Spon's Landscape and External Works Price Book

2004

Also available from Spon Press:

Spon's Manual for Educational Premises, 3rd edition – *Derek Barnsley* 0419233105

Spon's Railways Construction Price Book, 2nd edition – *Franklin + Andrews* 0415326230

Spon's Middle East Construction Costs Handbook – *Franklin + Andrews* 0415234387

Spon's Latin American Construction Costs Handbook – *Franklin + Andrews* 0415234379

Spon's African Construction Costs Handbook – *Franklin + Andrews* 0415234352

Spon's Asia Pacific Construction Costs Handbook, 3rd edition – *Davis Langdon & Seah* 0419254706

Spon's European Construction Costs Handbook, 3rd edition – *Davis Langdon & Everest* 0419254609

Spon's Irish Construction Price Book – *Franklin + Andrews* 0419242201

External Works, Roads and Drainage – *Phil Pitman* 0419257608

Guidelines for Landscape and Visual Impact Assessment – *The Landscape Institute and the Institute of Environmental Management and Assessment, Sue Wilson* 041523185X

The Regeneration of Public Parks – *Jan Woudstra & Ken Fieldhouse* 0419259007

Design for Outdoor Recreation – *Simon Bell* 0419203508

Manufactured Sites – *Niall Kirkwood* 0415243653

The Dynamic Landscape: Naturalistic Planting in an Urban Context – *Nigel Dunnett & James Hitchmough* 0415256208

Urban Open Spaces – *Helen Woolley* 0419256903

Shaping Neighbourhoods: A Guide for Health, Sustainability, Vitality – *Hugh Barton, Marcus Grant & Richard Guise* 041527852X HB 0415260094 PB

To order or obtain further information on the above or receive a full catalogue please contact:
The Marketing Department, Spon Press, 11 New Fetter Lane, London, EC4P 4EE.
Tel: 020 7583 9855 Fax 020 7842 2300
www.sponpress.com

Spon's Landscape and External Works Price Book

Edited by

DAVIS LANGDON & EVEREST

in association with
LANDSCAPE PROJECTS
Landscape Surveyors

2004

Twenty-third edition

Spon Press
Taylor & Francis Group

LONDON AND NEW YORK

First edition 1978
Twenty-third edition published 2004
by Spon Press
11 New Fetter Lane, London EC4P 4EE

Simultaneously published in the USA and Canada
by Spon Press
29 West 35th Street, New York, NY 10001

Spon Press is an imprint of the Taylor & Francis Group

© 2004 Spon Press

The right of Davis Langdon & Everest in association with Landscape Projects Landscape Surveyors to be identified as the Author of this Work has been asserted by them in accordance with the Copyright, Designs and Patents Act 1988

Printed and bound in Great Britain by
TJ International Ltd, Padstow, Cornwall

All rights reserved. No part of this book may be reprinted or reproduced or utilised in any form or by any electronic, mechanical, or other means, now known or hereafter invented, including photocopying and recording, or in any information storage or retrieval system, without permission in writing from the publishers.

The publisher makes no representation, express or implied, with regard to the accuracy of the information contained in this book and cannot accept any legal responsibility or liability for any errors or omissions that may be made.

Publisher's note
This book has been produced from camera-ready copy supplied by the authors.

British Library Cataloguing in Publication Data
A catalogue record for this book is available from the British Library

Library of Congress Cataloging in Publication Data
A catalogue record for this book has been requested

ISBN 0-415-32367-3
ISSN 0267-4181

Contents

Preface to the Twenty Third Edition	ix
List of Manufacturers and Suppliers	xi
Common Arrangement of Work Sections	xxvii

1 How to use this Book — 1
1.1 Introduction — 1
1.2 Important notes on the profit element of rates in this book — 1
1.3 Introductory notes on costing — 2
1.4 Adjustment and variation of the rates in this book — 4
1.5 Approximate Estimates — 4
1.6 How this book is updated each year — 5

2 Cost Information — 7
2.1 Labour Rates used in this edition — 7
2.2 Rates of Wages — 7
2.3 Daywork and Prime Cost — 11
2.4 Computation of Labour Rates, Cost of Materials and Plant — 13
2.5 Cost Indices — 22
2.6 Regional Variations — 24

3 Preliminaries — 25
3.1 Preliminaries/General Conditions — 25

4 Landscape Consultant's Appointment — 33

5 Fees for Professional Services — 43
Landscape Architects' Fees — 43
Architects' Fees — 47
Quantity Surveyors' Fees — 48

6 The Aggregates Tax — 65

7 Prices for Measured Works — 73

New Items for 2004 — 73

A Preliminaries — 77
A11 Tender and contract documents — 77
A32 Employers requirements: Management of works — 77
A34 Employers requirements: Security/safety/protection — 78
A41 Contractors general cost items: Site accommodation — 79
A43 Contractors general cost items: Mechanical plant — 79
A44 Contractors general cost items: Temporary roads — 80

B Complete buildings/structures/units — 81
B10 Prefabricated buildings/structures — 81

D Groundwork — 82
D11 Soil stabilization — 82
D20 Excavation and filling — 89

E In situ concrete/large precast concrete — 95
E10 In situ concrete — 95
E20 Formwork for in situ concrete — 98
E30 Reinforcement for in situ concrete — 98

F Masonry — 100
F10 Brick/block walling — 101
F20 Natural stone rubble walling — 104
F22 Cast stone walling — 105
F30 Accessories/Sundry items for brick/block/stone walling — 107
F31 Precast concrete sills/lintels/copings/features — 107

G Structural/Carcassing metal/timber — 108
G31 Prefabricated timber unit decking — 108

H Cladding/covering — 109
H51 Natural stone slab cladding/features — 109
H52 Cast stone slab cladding/features — 109

J Waterproofing — 110
J10 Specialist waterproof rendering — 110
J20 Mastic asphalt tanking/damp proofing — 110
J30 Liquid applied tanking/damp proofing — 110
J40 Flexible sheet tanking/damp proofing — 111
J50 Green Roof Systems — 111

M Surface finishes — 116
M10 Cement:sand/Concrete screeds/toppings — 116
M20 Plastered/Rendered/Roughcast coatings — 116
M40 Stone/Concrete/Quarry/Ceramic tiling — 116
M60 Painting/clear finishing — 117

P Building fabric sundries — 119
P30 Trenches/Pipeways/Pits for buried engineering services — 119

Q Paving/planting/fencing/site furniture — 120
Q10 Stone/concrete/brick kerbs/edgings/channels — 120
Q20 Bases and sub-bases to roads/pavings — 125
Q21 In situ concrete roads/pavings/bases — 126
Q22 Coated macadam/asphalt roads/pavings — 128
Q23 Gravel/hoggin roads/pavings — 130
Q24 Interlocking brick/block roads/pavings — 133
Q25 Slab/brick/sett/cobble pavings — 134
Q26 Special surfacings for sport — 145
Q30 Seeding/turfing — 148
Q31 Planting — 169
Q35 Landscape maintenance — 204
Q40 Fencing and gates — 211
Q50 Site/street furniture/equipment — 225

R Disposal systems		**243**
R12 Drainage below ground		243
R13 Land drainage		250
S Piped supply systems		**264**
S10 Cold water		264
S14 Irrigation		264
S15 Fountains/water features		265
V Electrical supply/power/lighting systems		**269**
V41 Street area floodlighting		269
8	**Approximate Estimates**	**279**
	8.1 Preliminaries	281
	8.2 Demolition and site clearance	282
	8.3 Groundwork	283
	8.4 Ground Stabilization	285
	8.5 Insitu Concrete	288
	8.6 Brick\Block Walling	289
	8.7 Roads and Pavings	291
	8.8 Special surfaces for Sport\Playgrounds	305
	8.9 Preparation for Planting\Turfing	308
	8.10 Seeding and Turfing	310
	8.11 Aftercare of Turf Areas	313
	8.12 Planting	314
	8.13 Fencing	320
	8.14 Street Furniture	326
	8.15 Drainage	327
	8.16 Irrigation	333
	8.17 Water features	334
	8.18 Timber decking	335
	8.19 Lighting	336
9	**Tables and Memoranda**	**337**
10	**Index**	**413**

SPON'S PRICEBOOKS 2004

Spon's Architects' & Builders' Price Book 2004
Davis Langdon & Everest

The most detailed professionally relevant source of construction price information currently available.

New features for 2004 include:
- Information on the new CIJC three year wage agreement, effective from Monday, 30th June 2003.
- Additional items for external wall, terracotta blockwork and rainscreen cladding.

HB & CD-ROM ◆ £120.00
0-415-32365-7 ◆ 1160pp

Spon's Landscape & External Works Price Book 2004
Davis Langdon & Everest

The only comprehensive source of information for detailed landscape and external works cost.

New features for 2004 include:
- More plant items to reflect the increased mechanisation of site handling of materials.
- Additional measured rate items for work in small gardens.

HB & CD-ROM ◆ £90.00
0-415-32367-3 ◆ 512pp

Spon's Mechanical & Electrical Services Price Book 2004
Mott Green & Wall

Still the only annual services engineering price book available.

New features for 2004 include:
- A complete revision of the Approximate Estimating section with more detail on a wider range of building types.
- Additional items for both Mechanical and Electrical services.

HB & CD-ROM ◆ £120.00
0-415-32368-1 ◆ 664pp

Spon's Civil Engineering & Highway Works Price Book 2004
Davis Langdon & Everest

More than just a price book, it is a comprehensive, work manual that all those in the civil engineering, surveying and construction business will find it hard to work without.

New features for 2004 include:
- An expansion of both the Civil Engineering Works and Highway Works sections to improve coverage of the two methods of measurement.

HB & CD-ROM ◆ £120.00
0-415-32366-5 ◆ 800pp

Receive a free cd-rom when you order any Spon 2004 Pricebook to help you produce tender documents, customise data, perform keyword searches and simple calculations.

Visit www.pricebooks.co.uk

To Order: Tel: +44 (0) 8700 768853, or +44 (0) 1264 343071 Fax: +44 (0) 1264 343005, or
Post: Spon Press Customer Services, Thomson Publishing Services, Cheriton House, Andover, Hants, SP10 5BE, UK
Email: book.orders@tandf.co.uk

For a complete listing of all our titles visit
www.sponpress.com

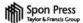
Spon Press
Taylor & Francis Group

Preface to the Twenty Third Edition

We are pleased to introduce new sections this year on *materials handling*, and *small gardens*. We anticipate expanding these sections in forthcoming editions. We have expanded and re-examined the disposal and filling sections. New products have been added or have replaced redundant products in their appropriate sections.

Labour Rates

Readers will note that this years book has increased the rate used for labour by a disproportionate amount (20%) over the annual increases reflected in previous editions. Our observations of market rates in the Landscape and External Works sector and the canvassing of professional contractors operating in the London area (see page 24 for regional variations) have influenced this decision. We believe that the base rate of £18.00 per man hour to be a competitive and profitable rate for a small company with overheads of £160,000.00 and employing 18 productive contract resources. A full break down of this labour rate is analysed on pages 17-18. Companies may request a copy of this spreadsheet electronically from us at the address at the end of this preface.

Disposal and Delivery of Bulk or Granular materials

We have become aware over the past year of practice amongst haulage companies and those handling granular materials which does impact on prices and profitability of parties buying these services or materials. This edition adjusts prices in these areas as follows:

Granular Materials in Bulk bags: This delivery format is often and mistakenly referred to as "One tonne bags". The suppliers of these products make no claim to supplying one tonne of material in these bags and confirm that the usual weight of materials here is 900 Kg. The tables in the memorandum section of this book referring to weights of materials have been used to calculate volumes in all instances.

Deliveries and Disposal costs: Most haulage is quoted at 20 tonnes in eight wheeled vehicles. Although these vehicles have capacity of 15 m3, vehicles would be contravening weight limits if vehicles were filled to the volume capacity of the vehicles. This and future editions will base all granular deliveries or disposal on volumes of between 11 and 12.5 m3 for granular materials and similarly for full loads of topsoil and the like. Grab lorries are calculated at a maximum volume of 7.25 m3 maximum volume and maximum weights of 12 tonnes. The section in this book on disposal reflects these ratios.

We furthermore would advise readers who operate in areas where materials are purchased and where therefore profitability is a factor, to ensure that their purchases are weight checked by production of weighbridge tickets or site measurement of the delivery weight.

This edition also includes a summary of the new fees guides for Landscape Architects. The front of the book also summarises the Landscape Architects appointment and fees guides for Quantity Surveyors. Readers should apply to the RICS or the Landscape Institute for the full texts of these documents. We acknowledge the cooperation by these two professional bodies afforded to us in compiling this edition.

We acknowledge the support provided to us by the suppliers of products and services who provide us with the base information used and whose contact details are published in the directory section at the front of this book. We advise that readers wishing to evaluate the cost of a product or service should approach these suppliers directly should they wish to confirm prices prior to submission of tenders or quotations.

Spon's Landscape and External Works prices do not allow for profit or site overhead. Readers should evaluate the market conditions in the sector or environment in which they operate and apply a percentage to these rates. Company overhead is allowed for within the labour rates used.

Market Conditions

The sector continues the buoyancy experienced over the last 3-4 years. Labour shortages still prevail. There is a hint of the easing of pressures due to the global economic and political situation. As usual the Landscape sector is the last to feel the brunt of these changes both in easing of works and in starting up of new works. Competitive commercial mark-up seems to be stable at 15% where company overhead is included elsewhere in the tender build-ups. Domestic work which is also very buoyant due to the property market enjoys profits of 20%-40% for the firms which operate in this sector. The higher end of profit mark-up is explained by the often very high specification or detailing on projects where an unusually high standard is required or if the degree of difficulty has impact.

Accuracy of Information in this Volume

Whilst every effort is made to ensure the accuracy of the information given in this publication, neither the editors nor the publishers in any way accept liability for loss of any kind resulting from the use made by any person of such information.

DAVIS LANGDON & EVEREST

MidCity Place

71 High Holborn,

London

WC1V 6QS

SAM HASSALL

LANDPRO LTD

Landscape Surveyors

6 Sandrock Hill Road

Farnham

Surrey

GU10 4NS

Tel: 020 7061 7000

e-mail: data.enquiry@davislangdon-uk.com

Tel: 01252 725513

e–mail: info@landpro.co.uk

List of Manufacturers and Suppliers

This list has been compiled from the latest information available but as firms frequently change their names, addresses and telephone numbers due to re-organization, users are advised to check this information before placing orders.

A Plant Acrow
Colnbrook By-Pass
Colnbrook
Slough
Berkshire
Plant Hire
Tel: 01753 680420
Fax: 01753 681338

A Plant Hire Co Ltd
Coothamlea Workshop
Cootham
Storrington
West Sussex RH20 4JN
Plant Hire
Tel: 01903 742348
Fax: 01903 742351

Abacus Municipal Ltd
Oddicroft Lane
Sutton-in-Ashfield
Notts. NG17 5FT
Lighting & Street Furniture
Tel: 01623 511111
Fax: 01623 552133

Ace Minimix
British Rail Goods Yard
York Way
Kings Cross
London N1 0AU
Ready Mixed Concrete
Tel: 0345 697690
Fax: 020 7278 5333

Addagrip Surface Treatments UK Ltd
Addagrip House
Bell Lane Industrial Estate
Uckfield
East Sussex TN22 1QL
Epoxy bound surfaces
Tel: 01825 761333
Fax: 01825 768566
E-mail: sales @addagrip.co.uk

AKZO Coatings Plc
Crown House
Hollin Lane
Darwin
Lancashire BB3 0BG
Sandtex matt paint
Tel: 01254 704951
Fax: 01254 774414

Alumasc Exterior Building Products Ltd
White House Works
Bold Road
Sutton, St.Helens
Merseyside WA9 4JG
Green Roof Systems
Tel: 01744 648400
Fax: 01744 648401
E-mail: info@alumasc-exterior.co.uk

Amberol Ltd
The Plantation
Spencer Road
Belper
Derbyshire DE56 1JW
Street furniture
Tel: 01773 830930
Fax: 01773 834191

List of Manufacturers and Suppliers

Amenity and Horticultural Services Ltd
Coppards Lane
Northiam
East Sussex TN31 6QP
Horticultural composts & fertilizers
Tel: 01797 252728
Fax: 01797 252724

Ancon Ltd
President Way
President Park
Sheffield S4 7UR
Fixings
Tel: 0114 275 5224
Fax: 0114 276 8543

Anderton Concrete Products Ltd
Anderton Wharf
Soot Hill
Anderton
Northwich
Cheshire CW9 6AA
Concrete fencing
Tel: 01606 79436
Fax: 01606 871590

Anglo Aquarium Plant Co Ltd
Strayfield Road
Enfield
Middlesex EN2 9JE
Aquatic plants
Tel: 020 8363 8548
Fax: 020 8363 8547
Website   www.anglo-aquarium.co.uk
E-mail: office@anglo-aquarium.co.uk

Arrow Chemicals Ltd
Stanhope Road
Swadlincote
Derbyshire DE11 9BE
Specialist coatings
Tel: 01283 221044
Fax: 01283 225731

Artificial Lawn Co.
Hartshill Nursery
Thong Lane
Gravesend
Kent DA12 4AD
Artificial Grass Suppliers
Tel: 01474 569576
Fax: 01474 321587
E-mail: landscape@baylis.freeserve.co.uk

Ashworth Ltd
Cray Avenue
St Mary Cray
Orpington
Kent BR5 3RH
Drainage Systems
Tel: 01689 891900
Fax: 01689 822372

Atlas Stone Products
Westington Quarry
Chipping Campden
Gloucestershire GL55 6EG
Paving manufacturer
Tel: 01386 840226
Fax: 01386 841356
Website: www.atlasstone.co.uk
E-mail: sales@atlasstone.co.uk

Autopa Ltd
Cottage Leap off Butlers Leap
Rugby
Warwickshire CV21 3XP
Street furniture
Tel: 01788 550556
Fax: 01788 550265
Website: www.autopa.co.uk
E-mail: info@autopa.co.uk

AVS Fencing Supplies Ltd
Unit 5F, Six Acre Works
Old Portsmouth Road
Peasmarsh
Guildford, Surrey GU3 1NE
Fencing
Tel: 01483 575212
Fax: 01483 306194
Website: www.avsfencing.co.uk
E-mail: sales@avsfencing.co.uk

Baylis Landscape Contractors
Hartshill Nursery
Thong Lane, Gravesend
Kent DA12 4AD
Sportsfield Constructon
Tel: 01474 569576
Fax: 01474 321587
E-mail: landscape@baylis.freeserve.co.uk

Bekaert Fencing Ltd
Woodhouse Lane
Wigan WN6 7NS
Security Fencing
Tel: 0870 1203252
Fax: 0870 1203242
Website: www.bekaert-fencing.com

Belwood Trees Ltd
Brigton of Ruthven
Meigle
Perthshire PH12 8RQ
Mature tree grower
Tel: 01828 640219
Fax: 01828 640623
Website: www.belwoodtrees.co.uk
E-mail: belwood@belwoodtrees.co.uk

Blanc De Bierges
Eastrea Road
Whittlesey
Peterborough
Cambridgeshire PE7 2AG
Concrete paving supplier
Tel: 01733 202566
Fax: 01733 205405
Website: www.blancedebierges.com
E-mail: mail@blancedebierges.com

Boughton Loam Ltd
Telford Way
Telford Way Industrial Estate
Kettering, Northants. NN16 8UN
Loams and Topsoils
Tel: 01536 510515
Fax: 01536 510691
Website: www.boughton-loam.co.uk
E-mail: enquiries@boughton-loam.co.uk

Breedon Plc
Breedon-on-the-Hill
nr. Melbourne
Derby DE73 1AP
Specialist gravels
Tel: 01332 862254
Fax: 01332 863149
Website: www.breedon.co.uk
E-mail: breedon@breedon.co.uk

British Seed Houses Ltd
Camp Road
Swinderby
Lincoln
LN6 9QJ
Grass and wildflower seed
Tel: 01522 868714
Fax: 01522 868095
Website: www.bshlincoln.co.uk
E-mail: simontaylor@bshlincoln.co.uk

Broxap Streetscene
Rowhurst Industrial Estate
Chesterton
Newcastle-under-Lyme
Staffordshire ST5 6BD
Street furniture
Tel: 01782 564411
Fax: 01782 565357
Website: www.broxap.com

Capital Garden Products Ltd
Gibbs Reed Barn
Pashley Road
Ticehurst
East Sussex TN5 7HE
Plant Containers
Tel: 01580 201092
Fax: 01580 201093
Website: www.capital-garden.com
E-mail: sales@capital-garden.com

CED Ltd
728 London Road
West Thurrock
Grays,
Essex RM20 3LU
Natural stone
Tel: 01708 867237
Fax: 01708 867230
Website: www.ced.ltd.uk
E-mail: sales@ced.ltd.uk

Champion Timber
205-209 Burlington Road
New Malden
Surrey KT3 4NB
Fencing
Tel: 020 8949 1621
Fax: 020 8942 9752
Website: www.championtimber.com

Charcon Hard Landscaping
Hulland Ward
Ashbourne
Derbyshire DE6 3ET
Paving and street furniture
Tel: 01335 372222
Fax: 01335 370074
Website: www.charcon.com
E-mail: charcon.sales@aggregate.com

Charles Morris (Fertilizers) Ltd
Longford House
Long Lane
Stanwell
Middlesex TW19 7AT
Topsoil
Tel: 01784 449144
Fax: 01784 449133

Clifton Nurseries Ltd
5a Clifton Villas
Little Venice
London W9 2PH
Nurserymen
Tel: 020 7286 6622
Fax: 020 7286 5655

Coblands Nurseries Ltd
Trench Road
Tonbridge
Kent TN10 3HQ
Nursery stock supplier
Tel: 01732 770999
Fax: 01732 770271
Website: www.coblands.co.uk

Colas Ltd
Rowfant
Crawley
West Sussex RH10 4NF
Road repair
Tel: 01342 711000
Fax: 01342 711199

Colourpave Ltd
Laymore
Forest Vale Industrial Estate
Cinderford
Gloucestershire GL14 2PH
Macadam Contractor
Tel: 01594 826768
Fax: 01594 826598

Columbia Cascade Ltd
Business Development Centre
Amanford Road
Tycroes
Ammanford
Carmarthenshire SA18 3QN
Street furniture
Tel: 01269 596396
Fax: 01269 596395

Cooper Clarke Group
Special Products Division
Bloomfield Road
Farnworth
Bolton BL4 9LP
Wall drainage and erosion control
Tel: 01204 862222
Fax: 01204 793856
Website: www.cooperclarke.com
E-mail: ccgmarketing@btinternet.com

Coverite Ltd
Palace Gates
Bridge Road
Wood Green
London N22 4SP
Waterproofing specialist
Tel: 020 8888 7821
Fax: 020 8889 0731

List of Manufacturers and Suppliers

Crowders Nurseries
Lincoln Road
Horncastle
Lincolnshire LN9 5LZ
Plant protection
Tel: 01507 525000
Fax: 01507 524000

CU Phosco Ltd
Lower Road
Great Amwell
Ware
Hertfordshire SG12 9TA
Lighting
Tel: 01920 860600
Fax: 01920 485915
Website: www.cuphosco.co.uk
E-mail: sales@cuphosco.co.uk

Dee-Organ Ltd
5 Sandyford Road
Paisley
Renfrewshire PA3 4HP
Street furniture
Tel: 0141 889 7000
Fax: 0141 889 7764
Website: www.dee-organ.co.uk
E-mail: signs@dee-organ.co.uk

DLF Perryfields Ltd
Thorn Farm
Inkberrow
Worcestershire WR7 4LJ
Grass and wild flower seed
Tel: 01386 793135
Fax: 01386 792715
Website: www.perryfields.co.uk
E-mail: amenity@perryfields.co.uk

Duracourt (Spade Oak) Ltd
Town Lane
Wooburn Green
High Wycombe
Buckinghamshire HP10 0PD
Tennis courts
Tel: 01628 528421
Fax: 01628 810509
Website: www.spadeoak.co.uk
E-mail: duracourt@spadeoak.co.uk

E.T. Clay Products
7 Fowler Road
Hainault
Essex IT6 3UT
Brick supplier
Tel: 020 8501 2100
Fax: 020 8500 9990
E-mail: eddie@etclay.freeserve.co.uk

Earth Anchors Ltd
15 Campbell Road
Croydon
Surrey CR0 2SQ
Anchors for site furniture
Tel: 020 8684 9601
Fax: 020 8684 2230
Website: www.earth-anchors.com

Edwards Sports Products Ltd
North Mills
Bridport
Dorset DT6 3AH
Sports equipment
Tel: 01308 424111
Fax: 01308 455800
Website: www.edsports.co.uk
E-mail: sales@edsports.co.uk

Elliot Workspace
The Fen
Baston
Peterborough
Cambridgeshire PE6 9PT
Site Offices
Tel: 01778 560891
Fax: 01778 560881

English Woodlands
Burrow Nursery
Cross in Hand
Heathfield
East Sussex TN21 0UG
Plant protection
Tel: 01435 862992
Fax: 01435 867742
Website: www.ewburrownursery.co.uk
E-mail: sales@ewburrownursery.co.uk

Ensor Rylance Ltd
Blackamoor Road
Guide
Blackburn
Lancashire BB1 2LQ
Drainage suppliers
Tel: 01254 52244
Fax: 01254 682371

Erisco-Bauder Ltd
Broughton House
Broughton Road
Ipswich
Suffolk IP1 3QR
Green Roof Systems
Tel: 01473 257671
Fax: 01473 230761
Website: www.erisco-bauder.com

Eve Trakway
Bramley Vale
Chesterfield
Derbyshire S44 5GA
Portable roads
Tel: 08700 767676
Fax: 08700 737373

Exclusive Leisure Ltd
28 Cannock Street
Leicester LE4 7HR
Artificial sports surfaces
Tel: 0116 233 2255
Fax: 0116 246 1561
Website: www.exclusiveleisure.co.uk
E-mail: info@exclusiveleisure.co.uk

Exxon Chemical Geopolymers Ltd
Mamhilad Park Industrial Estate
Pontypool
Gwent NP4 0YR
Geofabrics
Tel: 01495 757722
Fax: 01495 762383

Fairwater Water Gardens
Lodge Farm
Malthouse Lane
Ashington
West Sussex RH20 3BU
Water Feature Contractors
Tel: 01903 892522
Fax: 01903 892228

Farmura Ltd
Stone Hill
Egerton
Ashford,
Kent, TN27 9DU
Organic fertilizer suppliers
Tel: 01233 756241
Fax: 01233 756419
Website: www.farmura.com
www.alginure.co.uk
E-mail: info@farmura.com

Fleet (Line Markers) Limited
Fleet House
Spring Lane Industrial Estate
Malvern Link
Worcestershire WR14 1AT
Sports line marking
Tel: 01684 573535
Fax: 01684 892784

Forticrete Engineering Products Ltd
Hillhead Quarry
Harpur Hill
Buxton
Derbyshire SK17 9PS
Retaining wall systems
Tel: 01298 23333
Fax: 01298 23000

Fraser Ltd
Cray Avenue
St Mary Cray
Orpington
Kent BR5 3RH
Drainage Systems
Tel: 01689 891900
Fax: 01689 822372

List of Manufacturers and Suppliers

Furnitubes International
Meridian House
Royal Hill, Greenwich
London SE10 8RT
Street furniture
Tel: 020 8378 3200
Fax: 020 8378 3250
Website: www.furnitubes.com
E-mail: sales@furnitubes.com

Geometric Furniture Ltd
Birch Mill
Heywood Old Road
Heywood, Lancashire OL10 2QQ
Street furniture
Tel: 0161 653 2233
Fax: 0161 653 2299
Website: www.geometric-furniture.co.uk
E-Mail: geometric@geometric-furniture.co.uk

Grace Construction Products
Ajax Avenue
Slough
Berkshire SL1 4BH
Bitu-thene tanking
Tel: 01753 692929
Fax: 01753 691623

Grass Concrete Ltd
Walker House
22 Bond Street, Wakefield
West Yorkshire WF1 2QP
Grass block paving
Tel: 01924 379443
Fax: 01924 290289
Website: www.grasscrete.com
E-mail: info@grasscrete.com

Greenfix Ltd
Laverham House
77 St Georges Place
Cheltenham
Glos. GL50 3PP
Seeded erosion control mats
Tel: 01242 700092
Fax: 01242 700093
Websit: www.greenfix.co.uk
E-mail: cheltenham@greenfix.co.uk

Greenkeeper Ltd
Shelford Manor
Shelford
Nottingham NG12 1ER
Specialist turfs
Tel: 01949 21144
Fax: 01949 21144
Website: www.greenkeeperltd.co.uk

Greenleaf
The Old Dairy,
Croft Road
Hastings,
East Sussex TN34 4JU
Root Directors
Tel: 01424 717797
Website: www.ewburrownursery.co.uk
E-mail: sales@ewburrownursery.co.uk

H.S. Jackson & Son (Fencing) Ltd
Stowting Common
Ashford
Kent TN25 6BN
Fencing
Tel: 01233 750393
Fax: 01233 750403
Website: www.jacksons-fencing.co.uk
E-mail: sales@jacksonsfencing.co.uk

H.S.S. Hire Shops
Group Office
25 Willow Lane
Mitcham
Surrey CR4 4TS
Tool and plant hire
Tel: 020 8260 3100
Fax: 020 8687 5005

Haddonstone Ltd
The Forge House
East Haddon
Northampton NN6 8DB
Architectural stonework
Tel: 01604 770711
Fax: 01604 770027
Website: www.haddonstone.co.uk
E-mail: info@haddonstone.co.uk

Hanson Brick
Unicorn House
Wellington Street
Ripley
Derbyshire DE5 3DZ
Brick Manufacturer
Tel: 08705 258258
Fax: 01773 514040
Website: www.hanson-brickseurope.com
E-mail: info@hansonbrick.com

Harrison Flagpoles
Borough Road
Darlington
Co Durham DL1 1SW
Flagpoles
Tel: 01325 355433
Fax: 01325 461726
Website: www.flagpoles.co.uk
E-mail: sales@flagpoles.co.uk

Hepworth Plc
Hazlehead
Stocksbridge
Sheffield S30 5HG
Drainage
Tel: 0870 443 6000
Fax: 0870 443 8000
Website: hepworthdrainage.co.uk
E-mail:
   customerservice@hepworthdrainage.co.uk

Hill & Smith Ltd
Springvale Business & Industrial Park
Bilston
Wolverhampton WV14 0QZ
Safety barriers
Tel: 01902 499400
Fax: 01902 499419

Hills Industries Ltd
Pontygwindy Industrial Estate
Caerphilly
Mid Glamorgan CF8 3HU
Industrial washing lines
Tel: 029 2088 3951
Fax: 029 2088 6102
Website: www.hills-industries.co.uk
E-mail: info@hills-industries.co.uk

Hodkin & Jones (Sheffield) Ltd
Callywhite Lane
Dronfield
Sheffield S18 6XP
Drainage
Tel: 01246 290890
Fax: 01246 290292
Website: www.hodkin-jones.co.uk
E-mail: info@hodkin-jones.co.uk

Inturf
The Chestnuts
Wilberfoss
York YO4 5NT
Turf
Tel: 01759 321000
Fax: 01759 380130
Website: www.inturf.co.uk
E-mail: info@inturf.co.uk

J Toms Ltd
Grigg Lane
Headcorn
Ashford
Kent TN27 9XT
Plant Protection
Tel: 01622 891111
Fax: 016922 890783
Website: jtoms.co.uk
E-mail: tomsties@compuserve.com

List of Manufacturers and Suppliers

John Anderson Mobile Toilets
Smallford Lane
St. Albans
Herts. AL4 0LL
Mobile toilet facilities
Tel: 07000 822485
Fax: 01727 822886

Johnsons Wellfield Quarries Ltd
Crosland Hill
Huddersfield
West Yorkshire HD4 7AB
Natural Yorkstone pavings
Tel: 01484 652311
Fax: 01484 460007
Website: www.johnsons-wellfield.co.uk
E-mail: sales@johnsons-wellfield.co.uk

Jonathan James Ltd
15-17 New Road
Rainham
Essex RM13 8DJ
Specialist screeds and tanking
Tel: 01708 556921
Fax: 01708 520751

Jones of Oswestry
Whittington Road
Oswestry
Shropshire SY11 1HZ
Channels, gulleys, manhole covers
Tel: 01691 653251
Fax: 01691 658222

Keller Comtec
Barham Business Court
Teston, Maidstone
Kent ME18 5BZ
Hydro seeding specialist
Tel: 01622 618799
Fax: 01622 618790
Website: www.keller-ge.co.uk
E-mail: comtec@comtec-keller.co.uk

Kompan Ltd
Unit 20, Denbigh Hall
Bletchley
Milton Keynes
Buckinghamshire MK3 7QT
Play equipment
Tel: 01908 642466
Fax: 01908 270137
Website: www.kompan.com

Land and Water Services
3 Weston Farm Yard
The Street
Albury
nr. Guildford
Surrey
Specialist plant hire
Tel: 01483 202733
Fax: 01483 202510

Landline Ltd
1 Bluebridge Industrial Estate
Halstead
Essex CO9 2EX
Pond and lake installation
Tel: 01787 476699
Fax: 01787 472507

Landscapes By Design
The Studio
17 Church Road
Farnborough
Kent BR6 7DB
Trellis
Tel: 01689 851570
Fax: 01689 861151

Langdon & Son
51-53 2nd Avenue,
Chatham,
Kent, ME4 5BA
Railings
Tel: 01634 842485
Fax: 01634 831037

Lappset UK Ltd
Lappset House
Henson Way
Telford Way Industrial Estate
Kettering
Northamptonshire NN16 8PX
Play equipment
Tel: 01536 412612
Fax: 01536 521703

LDC Limited
Loampits Farm
99 Westfield Road
Woking
Surrey GU22 9QR
Willow walling
Tel: 01483 767488
Fax: 01483 764293

Leaky Pipe Systems Ltd
Frith Farm
Dean Street
East Farleigh
Maidstone, Kent ME15 0PR
Irrigation systems
Tel: 01622 746495
Fax: 01622 745118

Lister Lutyens Co Ltd
6 Alder Close
Eastbourne
East Sussex BN23 6QF
Street furniture
Tel: 01323 431177
Fax: 01323 639314
Website: www.listerteak.com
E-mail: sales@listerteak.com

Louis Poulsen UK Ltd
Surrey Business Park
Weston Road
Epsom
Surrey KT17 1JG
Outdoor lighting
Tel: 01372 848800
Fax: 01372 848801
E-mail: tcr-uk@lpmail.com

Maccaferri Ltd
7400 The Quorum
Oxford Business Park North
Garsington Road
Oxford OX4 2JZ
Gabions
Tel: 01865 770555
Fax: 01865 774550

Malcolm, Ogilvie & Co
Fowler Road
West Pitkerro Industrial Estate
Dundee DD5 3RU
Scotland
Geofabrics
Tel: 01382 735801
Fax: 01382 735808

Marlin Lighting
Hanworth Trading Estate
Hampton Road West
Feltham TW13 6DR
Lighting
Tel: 020 8894 5522
Fax: 020 8898 8480

Marshalls Mono Ltd
Southowram
Halifax
West Yorkshire HX3 9SY
Hard landscape materials and street furniture
Tel: 01422 306000
Fax: 01422 330185

Melcourt Industries Limited
Boldridge Brake
Long Newnton
Tetbury
Gloucestershire GL8 8RT
Mulch and compost
Tel: 01666 502711
Fax: 01666 504398
Website: www.melcourt.co.uk
E-mail: mail@melcourt.co.uk

Metro Sand
4 Leylands Business Park,
Nobbs Crook, Colden Common
Southampton,
Hants. SO21 1TH
Structural Tree Soil
02380 696957
02380 696958
info@metrosand.co.uk
www.metrosand.co.uk

Milton Pipes Ltd
Milton Regis
Sittingbourne
Kent ME10 2QF
Soakaway Rings
Tel: 01795 425191
Fax: 01795 420360
Website: www.miltonpipes.com
E-mail: sales@miltonpipes.com

Monarflex Geomembranes Ltd
Lyon Way
St Albans
Hertfordshire AL4 0LB
Lake liners
Tel: 01727 830116
Fax: 01727 858998

Neptune Outdoor Furniture Ltd
Thompsons Lane
Marwell
Winchester
Hampshire SO21 1JH
Street furniture
Tel: 01962 777799
Fax: 01962 777723
Website: www.nofl.co.uk
E-mail: info@nofl.co.uk

Netlon Professional
New Wellington Street
Blackburn
Lancashire BB2 4PJ
Erosion control, soil stabilisation, plant protection
Tel: 01254 262431
Fax: 01254 680008

Norbury Ltd
28 Marshgate Drive
Hertford
Herts. SG13 7AJ
Hardwood gates
Tel: 01992 554327
Fax: 01992 505978

Notcutts Nurseries Ltd
Woodbridge
Suffolk IP12 4AF
Nursery stock supplier
Tel: 01394 383344
Fax: 01394 445440

Orchard Street Furniture Ltd
Castle House, Castle Square
Benson
Oxfordshire OX10 6SD
Street furniture
Tel: 01491 642123
Fax: 01491 642126
Website: www.orchardstreet.co.uk
E-mail: sales@orchardstreet.co.uk

Outdoor Lighting
Surrey Business Park
Weston Road
Epsom
Surrey KT17 1JG
Outdoor lighting
Tel: 01372 848800
Fax: 01372 848801

Platipus Anchors Ltd
Kingsfield Business Centre
Philanthropic Road
Redhill
Surrey RH1 4DP
Tree anchors
Tel: 01737 762300
Fax: 01737 773395
Website: www.platipus-anchors.com
E-mail: info@platipus-anchors.com

Rawell Water Control Systems Ltd
Carr Lane
Hoylake, Wirral
Merseyside CH47 4FE
Lake liners
Tel: 0151 632 5771
Fax: 0151 632 0252

RCC
Barholm Road
Tallington
Stamford
Lincolnshire PE9 4RL
Precast concrete retaining units
Tel: 01778 344460
Fax: 01778 345949
Website: www.tarmacprecast.co.uk
E-mail: rcc@tarmac.co.uk

Recycled Materials Ltd
PO Box 519
Surbiton
Surrey KT6 4YL
Disposal Contractors
Tel: 020 8390 7010
Fax: 020 8390 7020
E-mail: rml@demon.co.uk

Rigby Taylor Ltd
The Riverway Estate
Portsmouth Road
Peasmarsh, Guildford
Surrey GU3 1LZ
Horticultural supply
Tel: 01483 446900
Fax: 01483 534058
Website: www.rigbytaylor.com
E-mail: sales@rigbytaylor.com

RIW Ltd
Arc House
Terrace Road South
Binfield, Bracknell
Berkshire RG42 4PZ
Waterproofing products
Tel: 01344 861988
Fax: 01344 862010
Website: www.riw.co.uk
E-mail: enquiries@riw.co.uk

RMC Concrete Products
London Road,
Wick,
Bristol, BS30 5SJ
Precast Concrete Paving
0117 937 3371
0117 9373002

Road Equipment Ltd
28/34 Feltham Road
Ashford
Middlesex TW15 1DL
Plant Hire
Tel: 01784256565
Fax: 01784 240398
E-mail: roadequipment@aol.com

Rolawn (Turf Growers) Ltd
Elvington
York YO41 4XR
Industrial turf
Tel: 01904 608661
Fax: 01904 608272
Website: www.rolawn.co.uk
E-mail: ian.elwick@rolawn.co.uk

Rom Ltd
Wheaton Road
Witham
Essex CM8 3BU
Reinforcement steel bars
Tel: 01376 514321
Fax: 01376 518628

Ruberoid Building Products Ltd
Tewin Road
Welwyn Garden City
Hertfordshire AL7 1BP
Synthaprufe waterproofing
Tel: 01707 822222
Fax: 01707 333763

Scotts UK Professional Ltd
Paper Mill Lane
Bramford
Ipswich
Suffolk IP8 4BZ
Fertilizers and chemicals
Tel: 01473 830492
Fax: 01473 830386

SMP Playgrounds Ltd
Ten Acre Lane
Thorpe
Egham
Surrey KT16 8EJ
Playground equipment
Tel: 01784 489100
Fax: 01784 431079
Website: www.smp.co.uk
E-mail: sales@smp.co.uk

Southern Conveyors
Unit 2, Denton Slipways Site
Wharf Road,
Gravesend, DAI2 2RU
Conveyors
Tel: 01474 564145
Fax: 01474 568036

Spade Oak Construction Co Ltd
Town Lane
Wooburn Green
High Wycombe
Bucks. HP10 0PD
Macadam Contractors
Tel: 01628 529421
Fax: 01628 810509
Website: www.spadeoak.co.uk
E-mail: email@spadeoak.co.uk

Sportsmark Group Ltd
Ealing Road
Brentford
Middlesex TW8 0LH
Sports surfaces
Tel: 020 8560 2010
Fax: 020 8568 2177

Steelway-Fensecure
Parkside Industrial Estate
Hickman Avenue
Wolverhampton
West Midlands WV1 2EN
Fencing
Tel: 01902 490919
Fax: 01902 490929

Sugg Lighting Ltd
Sussex Manor Business Park
Gatwick Road
Crawley
West Sussex RH10 2GD
Lighting
Tel: 01293 540111
Fax: 01293 540114
Website: www.sugglighting.co.uk
E-mail: sales@sugglighting.co.uk

Tarmac Southern
London Sales Office
24 Park Royal Road
London NW10 7JW
Ready mixed concrete
Tel: 020 8965 1864
Fax: 020 8965 1863

Tensar International (Netlon Group)
New Wellington Street
Blackburn
Lancashire BB2 4PJ
Erosion control, soil stabilisation
Tel: 01254 262431
Fax: 01254 266868
Website: www.tensar-international.com
E-mail: sales@tensar.co.uk

Terram Limited
Mamhilad Park Industrial Estate
Pontypool
Gwent NP4 0YR
Geofabrics
Tel: 01495 767402
Fax: 01495 767436
Website: www.terram.com
E-mail: info@terram.co.uk

List of Manufacturers and Suppliers

Terranova Lifting Ltd
Terranova House
Bennett Road
Reading,
Berks RG2 OQX
Crane Hire
0118 931 2345
0118 931 4114

Visqueen Agri
Yarm Road
Stockton-on-Tees
Cleveland TS18 3GE
Lake liners
Tel: 01642 677228
Fax: 01642 664325

Townscape Products Ltd
Fulwood Road South
Sutton-in-Ashfield
Nottinghamshire NG17 2JZ
Hard landscaping
Tel: 01623 513355
Fax: 01623 440267
Website: www.townscape-products.co.uk
E-mail: sales@townscape-products.co.uk

Wavin Building Products
Parsonage Way
Chippenham
Wiltshire SN15 5PN
Drainage products
Tel: 01249 766600
Fax: 01249 766671
Website: www.waving.co.uk

Tubex Ltd
Aberaman Park
Aberdare
Mid Glamorgan CF44 6DA
Plant Protection, tree guards
Tel: 01685 888000
Fax: 01685 888001
Website: www.tubex.com
E-mail: plantcare@tubex.com

The Western Log Group
Forest Product Centre
Afan Valley Road
Cwrnafan
Port Talbot
SA12 9AB
01639 861227
01639896715
sales@westernloggroup.com
www.westernloggroup.com

Turf Management Systems
Dromenach Farm
Seven Hills Road
Iver Heath
Bucks. SL0 0PA
Specialist turf systems
Tel: 01895 834411
Fax: 01895 834892

White Horse Contractors Ltd
Blakes Oak Farm
Lodge Hill
Abingdon
Oxfordshire OX14 2JD
Drainage
Tel: 01865 736272
Fax: 01865 326176
Website: www.whitehorsecontractors.co.uk
E-mail: whc@whitehorsecontractors.co.uk

List of Manufacturers and Suppliers

Wicksteed Leisure Ltd
Digby Street
Kettering
Northamptonshire NN16 8YJ
Play equipment
Tel: 01536 517028
Fax: 01536 410633
Website: www.wicksteed.co.uk
E-mail: sales@wicksteed.co.uk

Woodscape Ltd
Upfield, Pike Lowe
Brinscall
Chorley
Lancashire PR6 8SP
Street furniture
Tel: 01254 830886
Fax: 01254 831846
Website: www.woodscape.co.uk
E-mail: sales@woodscape.co.uk

Wybone Ltd
Mason Way
Platts Common Industrial Estate
Hoyland, Barnsley
South Yorkshire S74 9TF
Street furniture
Tel: 01226 744010
Fax: 01226 350105
Website: www.wybone.co.uk
E-mail: sales@wybone.co.uk

Wyckham-Blackwell
Old Station Road
Hampton in Arden,
Solihull,
West Midlands, B92 0HB
Timber Decking
Tel: 01675 442233
Fax: 01675 442227
Website: www.wyckham-blackwell.co.uk
E-mail: decking@wyckham-blackwell.co.uk

Yeoman Aggregates Ltd
Stone Terminal
Horn Lane
Acton
London W3 9EH
Aggregates
Tel: 020 8896 6820
Fax: 020 8896 6829
E-mail: yeoman.sales@ukonline.co.uk

LandPro Ltd
Landscape Surveyors

Contract Acquisition

Project Management

Cost feasibility

Quantity surveying services

for

Landscape Architects

Local Authorities

External Works and Landscape Contractors.

---

6 Sandrock Hill Road
Farnham
Surrey GU10 4NS
01252 725513
info@landpro.co.uk

Common Arrangement of Work Sections

The main work sections relevant to landscape work and their grouping:

A Preliminaries/General Conditions

- A10 Project particulars
- A11 Tender and contract documents
- A12 The site/existing buildings
- A13 Description of the work
- A20 The contract/sub-contract
- A30 Employers requirements: tendering/sub-letting/supply
- A31 Employers requirements: provision, content and use of documents
- A32 Employers requirements: management of the works
- A33 Employers requirements: quality standards/control
- A34 Employers requirements: security/safety/protection
- A35 Employers requirements: specific limitations on method/sequence/timing/use of site
- A36 Employers requirements: facilities/temporaryworks/services
- A37 Employers requirements: operation/maintenance of the finished building
- A40 Contractors general cost items: management and staff
- A41 Contractors general cost items: site accommodation
- A42 Contractors general cost items: services and facilities
- A43 Contractors general cost items: mechanical plant
- A44 Contractors general cost items: temporary works
- A50 Works/products by/on behalf of the employer
- A51 Nominated sub-contractors
- A52 Nominated suppliers
- A53 Work by statutory authorities/undertakers
- A54 Provisional work
- A55 Dayworks
- A60 Preliminaries/general conditions for demolition contract
- A61 Preliminaries/general conditions for investigation/survey contract
- A62 Preliminaries/general conditions for piling/embedded retaining wall contract
- A63 Preliminaries/general conditions for landscape contract
- A70 General specification requirements for work package

B Complete buildings/structures/units

- B10 Prefabricated buildings/structures
- B11 Prefabricated building units

C Existing site/buildings/services

- C10 Site survey
- C11 Ground investigation
- C12 Underground services survey

D Groundwork

- D11 Soil stabilisation
- D20 Excavating and filling
- D41 Crib walls/gabions/reinforced earth

E In situ concrete/large precast concrete

- E05 In situ concrete construction generally
- E10 Mixing/casting/curing in situ concrete
- E20 Formwork for in situ concrete
- E30 Reinforcement for in situ concrete

F Masonry

- F10 Brick/block walling
- F20 Natural stone rubble walling
- F21 Natural stone ashlar walling/dressings
- F22 Cast stone ashlar walling/dressings
- F30 Accessories/sundry items for brick/block/stone walling
- F31 Precast concrete sills/lintels/copings/features

G Structural/carcassing metal/timber

- G31 Prefabricated timber unit decking

H Cladding/covering

- H51 Natural stone slab cladding/features
- H52 Cast stone slab cladding/features

J Waterproofing

- J10 Specialist waterproof rendering
- J20 Mastic asphalt tanking/damp proofing
- J21 Mastic asphalt roofing/insulation/finishes
- J22 Proprietary roof decking with asphalt finish
- J30 Liquid applied tanking/damp proofing
- J31 Liquid applied waterproof roof coatings
- J40 Flexible sheet tanking/damp proofing
- J44 Sheet linings for pools/lakes/waterways

M Surface finishes

- M10 Cement:sand/concrete screeds
- M20 Plastered/rendered/roughcast coatings
- M40 Stone/concrete/quarry/ceramic tiling/mosaic
- M60 Painting/clear finishing

P Building fabric sundries

- P30 Trenches/pipeways/pits for buried engineering services

Q Paving/planting/fencing/site furniture

- Q10 Kerbs/edgings/channels/paving accessories
- Q20 Granular sub-bases to roads/pavings
- Q21 In situ concrete roads/pavings
- Q22 Coated macadam/asphalt roads/pavings
- Q23 Gravel/hoggin/woodchip roads/pavings
- Q24 Interlocking brick/block roads/pavings
- Q25 Slab/brick/sett/cobble pavings
- Q26 Special surfacings/pavings for sport/general amenity
- Q30 Seeding/turfing
- Q31 Planting
- Q32 Planting in special environments
- Q35 Landscape maintenance
- Q40 Fencing
- Q50 Site/street furniture/equipment

R   Disposal systems

    R12   Drainage below ground
    R13   Land drainage

S   Piped supply systems

    S10   Cold water
    S14   Irrigation
    S15   Fountains/water features

V   Electrical supply/power/lighting systems

    V41   Street/area floodlighting

Fast estimate

Advanced pricing tools for Spon's Price Book users

Everything you need to price work quickly and accurately using your own custom rates together with Spon's Price libraries.

- **Resource Library** - Keep your costs up-to-date and in one place
- **Suppliers** - Allocate resources to individual suppliers
- **Price Lists** - Access to the latest product prices from the UK's largest merchant
- **Composite Resources** - A real time saver for mortar; labour gangs etc.
- **Price Book Options** - Access industry standard Price books from Spon Press
- **Item Library** - Maintain your own rate library (with full build ups)
- **Composite Items** - Save time by selecting a single item covering multiple tasks
- **BQ Preparation** - Deliver professional documents to clients and colleagues
- **Resource Requirements** - Specify exactly what you need to complete the job
- **Purchase Ordering** - Convert job requirements into enquiries and orders
- **Flexibility** - Works for builders, contractors, sub-contractors and specialists
- **Output to MS Applications** - Export data to Word, Excel, Project etc.
- **Subcontractor Bills** - Generate bills from selected resources or tagged items.
- **Price CITE Bills** - Import, price and export bills supplied in this industry standard format.

Fast Estimate shares the same clear pricing processes as Spon Estimating, but with more pricing tools to give you outstanding control over your tender documents.

Try it now - Download your free trial copy:

www.estek.co.uk

1
How to use this Book

1.1 INTRODUCTION

First-time users of *Spon's Landscape and External Works Price Book* and others who may not be familiar with the way in which prices are compiled may find it helpful to read this section before starting to calculate the costs of landscape works.
The cost of an item of landscape construction or planting is made up of many components:
- the cost of the product;
- the labour and additional materials needed to carry out the job;
- the cost of running the contractor's business.

These are described more fully below.

1.2 IMPORTANT NOTES ON THE PROFIT ELEMENT OF RATES IN THIS BOOK

The rates shown in the Measured Works and Approximate Estimates sections of this book do not generally contain a profit element unless the rate has been provided by a subcontractor.

Analysed Rates versus Sub-Contractor Rates
As a general rule if a rate is shown as an analysed rate in the Measured Works section, i.e. it has figures shown in the columns other than the "Unit" and "Total Rate" column, it can be assumed that this has no profit or overhead element and that this calculation is shown as a direct labour/material supply/plant, task being performed by a contractor.

On the other hand if a rate is shown as a Total Rate only, this would normally be a sub-contractors rate and would contain the profit and overhead for the subcontractor.

The foregoing applies for the most part to the Approximate Estimates section. However in some items there may be an element of sub-contractor rates within direct works build-ups.

As an example of this, to excavate, lay a base and place a macadam surface; a general landscape contractor would normally perform the earthworks and base installation. The macadam surfacing would be performed by a specialist subcontractor to the landscape contractor.

The Approximate Estimate for this item uses rates from the Measured Works section to combine the excavation, disposal, base material supply and installation with all associated plant. There is no profit included on these elements. The cost of the surfacing, however, as supplied to us in the course of our annual price enquiry from the macadam subcontractor, would include the subcontractors profit element.

The landscape contractor would add this to his section of the work but would normally apply a mark up at a lower percentage to the subcontract element of the composite task.
Users of this book should therefore allow for the profit element at the prevailing rates. Please see the worked examples below and the notes on Overheads for further clarification.

1.3 INTRODUCTORY NOTES ON COSTING

The Prices for Measured Work are intended to apply to a medium-sized contract of about 6000 m2 in the outer London area, and assume that 50% is hard surfacing and 50% is soft landscaping and planting. Similarly it has been necessary to assume that the work is undertaken as a main contract, but if the work is let as a sub-contract, consideration should be given to the need for the addition of main contractor's discount and profit.

As explained in more detail later the prices are generally based on wage rates and material costs current at Spring 2003. They do not allow for preliminary items, which are dealt with in Section 2.3, or for any Value Added Tax which may be payable.

Adjustments should be made to standard rates for time, location, local conditions, site constraints and any other factors likely to affect the costs of a specific scheme.

Term contracts for general maintenance of large areas should be executed at rates somewhat lower than those given in this section.

There is now a facility available to readers that enables a comparison to be made between the level of prices given and those for projects carried out in regions other than outer London; this is dealt with in Section 2.6.

The units of measurement have been varied to suit the type of work and care should be taken when using any prices to ascertain the basis of measurement adopted.

The prices per unit of area for executing various mechanical operations are for work in the following areas and under the following conditions.

Prices per m^2 relate to areas not exceeding 100 m^2
(any plan configuration)

Prices per 100 m^2 relate to areas exceeding 100 m^2 but not exceeding 1/4 ha
(generally clear areas but with some sub-division)

Prices per ha relate to areas over 1/4 ha
(clear areas suitable for the use of tractors and tractor-operated equipment)

The prices per unit area for executing various operations by hand generally vary in direct proportion to the change in unit area.

Measured Works
Prime Cost: Commonly known as the 'PC'. Prime Cost is the actual price of the material item being addressed such as paving, shrubs, bollards or turves, as sold by the supplier. Prime Cost is given 'per square metre', 'per 100 bags' or 'each' according to the way the supplier sells his product. In researching the material prices for the book we requested that the suppliers price for full loads of their product delivered to a site close to the M25 in London. Spon's rates do not include VAT. Some companies may be able to obtain greater discounts on the list prices than those shown in the book. Prime Cost prices for those products and plants which have a wide cost range will be found under the heading of Market Prices in the main sections of this book, so that the user may select the product most closely related to his specification.

Materials: The PC material plus the additional materials required to fix the PC material.
Every job needs materials for its completion besides the product bought from the supplier. Paving needs sand for bedding, expansion joint strips and cement pointing; fencing needs concrete for post setting and nails or bolts; tree planting needs manure or fertilizer in the pit, tree stakes, guards and ties. If these items were to be priced out separately, Spon's Landscape and External Works Price Book (and the Bill of Quantities) would be impossibly unwieldy, so they are put together under the heading of Materials.

Labour. This figure covers the cost of planting shrubs or trees, laying paving, erecting fencing etc. and is calculated on the wage rate (skilled or unskilled) and the time needed for the job. Extras such as highly skilled craft work, difficult access, intermittent working and the need for labourers to back up the craftsman all add to the cost. Large regular areas of planting or paving are cheaper to install than smaller intricate areas, since less labour time is wasted moving from one area to another.

Labour Rates used in this edition
The rates for labour used in this edition have been based on surveys carried out on a cross section of external works contractors.
These rates include for company overheads such as employee administration, transport insurance, and on costs such as National Insurance. The rates do not include for profit.

Plant, consumable stores and services: This rather impressive heading covers all the work required to carry out the job which cannot be attributed exactly to any one item. It covers the use of machinery ranging from JCB's to shovels and compactors, fuel, static plant, water supply (which is metered on a construction site), electricity and rubbish disposal. The cost of transport to site is deemed to be included elsewhere and should be allowed for elsewhere or as a preliminary item. Hired plant is calculated on an average of 36 hours working time per week

Overheads. An allowance for this is included in the labour rates which are described above. The general overheads of the contract such as insurance, site huts, security, temporary roads and the statutory health and welfare of the labour force are not directly assignable to each item, so they are distributed as a percentage on each, or as a separate preliminary cost item. The contractor's and sub-contractor's profits are not included in this group of costs. Site overheads, which will vary from contract to contract according to the difficulties of the site, labour shortages, inclement weather or involvement with other contractors, have not been taken into account in the build up of these rates, while Overhead (or profit) may have to take into account losses on other jobs and the cost to the contractor of unsuccessful tendering.

Sub-contract rates
Where there is no analysis against an item, this is deemed to be a rate supplied by a sub-contractor. In most cases these are specialist items where most external works contractors would not have the expertise or the equipment to carry out the task described. It should be assumed that sub-contractor rates include for the sub contractors profit.
An example of this may be found for the Tennis court rates in section Q26

1.4 ADJUSTMENT AND VARIATION OF THE RATES IN THIS BOOK

It will be appreciated that a variation in any one item in any group will affect the final Measured Work price.

Any cost variation must be weighed against the total cost of the contract and a small variation in Prime Cost where the items are ordered in thousands may have more effect on the total cost than a large variation on a few items, while a change in design that necessitates the use of earth-moving equipment which must be brought to the site for that one job will cause a dramatic rise in the contract cost. Similarly, a small saving on multiple items will provide a useful reserve to cover unforeseen extras.

Worked examples

A variation in the Prime Cost of an item can arise from a specific quotation.
For example:,
The PC of 900 x 600 x 50mm precast concrete flags is given as £2.82 each which equates to £5.22/m^2; and to which the costs of bedding and pointing are added to give a total material cost of £9.26/m^2.
Further costs for labour and mechanical plant give a resultant price of £16.46/m^2.
If a quotation of £5.00/m^2 was received from a supplier of the flags the resultant price would be calculated as £16.46 less the original cost (£5.22) plus the revised cost (£5.00) to give £16.24/m^2.

A variation will also occur if, for example, the specification changes from 50 light standard trees to 50 extra heavy standard trees. In this case the Prime Cost will increase due to having to buy extra heavy standard trees instead of light standard stock.

Original price: The prime cost of an Acer platanoides light standard is given as £9.55, to which the cost of handling and planting, etc is added to give an overall cost of £15.55. Further costs for labour and mechanical plant for mechanical excavation of a 600 x 600 x 600 mm pit (£3.39), a single tree stake (£7.78), and importing "Topgrow" compost in 80 litre bags (£0.67) and backfilling (£6.30) give a resultant price of £33.69. Therefore the total cost of 50 light standard trees is £1684.50.

Revised price: The prime cost of an Acer platanoides extra heavy standard is given as £40.00, however, taking into account the additional staking, extra labour for handling and planting ,a larger tree pit, a larger stake and more compost, and this cost gives a total unit price of £75.27. This is significantly more than the light standard version. Therefore the total cost of 50 extra heavy trees is £3763.50 and the additional cost is £2079.00

This example of the effects of changing the tree size, illustrates that caution is needed when revising tender prices as merely altering the prime cost of an item is will not accurately reflect the total cost of the revised item.

1.5 APPROXIMATE ESTIMATES

These are combined Measured Work prices which give an approximate cost for a complete section of landscape work. For example, the construction of a car park comprises excavation, levelling, road-base, surfacing and marking parking bays. Each of these jobs is priced separately in the Measured Works section, but a comprehensive price is given in the Approximate Estimates section, which is intended to provide a quick guide to the cost of the job. It will be seen that the more items that go to make up an approximate estimate, the more possibilities there are for variations in the PC prices and the user should ensure that any PC price included in the estimate corresponds to his specification.

Worked example

In many instances a modular format has been use in order to enable readers to build up a rate for a required task. The following table describes the trench excavation pipe laying and backfilling operations as contained in section R12 of this book.

Pipe laying 100 m	£
Excavate for drain 150 mm wide x 450 mm deep inclusive of disposal on site; by machine	89.76
Lay flexible plastic pipe 100 mm wide	100.06
Backfilling with gravel rejects, blinding with sand and topping with 150 mm topsoil	238.02
Total	**427.84**

The figures given in Spon's Landscape and External Works Price Book are intended for general guidance, and if a significant variation appears in any one of the cost groups, the Price for Measured Work should be re-calculated.

1.6 HOW THIS BOOK IS UPDATED EACH YEAR

The basis for this book is a database of Material, Labour, Plant and Sub-Contractor resources each with its own area of the database.

Material, Plant and Sub-Contract Resources

Each year the suppliers of each material, plant or subcontract item are approached and asked to update their prices to those that will prevail in September of that year.

These resource prices are individually updated in the database. Each resource is then linked to one or many tasks. The tasks in the task library section of the database is automatically updated by changes in the resource library. A quantity of the resource is calculated against the task.

The calculation is generally performed once and the links remain in the database.
On occasions where new information or method or technology are discovered or suggested, these calculations would be revisited. A further source of information is simple time and production observations made during the course of the last year.

Labour Resource Update

Most tasks except those shown as sub-contractor rates (see above) employ an element of labour. The Data Department at Davis Langdon and Everest conducts ongoing research into the costs of labour in various parts of the country.

Tasks or entire sections would then be re-examined and recalculated. Comments on the rates published in this book are welcomed and may be submitted to the contact addresses shown in the preface.

The Dynamic Landscape
Naturalistic Planting in an Urban Context

Edited by Nigel Dunnett &
James Hitchmough

The last quarter of the twentieth-century witnessed a burgeoning of interest in ecological or naturally inspired use of vegetation in the designed landscape. More recently a strong aesthetic element has been added to what was formerly a movement aimed at creating nature-like landscapes.

This book advances a fusion of scientific and ecological planting design philosophy that can address the need for more sustainable designed landscapes. It is a major statement on the design, implementation and management of ecologically-inspired landscape vegetation. With contributions from people at the forefront of developments in this field, in both Europe and North America, it provides a valuable synthesis of current thinking.

Contents:
1. Introduction 2. The Historical Development of Naturalistic Planting 3. A Contemporary Overview of Naturalistic Planting 4. The Dynamic Nature of Plant Communities 5. A Naturalistic Design Process 6. Herbaceous Plantings 7. Exploring Woodland Design 8. Wetlands and Water Bodies 9. Communicating Naturalistic Plantings: Plans and Specifications 10. The Creative Management of Naturalistic Plantings 11. The Social and Cultural Context of Naturalistic Plantings. Index.

December 2003: 238x225: 384pp: Hb: 0-415-25620-8: £60.00

To Order: Tel: +44 (0) 8700 768853, or +44 (0) 1264 343071
Fax: +44 (0) 1264 343005, or **Post:** Spon Press Customer Services, ITPS Andover, Hants, SP10 5BE, UK

For a complete listing of all our titles visit: www.sponpress.com

2
Cost Information

2.1 LABOUR RATES USED IN THIS EDITION

Based on surveys carried out on a cross section of external works contractors, the following rates for labour have been used in this edition.

These rates include for company overheads such as employee administration, transport insurance, and on costs such as National Insurance. The rates do not include for profit.

The rates for all labour used in this edition is as follows

General contracting	£ 18.00 / hour
Maintenance contracting	£ 14.00 / hour

2.2 RATES OF WAGES

2.2.1 Building industry

Authorized rates of wages, etc., in the building industry in England and Wales and Scotland agreed by the Construction Industry Joint Council. Effective from 30 June 2003. Basic pay. Weekly rate based on 39 hours.

Craft Rate	Skill Rate 1	£284.70	General Operative	£225.03
£299.13	Skill Rate 2	£274.56		
	Skill Rate 3	£257.01		
	Skill Rate 4	£242.19		

The Prices for Measured Works used in this book are based upon the commercial wage rates currently used in the landscaping industry and are typical of the London area.

However, it is recognized that some contractors involved principally in soft landscaping and planting works may base their wages on the rates determined by the Agricultural Wages Board (England and Wales) and therefore the following information is given to assist readers to adjust the prices if necessary.

2.2.2 Agricultural Wages, England and Wales

Minimum payments agreed by the Agricultural Wages Board for standard 39 hour, five day week. Effective 1 October 2002.

Age	Unit	Basic pay	Craft Grade	Appointment Grade	
				Grade 2	Grade 1
Weekly pay:					
19+	£/ week	191.49	225.81	239.46	258.57
18	£/ week	162.63			
17	£/ week	134.16			
16	£/ week	115.05			
15 & under	£/ week	95.94			
Overtime pay:					
19+	£/ week	7.37	8.69	9.21	9.95
18	£/ week	6.26			
17	£/ week	5.16			
16	£/ week	4.43			
15 & under	£/ week	3.69			

Since 1997 there have been three main types of worker:
- Full time workers (standard and flexible)
- Part time workers (standard and flexible)
- Casual workers

Weekly rates apply to Full Time Standard Workers for 39 hours work.

Hourly rates apply to the hours worked in a week by a Part Time Standard worker.

Overtime rates apply to all overtime hours worked by Standard Workers -
- more than 39 hours in any week
- or more than 8 hours on any day
- or Saturday or Sunday
- or Public Holiday

From 1 June 2000 the minimum rate for holiday pay will be the same as the minimum rate for basic pay.

Copies of the Agricultural Wages Order 2002 (Number 1) may be obtained from
The Agricultural Wages Board for England and Wales

Nobel House
17 Smith Square
London SW1P 3JR

Telephone: 020 7238 5862

Grades and responsibilities

Craft Grade
Must have worked in agriculture for at least 3 out of last 4 years and
hold a valid craft certificate issued by the National Proficiency Tests Council in an approved subject or:
hold a valid Level 3 National Vocational Qualification (NVQ) in an approved subject or:
hold a Craft Certificate issued by an Agricultural Wages Committee based on an Employer's declaration or:
be a qualified former Council apprentice or:
hold a valid qualification as specified in sub-section 3.9 of the Agricultural Wages Order

Appointment Grade 2
Day to day responsibility for supervising the work on a farm or part of it, and implementing management decisions, or responsibility for the instruction and supervision of staff.

Appointment Grade 1
A management responsibility for an entire farm or part of it run as a separate operation or business, or responsibility for employing and disciplining staff.

2.2.3 Agricultural Wages, Scotland

New rules relating to agricultural wages and other conditions of service took effect from 1 January 2003. The new rules are contained in the Agricultural Wages (Scotland) Order No. 50 2002. The Order contains the detailed legal requirements for the calculations of minimum pay etc, but the Scottish Office has produced 'A Guide for Workers and Employers' which attempts to explain the new rules in simpler terms.

Both 'The Agricultural Wages (Scotland) Order No. 50 2002' and 'A Guide for Workers and Employers' may be obtained from:

Scottish Agricultural Wages Board
Pentland House
47 Robb's Loan, EDINBURGH EN14 1TY
Telephone: 0131 244 6195
Fax: 0131 244 6551

The new Order applies equally to all workers employed in agriculture in Scotland. It makes no distinction between full-time employees, part-time employees and students.

The old classifications of general farm worker, shepherd, stockworker, tractorman and supervisory grade have been removed and replaced by a single category covering all agricultural workers. The one exception to this is that special arrangements have been made for hill-shepherds.

Instead of minimum weekly rates of pay, the new Order works on minimum hourly rates of pay (with the exception hill-shepherds). It is very important to note that although wages are to be calculated on an hourly basis, a worker is still entitled to be paid for the full number of hours for which he is contracted (unless he is unavailable for work). In particular, this means that an employer cannot reduce the number of hours for which he is to be paid simply by e.g. sending him home early.

Cost information

Minimum rates of wages

There are two main pay scales. One is for workers who have been with their present employer for not more than 10 weeks and the other for workers who have been with their present employer for more than 10 weeks.

Within these two scales, there are varying minimum hourly rates of pay depending on the age of the worker.

The minimum hourly rates of pay are as follows:

	Minimum Hourly Rates (£)	
Age	Up to 10 weeks	After 10 weeks
	£	£
Under 16	1.76	2.36
16 and under 17	2.13	2.84
17 and under 18	2.49	3.31
18 and under 19	3.68	4.03
19 and over	4.31	4.74

Workers who have been with the same employer for more than 10 weeks and hold either:

(a) a Scottish, or National, Vocational Qualification in an agricultural subject at Level III or above, or

(b) an apprenticeship certificate approved by Lantra, NTO (formerly ATB Landbase), or a certificate of acquired experience issued by the ATB Landbase.

shall be paid, for each hour worked, an additional sum of not less than £0.71.

Minimum hourly overtime rate

The minimum hourly rate of wages payable to a worker :

(a) for each hour worked in excess of 8 hours on any day, and
(b) for each hour worked in excess of 39 hours in any week.

shall be calculated in accordance with the following formula:

$M \times 1.65$ (where M = the minimum hourly rate of pay to which the worker is entitled)

2.3 DAYWORK AND PRIME COST

The tables below analyse rates for daywork. The calculations used in the measured works and the derived approximate estimates use current commercial rates for the build-up of all items and do not employ the rates used below. Users should refer to Section 2.1 " Labour Rates used in this edition" at the beginning of this section for information on the actual hourly rates used.

When work is carried out which cannot be valued in any other way it is customary to assess the value on a cost basis with an allowance to cover overheads and profit. The basis of costing is a matter for agreement between the parties concerned, but definitions of prime cost for the building industry have been prepared and published jointly by The Royal Institution of Chartered Surveyors and the Construction Confederation (formerly the Building Employers Confederation, formerly the National Federation of Building Trades Employers) for the convenience of those who wish to use them. We publish these items with their kind permission. Reference should be made to *Spon's Architects' and Builders' Price Book* where these documents are reproduced by kind permission of the publishers together with the Schedule of Basic Plant Charges (January 1990 revision).

The calculations on page 12 shows an example of the calculation of typical standard hourly base rates for craft and general operatives prepared in accordance with the Definition of Prime Cost of Daywork carried out under a Building Contract. A building contract will normally provide for a contractor to tender a percentage adjustment for the base rates for incidental costs, overheads and profit.

Cost information

Daywork Rates - Building Craft & General Operatives

Rates effective from 30 June 2003

		Rate	Craftsman £	Rate	Labourer £
Guaranteed minimum weekly earnings					
Standard Basic Rate	46.2 wks	299.13	13819.81	225.03	10396.39
Guaranteed minimum bonus	46.2 wks	0.00	0.00	0.00	0.00
			13819.81		10396.39
Employer's National Insurance Contribution (12.8% after the first £89.01 per week)			1224.34		786.14
Employer's Contribution to:					
CITB annual levy at 0.50% of payroll			77.77		58.51
Holiday pay	4.2 wks	299.13	1256.35	225.03	945.13
Public holidays	1.6 wks	299.13	478.61	225.03	360.05
EasyBuild Stakeholder Pension (Death and accident cover is provided free)	52.0 wks	5.00	260.00	5.00	260.00
Annual labour cost as defined in section 3			£ 17116.87		£ 12806.21
Hourly rate of labour as defined in section 3, Clause 3.02 (Annual cost/hours per annum)			= £9.50		= £7.11

Note:

1. Calculated following Definition of Prime Cost of Daywork carried out under a Building Contract, published by the Royal Institution of Chartered Surveyors and the Construction Confederation.

2. Standard basic rate effective from 30 June 2003.

3. Standard working hours per annum calculated as follows

52 weeks @ 39 hours		2028.00
Less		
4.2 weeks holiday @ 39 hours	163.8	
8 days public holidays @ 7.8 hours	62.4	-226.2
		1801.8

4. All labour costs incurred by the Contractor in his capacity as an employer other than those contained in the hourly rate, are to be taken into account under Section 6 of the RICS definitions for labour costs.

5. The above examples is for guidance only and does not form part of the Definition; all the basic costs are subject to re-examination according to the time when and in the area where the daywork is executed.

6. All N.I. payments are at non-contracted out rates applicable from April 2003.

7. Basic rate & GMB number of weeks = 52.0 weeks - 4.2 weeks annual holidays – 1.6 weeks public holiday = 46.2 weeks

2.4 COMPUTATION OF LABOUR RATES, COST OF MATERIALS AND PLANT

Different organizations will have varying views on rates and costs, which will in any event be affected by the type of job, availability of labour and the extent to which mechanical plant can be used. However this information should assist the reader to:

(1) compare the prices to those used in his own organization;
(2) calculate the effect of changes in wage rates or prices of materials;
(3) calculate analogous prices for work similar to but differing in detail from the examples given.

Computation of labour rates

From 30 June 2003 basic weekly rates of pay for craft and general operatives are £299.13 and £225.03 respectively; to these rates have been added allowances for the items below in accordance with the recommended procedure of the Chartered Institute of Building in its Code of Estimating Practice. The resultant hourly rates are £10.63 and £7.96 for craft operatives and general operatives respectively.

The items referred to above for which allowances have been made are:

- Lost time
- Non-productive overtime
- Sick pay
- Construction Industry Training Board Levy
- Employer's contribution towards holidays with pay
- Employer's contribution towards pension and death benefit scheme
- Employer's contribution towards National Insurance
- Severance pay and sundry costs
- Employer's liability and third party insurance

Note: For travelling allowances and site supervision see Preliminaries, Section 2.3.

The tables which follow illustrate how the hourly rates referred to above have been calculated. Productive time has been based on a total of 1801.8 hours worked per year for daywork calculations above and for 1953.54 hours worked per year for the all-in labour rates (including 5 hours per week average overtime) for the all in labour rates below.

BUILDING CRAFT AND GENERAL OPERATIVES

Effective from 30 June 2003

		Craft Operatives		General Operatives	
		£	£	£	£
Wages at standard basic rate, productive time	44.30 wks 299.13	13250.23		225.03	9967.91
Lost time allowance	0.904 wks 299.13	270.41		225.03	203.43
Overtime	5.80 wks 448.70	2602.43		337.55	1957.76
			16123.08		12129.09
Extra payments under National Working Rules	45.20 wks	-		-	-
Sick pay	1.00 wk	-		-	-
CITB Allowance (0.50% of payroll)	1 year	90.96			68.43
Holiday Pay	4.20 wks 356.71	1498.16		268.34	1127.04
Public Holiday	1.60 wks 356.71	570.73		268.34	429.35
Travelling allowance (say 30 km)	226 dys	2.16		2.16	
Fare allowance (say 30 km)	226 dys	5.07		5.07	
Employer's contributions to EasyBuild Stakeholder Pension (Death and accident cover is provided free)	52 wks	5.00	260.00	5.00	260.00
National Insurance (average weekly payment)	48 wks	31.60	1516.88	20.95	1005.65
			20059.81		15019.56
Severance pay and sundry costs	Plus	1.5%	300.90	1.5%	225.29
			20360.71		15244.85
Employers liability and third-party insurance	Plus	2.0%	407.21	2.0%	304.90
Total cost per annum			20767.92		15549.75
Total cost per hour			**10.63**		**7.96**

NOTES:
1 Absence due to sickness has been assumed to be for periods not exceeding 3 days for which no payment is due. (Working Rule 20.7.3)
2 EasyBuild Stakeholder Pension effective from 1 July 2002. Death and accident benefit cover is provided free of charge. Taken as £5.00/week average as last year
3 All N.I. Payments are at not-contracted out rates applicable from April 2003
National Insurance is paid for 48 complete weeks(52wks-4.2wks) is based on employer making regular monthly payments into the Template holiday pay scheme and by doing so the employer achieves National Insurance savings on holiday wages.

Calculation of annual hours worked

Number of actual hours worked yearly

Normal working hours (52-6.8 =46.2wks x 39 hrs)   **1801.80**

Less Sick
1 week   **-39.00**
    1762.80

Less time lost for inclement weather @ 2%   **-35.26**
    1727.54

Overtime hours
52wksx5hrs.(overtime)   260.00
Less

Annual holidays	4.2wks x 5hrs	21.00	
Public holidays	1.6wks x 5hrs	8.00	
Sickness	1 week x 5hrs	5.00	**-34.00**

    -226.00

Number of actual hours worked yearly   **1953.54**

COMPUTATION OF LABOUR RATES IN PRICES FOR MEASURED WORK

A survey of typical landscape/ external works companies indicates that they are currently paying above average wages for multi skilled operatives regardless of specialist or supervisory capability. In our current overhaul of labour constants and costs we have departed from our previous policy of labour rates based on national awards and used instead a gross rate per hour for all rate calculations. This rate is tabled at the beginning of this section. Estimators can readily adjust this rate if they feel it inappropriate for their work.

The following section illustrates the calculation of the labour rate of £18.00 per hour used in this years book.

Calculation of Spon's Landscape and External Works Labour Rate

The productive labour of any organisation must return their salary plus the cost of the administration which supports the labour force.

The Labour rate used in this edition is calculated as follows:-

Team size
A three man team with annual salaries as shown and allowances for National insurance, uniforms and site tools plus a company vehicle returns a basic labour rate of £13.38 per hour.

Cost Information

Basic Labour Rate Calculation for Spon's 2004

Working Hours per year (2004) = 1953.4

	Number of	Nett Salary	NI	Nett Cost	Uniforms	Site Tools	TOTAL 3 Man Team
Foreman	1	21000.00	2499.00	23499.00	50.00	100.00	
Craftsman	1	18000.00	2142.00	20142.00	50.00	100.00	
Labourer	1	14000.00	1666.00	15666.00	50.00	100.00	
Overtime (Hrs) *1.5 times normal rate*	225	17.49	2.08	4403.62			
	225	14.99	1.78	3774.53			
	225	11.66	1.39	2935.74			
Total staff in Team	3.00	++	++	70420.89	150.00	300.00	£70,870.89
Vehicle Costs Inclusive of Fuel Insurances etc	Working Days		£ /Day				
	225.13		36.00			>	£ 8,104.50
						TOTAL	£78,975.39

The Basic Average labour rate excluding overhead costs (below) is

$$\frac{£78975.39}{(\text{Total staff in team}) \times (\text{Working hours per year})}$$

$$= £13.38$$

Overhead Costs

Add to the above basic rate the company overhead costs for a small company as per the table below. These costs are absorbed by the number of working men multiplied by the number of working hours supplied in the table on page 5. This then generates an hourly overhead rate which is added to the Nett cost rate above.

The table on the next page summarises these costs for a small to medium sized company

Cost information

Illustrative overhead for small company employing 9 -18 landscape operatives

Cost Centre	Number of	Cost	Total
MD Salary only; excludes profits;	1.0	40000.00	40000.00
Contracts managers Salaries and vehicles.	1.0	30000.00	30000.00
Secretary	1.0	15000.00	15000.00
Book Keeper	1.0	22000.00	22000.00
Rental	12.0	500.00	6000.00
Insurances	1.0	4000.00	4000.00
Telephone and Mobiles	12.0	200.00	2400.00
Office Equipment	1.0	2000.00	2000.00
Stationary	12.0	50.00	600.00
Advertising	12.0	100.00	1200.00
Other vehicles not allocated to contract teams	3.0	8000.00	24000.00
Other consultants	1.0	5000.00	5000.00
Accountancy	1.0	800.00	800.00
Lights water	1.0	1000.00	1000.00
Other	1.0	6000.00	6000.00
TOTAL OFFICE OVERHEAD			160000.00

The final labour rate used is then generated as follows:

Nett labour rate   +   Total Office Overhead
                         (Working Hours per year x Labour resources employed)

Total nr of site staff	Admin Cost per hour	Total Rate per Man Hour
6	13.65	27.13
9	9.10	22.58
12	6.83	20.31
15	5.46	18.94
18	4.55	18.03 (Rate used)

Hourly Labour Rates used for Spon's Landscape and External Works 2004

General contracting	£18.00
Maintenance contracting	£14.00

COMPUTATION THE COST OF MATERIALS

Percentages of default waste are placed against material resources within the supplier database. These range from 2.5% (for bricks) to 20% for topsoil

An allowance for the cost of unloading, stacking etc should be added to the cost of materials.

The following are typical hours of labour for unloading and stacking some of the more common building materials.

Material	Unit	Labourer hour
Cement	tonne	0.67
Lime	tonne	0.67
Common bricks	1000	1.70
Light engineering bricks	1000	2.00
Heavy engineering bricks	1000	2.40

COMPUTATION OF MECHANICAL PLANT COSTS

Plant used within resource calculations in this book have been used according to the following principles:

1. Plant which is non-pedestrian plant is fuelled by the Plant Hire company.
2. Operators of excavators above the size of 5 tonnes are provided by the Plant hire company and no labour calculation is allowed for these operators within the rate.
3. Three to Five tonne excavators and dumpers are operated by the landscape contractor and additional labour is shown within the calculation of the rate.
4. Small plant is operated by the landscape contractor.
5. No allowance for delivery or collection has been made within the rates shown.
6. Downtime is shown against each plant type either for non productive time or mechanical down time

The following tables lists the plant resources used in this years book

Plant Supplied by Road Equipment Ltd

Prices shown reflect 36 working hours per week

Description	Unit	Supply Quantity	Rate Used £	Down Time %
Excavator 360 Tracked 21 tonne- *fueled and operated*	hour	1	29.95	10
Excavator 360 Tracked 5 Ton *fueled only*	hour	1	12.01	10
Excavator 360 Tracked 7 Ton *fuelled and operated*	hour	1	22.61	10
Mini Excavator JCB 803 Rubber Tracks Self Drive *fuelled only*	hour	1	8.69	10
Mini Excavator JCB 803 Steel Tracks Self Drive *fuelled only*	hour	1	8.41	10
JCB 3Cx 4X4 Sitemaster, *fuelled and operated*	hour	1	28.10	10
JCB 3Cx 4 X 4 Sitemaster + Breaker *fuelled and operated*	hour	1	28.40	10
Dumper 5 Ton, Thwaites Self Drive *fuelled only*	hour	1	3.84	10
Dumper 5 Ton, Thwaites Self Drive *fuelled only*	hour	1	5.81	10
Manitou Telehandler *fuelled only*	hour	1	9.61	50

Cost Information

Plant Supplied by HSS Hire Shops

Description	Unit	Supply Quantity	Supply Cost £	Rate Used £	Down Time %
Auger Mechanical 2 Man Weekly Rate	hour	25	101.00	4.04	37.5
Engine Poker Vibrator	hour	25	90.00	3.60	37.5
Vibrating Plate Compactor	hour	25	58.00	2.32	37.5
Petrol Masonry Saw Bench	hour	25	120.00	4.80	35.5
Diamond Blade Consumable, 230 mm	mm	1	25.50	25.50	-
Access Platform To 5.2 M	day	5	132.00	26.40	-
Cutting Torch Oxy Acetylene	week	1	80.00	80.00	-
Cutting Torch Oxy Acetylene	hour	1	6.25	6.25	33.3
Bolt Croppers, 900 mm	hour	1	1.15	1.15	-
Hydraulic Breaker Diesel 5Hrs /Day	hour	1	2.30	3.90	37.5
Rotavator Howard Gem (Based On Weekly Rate) 25 Hrs/Week	hour	1	5.38	8.40	37.5
Rotavator Tractor Mounted 1200mm + Operator	hour	1	23.50	23.50	-
Compactor Bomag 136Kg/ 10.1Kn 4 Hrs /Day	hour	1	2.82	5.50	51
Compactor Bomag 470 Kg /18.7Kn 4 Hrs /Day	hour	1	3.58	7.00	51

2.5 COST INDICES

The purpose of this section is to show changes in the cost of carrying out landscape work (hard surfacing and planting) since 1990. It is important to distinguish between costs and tender prices: the following table reflects the change in cost to contractors but does not necessarily reflect changes in tender prices. In addition to changes in labour and material costs, which are reflected in the indices given below, tender prices are also affected by factors such as the degree of competition at the time of tender and in the particular area where the work is to be carried out, the availability of labour and materials, and the general economic situation. This can mean that in a period when work is scarce tender prices may fall despite the fact that costs are rising, and when there is plenty of work available, tender prices may increase at a faster rate than costs.

The Constructed Cost Index

A Constructed Cost Index based on PSA Price Adjustment Formulae for Construction Contracts (Series 2). Cost indices for the various trades employed in a building contract are published monthly by HMSO and are reproduced in the technical press.

The indices comprise 49 Building Work indices plus seven 'Appendices' and other specialist indices. The Building Work indices are compiled by monitoring the cost of labour and materials for each category and applying a weighting to these to calculate a single index.

Although the PSA indices are prepared for use with price-adjustment formulae for calculating reimbursement of increased costs during the course of a contract, they also present a time series of cost indices for the main components of landscaping projects. They can therefore be used as the basis of an index for landscaping costs.

The method used here is to construct a composite index by allocating weightings to the indices representing the usual work categories found in a landscaping contract, the weightings being established from an analysis of actual projects. These weightings totalled 100 in 1976 and the composite index is calculated by applying the appropriate weightings to the appropriate PSA indices on a monthly basis, which is then compiled into a quarterly index and rebased to 1976 = 100.

Constructed Landscaping (Hard Surfacing and Planting) Cost Index
Based on approximately 50% soft landscaping area and 50% hard external works.
1976 = 100

Year	First Quarter	Second Quarter	Third Quarter	Fourth quarter	Annual average
1990	330	334	353	357	344
1991	356	356	364	364	360
1992	364	365	376	378	371
1993	380	381	383	384	382
1994	387	388	394	397	391
1995	399	404	413	413	407
1996	417	419	428	430	424
1997	430	432	435	442	435
1998	442	445	466	466	455
1999	466	468	488	490	478
2000	493	496	513	514	504
2001	512	514	530	530	522
2002	531	540	573	574	555
2003	575*				

* *Provisional*

This index is updated every quarter in Spon's Price Book Update. The updating service is available, free of charge, to all purchasers of Spon's Price Books. (Complete the reply-paid card enclosed.)

2.6 REGIONAL VARIATIONS

Prices in Spon's Landscape and External Works Price Book are based upon conditions prevailing for a competitive tender in the outer London area. For the benefit of readers, this edition includes regional variation adjustment factors which can be used for an assessment of price levels in other regions.

Special further adjustment may be necessary when considering city centre or very isolated locations.

Region	Adjustment Factor
Outer London	1.00
Inner London	1.07
South East	0.93
South West	0.83
East Midlands	0.82
West Midlands	0.82
East Anglia	0.84
Northern	0.79
North West	0.82
Scotland	0.80
Wales	0.81
Northern Ireland	0.58
Channel Islands	1.24

The following example illustrates the adjustment of prices for regions other than outer London, by use of regional variation adjustment factors.

		£
A.	Value of items priced using Spon's Landscape and External Works Price Book	100 000
B.	Adjustment to value of A. to reflect East Midlands Region price level 100 000 x 0.82	82 000

ESSENTIAL READING FROM SPON PRESS

The Regeneration of Public Parks
Edited by Jan Woudstra & Ken Fieldhouse

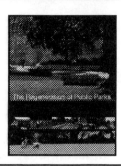

'Offers a timely overview of urban park design and discusses many issues common to park regeneration. It is an important reference document for many professions involved in public parks.'
- *Landscape Design*

The Urban Parks Programme, financed by the Heritage Lottery Fund, has sparked a new enthusiasm for the regeneration of Britain's parks. This unique reference book gives a valuable overview of all the elements of public park design. It emphasizes our parks' diversity and richness, and offers practical guidance as to their renovation and future care. It is essential reading for all those involved in the design, upkeep and regeneration of public parks.

Contents: Acknowledgements. Contributors. Foreword.INick Raynsford MP II Lorna McRobie III Alan Barber IV Stewart Harding. Introduction: The Regeneration of Public Parks -Jan Woudstra. Parks and People: The Social Functions - Hazel Conway. Memory and Value - David Jacques. Study Methods - Mary Lockwood. Buildings and Monuments - Hazel Conway and David Lambert. The Use of Iron - Edward Diestelkamp. Paths - Brent Elliott. Shrubbery Planting (1830-1900) - Kim Legate. Bedding Schemes - Brent Elliott. Late 20th Century Bedding Practice - Frazer Chapman. Lakes and Water Features - Hilary Taylor. Play and Sport - Brent Elliott and Ken Fieldhouse. Bibliography. Index.

June 2000: 276x219 mm: 200 pp.
132 b+w photos; 8 pages colour illustrations
PB: 0-419-25900-7: £32.00

To Order: Tel: +44 (0) 8700 768853, or +44 (0) 1264 343071 Fax: +44 (0) 1264 343005, or Post: Spon Press Customer Services, Thomson Publishing Services, Cheriton House, Andover, Hants, SP10 5BE, UK Email: book.orders@tandf.co.uk

For a complete listing of all our titles visit:
www.sponpress.com

Design for Outdoor Recreation

Simon Bell

Spon Press
Taylor & Francis Group

'very successful...nicely presented and easy to read...will provide a useful reference and stimulus to design professions, country managers and planners'
  - *Journal of Environmental Planning and Management*

This book takes a fresh, up-to-date look at all aspects of design of facilities needed by visitors to outdoor recreation destinations. The book follows the thought processes and physical requirements of visitors, from their perspective. This idea is followed step-by-step from setting off for a destination to the arrival and the visit itself, broken down into a range of different aspects, such as information needs, car parking, picnicking, hiking a trail, recreation by water, wildlife watching and camping. The book is fully illustrated with sketches, diagrams and photographs, many in colour, taken from a wide range of locations in the United Kingdom, Europe, North America and elsewhere.

Contents:
Introduction. Recreation planning. Design concepts for outdoor recreation. The journey to the destination. Providing visitor information. Parking the car. Toilet facilities. Picnicking. Children's play. Trails. Water based recreation. Wildlife viewing. Design for overnight visitors. Interpretation. Comprehensive site design. Bibliography. Index.

January 1997: 276x219: Pb: 0-419-20350-8: £52.00

To Order: Tel: +44 (0) 8700 768853, or +44 (0) 1264 343071
Fax: +44 (0) 1264 343005, or **Post:** Spon Press Customer Services, ITPS Andover, Hants, SP10 5BE, UK

For a complete listing of all our titles visit: www.sponpress.com

3
Preliminaries

3.1 PRELIMINARIES/GENERAL CONDITIONS

The number of items priced in the Preliminaries section of Tender Documents and the manner in which they are priced vary considerably between Contractors. Some Contractors, by modifying their percentage factor for overheads and profit, attempt to cover the costs of Preliminary items in their Prices for Measured Work. However, the cost of Preliminaries will vary widely according to job size and complexity, site location, accessibility, degree of mechanization practicable, position of the Contractor's head office and relationships with local labour/domestic sub-contractors. It is therefore usually far safer to price Preliminary items separately on their merits according to the job.

Reference should be made to *Spon's Architects' and Builders'* Price Book for the normal clause descriptions from the Preliminaries section of a bill of quantities where the *JCT Standard Form of Building Contract* 1980 Edition is to be used.

For the convenience of readers the clause descriptions which are likely to be found in the Preliminaries section of the tender documents for a project which consists of solely or mainly soft landscaping works and is therefore to be based on the *JCLI Form of Agreement* for Landscape Works January 1992 Edition, June 1996 revision, are given below together with further details against those items which are usually priced in tenders.

Note. The term 'Not priced' where used throughout this section means either that the cost implication is negligible or that it is usually included elsewhere in the tender.

1.	**Project, parties and consultants.**	Not priced
2.	**Description of site.**	Not priced
3.	**Drawings and other documents.**	Not priced
4.	**Form, type and conditions of contract.**	

Clause No.
1.0 *Intentions of the parties*
1.1	Contractor's obligation.	Not priced
1.2	Landscape Architect's duties.	Not priced
1.3	Contract Bills and SMM	Not priced
1.4	Reappointment of Planning Supervisor or Principal Contractor - notification to Contractor.	Not priced
1.5	Alternative B2 in the 5th recital - notification by Contractor – regulation 7(5) of the CDM Regulations.	Not priced

2.0 Commencement and completion
2.1	Commencement and completion.	Not priced
2.2	Extension of contract period.	Not priced
2.3	Damages for non-completion.	Not priced
2.4	Practical completion. Not priced	
2.5	Defects liability. Inevitably some defects will arise and an allowance will often be made to cover this either here or against Clause 2.7.	
2.6	Partial practical completion.	Not priced
2.7	Failures of plants A or B. Inevitably some replacements will be required and an allowance should be made here or in the rates. Plant replacements up to and including 10% loss are usually covered in the planting measured work prices.	

3.0 Control of the Works
3.1	Assignment.	Not priced
3.2	Sub-contracting.	Not priced
3.3	Contractor's representative. Any allowance for site supervision/administration which is not included in the rates should be priced here together with the cost of setting out the works.	
3.4	Exclusion from the Works.	Not priced
3.5	Landscape Architect's instructions.	Not priced
3.6	Variations. Not priced	
3.7	PC and Provisional sums. An amount should be added to the PC sum items, if required, for profit and a further sum, where applicable, for attendance.	
3.8	Objections to a Nomination.	Not priced

4.0 Payment
4.1	Correction of inconsistencies.	Not priced
4.2	Progress payments and retention.	Not priced
4.3	Penultimate certificate	Not priced
4.4	Final certificate.	Not priced
4.5	Contribution, levy and tax changes.	Not priced
4.6 A	Fixed price. The additional cost of providing a fixed price can be included here or in the rates.	
B	Fluctuations. An allowance can be made here for the estimated shortfall in reimbursement (if any) under a fluctuating price contract.	

5.0 Statutory obligations

5.1	Statutory obligations, notices, fees and charges. The cost of complying with this clause should be included here.	
5.2	Value Added Tax.	Not priced
5.3	Statutory tax deduction scheme.	Not priced
5.4	Number not used.	Not priced
5.5	Prevention of corruption.	Not priced
5.6	Employer's obligation - Planning Supervisor - Principal Contractor	Not priced
5.7	Duty of Principal Contractor.	Not priced
5.8	Successor appointed to the Contractor as Principal Contractor	Not priced
5.9	Health and Safety file.	Not priced

6.0 Injury, damage and insurance

6.1 Injury to or death of persons. The cost of maintaining insurance should be included here if not allowed for elsewhere.

6.2 Injury or damage to property. As above in 6.1.

6.3 A Insurance of the Works by Contractor - Fire, etc. If at the Contractor's risk the cost of maintaining insurance cover must be sufficient to include the full cost of reinstatement, all increases in cost, professional fees and any consequential costs such as clearing debris.

    B Insurance of the Works and any existing structures by Employer - Fire, etc. As above in 6.3 A.

6.4 Evidence of insurance.   Not priced

6.5 Malicious damage or theft A or B. If at the Contractor's risk the cost of complying with this clause should be included here.

7.0 Determination

7.1	Notices. Not priced	
7.2	Determination by Employer.	Not priced
7.3	Determination by Contractor.	Not priced

8.0 Supplementary Memorandum

8.1 Meaning of references in 4.5, 4.6, 5.2 and 5.3.   Not priced

9.0 Settlement of Disputes – Arbitration   Not priced

10. Contractor's liability.   Not priced

(If the JCLI Form of Agreement is used this item is covered under Item 6. If a different Form of Contract is used it may be necessary to include for Contractor's insurances, etc., under this item.)

Preliminaries

11. ***Employer's liability.*** Not priced
(If additional insurances are to be allowed for in the contract sum such cost will be given as a provisional sum.)

12. ***Local Authorities' fees and charges.*** Not priced
(If the JCLI Form of Agreement is used this item is covered under Item 5.
If a different Form of Contract is used it may be necessary to include for the fees and charges which the Contractor is required to pay under this Item unless given as provisional sums.)

13. ***Obligations and restrictions imposed by the employer.***
These include the following items and costs can only be assessed in the light of circumstances on a particular job.

- (a) Access to and possession of use of the site.
- (b) Limitations of working space.
- (c) Limitations of working hours.
- (d) The use or disposal of any materials found on site.
- (e) Hoardings, fences, and the like, temporary name boards and advertising rights.
- (f) The maintenance of existing live drainage, water, gas and other main or power services on or over the site.
- (g) The execution or completion of the work in any specific order or in sections or phases.
- (h) Temporary accommodation and facilities for the use of the Employer including heating, lighting, furnishing and attendance.
  This will include an office for the Clerk of Works if there is to be one on the site.
  Against this item the following should be priced:
  - i. Hire of Clerk of Works Office
  - ii. Transport to and from the site
  - iii. Erecting on a suitable base and later dismantling
  - iv. Lighting, heating and attendance on office
  - v. Local Authority rates
- (i) The installation of telephones for the use of the Employer. (The cost of his telephone calls should be covered by a provisional sum.)
- (k) Any other obligation or restriction.

Additional obligations may include the provision of a performance bond. If the Contractor is required to provide sureties for the fulfilment of the work the usual method of providing this is by a bond provided by one or more insurance companies. The cost of a performance bond depends largely on the financial standing of the applying contractor. Figures tend to range from 0.25 to 0.5% of the contract sum.

Preliminaries

Works by nominated sub-contractors, goods and materials from nominated suppliers and works by public bodies

14. **Works by nominated sub-contractors.**   Not priced
    (Work to be carried out by a nominated sub-contractor and given as a prime cost sum to which an amount should be added, if required, for profit and a further sum for attendance.)

15. **Goods and materials from nominated suppliers.**   Not priced
    (Goods and materials which are required to be obtained from a nominated supplier and given as a prime cost sum to which an amount should be added, if required, for profit.)

16. **Works by Public Bodies.**   Not priced
    (Works which are to be carried out by a Local Authority or public undertaking given as a provisional sum.)

17. **Works by others directly engaged by the Employer.**   Not priced
    (A description given for works by others directly engaged by the employer, and any attendance that is required shall be priced in the same way as works by nominated sub-contractors.)

General facilities and obligations

18. **Pricing**
    For convenience in pricing items will be listed which may include the following. The contractor should include maintaining any temporary works in connection with the items, adapting, clearing away and making good, and all notices and fees to local authorities and public undertakings.

    a) *Plant, tools and vehicles*

    The sixth edition of the Standard Method of Measurement of Building Works requires that items for plant be given at the beginning of each section, whereas the seventh edition provides for these items to be covered under A42 and A43. However, Contractors often include the cost of their own plant within the measured rates e.g. earthmoving.

b) *Site administration and security*

If the JCLI Form of Agreement is used the cost of administrative staff is often included against Item 3. When required allow for the provision of a watchman or inspection by a security organization.

c) *Transport for workpeople*

The labour rates per hour on which Prices for Measured Work have been based do not cover travel and lodging allowances which must be assessed according to the particular circumstances.

d) *Protecting the works from inclement weather*

In areas likely to suffer particularly inclement weather, some nominal allowance should be included for tarpaulins, polythene sheeting, etc., and the effect of any delays in concreting or brickwork by such weather.

e) *Water for the works*

Charges should properly be ascertained from the local Water Authority. If these are not readily available, an allowance of 0.33% of the value of the contract is probably adequate, providing water can be obtained directly from the mains. Failing this, each case must be dealt with on its merits. In all cases an allowance should also be made for temporary plumbing including site storage of water if required.

f) *Lighting and power for the works*

The Contractor is usually responsible for providing all temporary lighting and power for the works and all charges involved.

g) *Temporary roads, hardstandings, crossings and similar items*

Quite often consolidated bases of eventual site roads are used throughout a contract to facilitate movement of materials around the site. However, during the initial setting up of a site, with drainage works outstanding this is not always possible and occasionally temporary roadways have to be formed and ground levels later reinstated.

h) *Temporary accommodation for the use of the Contractor*

This includes all temporary offices and sheds for the Contractor and his domestic sub-contractors' use (temporary office for a Clerk of Works is covered under obligations and restrictions imposed by the employer).

Allowances must also be made for transport, erection, dismantling, etc., as previously shown under the item for Clerk of Works office.

i) *Temporary telephones for the use of the Contractor*

Against this item should be included the cost of installation, rental and an assessment of the cost of calls made during the contract.

j) *Traffic regulations*

Waiting and unloading restrictions can occasionally add considerably to costs, resulting in forced overtime or additional weekend working. Any such restrictions must be carefully assessed for the job in hand.

k) *Safety, health and welfare of workpeople*

The Contractor is required to comply with all relevant codes, regulations, agreements and statutes and in particular should allow for provision of the following:

1. Shelter from inclement weather
2. Accommodation for clothing
3. Accommodation and provision for meals
4. Provision of drinking water
5. Sanitary conveniences
6. Washing facilities
7. First aid
8. Site conditions

A variety of self-contained mobile or jack-type units are available for hire and allowance must be made in addition for transport costs to and from site, setting up costs, connections to mains, fuel supplies and attendance. A general provision to comply with the above code is often 0.50 to 0.75% of the contract value. The cost of safety supervisors (required for firms employing more than 20 people) is usually part of head office overhead costs.

l) *Disbursements arising from the employment of workpeople*

Travelling and lodging allowances have been dealt with under Transport for Workpeople and usually all other on-costs and disbursements are included in the all-in hourly rate used in the calculation of Prices for Measured Work. However, it is as well to check that all such disbursements have been included elsewhere.

m) *Maintenance of public and private roads*

Some additional insurance or value may be required against this item to insure against damage to entrance gates, kerbs or bridges caused by extraordinary traffic in the execution of the works.

n) *Removing rubbish, protective casings and coverings and cleaning the works on completion*

This includes removing surplus materials and final cleaning of the site prior to handover. Allow for sufficient 'bins' for the site throughout contract duration and for some operative time at the end of the contract for final clearing and cleaning ready for handover.

A general allowance of 0.20% of contract value is probably sufficient.

o) *Temporary fencing, hoardings and similar items*

This item must be considered in some detail as it is dependent on site perimeter, phasing of the work, etc.

p) *Control of noise, pollution and all other statutory obligations*

The Local Authority may impose restrictions on the timing of certain operations, particularly noisy or dust-producing operations, and may necessitate the carrying out of these outside normal working hours or using special tools and equipment.

The situation is most likely to occur in built up areas such as city centres, etc., where the site is likely to be in close proximity to offices or residential property.

19. Contingencies

Generally. Provision for contingencies is normally given as a provisional sum.

4
Landscape Consultant's Appointment

Note to Readers
The following information is extracted and summarized from the Landscape Architects Appointment publication which is obtainable in full from:
The Landscape Institute, 6/8 Barnard Mews, London SW11 1QU.
Tel: 020 7350 5200.   Fax: 020 7350 5201.

FOREWORD

'The Landscape Consultants Appointment' published by the Landscape Institute is a document designed to advise Landscape Consultants and their Clients in the execution of landscape commissions. It has been registered with the Office of Fair Trading under the terms of the Restrictive Trade Practices Act 1976.

The Landscape Institute is the Chartered Institute in the UK for Landscape Architects, incorporating Designers, Managers and Scientists. One of its objects is to promote the highest standard of professional service in the application of the arts and sciences of landscape architecture and management. All members of the Institute, collectively referred to in this document as Landscape Consultants, are governed by the Institute's Code of Professional Conduct, but the roles of the members of the three different landscape disciplines vary in accordance with their education and experience.

Landscape Architects (Design) identify and solve problems using design techniques based on an understanding of the external environment, knowledge of the functional and aesthetic characteristics of landscape materials, and of the organization of landscape elements, external spaces and activities.

Landscape Architects (Management) are concerned with the long term care and development of new and existing landscapes and also with policy and planning for future management and use. This involves them in the organization of manpower, machinery and materials and requires a working knowledge of statutory measures and Grant Aid Schemes in order to preserve and enhance the quality of the landscape.

Landscape Architects (Science) are specialists in the physical and biological aspects of landscape design and management. Relating scientific expertise to the practical problems of designers and managers, they may work in any of the many scientific subjects that are relevant to landscape, especially botany, ecology and soil surveys.

The Landscape Institute has produced three inter-related documents which assist clients and landscape consultants to reach clear agreement on the terms and conditions of an appointment. The documents are:

The Landscape Consultants Appointment which sets out the Memorandum of Agreement as well as the standard and additional services provided, conditions of service which apply and a schedule of services and fees.

Engaging a Landscape Consultant - Guidance to Clients on Fees: please refer to Chapter 5 of this book

Guide to Procedure for Competitive Tendering which guides clients through the initial considerations and the procedures needed to obtain comparable and valid fee tenders for projects.

Any question on or arising out of the information contained in any of these documents may be referred to the Director General, The Landscape Institute, 6/8 Barnard Mews, London SW11 1QU. Tel: 020 7350 5200. Fax: 020 7350 5201.

0 INTRODUCTION

0.1 The Landscape Consultant's professional responsibility is to act as the Client's adviser based upon a clear understanding of the Client's requirements and to administer any contract between Client and Contractor impartially and fairly. A good working relationship between the Client and the Landscape Consultant is therefore essential when proceeding with a commission. The Clients also have an important role. They must provide adequate information on the project, site and budget and fully understand and approve the Landscape Consultant's proposals at various stages of the work as it proceeds.

0.2 The most successful relationships are those which proceed in an atmosphere of mutual trust and goodwill. An understanding of the Client's and the Consultant's respective obligations by both parties is fundamental to the creation of such an atmosphere. The Landscape Institute therefore advises its members that before accepting a commission for professional services, they should agree with the Client the terms of the commission, including the scope of the services; the allocation of responsibilities and any limitation of liability; the payment of fees including the rates and methods of calculation and the provision for termination.

0.3 This document consists of the following parts:

0.3.1 **Memorandum of Agreement**
The Institute advises its members to use the Memorandum of Agreement and Schedule of Services and Fees. Alternatively, letters of appointment could also serve the purpose, provided that the services, responsibilities and fee basis are fully defined. If the agreement is not comprehensive it is likely to create uncertainties for either of both parties as the commission progresses.

0.3.2 The Landscape Consultant's Appointment

(Appendix I) in three parts:

Part 1 - Landscape Consultant's Services. The work which a Landscape Consultant normally undertakes after accepting a commission is primarily concerned with landscape design, construction (including planting), and management. It comprises Preliminary Services and Standard Services, sub-divided into work stages as follows:

Preliminary Services	Work Stage A - Inception
	Work Stage B - Feasibility
Standard Services	Work Stage C - Outline Proposals
	Work Stage D - Sketch Scheme Proposals
	Work Stage E - Detailed Proposals
	Work Stage F - Production Information
	Work Stage G - Bills of Quantities
	Work Stage H - Tender Action
	Work Stage J - Contract Preparation
	Work Stage K - Operations on site
	Work Stage L - Completion

The Landscape Consultant may also provide services outside the design, construction and management process, for example public inquiry commissions, landscape planning issues, environmental impact assessment, landscape appraisal and evaluation.

Part 2 - Other Services. Many commissions will require Other Services additional to the Preliminary and Standard Services, and it may be necessary to obtain and incorporate advice from various different consultants.

Part 3 - Conditions of Appointment. This part describes conditions which normally apply when a Landscape Consultant is commissioned.

0.4 There follows a summary of the subject headings in each of the Parts for ease of reference to the original document.

0.5 Guidance on the various types of fee arrangements and their appropriate use is currently the subject of further discussion with the Office of Fair Trading.

APPENDIX I

PART 1   **1**   **LANDSCAPE CONSULTANT'S SERVICES**

1.1 This Part describes the Preliminary and Standard Services which a Landscape Consultant will normally provide. These services are common to large and small commissions and none should be omitted if the commission is to be completed successfully. However, it may be prudent to vary the sequence of the Work Stages or to combine two or more stages to suit the particular circumstances. Where, for any reason, partial services only are to be provided, the agreement between the Client and the Consultant should indicate precisely the extent of those services. Preliminary Services comprise Work Stages A and B; Standard Services comprise Work Stages C to L inclusive. Other Services are described in Part 2.

Work Stage A   1.2   **Preliminary Services - Work Stage A: Inception**

1.2.1 The Brief
1.2.2 Information which may be provided by the Client
1.2.3 Site Appraisal
1.2.4 Advice on other Consultants, Specialist Firms and Site Staff

1.2.5 Programme

Work Stage B   1.3   **Preliminary Services - Work Stage B: Feasibility**

1.3.1 Feasibility Studies
1.3.2 Landscape Consultants Range of Services

Work Stage C   **1.4**   **Standard Services – Work Stage C: Outline Proposals**

1.4.1 Outline Proposals
1.4.2 CDM Designers Duties

Work Stage D   **1.5**   **Standard Services - Work Stage D: Sketch Scheme Proposals**

1.5.1 Sketch Scheme Proposals
1.5.2 Changes in Scheme Proposals
1.5.3 Outline Planning Application
1.5.4 Other Approvals

Work Stage E   **1.6**   **Standard Services - Work Stage E: Detailed Proposals**

1.6.1 Detailed Proposals
1.6.2 Cost Checks and Changes in Detailed Proposals
1.6.3 Detailed Statutory Approvals

Work Stages F & G	**1.7**	**Standard Services - Work Stages F and G: Production Information and Bills of Quantities**
	1.7.1	Production Information
	1.7.2	Bills of Quantities
Work Stages H & J	**1.8**	**Standard Services - Work Stages H and J: Tender Action and Contract Preparation**
	1.8.1	Other Contracts
	1.8.2	Tender Lists
	1.8.3	Tender Action or Negotiation
	1.8.4	Contract Document Preparation
Work Stage K	**1.9**	**Standard Services - Work Stage K: Operations on Site During Construction and 12 Months' Maintenance**
	1.9.1	Contract Administration
	1.9.2	Inspections
	1.9.3	Accounts
	1.9.4	Financial Appraisal and Programme
Work Stage L	**1.10**	**Standard Services - Work Stage L: Completion**
	1.10.1	Completion of Works
PART 2	**2**	**OTHER SERVICES**
	2.1	This part describes services which may be provided by the Landscape Consultant by prior agreement with the Client to augment the Preliminary and Standard Services described in Part 1 or which may be the subject of a separate appointment. The list of services is not exhaustive.
	2.2	**Surveys and Investigations**
	2.2.1	Site Evaluation
	2.2.2	Measured Surveys
	2.2.3	Site Investigation
	2.2.4	Maintenance and Management Cost in Use
	2.2.5	Environmental Assessment
	2.2.6	Development Plans

2.2.7	Demolition and Clearance
2.2.8	Special Drawings and Models
2.2.9	Prototype
2.2.10	Site Furniture and Equipment
2.2.11	Multi- Disciplinary Meetings
2.2.12	Public Meetings
2.2.13	Works of Art
2.2.14	Scientific Developments
2.2.15	Visits to Nurseries

2.3	**Cost Estimating and Financial Advisory Services**
2.3.1	Cost Plans and Cash Flow Requirements
2.3.2	Schedules of Rates and Quantities
2.3.3	Cost of Replacement and Reinstatement of Damaged Landscape
2.3.4	Grant Applications

2.4	**Planning Negotiations**
2.4.1	Planning Applications; Exceptional Negotiations
2.4.2	Planning Appeals and Public Inquiries
2.4.3	Royal Fine Art Commission
2.4.4	Building Regulations; Exceptional Negotiations
2.4.5	Landlords Approvals
2.4.6	Rights of Owners and Lessees
2.4.7	Statutory Bodies

2.5	**Additional Administration of Projects**
2.5.1	Site Staff
2.5.2	Extended Administration
2.5.3	Suspension
2.5.4	Project Management
2.5.5	Record Drawings
2.5.6	Landscape Management Plans
2.5.7	Contract Claims
2.5.8	Litigation and Arbitration

	2.5.9	Health and Safety Plan
	2.5.10	Planning Supervisor

	2.6	**Services normally provided by other Consultants**
	2.6.1	Other Consultants' Services

	2.7	**Additional Design Services**
	2.7.1	Alteration or modification

PART 3

	3	**CONDITIONS OF APPOINTMENT**
	3.1	This Part describes the conditions which apply to a Landscape Consultant's appointment. If different or additional services are required they should be set out in the Schedule of Services, Memorandum of Agreement or letter of appointment.
	3.2	Duty of Care
	3.3	Contamination or Pollution
	3.4	Landscape Consultant's Authority
	3.5	Modifications
	3.6	Terms for Payment
	3.7	Revisions to the Conditions of Appointment
	3.8	Project Control
	3.9	Impartiality
	3.10	Appointment of other Consultants
	3.11	Sub-contracting
	3.12	Liability of other Consultants
	3.13	Consultant Co-ordination
	3.14	Design work by Contractors/ Suppliers
	3.15	Contractor Responsibility
	3.16	Site Inspections
	3.17	Site Staff and Resident Landscape Staff
	3.18	Site Staff/Clerk of Works
	3.19	Information from Client
	3.20	Client Representative
	3.21	Copyright

3.22	Copyright Entitlement
3.23	Assignment
3.24	Suspension
3.25	Suspension by Consultant
3.26	Resumption of Service
3.27	Suspension of Obligations
3.28	Termination
3.29	Death or Incapacity
3.30	Adjudication
3.31	Appointment of Adjudicator
3.32	Ruling on a Joint Statement
3.33	Arbitration
3.34	Alternative for Scotland
3.35	Settlement of Disputes by Agreement
3.36	Governing laws England and Wales
	Alternative for Scotland
	Alternative for Northern Ireland
3.37	CDM Regulations 1994

ESSENTIAL READING FROM SPON PRESS

Urban Open Spaces
Helen Woolley

There is enormous interest in urban design and the regeneration of our urban areas, but current thinking often concentrates on the built form, forgetting the important role that open spaces play. *Urban Open Spaces* brings together extensive research and practical experience to prove the opportunities and benefits of different types of open space to society and individuals.

Focusing on the importance of open spaces in daily urban life, the book is divided into three sections. The first section describes the social, health, environmental and economic benefits and opportunities that open spaces can provide. The second section discusses the different types of urban open spaces that individuals or communities might use on a daily basis: from private gardens to commercial squares and waterway corridors. In the final section of the book best practice case-studies show how urban spaces are now being incorporated in new developments and community initiatives.

Not just for students and practitioners, this book will be of value for anyone interested in the design, development, regeneration, funding and use of open spaces in urban areas.

June 2003: 276x219 mm mm: 208 pp.
123 b+w photos
HB: 0-419-25690-3: £35.00

To Order: Tel: +44 (0) 8700 768853, or +44 (0) 1264 343071 Fax: +44 (0) 1264 343005, or Post: Spon Press Customer Services, Thomson Publishing Services, Cheriton House, Andover, Hants, SP10 5BE, UK Email: book.orders@tandf.co.uk

For a complete listing of all our titles visit:
www.sponpress.com

5
Fees for Professional Services

LANDSCAPE ARCHITECTS' FEES

The Landscape Institute no longer sets any fee scales for its members. A publication which offers guidance in determining fees for different types of project is available from

The Landscape Institute
6/8 Barnard Mews London SW11 1QY   Telephone: 020 7350 5200

The publication is entitled
"Engaging a Landscape Consultant – Guidance for Clients on Fees.

The document refers to suggested fee systems on the following basis:
- Time Charged Fee Basis
- Lump Sum Fee Basis
- Percentage Fee Basis
- Retainer Fee Basis

In regard to the 'Percentage Fee Basis' the publication lists various project types and relates them to complexity ratings for works valued at £22,500 and above. The suggested fee scales are dependent on the complexity rating of the project.

The following are samples of the Complexity Categories

Category 1:   Golf Courses, Country Parks and Estates, Planting Schemes

Category 2:   Coastal, River and Agricultural Works, Roads, Rural recreation schemes.

Category 3:   Hospital grounds, Sport stadia, Urban offices and Commercial properties, Housing

Category 4 : Domestic/Historic Garden Design, Urban rehabilitation and Environmental improvements

Table showing suggested percentage fees for various categories of complexity of Landscape Design or consultancy.

Project Value	Complexity			
	1	2	3	4
22,500	15.00%	16.50%	18.00%	21.00%
30,000	13.75	15.00	16.50	19.25
50,000	10.50	12.75	14.00	16.25
100,000	9.50	10.50	11.50	13.50
150,000	8.75	9.50	10.50	12.50
200,000	8.00	8.75	9.75	11.25
300,000	7.50	8.25	9.00	10.50
500,000	6.75	7.50	8.25	9.75
750,000	6.50	7.25	7.75	9.25
1,000,000	6.25	7.00	7.50	8.75
10,000,000	6.25	6.75	7.25	8.25

(extrapolated from the graph/curve chart in the aforementioned publication).

Guide to Stage Payments of Fees, Relevant Fee Basis and Proportion of Fee Applicable to Lump Sum and Percentage Fee Basis. Details of Preliminary, Standard and Other Services are set out in detail in the Landscape Consultant's Appointment.

Work Stage		Relevant Fee Basis			Proportion of Fee	
		Time	*Lump*	*%age*	*Proportion*	*Total*
Preliminary Services						
A	Inception	✓	✓	n/a	n/a	n/a
B	Feasibility	✓	✓	n/a	n/a	n/a
Standard Services						
C	Outline Proposals	✓	✓	✓	15%	15%
D	Sketch Scheme Proposals	✓	✓	✓	15%	30%
E	Detailed Proposals	✓	✓	✓	15%	45%
FG	Production Information	✓	✓	✓	20%	65%
HJ	Tender Action & Contract Preparation	✓	✓	✓	5%	70%
K	Operations on Site	✓	✓	✓	25%	95%
L	Completion	✓	✓	✓	5%	100%
Other Services		✓	✓	n/a		

Timing of Fee Payments

Percentage fees are normally paid at the end of each work stage. Time based fees are normally paid at monthly intervals. Lump sum fees are normally paid at intervals by agreement. Retainer or term commission fees are normally paid in advance, for predetermined periods of service.

WORKED EXAMPLES OF PERCENTAGE FEE CALCULATIONS

WORKED EXAMPLE 1
Project Type   Caravan Site
Services Required   To Detailed Proposals - Work Stages C to E
Budget   £120,000

Step 1 Decide on Work Type and therefore Complexity Rating - Complexity Rating 2
Step 2 Decide on Services required and Proportion of Fee - To Detailed Proposals, 45%
Step 3 Read off Graph, Complexity Rating 2, the % fee of £120,000 - Graph Fee 9.9%
Step 4 Multiply the Proportion of Fee (45%) by the Graph Fee (9.9%)- Adjusted Fee - 4.46%
Step 5 Calculate the Guide Fee (4.46% of £120,000) - Guide Fee - £5,352
Step 6 Agree fee with Client, complete Memorandum of Agreement & Schedule of Services & Fees

WORKED EXAMPLE 2
Project Type   New Housing
Services Required   Full Standard Services - Work Stages C to L
Budget   £350,000

Step 1 Decide on Work Type and therefore Complexity Rating - Complexity Rating 3
Step 2 Decide on Services required and Proportion of Fee - To Completion, 100%
Step 3 Read off Graph, Complexity Rating 3 the % fee of £350,000 -Graph Fee 8.8%
Step 4 Multiply the Proportion of Fee (100%) by the Graph Fee (8.8%)- Adjusted Fee - 8.8%
Step 5 Calculate the Guide Fee (8.8% of £350,000) - Guide Fee - £30,800
Step 6 Agree fee with Client, complete Memorandum of Agreement & Schedule of Services & Fees

WORKED EXAMPLE 3
Project Type   Urban Environmental Improvements
Services Required   To Production Information - Work Stages C to G
Budget   £1,250,000

Step 1 Decide on Work Type and therefore Complexity Rating - Complexity Rating 4
Step 2 Decide on Services required and Proportion of Fee - To Production Information, 65%
Step 3 Read off Graph, Complexity Rating 4 the % fee of £1,250,000 - Graph Fee 8.6%
Step 4 Multiply the Proportion of Fee (65%) by the Graph Fee (8.6%)- Adjusted Fee - 5.59%
Step 5 Calculate the Guide Fee (5.59% of £1,250,000) - Guide Fee - £69,875
Step 6 Agree fee with Client, complete Memorandum of Agreement & Schedule of Services & Fees

Fees for Other Professional Services

Extracts from the scales of fees for architects, quantity surveyors and consulting engineers are given together with extracts from the Town and Country Planning Regulations 1993 and Building Regulation Charges. These documents are referenced by kind permission of the bodies concerned, in the case of Building Regulation Charges, by kind permission of the London Borough of Ealing. Attention is drawn to the fact that the full scales are not reproduced here and that the extracts are given for guidance only. The full authority scales should be studied before concluding any agreement and the reader should ensure that the fees quoted here are still current at the time of reference.

ARCHITECTS' FEES

Standard Form of Agreement for the Appointment of an Architect (SFA/99),
Conditions of Engagement for the Appointment of an Architect (CE/99),
Small Works (SW/99),
Employer's Requirements (DB1/99),
Contractor's Proposals (DB2/99),
Form of Appointment as Planning Supervisor (PS/99)
Form of Appointment as Sub-Consultant (SC/99)
Form of Appointment as Project Manager (PM/99)

QUANTITY SURVEYORS' FEES

Author's Note:

The Royal Institution of Chartered Surveyors formally abolished the standard Quantity Surveyors fee scales with effect from 31 December 1998. To date the Institution has not published a guidance booklet to assist practitioners in compiling fee proposals and we believe that they are unlikely to do so. The last standard Quantity Surveyors fee scales, although now out of date, are nevertheless repeated below in order to assist the reader. Information considered to be relevant to Landscape and External works projects has been extracted from the published document and summarized. For the sake of clarity and brevity, scale descriptions are listed, followed by notes common to all, and finally by the scales themselves.

The following scale rates are addressed in this summary. Readers should refer to the full document for full information.

SCALE 36
Inclusive of professional charges for quantity surveying services for building works issued by the royal institution of chartered surveyors.

SCALE 37
Itemised scale of professional charges for quantity surveying services for building work issued by the royal institution of chartered surveyors.

Contracts based on bills of quantities: post-contract services

SCALE 40
Professional charges for quantity surveying services in connection with housing schemes for local authorities

SCALE 44
Professional charges for quantity surveying services in connection with improvements to existing housing and environmental improvement works

SCALE 45
Professional charges for quantity surveying services in connection with housing schemes financed by the housing corporation

NOTES:
In the following context "bona fide tender" shall be deemed to mean a tender submitted in good faith without major errors of computation and not subsequently withdrawn by the tenderer.

Fees for Professional Services

GENERALLY

The fees cover Quantity Surveying services such as:
- Budget estimating;
- Preparing tendering documents;
- Preparing recommendations for interim payments; periodic assessments of anticipated final cost and reporting thereon;.
- Providing a reasonable number of copies of bills of quantities and other documents
- Normal travelling and other expenses.

The full documentation provides for contingencies such as;
- Materials supplied by the employer
- Exceptional delays in building operations
- V.A.T.
- Copyright
- Exclusions and additional costs permissible
- Substantial variations in works
- Project abandonment / Termination of Contract
- No tender being accepted
- Liquidation, fire damage to the buildings, services in connection with arbitration, litigation etc.

In the full documentation provision is made for:
- When a contract comprises buildings which fall into more than one category,
- When a project is the subject of a number of contracts
- Works of alteration
- Works of redecoration and associated minor repairs
- Bills of Quantity prepared in special forms
- Reduction of tenders
- Provision of checkers on the site
- Re-use of old tender documents without amendment

In calculating the amount upon which fees are charged the total of any credits and the totals of any alternative bills shall be aggregated and added to the amounts described in the scales.

COST PLANNING
CONSULTATIVE SERVICES
COST REPORTING AND MONITORING SERVICES
PREPARING SCHEDULES OF PRICES

The full documentation provides for these additional costs and fees in respect of

The fee for providing the above services shall be based on the time involved .
Alternatively, the fee may be on a lump sum or percentage basis agreed between the employer and the quantity surveyor.
Full details in the original documentation.

TIME CHARGES

Time Charges are intended for use where other paras. of the Scale (not related to Time Charges) form a significant proportion of the overall fee.
In all other cases an increased time charge may be agreed.
Full details in the original documentation.

INSTALMENT PAYMENTS

Unless specified in the following texts and in the absence of agreement to the contrary, payments to the quantity surveyor shall be made by instalments by arrangement between the employer and the quantity surveyor.

In Scales 36 and 37 following, Building Works are categorized as follows:

Category A: Relatively complex works and/or works with little or no repetition.

Examples:
Ambulance and fire stations; banks; cinemas; clubs; computer buildings; council offices; crematoria; fitting out of existing buildings; homes for the elderly; hospitals and nursing homes; and the like.

Category B: Less complex works and/or works with some element of repetition.

Examples:
Adult education facilities; canteens; community centres; departmental stores; enclosed sports stadia and swimming baths; halls of residence; hostels; motels; schools; self-contained flats and maisonettes; shops and shopping centres; supermarkets and hypermarkets; and the like.

Category C: Simple works and/or works with a substantial element of repetition.

Examples:
Factories; garages; multi-storey car parks; open-air sports stadia; structural shell offices not fitted out; warehouses; workshops; and the like.

SCALE 36
INCLUSIVE OF PROFESSIONAL CHARGES FOR QUANTITY SURVEYING SERVICES FOR BUILDING WORKS ISSUED BY THE ROYAL INSTITUTION OF CHARTERED SURVEYORS.

1.0. GENERALLY

1.1   This scale is for the use when an inclusive scale of professional charges is considered to be appropriate..

1.2   This scale does not apply to civil engineering works, local authorities housing schemes and the like.

2.0 INCLUSIVE SCALE

    2.1.   The fees for the Quantity Surveying services outlined above shall be as follows:

Category A:

Value of work £	Category A fee		
	£		£
Up to 150,000	380 + 6.0%	(Minimum fee £3 380)	
150,000 - 300,000	9 380 + 5.0%	on balance over	150,000
300,000 - 600 000	16 880 + 4.3%	on balance over	300,000
600 000 - 1 500 000	29 780 + 3.4%	on balance over	600 000
1 500 000 - 3 000,000	60 380 + 3.0%	on balance over	1 500 000
3 000,000 - 6 000,000	105 380 + 2.8%	on balance over	3 000,000
Over 6 000,000	189 380 + 2.4%	on balance over	6 000,000

Category B:

Value of work £	Category B fee		
	£		£
Up to 150,000	360 + 5.8%	(Minimum fee £3 260)	
150,000 - 300,000	9 060 + 4.7%	on balance over	150,000
300,000 - 600 000	16 110 + 3.9%	on balance over	300,000
600 000 - 1 500 000	27 810 + 2.8%	on balance over	600 000
1 500 000 - 3 000,000	53 010 + 2.6%	on balance over	1 500 000
3 000,000 - 6 000,000	92 010 + 2.4%	on balance over	3 000,000
Over 6 000,000	164 010 + 2.0%	on balance over	6 000,000

Category C:

Value of work £	Category C fee		
	£		£
Up to 150,000	300 + 4.9%	(Minimum fee £2 750)	
150,000 - 300,000	7 650 + 4.1%	on balance over	150,000
300,000 - 600 000	13 800 + 3.3%	on balance over	300,000
600 000 - 1 500 000	23 700 + 2.5%	on balance over	600 000
1 500 000 - 3 000,000	46 200 + 2.2%	on balance over	1 500 000
3 000,000 - 6 000,000	79 200 + 2.0%	on balance over	3 000,000
Over 6 000,000	139 200 + 1.6%	on balance over	6 000,000

3.0 INSTALMENT PAYMENTS

3.1 In the absence of agreement to the contrary, fees shall be paid by instalments as follows:

(a) Upon acceptance by the employer of a tender for the works, one half of the fee calculated on the amount of the accepted tender.
(b) The balance by instalments at intervals to be agreed between the date of the first certificate and one month after final certification of the contractor's account.

SCALE 37
ITEMISED SCALE OF PROFESSIONAL CHARGES FOR QUANTITY SURVEYING SERVICES FOR BUILDING WORK ISSUED BY THE ROYAL INSTITUTION OF CHARTERED SURVEYORS.

CONTRACTS BASED ON BILLS OF QUANTITIES: PRE-CONTRACT SERVICES

1.0. BILLS OF QUANTITIES

1.1. Basic scale

For preparing bills of quantities and examining tenders received and reporting thereon.

Category A:

Value of work £	Category A fee		
	£		£
Up to 150,000	230 + 3.0%	(Minimum fee £1730)	
150,000 - 300,000	4 730 + 2.3%	on balance over	150,000
300,000 - 600 000	8 180 + 1.8%	on balance over	300,000
600 000 - 1 500 000	13 580 + 1.5%	on balance over	600 000
1 500 000 - 3 000,000	27 080 + 1.2%	on balance over	1 500 000
3 000,000 - 6 000,000	45 080 + 1.1%	on balance over	3 000,000
Over 6 000,000	78 080 + 1.0%	on balance over	6 000,000

Category B:

Value of work	Category B fee		
	£		£
Up to 150,000	210 + 2.8%	(Minimum fee £1 680)	
150,000 - 300,000	4 410 + 2.0%	on balance over	150,000
300,000 - 600 000	7 410 + 1.5%	on balance over	300,000
600 000 - 1 500 000	11 910 + 1.1%	on balance over	600 000
1 500 000 - 3 000,000	21 810 + 1.0%	on balance over	1 500 000
3 000,000 - 6 000,000	36 810 + 0.9%	on balance over	3 000,000
Over 6 000,000	63 810 + 0.8%	on balance over	6 000,000

Category C:

Value of work £	Category C fee		
	£		£
Up to 150,000	180 + 2.5%	(Minimum fee £1 430)	
150,000 - 300,000	3 930 + 1.8%	on balance over	150,000
300,000 - 600 000	6 630 + 1.2%	on balance over	300,000
600 000 - 1 500 000	10 230 + 0.9%	on balance over	600 000
1 500 000 - 3 000,000	18 330 + 0.8%	on balance over	1 500 000
3 000,000 - 6 000,000	30 330 + 0.7%	on balance over	3 000,000
Over 6 000,000	51 330 + 0.6%	on balance over	6 000,000

2.0. NEGOTIATING TENDERS

2.1. For negotiating and agreeing prices with a contractor:

Value of work £	Fee		
	£		£
Up to 150,000	0.5%		
150,000 - 600 000	750 + 0.3%	on balance over	150,000
600 000 - 1 200 000	2 100 + 0.2%	on balance over	600 000
Over 1 200 000	3 300 + 0.1%	on balance over	1 200 000

CONTRACTS BASED ON BILLS OF QUANTITIES: POST-CONTRACT SERVICES

Alternative scales (I and II) for post-contract services are set out below to be used at the quantity surveyor's discretion by prior agreement with the employer.

If the quantity surveyor appointed to carry out the post-contract services did not prepare the bills of quantities then the fees shall be increased to cover the additional services undertaken by the quantity surveyor.

3.0. ALTERNATIVE I: OVERALL SCALE OF CHARGES FOR POST-CONTRACT SERVICES

3.1. Basic scale

Category A:

Value of work £	Category A fee		
	£		£
Up to 150,000	150 + 2.0%	(Minimum fee £1 150)	
150,000 - 300,000	3 150 + 1.7%	on balance over	150,000
300,000 - 600 000	5 700 + 1.6%	on balance over	300,000
600 000 - 1 500 000	10 500 + 1.3%	on balance over	600 000
1 500 000 - 3 000,000	22 200 + 1.2%	on balance over	1 500 000
3 000,000 - 6 000,000	40 200 + 1.1%	on balance over	3 000,000
Over 6 000,000	73 200 + 1.0%	on balance over	6 000,000

Category B:

Value of work £	Category B fee		
	£		£
Up to 150,000	150 + 2.0%	(Minimum fee £1 150)	
150,000 - 300,000	3 150 + 1.7%	on balance over	150,000
300,000 - 600 000	5 700 + 1.5%	on balance over	300,000
600 000 - 1 500 000	10 200 + 1.1%	on balance over	600 000
1 500 000 - 3 000,000	20 100 + 1.0%	on balance over	1 500 000
3 000,000 - 6 000,000	35 100 + 0.9%	on balance over	3 000,000
Over 6 000,000	62 100 + 0.8%	on balance over	6 000,000

Category C:

Value of work £	Category C fee	
	£	£
Up to 150,000	120 + 1.6%	(Minimum fee £920)
150,000 - 300,000	2 520 + 1.5% on balance over	150,000
300,000 - 600,000	4 770 + 1.4% on balance over	300,000
600,000 - 1 500,000	8 970 + 1.1% on balance over	600,000
1 500,000 - 3 000,000	18 870 + 0.9% on balance over	1 500,000
3 000,000 - 6 000,000	32 370 + 0.8% on balance over	3 000,000
Over 6 000,000	56 370 + 0.7% on balance over	6 000,000

4.0. ALTERNATIVE II: SCALE OF CHARGES FOR SEPARATE STAGES OF POST-CONTRACT SERVICES

4.1. Valuations for interim certificates

(a) For taking particulars and reporting valuations for interim certificates for payments on account to the contractor.

Total of valuations £	Fee	
	£	£
Up to 300,000	0.5%	
300,000 - 1 000,000	1 500 + 0.4% on balance over	300,000
1 000,000 - 6 000,000	4 300 + 0.3% on balance over	1 000,000
Over 6 000,000	19 300 + 0.2% on balance over	6 000,000

NOTES:

The fees are to be calculated on the total of all interim valuations (i.e. the amount of the final account less only the net amount of the final valuation).

4.2 Preparing accounts of variation upon contracts

Value of work £	Fee	
	£	£
Up to 300,000	300 + 0.5%	
300,000 - 1 000,000	1 800 + 0.3% on balance over	300,000
Over 1 000,000	3 900 + 0.1% on balance over	1 000,000

5.0. BILLS OF APPROXIMATE QUANTITIES, INTERIM CERTIFICATES AND FINAL ACCOUNTS

5.1 Basic scale

Category A:

Value of work £	Category A fee		
	£		£
Up to 150,000	380 + 5.0%	(Minimum fee £2 880)	
150,000 - 300,000	7 880 + 4.0%	on balance over	150,000
300,000 - 600 000	13 880 + 3.4%	on balance over	300,000
600 000 - 1 500 000	24 080 + 2.8%	on balance over	600 000
1 500 000 - 3 000,000	49 280 + 2.4%	on balance over	1 500 000
3 000,000 - 6 000,000	85 280 + 2.2%	on balance over	3 000,000
Over 6 000,000	151 280 + 2.0%	on balance over	6 000,000

Category B:

Value of work £	Category B fee		
	£		£
Up to 150,000	360 + 4.8%	(Minimum fee £2 760)	
150,000 - 300,000	7 560 + 3.7%	on balance over	150,000
300,000 - 600 000	13 110 + 3.0%	on balance over	300,000
600 000 - 1 500 000	22 110 + 2.2%	on balance over	600 000
1 500 000 - 3 000,000	41 910 + 2.0%	on balance over	1 500 000
3 000,000 - 6 000,000	71 910 + 1.8%	on balance over	3 000,000
Over 6 000,000	125 910 + 1.6%	on balance over	6 000,000

Category C:

Value of work £	Category C fee		
	£		£
Up to 150,000	300 + 4.1%	(Minimum fee £2 350)	
150,000 - 300,000	6 450 + 3.3%	on balance over	150,000
300,000 - 600 000	11 400 + 2.6%	on balance over	300,000
600 000 - 1 500 000	19 200 + 2.0%	on balance over	600 000
1 500 000 - 3 000,000	37 200 + 1.7%	on balance over	1 500 000
3 000,000 - 6 000,000	62 700 + 1.5%	on balance over	3 000,000
Over 6 000,000	107 700 + 1.3%	on balance over	6 000,000

5.2 Where the quantity surveyor is specifically instructed to provide cost planning services, the fee shall be increased by a sum calculated in accordance with the following table and based upon the same value of work as that upon which the aforementioned fee has been calculated:

Categories A & B:

Value of work £	£		Fee £
Up to 600 000	0.70%		
600 000 - 3 000,000	4 200 + 0.40%	on balance over	600 000
3 000,000 - 6 000,000	13 800 + 0.35%	on balance over	3 000,000
Over 6 000,000	24 300 + 0.30%	on balance over	6 000,000

Category C: (as defined in para. 7.1 (c))

Value of work £	£		Fee £
Up to 600 000	0.50%		
600 000 - 3 000,000	3 000 + 0.30%	on balance over	600 000
3 000,000 - 6 000,000	10 200 + 0.25%	on balance over	3 000,000
Over 6 000,000	17 700 + 0.20%	on balance over	6 000,000

6.0. NEGOTIATING TENDERS

6.1 For negotiating and agreeing prices with a contractor:

Value of work £	£		Fee £
Up to 150,000	0.5%		
150,000 - 600 000	750 + 0.3%	on balance over	150,000
600 000 - 1 200 000	2 100 + 0.2%	on balance over	600 000
Over 1 200 000	3 300 + 0.1%	on balance over	1 200 000

7.0 PRICING BILLS OF APPROXIMATE QUANTITIES / APPROXIMATE ESTIMATES

7.1 The fee shall be one-third of the fee for negotiating tenders.

8.0 INSTALMENT PAYMENTS

8.1 For the purpose of instalment payments the fee for preparation of bills of approximate quantities only shall be the equivalent of forty per cent (40%) of the fees calculated in accordance with the appropriate sections of paras. 5.1 to 5.2,.

8.2 In the absence of agreement to the contrary, fees shall be paid by instalments as follows:

(a) Upon acceptance by the employer of a tender for the works the above defined fees for the preparation of bills of approximate quantities and for providing cost planning services.

(b) In the event of no tender being accepted, the aforementioned fees shall be paid within three months of completion of the bills of approximate quantities.

(c) The balance by instalments at intervals to be agreed between the date of the first certificate and one month after certification of the contractor's account.

CONTRACTS BASED ON SCHEDULES OF PRICES: POST-CONTRACT SERVICES

9.0. FINAL ACCOUNTS

9.1. Basic Scale

(a) For taking particulars and reporting valuations for interim certificates for payments on account to the contractor, preparing periodic assessments of anticipated final cost and reporting thereon etc., the fee shall be equivalent to sixty per cent (60%) of the fee calculated in accordance with para. 5.1.

(b) When the quantity surveyor is required to prepare, an additional fee shall be charged on the basis of the time involved.

(c) The basic scale for post-contract services includes for a simple routine of periodically estimating final costs.

PRIME COST CONTRACTS: PRE-CONTRACT AND POST-CONTRACT SERVICES

COST PLANNING

10.0. ESTIMATES OF COST

10.1.
(a) For preparing an approximate estimate:

Value of work £	£		Fee £
Up to 30 000	1.25%		
30 000 - 150,000	375 + 1.00%	on balance over	30 000
150,000 - 600 000	1 575 + 0.75%	on balance over	150,000
Over 600 000	4 950 + 0.50%	on balance over	600 000

(b) The fee shall be calculated upon the total of the approved estimates.

11.0. FINAL ACCOUNTS

11.1
(a) For checking prime costs, reporting for interim certificates for payment on account to the contractor and preparing final accounts:

Value of work £	£		Fee £
Up to 30 000	2.50%		
150,000 - 150,000	750 + 2.00%	on balance over	30 000
150,000 - 600 000	3 150 + 1.50%	on balance over	150,000
Over 600 000	9 900 + 1.25%	on balance over	600 000

SCALE 40
PROFESSIONAL CHARGES FOR QUANTITY SURVEYING SERVICES IN CONNECTION WITH HOUSING SCHEMES FOR LOCAL AUTHORITIES

1.0 GENERALLY

1.1 The scale is applicable to housing schemes of self-contained dwellings regardless of type but not to improvement to existing dwellings.

2.0 BASIC SCALE

2.1 The basic fee for the services outlined in para. 1.2 shall be as follows:-

Value of work £	Fee		
	£		£
Up to 75 000	250 + 4.6%		
75 000 - 150,000	3 700 + 3.6%	on balance over	75 000
150,000 - 750 000	6 400 + 2.3%	on balance over	150,000
750 000 - 1 500 000	20 200 + 1.7%	on balance over	750 000
Over 1 500 000	32 950 + 1.5%	on balance over	1 500 000

3.0 COST PLANNING

3.1 When the quantity surveyor is specifically instructed to provide cost planning services, the fee calculated in accordance with paras. 2.1 and 2.3 shall be increased by a sum calculated in accordance with the following table and based upon the amount of the accepted tender.

Value of work £	Fee		
	£		£
Up to 150,000	0.45%		
150,000 - 750 000	675 + 0.35%	on balance over	150,000
Over 750 000	2 775 + 0.25%	on balance over	750 000

4.0 INSTALMENT PAYMENTS

4.1 In the absence of agreement to the contrary, fees shall be paid by instalments as follows:

(a) 60% of the fees upon receipt by the employer of a tender for the works plus the appropriate recoverable expenses and the full amount of the fee for cost planning services if such services have been instructed by the employer.

(b) The balance of fees and expenses by instalments at intervals to be agreed between the date of the first certificate and one month after final certification of the contractor's account.

SCALE 44
PROFESSIONAL CHARGES FOR QUANTITY SURVEYING SERVICES IN CONNECTION WITH IMPROVEMENTS TO EXISTING HOUSING AND ENVIRONMENTAL IMPROVEMENT WORKS

1. This scale of charges is applicable to all works of improvement to existing housing for local authorities, development corporations, housing associations and the like.

2. The fees cover such quantity surveying services as may be required in connection with an improvement project irrespective of the type of contract or contract documentation from initial appointment to final certification of the contractor's account such as:

3. Such services do not include the carrying out of structural surveys.

4. These scales shall not be abated if, by agreement, any of the services are not required to be provided by the quantity surveyor.

IMPROVEMENT WORKS TO HOUSING

5. The fee for quantity surveying services in connection with improvement works to existing housing and external works in connection therewith shall be calculated from a sliding scale based upon the total number of houses or flats in a project divided by the total number of types substantially the same in design and plan as follows:

Total number of houses or flats divided by total number of types substantially the same in design and plan	Fee
not exceeding 1	see note below
exceeding 1 but not exceeding 2	7.0%
exceeding 2 but not exceeding 3	5.0%
exceeding 3 but not exceeding 4	4.5%
exceeding 4 but not exceeding 20	4.0%
exceeding 20 but not exceeding 50	3.6%
exceeding 50 but not exceeding 100	3.2%
exceeding 100	3.0%

and to the result of the computation shall be added 12.5%

NOTE: For schemes of only one house or flat per type an appropriate fee is to be agreed between the employer and the quantity surveyor on a percentage, lump sum or time basis.

ENVIRONMENTAL IMPROVEMENT WORKS

6. The fee for quantity surveying services in connection with environmental improvement works associated with improvements to existing housing or environmental improvement works of a similar nature shall be as follows:

Value of work		Fee		
£		£		£
Up to	50 000	4.5%		
50 000 -	200 000	2 250 + 3.0%	on balance over	50 000
200 000 -	500 000	6 750 + 2.1%	on balance over	200 000
Over	500 000	13 050 + 2.0%	on balance over	500 000

and to the result of that computation shall be added 12.5%.

INSTALMENT PAYMENTS

7.0 In the absence of agreement to the contrary, fees shall be paid by instalments as follows:

One half of the fee upon acceptance of the tender
The balance by instalments to be agreed.

SCALE 45
PROFESSIONAL CHARGES FOR QUANTITY SURVEYING SERVICES IN CONNECTION WITH HOUSING SCHEMES FINANCED BY THE HOUSING CORPORATION

1. (a) This scale of charges has been agreed between The Royal Institution of Chartered Surveyors and the Housing Corporation and shall apply to housing schemes of self-contained dwellings financed by the Housing Corporation regardless of type (e.g. houses, maisonettes, bungalows or flats)
   (b) This scale does not apply to services in connection with improvements to existing dwellings.
2. (a) The basic fee shall be as follows:

Value of work		Fee		
£		£		£
Up to	75 000	210 + 3.8%		
75 000 -	150,000	3 060 + 3.0%	on balance over	75 000
150,000 -	750 000	5 310 + 2.0%	on balance over	150,000
750 000 -	1 500 000	17 310 + 1.5%	on balance over	750 000
Over	1 500 000	28 560 + 1.3%	on balance over	1 500 000

Fees for Professional Services

(b) (i) Additional fees are chargeable for services in connection with Categories 1 and 2 Accommodation designed for Old People

(ii) For services in connection with Accommodation designed for the Elderly in Scotland in accordance with the standards described in Scottish Housing Handbook Part 5, Housing for the Elderly, the fee shall be calculated as follows:

Mainstream and Amenity Housing	Basic fee in accordance with para. 2 (a)
Basic Sheltered Housing (i.e. Amenity Housing plus Warden=s accommodation and alarm system)	An addition of five per cent (5%) to the basic fee calculated in accordance with para. 2 (a)
Sheltered Housing, including optional facilities	An addition of twelve and a half per cent (12.5%) of the basic fee calculated in accordance with para. 2 (a)

(c) (i) For services in connection with Accommodation designed for Disabled and Disabled Old People, there shall be an addition of fifteen per cent (15%).

(ii) For services in connection with Accommodation designed for the Disabled in Scotland, there shall be an addition of fifteen per cent (15%).

(d) For services in connection with Subsidised Fair Rent New Build Housing, there shall be a fee, in addition to that in accordance with paras. 2 (a) to (c), calculated as follows:

Value of work £	Category A fee		
	£		£
Up to 75 000	20 + 0.40%		
75 000 - 150,000	320 + 0.20%	on balance over	75 000
150,000 - 500 000	470 + 0.07%	on balance over	150,000
Over 500 000	715		

3. (a) Where with the approval of the Housing Corporation the employer instructs the quantity surveyor to carry out cost planning services there shall be a fee additional to that charged under para. 2 as follows:

Value of work £	Category A fee		
	£		£
Up to 150,000	0.45%		
150,000 - 750 000	675 + 0.35%	on balance over	150,000
Over 750 000	2 775 + 0.25%	on balance over	750 000

EXPLANATORY NOTE:

Further information and clarification regarding fees in connection with rehabilitation projects and shared housing schemes may be found in the August 1986 edition of Chartered Quantity Surveyor

SPON'S PRICEBOOKS 2004

Spon's Architects' & Builders' Price Book 2004
Davis Langdon & Everest

The most detailed professionally relevant source of construction price information currently available.

New features for 2004 include:
- Information on the new CIJC three year wage agreement, effective from Monday, 30th June 2003.
- Additional items for external wall, terracotta blockwork and rainscreen cladding.

HB & CD-ROM ♦ £120.00
0-415-32365-7 ♦ 1160pp

Spon's Landscape & External Works Price Book 2004
Davis Langdon & Everest

The only comprehensive source of information for detailed landscape and external works cost.

New features for 2004 include:
- More plant items to reflect the increased mechanisation of site handling of materials.
- Additional measured rate items for work in small gardens.

HB & CD-ROM ♦ £90.00
0-415-32367-3 ♦ 512pp

Spon's Mechanical & Electrical Services Price Book 2004
Mott Green & Wall

Still the only annual services engineering price book available.

New features for 2004 include:
- A complete revision of the Approximate Estimating section with more detail on a wider range of building types.
- Additional items for both Mechanical and Electrical services.

HB & CD-ROM ♦ £120.00
0-415-32368-1 ♦ 664pp

Spon's Civil Engineering & Highway Works Price Book 2004
Davis Langdon & Everest

More than just a price book, it is a comprehensive, work manual that all those in the civil engineering, surveying and construction business will find it hard to work without.

New features for 2004 include:
- An expansion of both the Civil Engineering Works and Highway Works sections to improve coverage of the two methods of measurement.

HB & CD-ROM ♦ £120.00
0-415-32366-5 ♦ 800pp

Receive a free cd-rom when you order any Spon 2004 Pricebook to help you produce tender documents, customise data, perform keyword searches and simple calculations.

Visit www.pricebooks.co.uk

To Order: Tel: +44 (0) 8700 768853, or +44 (0) 1264 343071 Fax: +44 (0) 1264 343005, or
Post: Spon Press Customer Services, Thomson Publishing Services, Cheriton House, Andover, Hants, SP10 5BE, UK
Email: book.orders@tandf.co.uk

For a complete listing of all our titles visit www.sponpress.com

Spon Press
Taylor & Francis Group

ESSENTIAL READING FROM SPON PRESS

Manufactured Sites
Rethinking the Post-Industrial Landscape

Edited by Niall Kirkwood

'Provide(s) a thoughtful overview of current approaches and issues that include numerous case studies of actual sites.'
— *Landscape Architecture*

'*Manufactured Sites* stands as an important contribution to the field of brownfield reclamation, both for documenting inventive work on a range of professional fronts and for its successful attempts to broaden the cultural and critical discussions surrounding brownfield reuse.'
— *Land Forum*

Manufactured Sites focuses on the legacy of industrial production and pollutants on the contemporary landscape and their influence on new scientific research, innovative site technologies and progressive site design. It presents innovative environmental, engineering and design approaches along with ongoing research and built projects of international significance. Contributions range from innovative scientific engineering research from industry and federal agencies to contemporary international and regional professional reclamation and redevelopment projects such as the 2000 Olympic Games in Sydney, Australia and the A.G. Thyssen steelworks and blast furnace planning in Germany's Ruhr region.

April 2001: 246x189 mm: 272 pp.
154 b+w photos
HB: 0-415-24365-3: £47.50

To Order: Tel: +44 (0) 8700 768853, or +44 (0) 1264 343071 Fax: +44 (0) 1264 343005, or Post: Spon Press Customer Services, Thomson Publishing Services, Cheriton House, Andover, Hants, SP10 5BE, UK Email: book.orders@tandf.co.uk

For a complete listing of all our titles visit:
www.sponpress.com

6
The Aggregates Tax

The Tax

The Aggregates Tax came into operation on 1 April 2002 in the UK, except for Northern Ireland where it will be phased in over five years commencing in 2003.

It is currently levied at the rate of £1.60 per tonne on anyone considered to be responsible for commercially exploiting 'virgin' aggregate in the UK and should naturally be passed by price increase to the ultimate user.

All materials falling within the definition of 'aggregate' is liable for the levy unless it is specifically exempted.

It does not apply to clay, soil, vegetable or other organic matter.

The hope is that this will :-

- Encourage the use of alternative materials that would otherwise be disposed of to landfill sites.
- promote development of new recycling processes, such as using waste tyres and glass
- promote greater efficiency in the use of virgin aggregates
- reduce noise + vibration, dust + other emissions to air, visual intrusion, loss of amenity + damage to wildlife habitats

The intention is for part of the revenue from the levy to be recycled to business and communities affected by aggregates extraction, through :-

- a 0.1 percentage point cut in employers' national insurance contributions
- a new £35 million per annum 'Sustainability Fund' to reduce the need for virgin materials and to limit the effects of extraction on the environmental where it takes place. A number of key priorities have been identified with a promise to give effect to them through existing programmes. The intention was to publish the results early in 2003.

Definition

'Aggregate' means any rock, gravel or sand which is extracted or dredged in the UK for aggregates use. It includes whatever substances are for the time being incorporated in it or naturally occur mixed with it.

'Exploitation' is defined as involving any one or a combination of any of the following :-

- being removed from its original site
- becoming subject to a contract or other agreement to supply to any person
- being used for construction purposes
- being mixed with any material or substance other than water, except in permitted circumstances

Aggregates Tax

Incidence

It is a tax on primary aggregate production – i.e. 'virgin' aggregate won from a source and used in a location within the UK territorial boundaries (land or sea). The tax is not levied on aggregate which is exported nor on aggregate imported from outside the UK territorial boundaries.

It is levied at the point of sale.

Exemption from tax

An 'aggregate' is exempt from the levy if it is :-

- material which has previously been used for construction purposes
- aggregate that has already been subject to a charge to the aggregates levy
- aggregate which was previously removed from its originating site before the commencement date of the levy
- aggregate which is being returned to the land from which it was won
- aggregate won from a farm land or forest where used on that farm or forest
- rock which has not been subjected to an industrial crushing process
- aggregate won by being removed from the ground on the site of any building or proposed building in the course of excavations carried out in connection with the modification or erection of the building and exclusively for the purpose of laying foundations or of laying any pipe or cable
- aggregate won by being removed from the bed of any river, canal or watercourse or channel in or approach to any port or harbour (natural or artificial), in the course of carrying out any dredging exclusively for the purpose of creating, restoring, improving or maintaining that body of water
- aggregate won by being removed from the ground along the line of any highway or proposed highway in the course of excavations for improving, maintaining or constructing the highway + otherwise than purely to extract the aggregate
- drill cuttings from petroleum operations on land and on the seabed
- aggregate resulting from works carried out in exercise of powers under the New Road and Street Works Act 1991, the Roads (Northern Ireland) Order 1993 or the Street Works (Northern Ireland) Order 1995
- aggregate removed for the purpose of cutting of rock to produce dimension stone, or the production of lime or cement from limestone.
- Aggregate arising as a waste material during the processing of the following industrial minerals:
    - ball clay
    - barytes
    - calcite
    - china clay
    - coal, lignite, slate or shale
    - feldspar
    - flint
    - fluorspar
    - fuller's earth
    - gems and semi-precious stones
    - gypsum
    - any metal or the ore of any metal
    - muscovite
    - perlite

- potash
- pumice
- rock phosphates
- sodium chloride
- talc
- vermiculite

However, the levy is still chargeable on any aggregate arising as the spoil or waste from or the by-products of the above exempt processes. This includes quarry overburden.

Anything that consists 'wholly or mainly' of the following is exempt from the levy (note that 'wholly' is defined as 100% but 'mainly' as more than 50%, thus exempting any contained aggregates amounting to less than 50% of the original volumes :

- clay, soil, vegetable or other organic matter
- coal, slate or shale
- china clay waste and ball clay waste

Relief from the levy either in the form of credit or repayment is obtainable where :

- it is subsequently exported from the UK in the form of aggregate
- it is used in an exempt process
- where it is used in a prescribed industrial or agricultural process
- it is waste aggregate disposed of by dumping or otherwise, e.g. sent to landfill or returned to the originating site

BACKGROUND

Environment Act 1995

Provides for planning permissions for minerals extraction (which includes aggregates) to be reviewed and updated on a regular basis by the mineral planning authorities.

Minerals planning guidance notes were to be updated to ensure that all mineral operators provide local benefits that clearly outweigh the likely impacts of mineral extraction.

The Rural White Paper published in November 2000

This set out the government's 'vision of a countryside that is sustainable economically, socially and environmentally' apart from economic support, action was proposed to deal with a number of environmental threats, one is to reduce the environmental impact (such as damage to biodiversity and visual intrusion) of aggregate extraction by imposing an aggregates levy.

Finance Act 2001

This provides for a levy being charged on aggregate which is subject to commercial exploitation. It places the levy under the care and management of the Commissioners of Customs & Excise.

Anticipated impact

The British Aggregates Association has suggested that the additional cost imposed by quarries is more likely to be in the order of £2.65 per tonne on mainstream products, applying an above average rate on these in order that by-products and low grade waste products can be held at competitive rates, as well as making some allowance for administration and increased finance charges.

With many gravel aggregates costing in the region of £7.50 to £9.50 per tonne, there will be a significant impact on construction costs.

The Civil Engineering Contractors Association has predicted that the tax will push up total construction prices in the civil sector by 1.3%. Other models have come up with figures much higher than this, but the level for any particular project will of course depend on the constituents and type of work involved. It has been estimated that between 20 and 25% of aggregate production is used for road maintenance and construction. Obviously there will be a substantial cost impact on civil engineering schemes involving large quantities of aggregates as fill materials etc. in schemes such as roads, railways and airports, but there will also be a significant impact on any building work which employs large quantities of aggregates not only in their primary state but also as concrete or concrete based products. An average building project could attract an increase of approximately 0.8 to 1%.

Individual typical materials could be affected as follows :-

- Concrete: given that a cubic metre contains 1.9 - 2.4 tonnes of sand and aggregate, the levy could push up prices by £ 3.04 - 3.84 per m^3 at the £1.60 rate, were this to be £2.65 as reasoned above, the resulting increase would be £5.04 - 6.36 per m^3. On a ready mix supply price of say £65 per m^3, this would represent an average rise of 5.29% or 8.77% respectively.
- Precast concrete blocks: anticipated at 10 p per m^2 on dense aggregate
- Concrete roof tiles: anticipated at 10 p to 50 p per m^2 of roofing
- Precast concrete pavings: anticipated at 2 to 5% on the cost of concrete pavings, flags, blocks, kerbs etc.
- Bitumen macadam paving: anticipated to be in the region of 2.5% on the cost of materials

Avoidance

An alternative to using new aggregate in filling operations is to crush and screen rubble which may become available during the process of demolition and site clearance as well as removal of obstacles during the excavation processes.

Example : Assuming that the material would be suitable for fill material under buildings or roads, a simple cost comparison would be as follows (note that for the purpose of the exercise, the material is taken to be 1.80 t/m^3 and the total quantity involved less than 1,000 m^3) :-

Importing fill material :

	£/m³	£/tonne
Cost of 'new' aggregate delivered to site	17.10	9.50
Addition for Aggregates Tax	2.88	1.60
Total cost of importing fill materials	19.98	11.10

Disposing of site material :

	£/m³	£/tonne
Cost of removing materials from site	9.50	5.28
Total cost of disposal of site material	9.50	5.28

Crushing site materials :

	£/m³	£/tonne
Transportation of material from excavations or demolition places to temporary stockpiles	1.25	.69
Transportation of material from temporary stockpiles to the crushing plant	1.00	.56
Establishing plant and equipment on site; removing on completion	.75	.42
Maintain and operate plant	3.75	2.08
Crushing hard materials on site	5.50	3.05
Screening material on site	.75	.42
Total cost of crushing site materials	13.00	7.22

From the above it can be seen that potentially there is a great benefit in crushing site materials for filling rather than importing fill materials.

Setting the cost of crushing against the import price would produce a saving of £6.98 per m³. If the site materials were otherwise intended to be removed from the site, then the cost benefit increases by the saved disposal cost to £16.48 per m³.

Even if there is no call for any or all of the crushed material on site, it ought to be regarded as a useful asset and either sold on in crushed form or else sold with the prospects of crushing elsewhere.

Specimen Unit rates

Establishing plant and equipment on site; removing on completion
    Crushing plant   £500
    Screening plant   £250

Maintain and operate plant
    Crushing plant   £3,000 per week
    Screening plant   £750 per week

Transportation of material from excavations or demolition places to temporary stockpiles   £1.25 per m³

Transportation of material from temporary stockpiles to the crushing plant   £1.00 per m³

Breaking up material on site using impact breakers
    mass concrete   £5.50 per m³
    reinforced concrete   £6.50 per m³
    brickwork   £2.50 per m³

Crushing material on site
    mass concrete ne 1000m³   £5.50 per m³
    mass concrete 1000 - 5000m³   £5.00 per m³
    mass concrete over 5000m³   £4.50 per m³
    reinforced concrete ne 1000m³   £6.50 per m³
    reinforced concrete 1000 - 5000m³   £6.00 per m³
    reinforced concrete over 5000m³   £5.50 per m³
    brickwork ne 1000m³   £5.00 per m³
    brickwork 1000 - 5000m³   £4.50 per m³
    brickwork over 5000m³   £4.00 per m³

Screening material on site   £0.75 per m³

7
Prices for Measured Works

ESSENTIAL READING FROM SPON PRESS

Shaping Neighbourhoods
A Guide for Health, Sustainability, Vitality

Hugh Barton, Marcus Grant, Richard Guise

Current policies in planning emphasise the importance of rejuvenating neighbourhoods. This new guide seeks to bridge the gap between rhetoric and reality, promoting an interprofessional and collaborative approach to making localities work. The objective is to design neighbourhoods that are healthy, safe, friendly and attractive for residents and users, whilst being more self-sustaining in terms of water, energy, local services and work. Social inclusion, accessibility and the quality of the public realm will be central tenets. The guide is distinctive in showing not only what should be done but also how it could be done. The book provides a practical guide with illustrations and case studies.

Contents: 1. Orientation and Principles. Neighbourhood as Habitat. Neighbourhood in Focus. 2. Neighbourhood Checklists. Community Checklist. Investors Checklist. 3. A Neighbourhood Planning Process. Overview. The Collaborative Approach. Getting Going. Creating a Strategy. Making it Happen . Case Studies: 4. Providing for Local Needs. People and Community. Access to Jobs. Access to Local Facilities. Planning Movement. Case Studies. 5. Use of Local Resources. Energy. Water. Food, Waste. Biodiversity. Case Studies. 6. Neighbourhood Design. Neighbourhood Character. Key Structuring Elements. Shaping Towns and Townships. Designing Places. Case Studies. Sources. Index.

November 2002: 276x219 mm mm: 256 pp.
54 line drawings and 140 b+w photos
HB: 0-415-27852-X: £65.00 : Pb: 0-415-26009-4: £27.50

To Order: Tel: +44 (0) 8700 768853, or +44 (0) 1264 343071 Fax: +44 (0) 1264 343005, or Post: Spon Press Customer Services, Thomson Publishing Services, Cheriton House, Andover, Hants, SP10 5BE, UK Email: book.orders@tandf.co.uk

For a complete listing of all our titles visit:
www.sponpress.com

Prices for Measured Works
NEW ITEMS FOR THIS EDITION

Item Excluding site overheads and profit	PC £	Labour hours	Labour £	Plant £	Material £	Unit	Total rate £
MOVEMENT OF MATERIALS							
Southern Conveyors; Moving only of granular material by belt conveyor; Conveyors fitted with troughed belts, receiving hopper, front and rear undercarriages and driven by Electrical and Air motors; support work for conveyor installation (scaffolding), delivery, collection and installation all excluded.							
Conveyor belt width 400 mm; Mechanically loaded and removed at offload point; Conveyor length;							
Up to 5 m	-	1.50	27.00	7.00	-	m³	34.00
10 m	-	1.50	27.00	7.70	-	m³	34.70
12.5 m	-	1.50	27.00	8.02	-	m³	35.02
15 m	-	1.50	27.00	8.41	-	m³	35.41
20 m	-	1.50	27.00	9.11	-	m³	36.11
25 m	-	1.50	27.00	9.74	-	m³	36.74
30 m	-	1.50	27.00	10.44	-	m³	37.44
Conveyor belt width 600 mm; Mechanically loaded and removed at offload point; Conveyor length;							
Up to 5 m	-	1.00	18.00	5.03	-	m³	23.04
10 m	-	1.00	18.00	5.61	-	m³	23.61
12.5 m	-	1.00	18.00	5.92	-	m³	23.92
15 m	-	1.00	18.00	6.23	-	m³	24.23
20 m	-	1.00	18.00	6.86	-	m³	24.86
25 m	-	1.00	18.00	7.46	-	m³	25.46
30 m	-	1.00	18.00	8.05	-	m³	26.06
Conveyor belt width 400 mm; Mechanically loaded and removed by hand at offload point; Conveyor length;							
Up to 5 m	-	3.19	57.43	5.25	-	m³	62.68
10 m	-	3.19	57.43	6.19	-	m³	63.62
12.5 m	-	3.19	57.43	6.60	-	m³	64.03
15 m	-	3.19	57.43	7.13	-	m³	64.55
20 m	-	3.19	57.43	8.06	-	m³	65.49
25 m	-	3.19	57.43	8.90	-	m³	66.32
30 m	-	3.19	57.43	9.83	-	m³	67.26

Prices for Measured Works
NEW ITEMS FOR THIS EDITION

Item Excluding site overheads and profit	PC £	Labour hours	Labour £	Plant £	Material £	Unit	Total rate £
MOVEMENT OF MATERIALS – cont'd							
Terranova Cranes Ltd. Crane Hire; Materials handling and lifting; Telescopic Cranes supply and management. Exclusive of roadway management or planning applications. Prices below illustrate lifts of 1 tonne at maximum crane reach; Lift cycle of 0.25 hours; Mechanical filling of material skip if appropriate							
35 tonne mobile crane; Lifting capacity of 1 tonne at 25 meters; C.P.A hire (all management by hirer)							
granular materials including concrete	-	0.75	13.50	15.03	-	tonne	28.53
palletised or packed materials	-	0.50	9.00	12.03	-	tonne	21.03
35 tonne mobile crane; Lifting capacity of 1 tonne at 25 meters; contract lift							
granular materials or concrete	-	0.25	4.50	40.31	-	tonne	44.81
palletised or packed materials	-	-	-	37.31	-	tonne	37.31
50 tonne mobile crane; Lifting capacity of 1 tonne at 30 meters; C.P.A hire (all management by hirer)							
granular materials including concrete	-	0.75	13.50	20.19	-	tonne	33.69
palletised or packed materials	-	0.50	9.00	17.19	-	tonne	26.19
50 tonne mobile crane; Lifting capacity of 1 tonne at 30 meters; contract lift							
granular materials or concrete	-	0.25	4.50	46.50	-	tonne	51.00
palletised or packed materials	-	-	-	43.50	-	tonne	43.50
80 tonne mobile crane; Lifting capacity of 1 tonne at 38 meters; C.P.A hire (all management by hirer)							
granular materials including concrete	-	0.75	13.50	28.75	-	tonne	42.25
palletised or packed materials	-	0.50	9.00	25.75	-	tonne	34.75
80 tonne mobile crane; Lifting capacity of 1 tonne at 38 meters; contract lift							
granular materials or concrete	-	0.25	4.50	56.75	-	tonne	61.25
palletised or packed materials	-	-	-	53.75	-	tonne	53.75
Q25 SLAB/BRICK/SETT/COBBLE PAVINGS							
Marshalls; La Linia Pavings; Fine textured exposed aggregate 80 mm thick pavings in various sizes to designed laying patterns; laid to 50mm sharp sand bed on Type 1 base all priced separately							
Bonded laying patterns 300 x 300							
Light Granite / Anthracite basalt	23.82	2.00	36.00	-	24.42	m²	60.42
Indian granite / Yellow	-	2.00	36.00	-	26.55	m²	62.55
Random Scatter pattern incorporating 100 x 200, 200 x 200 and 300 x 200 units							
Light Granite / Anthracite basalt	23.82	3.00	54.05	-	23.82	m²	77.87
Indian granite / Yellow	-	3.00	54.05	-	25.90	m²	79.95

Prices for Measured Works
NEW ITEMS FOR THIS EDITION

Item Excluding site overheads and profit	PC £	Labour hours	Labour £	Plant £	Material £	Unit	Total rate £
Marshalls; "La Linia Grande Paving"; as above but 140 mm thick							
Bonded laying patterns 300 x 300							
All colours	46.54	3.00	54.05	-	46.54	m²	**100.59**
Random Scatter pattern incorporating 100 x 200, 200 x 200 and 300 x 200 units							
Light Granite / Anthracite basalt	46.54	4.00	72.00	-	46.54	m²	**118.54**
Extra over to the above for incorporating inlay stones to patterns; blue, light granite or anthracite basalt							
to prescribed patterns	100.89	1.50	27.00	-	100.89	m²	**127.89**
as individual units; triangular 200 x 200 x 282 mm	0.57	0.05	0.90	-	0.57	nr	**1.47**
Q31 TREE PLANTING							
Greenleaf Tree Planting - Irrigation Systems; "Root Rain" tree pit irrigation systems							
"Metro" - "Small"; 35 mm pipe diameter 1.25 m long for specimen shrubs and standard trees							
Plastic	5.09	0.25	4.50	-	5.09	nr	**9.59**
Plastic with chain	6.65	0.25	4.50	-	6.65	nr	**11.15**
Metal with chain	8.24	0.25	4.50	-	8.24	nr	**12.74**
"Medium" 35 mm pipe diameter 1.75 m long for s standard and selected standard trees							
Plastic	5.40	0.29	5.14	-	5.40	nr	**10.54**
Plastic with chain	6.98	0.29	5.14	-	6.98	nr	**12.12**
Metal with chain	8.55	0.29	5.14	-	8.55	nr	**13.69**
"Large" 35 mm pipe diameter 2.50 m long for selected standards and extra heavy standards							
Plastic	6.24	0.33	6.00	-	6.24	nr	**12.24**
Plastic with chain	7.82	0.33	6.00	-	7.82	nr	**13.82**
Metal with chain	10.44	0.33	6.00	-	10.44	nr	**16.44**
"Large" 35 mm pipe diameter 2.50 m long for selected standards and extra heavy standards							
Plastic	6.24	0.33	6.00	-	6.24	nr	**12.24**
Plastic with chain	7.82	0.33	6.00	-	7.82	nr	**13.82**
Metal with chain	10.44	0.33	6.00	-	10.44	nr	**16.44**
"Urban" Irrigation and aeration systems; for large capacity general purpose irrigation to parkland and street verge planting							
RRUrb1; 3.0 m pipe	-	0.29	5.14	-	12.60	nr	**17.74**
RRUrb2; 5.0 m pipe	14.60	0.33	6.00	-	14.60	nr	**20.60**
RRUrb3; 8.0 m pipe	15.90	0.40	7.20	-	15.90	nr	**23.10**
"Precinct" Irrigation and aeration systems; heavy cast aluminium inlet for heavily trafficked locations							
"5 m pipe"	27.50	0.33	6.00	-	27.50	nr	**33.50**
"8 m pipe"	30.00	0.40	7.20	-	30.00	nr	**37.20**

Prices for Measured Works
NEW ITEMS FOR THIS EDITION

Item Excluding site overheads and profit	PC £	Labour hours	Labour £	Plant £	Material £	Unit	Total rate £
SMALL GARDENS							
Notes; In calculating the costs and outputs of work in small garden, the following factors are taken into consideration.							
Access More difficult access decreases production due to longer time periods to move materials in and out of working areas. Often steps, ramps driveway or garden gradients and narrow access gates impact on the labour outputs							
Production rates Slower production due to lower motivation on working teams. This is assumed for in cases where the team foreman is not the company owner or operator							
Ground conditions Compaction and working room in smaller spaces:- The smaller space has been trafficked relatively more often and excavation is hence more compact and more difficult							
Additional care and attention to detail In domestic situations, working conditions are generally slower due to the presence of the employer, confined working, and additional care required due to quality considerations							
Excavating; hand Topsoil for preservation							
average depth 100 mm	-	0.46	8.23	-	-	m²	8.23
average depth 150 mm	-	0.69	12.34	-	-	m²	12.34
average depth 200 mm	-	0.91	16.46	-	-	m²	16.46
average depth 250 mm	-	1.14	20.57	-	-	m²	20.57
average depth 300 mm	-	1.37	24.69	-	-	m²	24.69
average depth 400 mm	-	2.01	36.21	-	-	m²	36.21
average depth 500 mm	-	2.29	41.14	-	-	m²	41.14
Over 500 mm	-	6.86	123.40	-	-	m³	123.40
Excavating; hand Topsoil to reduce levels							
average depth 100 mm	-	0.50	9.05	-	-	m²	9.05
average depth 150 mm	-	0.75	13.58	-	-	m²	13.58
average depth 200 mm	-	1.01	18.10	-	-	m²	18.10
average depth 250 mm	-	1.26	22.63	-	-	m²	22.63
average depth 300 mm	-	1.51	27.15	-	-	m²	27.15
average depth 400 mm	-	2.21	39.83	-	-	m²	39.83
average depth 500 mm	-	2.51	45.26	-	-	m²	45.26
Over 500 mm	-	7.54	135.74	-	-	m³	135.74
Movement of materials in Small gardens filling of wheel barrows and moving to remote location average 25 m distant; tipping							
Excavated material; on site; in spoil heaps							
average 25 m distance	-	4.80	86.40	-	-	m³	86.40
average 50 m distance	-	6.00	108.00	-	-	m³	108.00

Prices for Measured Works
A PRELIMINARIES

Item Excluding site overheads and profit	PC £	Labour hours	Labour £	Plant £	Material £	Unit	Total rate £
A11 TENDER AND CONTRACT DOCUMENTS							
Health and Safety							
Produce health and safety file including preliminary meeting and subsequent progress meetings with external planning officer in connection with health and safety: Project value:							
£35,000	-	8.00	200.00	-	-	nr	200.00
£75000	-	12.00	300.00	-	-	nr	300.00
£100000	-	16.00	400.00	-	-	nr	400.00
£200,000 to £500,000	-	40.00	1000.00	-	-	nr	1000.00
Maintain health and safety file for project duration							
£35,000	-	4.00	100.00	-	-	week	100.00
£75000	-	4.00	100.00	-	-	week	100.00
£100000	-	8.00	200.00	-	-	week	200.00
£200,000 to £500,000	-	8.00	200.00	-	-	week	200.00
Produce written risk assessments on all areas of operations within the scope of works of the contract. Project value:							
£35,000	-	2.00	50.00	-	-	week	50.00
£75000	-	3.00	75.00	-	-	week	75.00
£100000	-	5.00	125.00	-	-	week	125.00
£200,000 to £500,000	-	8.00	200.00	-	-	week	200.00
Produce Coshh assessments on all substances to be used in connection with the contract. Project value:							
£35,000	-	1.50	37.50	-	-	nr	37.50
£75000	-	2.00	50.00	-	-	nr	50.00
£100000	-	3.00	75.00	-	-	nr	75.00
£200,000 to £500,000	-	3.00	75.00	-	-	nr	75.00
Method Statements							
Provide detailed method statements on all aspects of the works; Project value:							
£ 30,000	-	2.50	62.50	-	-	nr	62.50
£ 50,000	-	3.50	87.50	-	-	nr	87.50
£ 75,000	-	4.00	100.00	-	-	nr	100.00
£ 100,000	-	5.00	125.00	-	-	nr	125.00
£ 200,000	-	6.00	150.00	-	-	nr	150.00
A32 EMPLOYERS REQUIREMENTS: MANAGEMENT OF WORKS							
Programmes							
Allow for production of works programmes prior to the start of the works; Project value:							
£ 30,000	-	3.00	75.00	-	-	nr	75.00
£ 50,000	-	6.00	150.00	-	-	nr	150.00
£ 75,000	-	8.00	200.00	-	-	nr	200.00
£ 100,000	-	10.00	250.00	-	-	nr	250.00
£ 200,000	-	14.00	350.00	-	-	nr	350.00
Allow for updating the works programme during the course of the works; Project value							
£ 30,000	-	1.00	25.00	-	-	nr	25.00
£ 50,000	-	1.50	37.50	-	-	nr	37.50
£ 75,000	-	2.00	50.00	-	-	nr	50.00
£ 100,000	-	3.00	75.00	-	-	nr	75.00
£ 200,000	-	5.00	125.00	-	-	nr	125.00

Prices for Measured Works
A PRELIMINARIES

Item Excluding site overheads and profit	PC £	Labour hours	Labour £	Plant £	Material £	Unit	Total rate £
A32 EMPLOYERS REQUIREMENTS: **MANAGEMENT OF WORKS** – cont'd							
Setting out Setting out for external works operations comprising hard and soft works elements. Placing of pegs and string lines to Landscape Architects drawings; surveying levels and placing level pegs; Obtaining approval from the Landscape Architect to commence works; areas of entire site							
1,000 m²	-	5.00	90.00	-	5.60	nr	95.60
2,500 m²	-	8.00	144.00	-	11.20	nr	155.20
5,000 m²	-	8.00	144.00	-	11.20	nr	155.20
10,000 m²	-	32.00	576.00	-	28.00	nr	604.00
A34 SECURITY/SAFETY /PROTECTION							
Jacksons Fencing; Express fence; **Framed mesh unclimbable fencing;** **including precast concrete supports and couplings** weekly hire							
2.0 m high; weekly hire rate	-	-	-	0.50	-	m	0.50
erection of fencing; labour only	-	0.10	1.80	-	-	m	1.80
delivery charge	-	-	-	0.80	-	m	0.80
return haulage charge	-	-	-	0.60	-	m	0.60
A41 CONTRACTORS GENERAL COST ITEMS: **SITE ACCOMMODATION**							
General The following items are instances of the commonly found preliminary costs associated with external works contracts. The assumption is made that the external works contractor is sub-contracted to a main contractor.							
Erect temporary office on concrete base measured separately. Elliot Workspace; Prefabricated office hire; jackleg; open plan							
3.0 x 2.4 m	31.76	-	-	-	31.76	week	31.76
4.8 x 2.4 m	33.80	-	-	-	33.80	week	33.80
Delivery and collection charges on site offices							
Delivery charge	-	-	-	-	130.00	load	130.00
Collection charge	-	-	-	-	130.00	load	130.00
Erect temporary secure storage container for tools and equipment Armoured store							
3.0 x 2.4 m	-	-	-	-	11.00	week	11.00
2.4 x 2.4 m	-	-	-	-	11.00	week	11.00

Prices for Measured Works
A PRELIMINARIES

Item Excluding site overheads and profit	PC £	Labour hours	Labour £	Plant £	Material £	Unit	Total rate £
Toilet facilities							
John Anderson Ltd. Serviced self contained toilet delivered to and collected from site. Maintained by toilet supply company.							
single chemical toilet	-	-	-	-	20.00	week	20.00
delivery and collection; each way	-	-	-	-	17.00	nr	17.00
A43 CONTRACTORS GENERAL COST ITEMS: MECHANICAL PLANT							
Southern Conveyors; Moving only of granular material by belt conveyor; Conveyors fitted with troughed belts, receiving hopper, front and rear undercarriages and driven by Electrical and Air motors; support work for conveyor installation (scaffolding), delivery, collection and installation all excluded.							
Conveyor belt width 400 mm; Mechanically loaded and removed at offload point; Conveyor length;							
Up to 5 m	-	1.50	27.00	7.00	-	m³	34.00
10 m	-	1.50	27.00	7.70	-	m³	34.70
12.5 m	-	1.50	27.00	8.02	-	m³	35.02
15 m	-	1.50	27.00	8.41	-	m³	35.41
20 m	-	1.50	27.00	9.11	-	m³	36.11
25 m	-	1.50	27.00	9.74	-	m³	36.74
30 m	-	1.50	27.00	10.44	-	m³	37.44
Conveyor belt width 600 mm; Mechanically loaded and removed at offload point; Conveyor length;							
Up to 5 m	-	1.00	18.00	5.03	-	m³	23.04
10 m	-	1.00	18.00	5.61	-	m³	23.61
12.5 m	-	1.00	18.00	5.92	-	m³	23.92
15 m	-	1.00	18.00	6.23	-	m³	24.23
20 m	-	1.00	18.00	6.86	-	m³	24.86
25 m	-	1.00	18.00	7.46	-	m³	25.46
30 m	-	1.00	18.00	8.05	-	m³	26.06
Conveyor belt width 400 mm; Mechanically loaded and removed by hand at offload point; Conveyor length;							
Up to 5 m	-	3.19	57.43	5.25	-	m³	62.68
10 m	-	3.19	57.43	6.19	-	m³	63.62
12.5 m	-	3.19	57.43	6.60	-	m³	64.03
15 m	-	3.19	57.43	7.13	-	m³	64.55
20 m	-	3.19	57.43	8.06	-	m³	65.49
25 m	-	3.19	57.43	8.90	-	m³	66.32
30 m	-	3.19	57.43	9.83	-	m³	67.26

Prices for Measured Works
A PRELIMINARIES

Item Excluding site overheads and profit	PC £	Labour hours	Labour £	Plant £	Material £	Unit	Total rate £
A43 CONTRACTORS GENERAL COST ITEMS: MECHANICAL PLANT – cont'd							
Terranova Cranes Ltd. Crane Hire; Materials handling and lifting; Telescopic Cranes supply and management. Exclusive of roadway management or planning applications. Prices below illustrate lifts of 1 tonne at maximum crane reach; Lift cycle of 0.25 hours; Mechanical filling of material skip if appropriate							
35 tonne mobile crane; Lifting capacity of 1 tonne at 25 meters; C.P.A hire (all management by hirer)							
granular materials including concrete	-	0.75	13.50	15.03	-	tonne	28.53
palletised or packed materials	-	0.50	9.00	12.03	-	tonne	21.03
35 tonne mobile crane; Lifting capacity of 1 tonne at 25 meters; contract lift							
granular materials or concrete	-	0.25	4.50	40.31	-	tonne	44.81
palletised or packed materials	-	-	-	37.31	-	tonne	37.31
50 tonne mobile crane; Lifting capacity of 1 tonne at 30 meters; C.P.A hire (all management by hirer)							
granular materials including concrete	-	0.75	13.50	20.19	-	tonne	33.69
palletised or packed materials	-	0.50	9.00	17.19	-	tonne	26.19
50 tonne mobile crane; Lifting capacity of 1 tonne at 30 meters; contract lift							
granular materials or concrete	-	0.25	4.50	46.50	-	tonne	51.00
palletised or packed materials	-	-	-	43.50	-	tonne	43.50
80 tonne mobile crane; Lifting capacity of 1 tonne at 38 meters; C.P.A hire (all management by hirer)							
granular materials including concrete	-	0.75	13.50	28.75	-	tonne	42.25
palletised or packed materials	-	0.50	9.00	25.75	-	tonne	34.75
80 tonne mobile crane; Lifting capacity of 1 tonne at 38 meters; contract lift							
granular materials or concrete	-	0.25	4.50	56.75	-	tonne	61.25
palletised or packed materials	-	-	-	53.75	-	tonne	53.75
A44 TEMPORARY ROADS							
Eve Trackway; portable roadway systems Temporary roadway system laid directly onto existing surface or onto PVC matting to protect existing surface. Most systems are based on a weekly hire charge with transportation, installation and recovery charges included.							
Heavy Duty Trakpanel - per panel (3.05m x 2.59m) per week	-	-	-	-	-	m²	5.06
Outrigger Mats for use in conjunction with Heavy Duty Trakpanels - per set of 4 mats per week	-	-	-	-	-	set	82.40
Medium Duty Trakpanel - per panel (2.44m x 3.00m) per week	-	-	-	-	-	m²	5.46
LD20 Eveolution - Light Duty Trakway - roll out system minimum delivery 50 m	-	-	-	-	-	m²	4.86
Terraplas Walkways - turf protection system - per section (1m x 1m) per week	-	-	-	-	-	m²	3.50

Prices for Measured Works
B COMPLETE BUILDINGS/STRUCTURES/UNITS

Item Excluding site overheads and profit	PC £	Labour hours	Labour £	Plant £	Material £	Unit	Total rate £
B10 PREFABRICATED BUILDINGS/ STRUCTURES							
Cast stone buildings; Haddonstone Ltd; Ornamental garden buildings in Portland Bath or Terracotta finished cast stone; Prices for stonework and facades only; excavations, foundations, reinforcement, concrete infill, roofing and floors all priced separately							
Pavilion Venetian Folly L9400; Tuscan columns, pedimented arch, quoins and optional balustrading							
4184 high x 4728 mm wide x 3147 deep	7983.00	175.00	3150.00	184.80	8086.96	nr	**11421.76**
Pavilion L9300;Tuscan columns							
3496 mm high x 3634 wide	4570.00	144.00	2592.00	158.40	4673.96	nr	**7424.36**
Small Classical Temple L9250; 6 column with fibre glass lead effect finish dome roof.							
Overall height 3610 mm, diameter 2540 mm	5430.00	130.00	2340.00	132.00	5533.96	nr	**8005.96**
Large Classical Temple L9100; 8 column with fibre glass lead effect finish dome roof.							
overall height 4664 mm, diameter 3190 mm	8690.00	165.00	2970.00	132.00	8793.96	nr	**11895.96**
Stepped floors to temples							
Single step; Large Classical Temple	1003.00	26.00	468.00	-	1068.51	nr	**1536.51**
Single step; Small Classical Temple	709.00	24.00	432.00	-	763.60	nr	**1195.60**

D GROUNDWORK

Item Excluding site overheads and profit	PC £	Labour hours	Labour £	Plant £	Material £	Unit	Total rate £
D11 SOIL STABILIZATION							
Soil Stabilization - General							
Preamble - Earth-retaining and stabilizing materials are often specified as part of the earth-forming work in landscape contracts, and therefore this section lists a number of products specially designed for large-scale earth control. There are two types: rigid units for structural retention of earth on steep slopes; and flexible meshes and sheets for control of soil erosion where structural strength is not required. Prices for these items depend on quantity, difficulty of access to the site and availability of suitable filling material: estimates should be obtained from the manufacturer when the site conditions have been determined.							
Crib Walls; Keller Comtec							
"Timbercrib" timber crib walling system; machine filled with crushed rock inclusive of reinforced concrete footing cribfill stone, rear wall land drain and rear wall drainage/ separation membrane; excluding excavation							
ref 450/38; for retaining walls up to 1.20 m high	-	-	-	-	-	m²	110.00
ref 600/38; for retaining walls up to 2.20 m high	-	-	-	-	-	m²	115.00
ref 750/38; for retaining walls up to 3.10 m high	-	-	-	-	-	m²	120.00
ref 900/48; for retaining walls up to 3.60 m high	-	-	-	-	-	m²	145.00
ref 1050/48; for retaining walls up to 4.40 m high	-	-	-	-	-	m²	155.00
ref 1200/48; for retaining walls up to 5.20 m high	-	-	-	-	-	m²	170.00
ref 1500/48; for retaining walls up to 6.50 m high	-	-	-	-	-	m²	190.00
ref 1800/48; for retaining walls up to 8.20 m high	-	-	-	-	-	m²	215.00
Retaining Walls; Keller Comtec							
"Textomur" reinforced soil system embankments reinforced soil slopes at angles of 60 - 70 degrees to the horizontal; as an alternative to re-inforced concrete or gabion solutions, finishing with excavated material/grass or shrubs.	-	-	-	-	-	m²	90.00
Retaining Walls; RCC Ltd							
Retaining walls of units with plain concrete finish; prices based on 24 tonne loads but other quantities available (excavation, temporary shoring, foundations and backfilling not included)							
1000 wide x 1250 mm high	87.71	1.27	22.87	12.44	90.29	m	125.59
1000 wide x 1750 mm high	105.22	1.27	22.87	12.44	108.78	m	144.08
1000 wide x 2400 mm high	154.02	1.27	22.87	12.44	159.19	m	194.49
1000 wide x 2690 mm high	193.24	1.27	22.87	31.45	199.88	m	254.19
1000 wide x 3000 mm high	203.72	1.47	26.46	39.31	212.19	m	277.96
1000 wide x 3750 mm high	304.43	1.58	28.35	47.17	317.89	m	393.41
Retaining Walls; Maccaferri Ltd							
Wire mesh gabions; galvanized mesh 80 mm x 100 mm; filling with broken stones 125 mm - 200 mm size; wire down securely to manufacturer's instructions; filling front face by hand							

Prices for Measured Works
D GROUNDWORK

Item Excluding site overheads and profit	PC £	Labour hours	Labour £	Plant £	Material £	Unit	Total rate £
2 x 1 x 0.50 m	22.79	2.00	36.00	8.29	86.63	nr	130.92
2 x 1 x 1.00 m	31.93	4.00	72.00	16.58	159.61	nr	248.19
3 x 1 x 0.50 m	32.10	3.00	54.00	6.22	127.86	nr	188.08
3 x 1 x 1.00 m	45.00	6.00	108.00	12.44	236.52	nr	356.96
"Reno" mattress gabions							
6 x 2 x 0.17 m	69.39	3.00	54.00	12.44	199.62	nr	266.06
6 x 2 x 0.23 m	76.31	4.50	81.00	16.58	252.51	nr	350.09
6 x 2 x 0.30 m	82.96	6.00	108.00	18.65	312.78	nr	439.44
Retaining Walls; Tensar International "Tensar" retaining wall system; modular dry laid concrete blocks; 220 mm x 400 mm long x 150 mm high connected to "Tensar SR" geogrid with proprietary connectors; geogrid laid horizontally within the fill at 300 mm centres; on 150 x 450 concrete foundation; filling with imported granular material.							
1.00 m high	70.00	3.00	54.00	7.73	77.47	m²	139.20
2.00 m high	70.00	4.00	72.00	7.73	77.47	m²	157.20
3.00 m high	70.00	4.50	81.00	7.73	77.47	m²	166.20
Retaining Walls; Grass Concrete International Ltd "Betoflor" precast concrete landscape retaining walls including soil filling to pockets (excavation, concrete foundations, backfilling stones to rear of walls and planting not included)							
"Betoflor" interlocking units; 250 mm long x 250 mm x 200 mm modular deep; in walls 250 mm wide	-	-	-	-	-	m²	70.16
extra over "Betoflor" interlocking units for colours	-	-	-	-	-	m²	1.67
"Betoatlas" interlocking units; 250 mm long x 500 mm wide x 200 mm modular deep; in walls 500 mm wide	-	-	-	-	-	m²	90.65
extra over "Betoatlas" interlocking units for colours	-	-	-	-	-	m²	3.44
Excavating existing banks							
1000 x 1000 mm deep	-	0.67	12.00	3.36	-	m	15.36
2000 x 2000 mm deep	-	1.50	27.00	7.57	-	m	34.57
3000 x 2000 mm deep	-	1.50	27.00	18.65	-	m	45.65
Setting out; grading and levelling; compacting bottoms of excavations	-	0.13	2.40	0.44	-	m	2.84
Extra over "Betoflor" retaining walls for C7P concrete foundations							
700 mm x 300 mm deep	14.29	0.27	4.80	-	14.29	m	19.10
Retaining Walls; Forticrete Engineering Systems Ltd "Keystone" pc concrete block retaining wall; geogrid included for walls over 1.00 m high; excavation, concrete foundation, stone backfill to rear of wall all measured separately							
1.0m high	52.00	2.40	43.20	-	53.30	m²	96.50
2.0m high	52.00	2.40	43.20	6.60	66.28	m²	116.08
3.0m high	52.00	2.40	43.20	8.80	75.30	m²	127.30
4.0m high	52.00	2.40	43.20	10.56	82.29	m²	136.05

Prices for Measured Works
D GROUNDWORK

Item Excluding site overheads and profit	PC £	Labour hours	Labour £	Plant £	Material £	Unit	Total rate £
D11 SOIL STABILIZATION – cont'd							
Retaining Walls; Forticrete Architectural Masonry;							
Stepoc Blocks; interlocking blocks; 10 mm reinforcing laid loose horizontally to preformed notches and vertical reinforcing nominal size 10 mm fixed to starter bars; infilling with concrete; foundations and starter bars measured separately							
Type 256 400 x 225 x 256 mm	33.50	1.20	21.60	-	48.94	m²	70.54
Type 190 400 x 225 x 190 mm	25.00	1.00	18.00	-	37.11	m²	55.11
Embankments; Tensar International							
Embankments; reinforced with "Tensar Uniaxial Geogrid"; ref 40 RE; 40 kN/m width; Geogrid laid horizontally within fill to 100% of vertical height of slope at 1.00 m centres; filling with excavated material							
slopes less than 45 degrees	2.15	0.19	3.34	2.86	2.26	m³	8.46
slopes exceeding 45 degrees	4.62	0.19	3.34	4.44	4.85	m³	12.64
Embankments; reinforced with "Tensar Uniaxial Geogrid"; ref 55 RE; 55 kN/m width; Geogrid laid horizontally within fill to 100% of vertical height of slope at 1.00 m centres; filling with excavated material							
slopes less than 45 degrees	2.75	0.19	3.34	2.86	2.89	m³	9.09
slopes exceeding 45 degrees	5.91	0.19	3.34	4.44	6.21	m³	13.99
Extra for "Tensar Mat"; erosion control mats; to faces of slopes of 45 degrees or less; filling with 20 mm fine topsoil; seeding	3.00	0.02	0.36	0.15	4.01	m²	4.53
Embankments; reinforced with "Tensar Uniaxial Geogrid" ref "55 RE"; 55 kN/m width; Geogrid laid horizontally within fill to 50% of horizontal length of slopes at specified centres; filling with imported fill PC £16.64/m³							
300 mm centres; slopes up to 45 degrees	4.57	0.23	4.20	3.31	24.76	m³	32.28
600 mm centres; slopes up to 45 degrees	2.28	0.19	3.34	4.44	22.36	m³	30.15
Extra for "Tensar Mat"; erosion control mats; to faces of slopes of 45 degrees or less; filling with 20 mm fine topsoil; seeding	3.00	0.02	0.36	0.15	4.01	m²	4.53
Embankments; reinforced with "Tensar Uniaxial Geogrid" ref "55 RE"; 55 kN/m width; Geogrid laid horizontally within fill; filling with excavated material; wrapping around at faces							
300 mm centres; slopes exceeding 45 degrees	9.16	0.19	3.34	2.86	9.62	m³	15.82
600 mm centres; slopes exceeding 45 degrees	4.57	0.19	3.34	2.86	4.79	m³	10.99
Extra over embankments for bagwork face supports for slopes exceeding 45 degrees	-	-	-	-	12.50	m²	12.50
Extra over embankments for seeding of bags	-	-	-	-	1.00	m²	1.00
Extra over embankments for "Bodkin" joints	-	-	-	-	1.00	m	1.00
Extra over embankments for temporary shuttering to slopes exceeding 45 degrees	-	-	-	-	10.00	m²	10.00

Prices for Measured Works
D GROUNDWORK

Item Excluding site overheads and profit	PC £	Labour hours	Labour £	Plant £	Material £	Unit	Total rate £
Anchoring Systems; Surface stabilization; Platipus Engineering Systems; Centres shown should be verified with a design engineer. They may vary either way dependent on circumstances							
Concrete Revetments; Hard anodised stealth anchors with stainless steel accessories installed to a depth of 1 m							
"SO4" anchors to a depth of 1.00 - 1.50 m at 1.00 m ccs	15.00	0.33	6.00	0.90	15.00	m²	21.90
Loadlocking and crimping tool; for stressing and crimping the anchors							
Purchase price	350.00	-	-	-	350.00	nr	350.00
Hire rate per week	60.00	-	-	-	60.00	nr	60.00
Surface erosion; geotextile anchoring on slopes; Aluminium alloy Stealth anchors							
"SO4", "SO6" to a depth of 1.0 - 2.0 m at 2.00 m ccs	10.00	0.20	3.60	0.90	10.00	m²	14.50
Brick block or in-situ concrete, distressed retaining walls; anchors installed in two rows at varying tensions between 18 kN - 50 kN along the length of the retaining wall. Core drilling of wall not included							
"SO8", "BO6", "BO8" bronze Stealth and Bat anchors installed in combination with stainless steel accessories to a depth of 6 m. Average cost base on 1.50 m ccs; long term solution	300.00	0.50	9.00	0.90	300.00	m²	309.90
"SO8", "BO6", "BO8" bronze Stealth and Bat anchors installed in combination with stainless steel accessories to a depth of 6 m. Average cost base on 1.50 m ccs; short term solution	112.50	0.50	9.00	0.90	112.50	m²	122.40
Timber retaining wall support; anchors installed 19 kN along the length of the retaining wall.							
"SO6", "SO8", bronze Stealth anchors installed in combination with stainless steel accessories to a depth of 3.0 m. 1.50 centres; Average cost;	45.00	0.50	9.00	0.72	45.00	m²	54.72
Gabions; anchors for support and stability to moving, overturning or rotating gabion retaining walls							
"SO8", "BO6" anchors including base plate at 1.00 m centres to a depth of 4.00 m	100.00	0.33	6.00	0.90	100.00	m²	106.90
Soil nailing of geofabrics; anchors fixed through surface of fabric or erosion control surface treatment (not included)							
"SO8", "BO6" anchors including base plate at 1.50 m centres to a depth of 3.00 m	100.00	0.17	3.00	0.60	100.00	m²	103.60

Prices for Measured Works
D GROUNDWORK

Item Excluding site overheads and profit	PC £	Labour hours	Labour £	Plant £	Material £	Unit	Total rate £
D11 SOIL STABILIZATION – cont'd							
Embankments; Grass Concrete International Ltd							
"Grasscrete"; in situ reinforced concrete surfacing; to 20 mm thick sand blinding layer (not included); including soiling and seeding							
ref GC1; 100 mm thick	-	-	-	-	-	m²	24.07
ref GC2; 150 mm thick	-	-	-	-	-	m²	28.81
"Grassblock 103"; solid matrix precast concrete blocks; to 20 mm thick sand blinding layer; excluding edge restraint; including soiling and seeding							
406 mm x 406 mm x 103 mm; fully interlocking	-	-	-	-	-	m²	25.75
Embankments; Cooper Clarke Group							
"Ecoblock" polyethylene; 925 mm x 310 mm x 50 mm; heavy duty for car parking and fire paths							
to firm sub-soil (not included)	11.40	0.04	0.72	-	11.74	m²	12.46
to 100 mm granular fill and "Geotextile"	11.40	0.08	1.44	0.16	13.74	m²	15.34
to 250 mm granular fill and "Geotextile"	11.40	0.20	3.60	0.26	15.92	m²	19.77
Extra for filling "Ecoblock" with topsoil; seeding with rye grass at 50 g/m²	0.88	0.02	0.36	0.20	1.05	m²	1.62
"Geoweb" polyethylene soil-stabilizing panels; to soil surfaces brought to grade (not included); filling with excavated material							
panels; 2.40 x 6.10 m x 100 mm deep	5.86	0.10	1.80	2.94	5.98	m²	10.72
panels; 2.40 x 6.10 m x 200 mm deep	11.79	0.13	2.40	3.50	12.03	m²	17.93
"Geoweb" polyethylene soil-stabilizing panels; to soil surfaces brought to grade (not included); filling with ballast							
panels; 2.40 x 6.10 m x 100 mm deep	5.86	0.11	2.00	2.94	7.18	m²	12.13
panels; 2.40 x 6.10 m x 200 mm deep	11.79	0.15	2.77	3.50	14.44	m²	20.71
"Geoweb" polyethylene soil-stabilizing panels; to soil surfaces brought to grade (not included); filling with ST2 10 N/mm2 concrete							
panels; 2.40 x 6.10 m x 100 mm deep	5.86	0.16	2.88	-	12.49	m²	15.37
panels; 2.40 x 6.10 m x 200 mm deep	11.79	0.20	3.60	-	25.05	m²	28.65
Extra over for filling "Geoweb" with imported topsoil; seeding with rye grass at 50 g/m²	1.83	0.07	1.20	0.34	2.18	m²	3.71
Timber log retaining walls; Western Log Company							
Machine rounded softwood logs to trenches priced separately; disposal of excavated material priced separately; inclusive of 75 mm hardcore blinding to trench and backfilling trench with site mixed concrete 1:3:6; geofabric pinned to rear of logs; Heights of logs above ground							
500 mm (constructed from 1.80 m lengths)	22.45	1.50	27.00	-	30.66	m	57.66
1.20 m (constructed from 1.80 m lengths)	44.90	1.30	23.40	-	63.81	m	87.21
1.60 m (constructed from 2.40 m lengths)	60.60	2.50	45.00	-	84.29	m	129.29
2.00 m (constructed from 3.00 m lengths)	88.50	3.50	63.00	-	116.97	m	179.97
As above but with 150 mm machine rounded timbers							
500 mm (constructed from 1.80 m lengths)	-	2.50	45.00	-	80.21	m	125.21
1.20 m (constructed from 1.80 m lengths)	96.00	1.75	31.50	-	114.91	m	146.41
1.60 m (constructed from 2.40 m lengths)	128.00	3.00	54.00	-	151.69	m	205.69

Prices for Measured Works
D GROUNDWORK

Item Excluding site overheads and profit	PC £	Labour hours	Labour £	Plant £	Material £	Unit	Total rate £
2.00 m (constructed from 3.00 m lengths)	160.33	4.00	72.00	-	188.81	m	260.81
2.40 m (constructed from 3.60 m lengths)	-	4.50	81.00	-	223.86	m	304.86
Grass Reinforcement; Farmura Environmental							
"Matrix" Grass Paver; recycled polyethylene and polypropylene mixed interlocking erosion control and grass reinforcement system laid to rootzone prepared separately and filled with screened topsoil and seeded with grass seed; green							
640 x 330 x38 mm	11.00	0.13	2.25	-	13.30	m²	15.55
330 x 330 x 38 mm	11.00	0.17	3.00	-	13.30	m²	16.30
extra over for coloured material	4.00	-	-	-	4.00	m²	4.00
Willow Walling to River banks; L.D.C. Ltd							
Woven willow walling as retention to river banks; driving or concreting posts in at 2 m centres; intermediate posts at 500 mm centres							
1.20 m high	-	-	-	-	-	m	70.15
1.50 m high	-	-	-	-	-	m	95.45
Flexible sheet materials; Tensar International							
"Tensar Mat"; erosion mats; 3 m - 4.50 m wide; securing with "Tensar" pegs; lap rolls 100 mm; anchors at top and bottom of slopes; in trenches	3.13	0.02	0.30	-	3.44	m²	3.74
Topsoil filling to "Tensar Mat"; including brushing and raking	0.43	0.02	0.30	0.11	0.54	m²	0.95
"Tensar Bi-axial Geogrid"; to graded compacted base; filling with 200 mm granular fill; compacting (turf or paving to surfaces not included); 400 mm laps							
ref SS20; 39 mm x 39 mm mesh	1.40	0.01	0.24	0.25	5.39	m²	5.88
ref SS30; 39 mm x 39 mm mesh	2.20	0.01	0.24	0.25	6.19	m²	6.68
Flexible sheet materials; Malcolm Ogilvie							
"Wiretex" polypropylene/wire-reinforced fabric; to graded base (not included); fixing with steel pegs at 500 mm centres; 25 mm imported topsoil (seeding not included)							
ref Nr.8 (roll 150 m²)	2.57	0.08	1.44	0.06	3.93	m²	5.42
ref Nr.9 (roll 150 m²)	1.72	0.08	1.44	0.18	3.08	m²	4.69
Flexible sheet materials; Terram Ltd							
"Terram" synthetic fibre filter fabric; to graded base (not included)							
"Terram 1000", 0.70 mm thick; mean water flow 50 litre/m²/s	0.43	0.02	0.30	-	0.43	m²	0.73
"Terram 2000"; 1.00 mm thick; mean water flow 33 litre/m²/s	0.87	0.02	0.30	-	1.04	m²	1.34
"Terram Minipack"	1.32	-	0.06	-	1.32	m²	1.38

Prices for Measured Works
D GROUNDWORK

Item Excluding site overheads and profit	PC £	Labour hours	Labour £	Plant £	Material £	Unit	Total rate £
D11 SOIL STABILIZATION – cont'd							
Flexible sheet materials; Greenfix Ltd							
"Greenfix"; erosion control mats; 10 mm -15 mm thick; fixing with 4 nr crimped pins in accordance with manufacturer's instructions; to graded surface (not included)							
unseeded "Eromat Light"; 2.40 m wide	135.00	2.00	36.00	-	200.50	100 m²	236.50
unseeded "Eromat Standard"; 2.40 m wide	145.00	2.00	36.00	-	211.50	100 m²	247.50
unseeded "Eromat Coco"; 2.40 m wide	165.00	2.00	36.00	-	233.50	100 m²	269.50
seeded "Covamat Standard"; 2.40 m wide	200.00	2.00	36.00	-	262.00	100 m²	298.00
seeded "Covamat Special"; 2.40 m wide	216.67	2.00	36.00	-	279.50	100 m²	315.50
seeded "Covamat Coco"; 2.40 m wide	216.67	2.00	36.00	-	279.50	100 m²	315.50
"Bio Roll" 300 mm diameter to river banks and revetments	10.00	0.13	2.25	-	11.54	m	13.79
Extra over "Greenfix" erosion control mats for fertilizer applied at 70 g/m²	3.46	0.20	3.60	-	3.46	100 m²	7.06
Extra over "Greenfix" erosion control mats for polymer mesh; fixing with steel pins	238.00	1.33	24.00	-	238.00	100 m²	262.00
Extra over "Greenfix" erosion control mats for laying to slopes exceeding 30 degrees	-	-	-	-	-	25%	-
Extra for the following operations							
Spreading 25 mm approved topsoil							
by machine	0.43	0.01	0.11	0.10	0.52	m²	0.73
by hand	0.43	0.04	0.63	-	0.52	m²	1.15
Grass seed; PC £2.46/kg; spreading in two operations; by hand							
35 g/m²	8.61	0.17	3.00	-	8.61	100 m²	11.61
50 g/m²	12.30	0.17	3.00	-	12.30	100 m²	15.30
70 g/m²	17.22	0.17	3.00	-	17.22	100 m²	20.22
100 g/m²	24.60	0.20	3.60	-	24.60	100 m²	28.20
125 g/m²	30.75	0.20	3.60	-	30.75	100 m²	34.35
Extra over seeding by hand for slopes over 30 degrees (allowing for the actual area but measured in plan)							
35 g/m²	1.28	-	0.07	-	1.28	100 m²	1.35
50 g/m²	1.85	-	0.07	-	1.85	100 m²	1.91
70 g/m²	2.58	-	0.07	-	2.58	100 m²	2.65
100 g/m²	3.69	-	0.08	-	3.69	100 m²	3.77
125 g/m²	4.60	-	0.08	-	4.60	100 m²	4.68
Grass seed; PC £2.46/kg; spreading in two operations; by machine							
35 g/m²	8.61	-	-	0.49	8.61	100 m²	9.10
50 g/m²	12.30	-	-	0.49	12.30	100 m²	12.79
70 g/m²	17.22	-	-	0.49	17.22	100 m²	17.71
100 g/m²	24.60	-	-	0.49	24.60	100 m²	25.09
125 kg/ha	307.50	-	-	49.03	307.50	ha	356.53
150 kg/ha	369.00	-	-	49.03	369.00	ha	418.03
200 kg/ha	492.00	-	-	49.03	492.00	ha	541.03
250 kg/ha	615.00	-	-	49.03	615.00	ha	664.03
300 kg/ha	738.00	-	-	49.03	738.00	ha	787.03
350 kg/ha	861.00	-	-	49.03	861.00	ha	910.03
400 kg/ha	984.00	-	-	49.03	984.00	ha	1033.03
500 kg/ha	1230.00	-	-	49.03	1230.00	ha	1279.03
700 kg/ha	1722.00	-	-	49.03	1722.00	ha	1771.03
1400 kg/ha	3444.00	-	-	49.03	3444.00	ha	3493.03

Prices for Measured Works
D GROUNDWORK

Item Excluding site overheads and profit	PC £	Labour hours	Labour £	Plant £	Material £	Unit	Total rate £
Extra over seeding by machine for slopes over 30 degrees (allowing for the actual area but measured in plan)							
35 g/m²	8.61	-	-	0.07	1.29	100 m²	1.37
50 g/m²	12.30	-	-	0.07	1.85	100 m²	1.92
70 g/m²	17.22	-	-	0.07	2.58	100 m²	2.66
100 g/m²	24.60	-	-	0.07	3.69	100 m²	3.76
125 kg/ha	307.50	-	-	7.36	46.13	ha	53.48
150 kg/ha	369.00	-	-	7.36	55.35	ha	62.71
200 kg/ha	492.00	-	-	7.36	73.80	ha	81.16
250 kg/ha	615.00	-	-	7.36	92.25	ha	99.61
300 kg/ha	738.00	-	-	7.36	110.70	ha	118.06
350 kg/ha	861.00	-	-	7.36	129.15	ha	136.51
400 kg/ha	984.00	-	-	7.36	147.60	ha	154.96
500 kg/ha	1230.00	-	-	7.36	184.50	ha	191.86
700 kg/ha	1722.00	-	-	7.36	258.30	ha	265.66
1400 kg/ha	3444.00	-	-	7.36	516.60	ha	523.96
D20 EXCAVATION AND FILLING							
Market Prices of Topsoil; Charles Morris Fertilizers Ltd							
Multiple source screened topsoil	-	-	-	-	11.66	m³	11.66
Single source screened topsoil	-	-	-	-	17.32	m³	17.32
"P30" high grade topsoil	-	-	-	-	29.28	m³	29.28
Site preparation							
Removing trees							
girth 600 mm - 1.50 m	-	5.00	90.00	18.30	-	nr	108.30
girth 1.50 m - 3.00 m	-	17.00	306.00	73.20	-	nr	379.20
girth over 3.00 m girth	-	48.53	873.50	117.12	-	nr	990.62
Removing tree stumps							
girth 600 mm - 1.50 m	-	2.00	36.00	61.82	-	nr	97.82
girth 1.50 m - 3.00 m	-	7.00	126.00	174.10	-	nr	300.10
girth over 3.00 m	-	12.00	216.00	298.45	-	nr	514.45
Stump grinding; disposing to spoil heaps							
girth 600 mm - 1.50 m	-	2.00	36.00	31.00	-	nr	67.00
girth 1.50 m - 3.00 m	-	2.50	45.00	62.00	-	nr	107.00
girth over 3.00 m	-	4.00	72.00	54.25	-	nr	126.25
Clearing site vegetation							
mechanical clearance	-	0.25	4.50	8.94	-	100 m²	13.44
hand clearance	-	2.00	36.00	-	-	100 m²	36.00
Lifting turf for preservation							
machine lift and stack	-	0.75	13.50	9.71	-	100 m²	23.21
hand lift and stack	-	8.33	149.99	-	-	100 m²	149.99
Excavating; mechanical							
Topsoil for preservation; JCB Sitemaster 3CX							
average depth 100 mm	-	0.70	12.60	21.64	-	100 m²	34.24
average depth 150 mm	-	1.00	18.00	30.91	-	100 m²	48.91
average depth 200 mm	-	1.40	25.20	43.27	-	100 m²	68.47
average depth 250 mm	-	1.68	30.24	51.93	-	100 m²	82.17
average depth 300 mm	-	2.00	36.00	61.82	-	100 m²	97.82

Prices for Measured Works
D GROUNDWORK

Item Excluding site overheads and profit	PC £	Labour hours	Labour £	Plant £	Material £	Unit	Total rate £
D20 EXCAVATION AND FILLING – cont'd							
To reduce levels							
maximum depth not exceeding 0.25 m; JCB sitemaster	-	0.07	1.26	2.16	-	m³	**3.42**
maximum depth not exceeding 1.00 m; JCB sitemaster	-	0.05	0.90	1.55	-	m³	**2.45**
maximum depth not exceeding 1.00 m; 360 Tracked excavator	-	0.04	0.72	1.20	-	m³	**1.92**
maximum depth not exceeding 2.00 m; 360 Tracked excavator	-	0.06	1.08	1.80	-	m³	**2.88**
Pits; 3 ton tracked excavator							
maximum depth not exceeding 0.25 m	-	0.58	10.44	6.05	-	m³	**16.49**
maximum depth not exceeding 1.00 m	-	0.50	9.00	5.21	-	m³	**14.21**
maximum depth not exceeding 2.00 m	-	0.60	10.80	6.26	-	m³	**17.06**
Trenches; width not exceeding 0.30 m							
maximum depth not exceeding 0.25 m	-	0.64	11.52	3.23	-	m³	**14.75**
maximum depth not exceeding 1.00 m	-	0.54	9.72	2.72	-	m³	**12.44**
maximum depth not exceeding 2.00 m	-	0.60	10.80	3.03	-	m³	**13.83**
Trenches; width exceeding 0.30 m							
maximum depth not exceeding 0.25 m	-	0.60	10.80	3.03	-	m³	**13.83**
maximum depth not exceeding 1.00 m	-	0.50	9.00	2.52	-	m³	**11.52**
maximum depth not exceeding 2.00 m	-	0.56	10.08	2.83	-	m³	**12.91**
Extra over any types of excavating irrespective of depth for breaking out existing materials; JCB with breaker attachment							
hard rock	-	0.50	9.00	56.80	-	m³	**65.80**
concrete	-	0.50	9.00	21.30	-	m³	**30.30**
reinforced concrete	-	1.00	18.00	28.40	-	m³	**46.40**
brickwork, blockwork or stonework	-	0.25	4.50	21.30	-	m³	**25.80**
Extra over any types of excavating irrespective of depth for breaking out existing hard pavings; JCB with breaker attachment							
concrete; 100 mm thick	-	-	-	1.56	-	m²	**1.56**
concrete; 150 mm thick	-	-	-	2.60	-	m²	**2.60**
concrete; 200 mm thick	-	-	-	3.12	-	m²	**3.12**
concrete; 300 mm thick	-	-	-	4.69	-	m²	**4.69**
reinforced concrete; 100 mm thick	-	0.08	1.50	2.20	-	m²	**3.70**
reinforced concrete; 150 mm thick	-	0.08	1.35	2.83	-	m²	**4.18**
reinforced concrete; 200 mm thick	-	0.10	1.80	3.78	-	m²	**5.58**
reinforced concrete; 300 mm thick	-	0.15	2.70	5.67	-	m²	**8.37**
tarmacadam; 75 mm thick	-	-	-	1.56	-	m²	**1.56**
tarmacadam and hardcore; 150 mm thick	-	-	-	2.50	-	m²	**2.50**
Extra over any types of excavating irrespective of depth for taking up							
precast concrete paving slabs	-	0.07	1.20	0.36	-	m²	**1.56**
natural stone paving	-	0.10	1.80	0.54	-	m²	**2.34**
cobbles	-	0.13	2.25	0.68	-	m²	**2.93**
brick paviors	-	0.13	2.25	0.68	-	m²	**2.93**
Filling to make up levels; mechanical							
Arising from the excavations							
average thickness 100 mm maximum thickness	-	-	0.07	0.12	-	m²	**0.20**
average thickness 150 mm maximum thickness	-	0.01	0.11	0.19	-	m²	**0.29**
average thickness 200 mm maximum thickness	-	0.01	0.14	0.25	-	m²	**0.39**
average thickness 250 mm maximum thickness	-	0.01	0.18	0.31	-	m²	**0.49**
average thickness 300 mm maximum thickness	-	0.01	0.24	0.41	-	m²	**0.65**

Prices for Measured Works
D GROUNDWORK

Item Excluding site overheads and profit	PC £	Labour hours	Labour £	Plant £	Material £	Unit	Total rate £
average thickness 400 mm maximum thickness	-	0.02	0.29	0.49	-	m²	0.78
average thickness 500 mm maximum thickness	-	0.02	0.36	0.62	-	m²	0.98
average thickness 750 mm maximum thickness	-	0.03	0.54	0.93	-	m²	1.47
Obtained from on site spoil heaps; average 25 m distance; multiple handling							
average thickness 250 mm	-	0.02	0.36	0.73	-	m²	1.09
average thickness 300 mm	-	0.02	0.43	0.88	-	m²	1.31
average thickness 400 mm	-	0.03	0.58	1.18	-	m²	1.75
average thickness 500 mm	-	0.04	0.75	1.53	-	m²	2.28
average thickness 750 mm	-	0.06	1.13	2.30	-	m²	3.42
Obtained off site; soil PC £17.32 / m³							
average thickness 100 mm	1.73	0.01	0.12	0.21	2.08	m²	2.41
average thickness 200 mm	3.46	0.01	0.24	0.41	4.16	m²	4.81
average thickness 250 mm	4.33	0.02	0.30	0.52	5.20	m²	6.01
average thickness 300 mm	5.20	0.02	0.27	0.46	6.24	m²	6.97
average thickness 400 mm	6.93	0.02	0.36	0.62	8.31	m²	9.29
average thickness 500 mm	8.66	0.03	0.45	0.77	10.39	m²	11.61
average thickness 750 mm	12.99	0.04	0.68	1.16	15.59	m²	17.42
Obtained off site; hardcore; PC £16.00 / m³							
average thickness 100 mm	1.60	0.01	0.12	0.21	1.76	m²	2.09
average thickness 200 mm	3.20	0.01	0.24	0.41	3.52	m²	4.17
average thickness 250 mm	4.00	0.02	0.30	0.52	4.40	m²	5.22
average thickness 300 mm	4.80	0.02	0.27	0.46	5.28	m²	6.01
average thickness 400 mm	6.40	0.02	0.36	0.62	7.04	m²	8.02
average thickness 500 mm	8.00	0.03	0.45	0.77	8.80	m²	10.02
average thickness 750 mm	12.00	0.04	0.68	1.16	13.20	m²	15.04
Excavating; hand							
Topsoil for preservation; loading to barrows							
average depth 100 mm	-	0.24	4.32	-	-	m²	4.32
average depth 150 mm	-	0.36	6.48	-	-	m²	6.48
average depth 200 mm	-	0.58	10.37	-	-	m²	10.37
average depth 250 mm	-	0.72	12.96	-	-	m²	12.96
average depth 300 mm	-	0.86	15.55	-	-	m²	15.55
Excavating; hand							
Topsoil to reduce levels							
average depth 100 mm	-	0.33	6.00	-	-	m²	6.00
average depth 150 mm	-	0.53	9.45	-	-	m²	9.45
average depth 200 mm	-	0.73	13.20	-	-	m²	13.20
average depth 250 mm	-	1.00	18.00	-	-	m²	18.00
average depth 300 mm	-	1.30	23.40	-	-	m²	23.40
average depth 400 mm	-	1.87	33.60	-	-	m²	33.60
average depth 600 mm	-	3.00	54.00	-	-	m²	54.00
average depth 750 mm	-	4.00	72.00	-	-	m²	72.00
average depth 1.00 mm	-	6.67	119.99	-	-	m²	119.99
Pits							
maximum depth not exceeding 0.25 m	-	2.67	48.00	-	-	m³	48.00
maximum depth not exceeding 1.00 m	-	3.47	62.40	-	-	m³	62.40
maximum depth not exceeding 2.00 m	-	6.93	124.80	-	-	m³	124.80
Trenches; width not exceeding 0.30 m							
maximum depth not exceeding 0.25 m	-	2.86	51.43	-	-	m³	51.43
maximum depth not exceeding 1.00 m	-	3.72	66.98	-	-	m³	66.98
Trenches; width exceeding 0.30 m wide							
maximum depth not exceeding 0.25 m	-	2.86	51.43	-	-	m³	51.43
maximum depth not exceeding 1.00 m	-	4.00	72.00	-	-	m³	72.00
maximum depth not exceeding 2.00 m	-	8.00	144.00	-	-	m³	144.00

Prices for Measured Works
D GROUNDWORK

Item Excluding site overheads and profit	PC £	Labour hours	Labour £	Plant £	Material £	Unit	Total rate £
D20 EXCAVATION AND FILLING – cont'd							
Extra over any types of excavating irrespective of depth for breaking out existing materials; hand held pneumatic breaker							
rock	-	5.00	90.00	18.00	-	m³	108.00
concrete	-	2.50	45.00	9.00	-	m³	54.00
reinforced concrete	-	4.00	72.00	20.65	-	m³	92.65
brickwork, blockwork or stonework	-	1.50	27.00	5.40	-	m³	32.40
Disposal; mechanical; Recycled Materials Ltd							
Note to users. Most commercial site disposal is carried out by 20 tonne - 8 wheeled vehicles. It has been customary to calculate disposal from construction sites in terms of full 15 m³ loads. Spon's research has found that based on weights of common materials such as clean hardcore and topsoil, that vehicles could not load more than 12 m³ at a time. Most hauliers do not make it apparent that their loads are calculated by weight and not by volume. The rates below reflect these lesser volumes which are limited by the 20 tonne limit							
The volumes shown below are based on volumes "in the solid". weight and bulking factors have been applied. For further information please see the weights of typical materials in the Earthworks section of the Memoranda at the back of this book.							
Light soils and loams (bulking factor - 1.25); 40 tonnes (2 loads per hour)							
Excavated material; off site; to tip; mechanically loaded. (JCB)							
inert material	-	0.04	0.75	1.17	14.85	m³	16.77
soil (sandy and loam) dry	-	0.04	0.75	1.17	14.85	m³	16.77
soil (sandy and loam) wet	-	0.04	0.75	1.17	15.75	m³	17.67
broken out compacted materials such as road bases and the like	-	0.04	0.75	1.17	16.50	m³	18.42
soil (clay) dry	-	0.04	0.75	1.17	19.27	m³	21.20
soil (clay) wet	-	0.04	0.75	1.17	20.03	m³	21.95
slightly contaminated	-	0.04	0.75	1.17	20.63	m³	22.55
rubbish	-	0.04	0.75	1.17	24.75	m³	26.67
As above but allowing for 3 loads (36 m³ - 60 tonne) per hour removed							
Inert material	-	0.03	0.50	0.78	14.85	m³	16.13
Excavated material; off site; to tip; mechanically loaded by grab; capacity of load 7.25 m³							
inert material	-	-	-	-	-	m³	36.30
soil (sandy and loam) dry	-	-	-	-	-	m³	36.30
soil (sandy and loam) wet	-	-	-	-	-	m³	38.50
broken out compacted materials such as road bases and the like	-	-	-	-	-	m³	40.33
soil (clay) dry	-	-	-	-	-	m³	47.12
soil (clay) wet	-	-	-	-	-	m³	48.95

Prices for Measured Works
D GROUNDWORK

Item Excluding site overheads and profit	PC £	Labour hours	Labour £	Plant £	Material £	Unit	Total rate £
Disposal by skip; 7 yd3 (5.35 m³)							
Excavated material; off site; to tip							
by machine	-	0.13	2.25	1.26	21.80	m³	25.31
by hand	-	3.00	54.00	-	32.70	m³	86.70
Excavated material; on site; in spoil heaps							
average 25 m distance	-	0.02	0.41	0.85	-	m³	1.26
average 50 m distance	-	0.04	0.70	1.48	-	m³	2.18
average 100 m distance	-	0.06	1.12	2.35	-	m³	3.46
average 200 m distance	-	0.12	2.16	4.55	-	m³	6.71
Excavated material; spreading on site							
average 25 m distance	-	0.02	0.41	3.84	-	m³	4.24
average 50 m distance	-	0.04	0.70	4.34	-	m³	5.04
average 100 m distance	-	0.06	1.12	5.34	-	m³	6.45
average 200 m distance	-	0.12	2.16	5.16	-	m³	7.32
Disposal; hand							
Excavated material; on site; in spoil heaps							
average 25 m distance	-	2.40	43.20	-	-	m³	43.20
average 50 m distance	-	2.64	47.52	-	-	m³	47.52
average 100 m distance	-	3.00	54.00	-	-	m³	54.00
average 200 m distance	-	3.60	64.80	-	-	m³	64.80
Excavated material; spreading on site							
average 25 m distance	-	2.64	47.52	-	-	m³	47.52
average 50 m distance	-	3.00	54.00	-	-	m³	54.00
average 100 m distance	-	3.60	64.80	-	-	m³	64.80
average 200 m distance	-	4.20	75.60	-	-	m³	75.60
Cultivating							
Ripping up subsoil; using approved subsoiling machine; minimum depth 250 mm below topsoil; at 1.20 m centres; in							
gravel or sandy clay	-	-	-	2.31	-	100m2	2.31
soil compacted by machines	-	-	-	2.70	-	100m2	2.70
clay	-	-	-	2.89	-	100m2	2.89
chalk or other soft rock	-	-	-	5.78	-	100m2	5.78
Extra for subsoiling at 1 m centres	-	-	-	0.58	-	100m2	0.58
Breaking up existing ground; using pedestrian operated tine cultivator or rotavator; loam or sandy soil							
100 mm deep	-	0.22	3.96	2.24	-	100m2	6.20
150 mm deep	-	0.28	4.95	2.80	-	100m2	7.75
200 mm deep	-	0.37	6.60	3.73	-	100m2	10.33
As above but in heavy clay or wet soils							
100 mm deep	-	0.44	7.92	4.48	-	100m2	12.40
150 mm deep	-	0.66	11.88	6.72	-	100m2	18.60
200 mm deep	-	0.83	14.85	8.40	-	100m2	23.25
Breaking up existing ground; using tractor drawn tine cultivator or rotavator							
100 mm deep	-	-	-	0.52	-	100 m²	0.52
150 mm deep	-	-	-	0.65	-	100 m²	0.65
200 mm deep	-	-	-	0.86	-	100 m²	0.86
400 mm deep	-	-	-	2.59	-	100 m²	2.59
Cultivating ploughed ground; using disc, drag, or chain harrow							
4 passes	-	-	-	3.10	-	100 m²	3.10

D GROUNDWORK

Item Excluding site overheads and profit	PC £	Labour hours	Labour £	Plant £	Material £	Unit	Total rate £
D20 EXCAVATION AND FILLING - cont'd							
Rolling cultivated ground lightly; using self-propelled agricultural roller	-	0.06	1.00	0.58	-	100 m²	1.58
Importing and storing selected and approved topsoil; to BS 3882; from source not exceeding 13 km from site							
1 - 14 m³	17.32	-	-	-	51.96	m³	51.96
over 15 m³	17.32	-	-	-	17.32	m³	17.32
Filling to make up levels; mechanical							
Arising from the excavations							
average thickness not exceeding 0.25 m; depositing in layers 150 mm maximum thickness	-	0.04	0.72	1.24	-	m³	1.96
average thickness exceeding 0.25 m; depositing in layers 150 mm maximum thickness	-	0.03	0.60	1.03	-	m³	1.63
Obtained off site; soil							
average thickness not exceeding 0.25 m; depositing in layers 150 maximum thickness	17.32	0.07	1.20	2.06	20.78	m³	24.05
Filling to make up levels; hand							
Arising from the excavations							
average thickness exceeding 0.25 m; depositing in layers 150 mm maximum thickness	-	0.60	10.80	-	-	m³	10.80
Obtained from on site spoil heaps; average 25 m distance; multiple handling							
average thickness exceeding 0.25 m thick; depositing in layers 150 mm maximum thickness	-	1.00	18.00	-	-	m³	18.00
Surface treatments							
Compacting							
bottoms of excavations	-	0.01	0.09	0.03	-	m²	0.12
Grading							
Trimming surfaces of cultivated ground to final levels, removing roots stones and debris exceeding 50 mm in any direction to tip offsite; slopes less than 15 degrees							
clean ground with minimal stone content	-	0.25	4.50	-	-	100 m²	4.50
slightly stony - 0.5 kg stones per m²	-	0.33	5.99	-	-	100 m²	6.00
very stony - 1.0 - 3.00 kg stones per m²	-	0.50	9.00	-	0.01	100 m²	9.01
clearing mixed slightly contaminated rubble inclusive of roots and vegetation	-	0.50	9.00	-	0.03	100 m²	9.03
clearing brick-bats stones and clean rubble	-	0.60	10.80	-	0.02	100 m²	10.82

Prices for Measured Works
E IN SITU CONCRETE/LARGE PC CONCRETE

Item Excluding site overheads and profit	PC £	Labour hours	Labour £	Plant £	Material £	Unit	Total rate £
E10 IN SITU CONCRETE							
General The concrete mixes used here are referred to as "Design" "Standard" and "Designated" mixes. The BS references on these are used to denote the concrete strength and mix volumes. Please refer to the definitions and the tables in the Memoranda for each Trade, Concrete Work, at the back of this book							
Designed mix User specified performance of the concrete. Producer responsible for selecting appropriate mix. Strength testing is essential							
Prescribed mix User specified mix constituents and is responsible for ensuring that the concrete meets performance requirements. The mix proportion is essential							
Standard mix Specified from the list in BS 5328 Pt 2 1991 s.4. Made with a restricted range of materials. Specification to include the proposed use of the material as well as; the standard mix reference, the type of cement, type and size of aggregate, slump (workability). Quality assurance required							
Designated mix Mix specified in BS 5328 Pt 2: 1991 s.5. Producer to hold current product conformity certification and quality approval to BS 5750 Pt 1 (EN 29001). Quality assurance essential. The mix may not be modified.							
Concrete mixes; mixed on site; costs for producing concrete; prices for commonly used mixes for various types of work; based on bulk load 20 tonne rates for aggregates; Roughest type mass concrete such as footings, road haunchings 300 thick							
1:3:6	48.72	1.00	18.00	-	48.72	m³	66.72
1:3:6 sulphate resisting	52.59	1.00	18.00	-	52.59	m³	70.59
As above but aggregated delivered in 10 tonne loads							
1:3:6	54.12	1.00	18.00	-	54.12	m³	72.12
1:3:6 sulphate resisting	57.99	1.00	18.00	-	57.99	m³	75.99
As above but aggregates delivered in 900 Kg bulk bags							
1:3:6	76.64	1.00	18.00	-	76.64	m³	94.64
1:3:6 sulphate resisting	80.51	1.00	18.00	-	80.51	m³	98.51
Most ordinary use of concrete such as mass walls above ground, road slabs etc. and general reinforced concrete work							
1:2:4	57.26	1.00	18.00	-	57.26	m³	75.26
1:2:4 sulphate resisting	62.74	1.00	18.00	-	62.74	m³	80.74

Prices for Measured Works
E IN SITU CONCRETE/LARGE PC CONCRETE

Item Excluding site overheads and profit	PC £	Labour hours	Labour £	Plant £	Material £	Unit	Total rate £
E10 IN SITU CONCRETE – cont'd							
Concrete mixes; mixed on site; - cont'd							
As above but aggregates delivered in 10 tonne loads							
1:2:4	62.66	1.00	18.00	-	62.66	m3	**80.66**
1:2:4 sulphate resisting	68.14	1.00	18.00	-	68.14	m3	**86.14**
As above but aggregates delivered in 900 kg bulk bags							
1:2:4	85.18	1.00	18.00	-	85.18	m3	**103.18**
1:2:4 sulphate resisting	90.66	1.00	18.00	-	90.66	m3	**108.66**
Watertight floors, pavements and walls, tanks pits steps paths surface of two course roads, reinforced concrete where extra strength is required.							
1:1.5:3	65.42	1.00	18.00	-	65.42	m3	**83.42**
As above but aggregates delivered in 10 tonne loads							
1:1.5:3	70.82	1.00	18.00	-	70.82	m3	**88.82**
As above but aggregates delivered in 900 kg bulk bags							
1:1.5:3	93.34	1.00	18.00	-	93.34	m3	**111.34**
Plain in situ concrete; site mixed; 10 N/mm2 - 40 aggregate (1:3:6); aggregates delivered in 900 kg bulk bags							
Foundations							
ordinary portland cement	94.64	1.65	29.70	-	97.01	m3	**126.71**
sulphate resistant cement	98.31	1.65	29.70	-	100.77	m3	**130.47**
Foundations; poured on or against earth or unblinded hardcore							
1:3:6	94.64	1.65	29.70	-	97.01	m3	**126.71**
1:3:6 sulphate resisting	98.31	1.65	29.70	-	100.77	m3	**130.47**
Isolated foundations							
1:3:6	94.64	2.20	39.60	-	97.01	m3	**136.61**
1:3:6 sulphate resisting	98.31	2.20	39.60	-	100.77	m3	**140.37**
Plain in situ concrete; site mixed; 21 N/mm2 - 20 aggregate (1:2:4); aggregates delivered in 900 kg bulk bags							
Foundations							
ordinary portland cement	85.18	1.65	29.70	-	87.31	m3	**117.01**
sulphate resistant cement	108.66	1.65	29.70	-	111.38	m3	**141.08**
Foundations; poured on or against earth or unblinded hardcore							
ordinary portland cement	85.18	1.65	29.70	-	87.31	m3	**117.01**
sulphate resistant cement	108.66	1.65	29.70	-	111.38	m3	**141.08**
Isolated foundations							
ordinary portland cement	85.18	2.20	39.60	-	87.31	m3	**126.91**
sulphate resistant cement	108.66	2.20	39.60	-	111.38	m3	**150.98**

E IN SITU CONCRETE/LARGE PC CONCRETE

Item Excluding site overheads and profit	PC £	Labour hours	Labour £	Plant £	Material £	Unit	Total rate £
Reinforced insitu concrete; site mixed; 21 N/mm2 - 20 aggregate (1:2:4); aggregates delivered in 10 tonne loads							
Foundations							
ordinary portland cement	80.66	2.20	39.60	-	82.68	m3	122.28
sulphate resistant cement	86.14	2.20	39.60	-	88.29	m3	127.89
Foundations; poured on or against earth or unblinded hardcore							
ordinary portland cement	80.66	2.20	39.60	-	82.68	m3	122.28
sulphate resistant cement	108.66	2.20	39.60	-	111.38	m3	150.98
Isolated foundations							
ordinary portland cement	80.66	2.75	49.50	-	82.68	m3	132.18
sulphate resistant cement	108.66	2.75	49.50	-	111.38	m3	160.88
Plain in situ concrete; ready mixed; Tarmac Southern Ltd., 10 N/mm mixes; suitable for mass concrete fill and blinding							
Foundations							
GEN1; Designated mix	62.45	1.50	27.00	-	62.45	m3	89.45
ST2; Standard mix	65.13	1.50	27.00	-	68.39	m3	95.39
Foundations; poured on or against earth or unblinded hardcore							
GEN1; Designated mix	62.45	1.57	28.35	-	62.45	m3	90.80
ST2; Standard mix	65.13	1.57	28.35	-	70.01	m3	98.36
Isolated foundations							
GEN1; Designated mix	62.45	2.00	36.00	-	62.45	m3	98.45
ST2; Standard mix	65.13	2.00	36.00	-	68.39	m3	104.39
Plain in situ concrete; ready mixed; Tarmac Southern Ltd., 15 N/mm mixes; suitable for oversite below suspended slabs and strip footings in non aggressive soils							
Foundations							
GEN 2; Designated mix	64.23	1.50	27.00	-	67.44	m3	94.44
ST3; Standard mix	66.48	1.50	27.00	-	69.80	m3	96.80
Foundations; poured on or against earth or unblinded hardcore							
GEN2; Designated mix	64.23	1.57	28.35	-	64.23	m3	92.58
ST3; Standard mix	66.48	1.57	28.35	-	66.48	m3	94.83
Isolated foundations							
GEN2; Designated mix	64.23	2.00	36.00	-	64.23	m3	100.23
ST3; Standard mix	66.48	2.00	36.00	-	66.48	m3	102.48
Plain in situ concrete; ready mixed; Tarmac Southern Ltd.; Air entrained mixes suitable for paving							
Beds or slabs; House drives parking and external paving							
PAV 1; 35 N/mm2; Designated mix	73.73	1.50	27.00	-	77.42	m3	104.42
Beds or slabs; Heavy duty external paving							
PAV 2; 40 N/mm2; Designated mix	75.68	1.50	27.00	-	79.46	m3	106.46

Prices for Measured Works
E IN SITU CONCRETE/LARGE PC CONCRETE

Item Excluding site overheads and profit	PC £	Labour hours	Labour £	Plant £	Material £	Unit	Total rate £
E10 IN SITU CONCRETE – cont'd							
Reinforced in situ concrete; ready mixed; Tarmac Southern Ltd.; 35N/mm2 mix; suitable for foundations in class 2 sulphate conditions							
Foundations							
RC 35; Designated mix	69.27	2.00	36.00	-	72.73	m³	**108.73**
Foundations; poured on or against earth or unblinded hardcore							
RC 35; Designated mix	69.27	2.10	37.80	-	74.47	m³	**112.27**
Isolated foundations							
RC 35; Designated mix	69.27	2.00	36.00	-	69.27	m³	**105.27**
Ready mix concrete; Extra for loads of less than 6 m³; Ace Minimix							
Ready mix concrete small load surcharge	-	-	-	-	15.00	m³	**15.00**
E20 FORMWORK FOR IN SITU CONCRETE							
Plain vertical formwork; basic finish							
Sides of foundations							
height exceeding 1.00 m	-	2.00	36.00	-	9.07	m²	**45.08**
height not exceeding 250 mm	-	1.00	18.00	-	2.42	m	**20.42**
height 250 - 500 mm	-	1.00	18.00	-	4.54	m	**22.54**
height 500 mm - 1.00 m	-	1.50	27.00	-	9.07	m	**36.08**
Sides of foundations; left in							
height over 1.00 m	-	2.00	36.00	-	25.46	m²	**61.46**
height not exceeding 250 mm	-	1.00	18.00	-	9.56	m	**27.56**
height 250 - 500 mm	-	1.00	18.00	-	18.83	m	**36.83**
height 500 mm - 1.00 m	-	1.50	27.00	-	37.65	m	**64.65**
E30 REINFORCEMENT FOR IN SITU CONCRETE							
The rates for reinforcement shown for steel bar below are based on prices which would be supplied on a typical landscape contract. The steel prices shown have been priced on a selection of steel delivered to site where the total order quantity is in the region of 2 tonnes. The assumption is that should larger quantities be required, the work would fall outside the scope of the typical landscape contract defined in the front of this book. Keener rates can be obtained for larger orders.							
Reinforcement bars; BS 4449; hot rolled plain round mild steel; straight							
Bars							
8 mm nominal size	426.00	27.00	486.00	-	426.00	tonne	**912.00**
10 mm nominal size	412.50	26.00	468.00	-	412.50	tonne	**880.50**
12 mm nominal size	409.50	25.00	450.00	-	409.50	tonne	**859.50**
16 mm nominal size	405.00	24.00	432.00	-	405.00	tonne	**837.00**
20 mm nominal size	405.00	23.00	414.00	-	405.00	tonne	**819.00**

Prices for Measured Works
E IN SITU CONCRETE/LARGE PC CONCRETE

Item Excluding site overheads and profit	PC £	Labour hours	Labour £	Plant £	Material £	Unit	Total rate £
Reinforcement bars; BS 4449; hot rolled plain round mild steel; bent							
Bars							
8 mm nominal size	469.50	27.00	486.00	-	469.50	tonne	**955.50**
10 mm nominal size	454.50	26.00	468.00	-	454.50	tonne	**922.50**
12 mm nominal size	451.50	25.00	450.00	-	451.50	tonne	**901.50**
16 mm nominal size	445.50	24.00	432.00	-	445.50	tonne	**877.50**
20 mm nominal size	445.50	23.00	414.00	-	445.50	tonne	**859.50**
Reinforcement fabric; BS 4483; lapped; in beds or suspended slabs							
Fabric							
ref A98 (1.54 kg/m2)	3.13	0.22	3.96	-	3.44	m2	**7.40**
ref A142 (2.22 kg/m2)	3.13	0.22	3.96	-	3.44	m2	**7.40**
ref A193 (3.02 kg/m2)	4.17	0.22	3.96	-	4.58	m2	**8.54**
ref A252 (3.95 kg/m2)	5.38	0.24	4.32	-	5.92	m2	**10.24**
ref A393 (6.16 kg/m2)	8.43	0.28	5.04	-	9.28	m2	**14.32**

Prices for Measured Works
F MASONRY

Item Excluding site overheads and profit	PC £	Labour hours	Labour £	Plant £	Material £	Unit	Total rate £
F MARKET PRICES OF MATERIALS							
Cement; Builder Centre							
Portland cement	-	-	-	-	2.40	25kg	2.40
Sulphate resistant cement	-	-	-	-	2.85	25kg	2.85
White cement	-	-	-	-	4.90	25kg	4.90
Sand; Builder Centre							
Building sand							
loose	-	-	-	-	49.24	m³	49.24
900 kg bulk bags	-	-	-	-	27.36	nr	27.36
Sharp sand							
loose	-	-	-	-	49.24	m³	49.24
900 kg bulk bags	-	-	-	-	27.36	nr	27.36
Sand; Yeoman Aggregates							
Sharp sand	-	-	-	-	14.56	tonne	14.56
Bricks; E.T. Clay Products							
Ibstock; facing bricks; 215 x 102.5 x 65 mm							
Leicester Red Stock	-	-	-	-	404.25	1000	404.25
Leicester Yellow Stock	-	-	-	-	341.25	1000	341.25
Himley Mixed Russet	-	-	-	-	352.27	1000	352.27
Himley Worcs. Mixture	-	-	-	-	341.25	1000	341.25
Roughdales Red Multi Rustic	-	-	-	-	375.38	1000	375.38
Ashdown Cottage Mixture	-	-	-	-	415.80	1000	415.80
Ashdown Crowborough Multi	-	-	-	-	482.79	1000	482.79
Ashdown Pevensey Multi	-	-	-	-	433.13	1000	433.13
Chailey Stock	-	-	-	-	471.45	1000	471.45
Dorking Multi Coloured	-	-	-	-	358.05	1000	358.05
Holbrook Smooth Red	-	-	-	-	444.68	1000	444.68
Stourbridge Kenilworth Multi	-	-	-	-	340.73	1000	340.73
Stourbridge Pennine Pastone	-	-	-	-	309.75	1000	309.75
Stratford Red Rustic	-	-	-	-	323.40	1000	323.40
Swanage Restoration Red	-	-	-	-	675.67	1000	675.67
Laybrook Sevenoaks Yellow	-	-	-	-	336.00	1000	336.00
Laybrook Arundel Yellow	-	-	-	-	357.00	1000	357.00
Laybrook Thakeham Red	-	-	-	-	304.50	1000	304.50
Funton Second Hard Stock	-	-	-	-	409.50	1000	409.50
Hanson Brick Ltd, London Brand; facing bricks; 215 x 102.5 x 65 mm							
Capel Multi Stock	-	-	-	-	290.00	1000	290.00
Rusper Stock	-	-	-	-	295.00	1000	295.00
Delph Autumn	-	-	-	-	235.00	1000	235.00
Regency	-	-	-	-	266.00	1000	266.00
Sandfaced	-	-	-	-	266.00	1000	266.00
Saxon Gold	-	-	-	-	222.00	1000	222.00
Tudor	-	-	-	-	245.00	1000	245.00
Windsor	-	-	-	-	239.00	1000	239.00
Autumn Leaf	-	-	-	-	185.00	1000	185.00
Claydon Red Multi	-	-	-	-	205.00	1000	205.00
Other facing bricks; 215 x 102.5 x 65 mm							
Soft Reds - Milton Hall	-	-	-	-	375.00	1000	375.00
Soft Reds - Blenheim Reds	-	-	-	-	530.00	1000	530.00
Staffs Blues - Blue Smooth	-	-	-	-	465.00	1000	465.00
Staffs Blues - Blue Brindle	-	-	-	-	425.00	1000	425.00

Prices for Measured Works
F MASONRY

Item Excluding site overheads and profit	PC £	Labour hours	Labour £	Plant £	Material £	Unit	Total rate £
Reclaimed (second hand) bricks							
Yellows	-	-	-	-	850.00	1000	850.00
Yellow Multi	-	-	-	-	850.00	1000	850.00
Mixed London Stock	-	-	-	-	650.00	1000	650.00
Red Multi	-	-	-	-	530.00	1000	530.00
Gaults	-	-	-	-	460.00	1000	460.00
Red Rubbers	-	-	-	-	740.00	1000	740.00
Common bricks	-	-	-	-	185.00	1000	185.00
Engineering bricks	-	-	-	-	180.00	1000	180.00
Kempston facing bricks; 215 x 102.5 x 65 mm							
Melford Yellow	-	-	-	-	269.00	1000	269.00
Anglian Blend	-	-	-	-	280.00	1000	280.00
F10 BRICK/BLOCK WALLING							
Mortar mixes; Common mixes for various types of work; Mortar mixed on site; Prices based on builders merchant rates for cement; aggregates delivered in 900 kg bulk bags; mechanically mixed; Batching quantities for these mortar mixes may be found in the memorandum section of this book							
1:3	-	0.75	13.50	-	112.03	m³	125.53
1:4	-	0.75	13.50	-	90.00	m³	103.50
1:1:6	-	0.75	13.50	-	96.28	m³	109.78
1:1:6 Sulphate resisting	-	0.75	13.50	-	101.14	m³	114.64
Mortar mixes; Common mixes for various types of work; Mortar mixed on site; Prices based on builders merchant rates for cement; aggregates delivered in 900 kg bulk bags; hand mixed							
1:3	-	1.00	18.00	-	112.03	m³	130.03
1:4	-	1.00	18.00	-	88.74	m³	106.74
1:1:6	-	1.00	18.00	-	96.28	m³	114.28
1:1:6 Sulphate resisting	-	1.00	18.00	-	101.14	m³	119.14
Mortar mixes; Common mixes for various types of work; Mortar mixed on site; Prices based on builders merchant rates for cement; aggregates delivered in 10 tonne loads; mechanically mixed; Batching quantities for these mortar mixes may be found in the memorandum section of this book							
1:3	-	0.75	13.50	-	88.47	m³	101.97
1:4	-	0.75	13.50	-	69.04	m³	82.54
1:1:6	-	0.75	13.50	-	76.58	m³	90.08
1:1:6 Sulphate resisting	-	0.75	13.50	-	81.44	m³	94.94

Prices for Measured Works
F MASONRY

Item Excluding site overheads and profit	PC £	Labour hours	Labour £	Plant £	Material £	Unit	Total rate £
F10 BRICK/BLOCK WALLING – cont'd							
Common bricks; PC £180.00 /1000; **English garden wall bond; in gauged** **mortar (1:1:6)** Mechanically offloading; maximum 25 m distance; loading to wheel barrows and transporting to location; per 215 mm thick walls	-	0.42	7.50	-	-	m²	7.50
Walls							
half brick thick	-	1.80	32.40	-	12.90	m²	45.30
one brick thick	-	2.40	43.20	-	25.37	m²	68.57
one and a half brick thick	-	3.60	64.80	-	38.05	m²	102.85
two brick thick	-	4.80	86.39	-	50.73	m²	137.13
Walls; curved; mean radius 6 m							
half brick thick	-	1.67	30.00	-	13.96	m²	43.96
one brick thick	-	3.36	60.48	-	26.88	m²	87.36
Walls; curved; mean radius 1.50 m							
half brick thick	-	1.89	33.96	-	14.50	m²	48.46
one brick thick	-	4.00	72.00	-	27.96	m²	99.96
Extra for cement mortar (1:3) in lieu of gauged mortar							
half brick thick	-	-	-	-	-0.12	m²	-0.12
one brick thick	-	-	-	-	-0.24	m²	-0.24
one and a half brick thick	-	-	-	-	-0.36	m²	-0.36
two brick thick	-	-	-	-	-0.48	m²	-0.48
Walls; stretcher bond; wall ties at 450 centres vertically and horizontally							
one brick thick	-	1.71	30.86	-	25.81	m²	56.67
one and a half brick thick	-	2.56	46.08	-	38.95	m²	85.03
two brick thick	-	3.43	61.72	-	52.08	m²	113.80
Facing bricks; PC £300.00 /1000; **English garden wall bond; in gauged** **mortar (1:1:6); facework one side** Mechanically offloading; maximum 25 m distance; loading to wheel barrows and transporting to location; per 215 mm thick walls	-	0.42	7.50	-	-	m²	7.50
Walls							
half brick thick	-	1.40	25.20	-	20.46	m²	45.66
one brick thick	-	2.82	50.76	-	40.92	m²	91.68
one and a half brick thick	-	4.24	76.32	-	61.38	m²	137.70
two brick thick	-	5.65	101.70	-	81.84	m²	183.54
Walls; curved; mean radius 6 m							
half brick thick	-	2.10	37.80	-	21.88	m²	59.68
one brick thick	-	4.21	75.79	-	42.72	m²	118.51
Walls; curved; mean radius 1.50 m							
half brick thick	-	2.40	43.20	-	21.43	m²	64.63
one brick thick	-	4.80	86.41	-	41.82	m²	128.23
Walls; tapering; one face battering; average							
one and a half brick thick	-	5.60	100.81	-	64.08	m²	164.89
two brick thick	-	7.47	134.38	-	85.44	m²	219.81
Walls; battering (retaining)							
one and a half brick thick	-	5.60	100.81	-	64.08	m²	164.89
two brick thick	-	7.47	134.38	-	85.44	m²	219.81
Isolated piers; English bond; facework all round							
one brick thick	-	7.00	126.00	-	43.56	m²	169.56
one and a half brick thick	-	9.00	162.00	-	65.34	m²	227.34

Prices for Measured Works
F MASONRY

Item Excluding site overheads and profit	PC £	Labour hours	Labour £	Plant £	Material £	Unit	Total rate £
two brick thick	-	10.00	180.00	-	88.08	m²	268.08
three brick thick	-	12.40	223.20	-	130.68	m²	353.88
Projections; vertical							
one brick x half brick	-	0.70	12.60	-	4.56	m	17.16
one brick x one brick	-	1.40	25.20	-	9.13	m	34.33
one and a half brick x one brick	-	2.10	37.80	-	13.69	m	51.49
two brick x one brick	-	2.30	41.40	-	18.25	m	59.65
Brickwork fair faced both sides; facing bricks in gauged mortar (1:1:6)							
Extra for fair face both sides; flush, struck, weathered, or bucket-handle pointing	-	0.67	12.00	-	-	m²	12.00
Extra for cement mortar (1:3) in lieu of gauged mortar							
half brick thick	-	-	-	-	-0.12	m²	-0.12
one brick thick	-	-	-	-	-0.24	m²	-0.24
one and a half brick thick	-	-	-	-	-0.36	m²	-0.36
two brick thick	-	-	-	-	-0.48	m²	-0.48
Class B engineering bricks; PC £260.00 /1000; double Flemish bond in cement mortar (1:3)							
Mechanically offloading; maximum 25 m distance; loading to wheel barrows; transporting to location; per 215 mm thick walls	-	0.42	7.50	-	-	m²	7.50
Walls							
half brick thick	-	1.41	25.35	-	17.82	m²	43.17
one brick thick	-	2.82	50.82	-	35.64	m²	86.46
one and a half brick thick	-	4.24	76.24	-	62.73	m²	138.97
two brick thick	-	5.65	101.64	-	71.28	m²	172.92
Walls; curved; mean radius 6 m							
half brick thick	-	2.10	37.80	-	18.78	m²	56.58
one brick thick	-	4.21	75.79	-	35.64	m²	111.43
Walls; curved; mean radius 1.50 m							
half brick thick	-	2.40	43.20	-	18.30	m²	61.50
one brick thick	-	4.80	86.41	-	35.64	m²	122.05
Walls; tapering; one face battering; average							
one and a half brick thick	-	5.60	100.81	-	53.46	m²	154.27
two brick thick	-	7.47	134.38	-	71.28	m²	205.66
Walls; battering (retaining)							
one and a half brick thick	-	5.60	100.81	-	53.46	m²	154.27
two brick thick	-	7.47	134.38	-	71.28	m²	205.66
Isolated piers							
one brick thick	-	7.00	126.00	-	38.52	m²	164.52
one and a half brick thick	-	9.00	162.00	-	57.78	m²	219.78
two brick thick	-	10.00	180.00	-	78.00	m²	258.00
three brick thick	-	12.40	223.20	-	115.56	m²	338.76
Projections; vertical							
one brick x half brick	-	0.70	12.60	-	3.98	m	16.58
one brick x one brick	-	1.40	25.20	-	7.95	m	33.15
one and a half brick x one brick	-	2.10	37.80	-	11.93	m	49.73
two brick by one brick	-	2.30	41.40	-	15.90	m	57.30
Walls; half brick thick							
in honeycomb bond	-	1.80	32.40	-	12.58	m²	44.98
in quarter bond	-	1.67	30.00	-	17.43	m²	47.43
Brickwork fair faced both sides; facing bricks in gauged mortar (1:1:6)							
Extra for fair face both sides; flush, struck, weathered, or bucket-handle pointing	-	0.67	12.00	-	-	m²	12.00

Prices for Measured Works
F MASONRY

Item Excluding site overheads and profit	PC £	Labour hours	Labour £	Plant £	Material £	Unit	Total rate £
Brick copings							
Copings; all brick headers-on-edge; to BS 4729; two angles rounded 53 mm radius; flush pointing top and both sides as work proceeds; one brick wide; horizontal							
machine-made specials	19.33	0.16	2.81	-	20.48	m	**23.30**
hand-made specials	19.33	0.16	2.81	-	20.49	m	**23.30**
Extra over copings for two courses machine-made tile creasings, projecting 25 mm each side; 260 mm wide copings; horizontal	5.26	0.50	9.00	-	5.81	m	**14.81**
Copings; all brick headers-on-edge; flush pointing top and both sides as work proceeds; one brick wide; horizontal							
facing bricks PC £300.00/1000	4.00	0.31	5.63	-	4.30	m	**9.93**
engineering bricks PC £260.00/1000	3.47	0.31	5.63	-	3.64	m	**9.27**
Variation in brick prices							
Add or subtract the following amounts for every £1.00/1000 difference in the PC price							
half brick thick	-	-	-	-	0.06	m²	**0.06**
one brick thick	-	-	-	-	0.13	m²	**0.13**
one and a half brick thick	-	-	-	-	0.19	m²	**0.19**
two brick thick	-	-	-	-	0.25	m²	**0.25**
Dense aggregate concrete blocks; "Tarmac Topblock" or other equal and approved; in gauged mortar (1:2:9)							
Walls							
Solid Blocks 7N/mm2							
440 x 215 x 100 mm thick	-	1.20	21.60	-	8.02	m²	**29.62**
440 x 215 x 140 mm thick	10.21	1.30	23.40	-	11.07	m²	**34.47**
440 x 215 x 215 mm thick	-	1.50	27.00	-	16.94	m²	**43.94**
Solid Blocks 7N/mm2 laid flat							
440 x 100 x 215 mm thick	14.79	3.22	57.96	-	16.85	m²	**74.81**
Hollow Concrete blocks							
440 x 215 x 215 mm thick	10.25	1.30	23.40	-	11.11	m²	**34.51**
Filling of Hollow concrete blocks with concrete as work proceeds; tamping and compacting							
440 x 215 x 215 mm thick	-	0.20	3.60	-	17.55	m²	**21.15**
F20 NATURAL STONE RUBBLE WALLING							
Granite walls							
Granite random rubble walls; laid dry							
200 mm thick; single faced	33.33	6.66	119.88	-	33.33	m²	**153.21**
Granite walls; one face battering to 50 degrees; pointing faces							
450 mm (average) thick	75.00	10.00	180.00	-	81.23	m²	**261.23**
Dry stone walling - General							
Preamble: In rural areas where natural stone is a traditional material, it may be possible to use dry stone walling or dyking as an alternative to fences or brick walls. Many local authorities are willing to meet the extra cost of stone walling in areas of high landscape value, and they may hold lists of available craftsmen.							

Prices for Measured Works
F MASONRY

Item Excluding site overheads and profit	PC £	Labour hours	Labour £	Plant £	Material £	Unit	Total rate £
Dry Stone Association, YFC Centre, National Agricultural Centre, Stoneleigh Park, Kenilworth, Warwickshire. CV8 2LG. Tel: (01203) 696 544.							
Rockery stone - General Preamble: Rockery stone prices vary considerably with source, carriage, distance and load. Typical garden centre prices for small quantities are in the range of £30 - £40 per tonne collected.							
Rockery stone; Breedon Lumping limestone 9 - 12 inches; minimum 15/20 tonne loads							
boulders; maximum diameter 400 mm; to positions maximum 25 m distance from offload	56.50	2.60	46.80	-	56.50	tonne	103.30
Rockery stone; Civil Engineering Developments Glacial boulders; maximum distance 25 m; by machine							
750 mm diameter	115.70	0.90	16.20	12.44	115.70	nr	144.34
1 m diameter	230.00	2.00	36.00	24.87	230.00	nr	290.87
1.5 m diameter	765.00	2.00	36.00	42.00	765.00	nr	843.00
2 m diameter	1735.00	2.00	36.00	42.00	1735.00	nr	1813.00
Glacial boulders; maximum distance 25 m; by hand							
not exceeding 750 mm diameter	48.50	0.75	13.50	-	48.50	nr	62.00
not exceeding 1 m diameter	115.70	1.89	33.94	-	115.70	nr	149.64
F22 CAST STONE WALLING							
Atlas Stone Products; dwarf walls; in Cotswold stone; grey, brown or red "Campden Pitched" stone walls							
uniform size precast units	12.07	2.50	45.00	-	17.83	m²	62.83
random size precast units	12.07	4.00	72.00	-	17.83	m²	89.83
Atlas Stone Products; cast stone characteristic walls with irregular size stone facing "Weathered Cotswold" drystone walls							
600 x 100 x 125 mm	39.48	0.67	12.00	-	42.36	m²	54.36
600 x 100 x 125 mm	59.99	1.50	27.00	-	62.87	m²	89.87
Copings 400 x 150 x 150	3.45	0.40	7.20	-	6.33	m	13.53
Copings							
flat	3.45	0.20	3.60	-	3.52	m	7.12

Prices for Measured Works
F MASONRY

Item Excluding site overheads and profit	PC £	Labour hours	Labour £	Plant £	Material £	Unit	Total rate £
F22 CAST STONE WALLING cont'd							
Haddonstone Ltd; Cast stone piers; **Ornamental masonry in Portland Bath or** **Terracotta finished cast stone**							
Gate Pier S120; to foundations and underground work measured separately; concrete infill							
S 120G base unit to pier; 699 x 699 x 172 mm	109.00	0.75	13.50	-	112.94	nr	**126.44**
S 120F shaft base unit; 533 x 533 x 280 mm	87.00	1.00	18.00	-	91.06	nr	**109.06**
S 120E main shaft unit; 533 x 533 x 280 mm; nr of units required dependent on height of pier	87.00	1.00	18.00	-	91.06	nr	**109.06**
S 120D top shaft unit; 33 x 533 x 280 mm	87.00	1.00	18.00	-	91.06	nr	**109.06**
S 120C pier cap unit; 737 x 737 x 114 mm	104.00	0.50	9.00	-	108.06	nr	**117.06**
S 120B pier block unit; base for finial 533 x 533 x 64 mm	45.00	0.33	5.94	-	45.12	nr	**51.06**
Pier blocks; flat to receive gate finial							
S100B; 440 x 440 x 63 mm	29.00	0.33	6.00	-	29.29	nr	**35.29**
S120B; 546 x 546 x 64 mm	45.00	0.33	6.00	-	45.29	nr	**51.29**
S150B; 330 x 330 x 51 mm	15.00	0.33	6.00	-	15.29	nr	**21.29**
Pier caps; part weathered							
S100C; 915 x 915 x 150 mm	246.00	0.50	9.00	-	246.58	nr	**255.58**
S120C; 737 x 737 x 114 mm	104.00	0.50	9.00	-	104.29	nr	**113.29**
S150C; 584 x 584 x 120 mm	758.00	0.50	9.00	-	758.29	nr	**767.29**
Pier caps; weathered							
S230C; 1029 x 1029 x 175 mm	323.00	0.50	9.00	-	323.29	nr	**332.29**
S215C; 687 x 687 x 175 mm	141.00	0.50	9.00	-	141.29	nr	**150.29**
S210C; 584 x 584 x 175 mm	89.00	0.50	9.00	-	89.29	nr	**98.29**
Pier strings							
S100S; 800 x 800 x 55 mm	84.00	0.50	9.00	-	84.18	nr	**93.18**
S120S; 555 x 555 x 44 mm	38.00	0.50	9.00	-	38.18	nr	**47.18**
S150S; 457 x 457 x 48 mm	24.00	0.50	9.00	-	24.18	nr	**33.18**
Balls and Bases							
E150A Ball 535 mm and E150C collared base	216.17	0.50	9.00	-	216.35	nr	**225.35**
E120A Ball 330 mm and E120C collared base	75.74	0.50	9.00	-	75.92	nr	**84.92**
E110A Ball 230 mm and E110C collared base	54.47	0.50	9.00	-	54.65	nr	**63.65**
E100A Ball 170 mm and E100B plain base	34.89	0.50	9.00	-	35.07	nr	**44.07**
Haddonstone Ltd; Cast stone copings; **Ornamental masonry in Portland Bath or** **Terracotta finished cast stone**							
Copings for walls; bedded, jointed and pointed in approved coloured cement-lime mortar 1:2:9							
T100; weathered coping 102 mm high 178 mm wide x 914 mm	27.25	0.33	5.99	-	27.50	m	**33.49**
T140; weathered coping 102 mm high 337 mm wide x 914 mm	44.69	0.33	6.00	-	44.94	m	**50.94**
T200; weathered coping 127 mm high 508 mm wide x 750 mm	68.67	0.33	6.00	-	68.92	m	**74.92**
T170; weathered coping 108 mm high 483 mm wide x 914	73.03	0.33	6.00	-	73.28	m	**79.28**
T340 raked coping 100 -75 mm high 290 wide x 900 mm	41.42	0.33	6.00	-	41.67	m	**47.67**
T310 raked coping 89 -76 mm high 381 wide x 914 mm	49.05	0.33	6.00	-	49.30	m	**55.30**

Prices for Measured Works
F MASONRY

Item Excluding site overheads and profit	PC £	Labour hours	Labour £	Plant £	Material £	Unit	Total rate £
F30 ACCESSORIES/SUNDRY ITEMS FOR BRICK/BLOCK STONE WALLING							
Damp proof courses; bitumen hessian based; 200 mm laps							
Horizontal							
width not exceeding 225 mm	3.51	0.58	10.44	-	3.77	m²	14.21
width exceeding 225 mm	3.51	0.60	10.80	-	4.39	m²	15.19
Vertical							
not exceeding 225 mm	3.51	1.72	30.96	-	3.77	m²	34.73
Damp proof courses; "Ledkore" grade A; bitumen based; lead cored; 200 mm laps							
Horizontal							
width not exceeding 225 mm	18.00	1.00	18.00	-	19.35	m²	37.35
width exceeding 225 mm	18.00	0.76	13.68	-	22.50	m²	36.18
Vertical							
width not exceeding 225 mm wide	18.00	1.72	30.96	-	19.35	m²	50.31
Damp proof courses; pitch polymer; 150 mm laps							
Horizontal							
width not exceeding 225 mm	3.90	1.14	20.52	-	5.34	m²	25.86
width exceeding 225 mm	3.90	0.58	10.44	-	5.82	m²	16.26
Vertical							
width not exceeding 225 mm	3.90	1.72	30.96	-	5.34	m²	36.30
Two courses slates in cement mortar (1:3)							
Horizontal							
width exceeding 225 mm	8.43	3.46	62.28	-	11.31	m²	73.59
Vertical							
width exceeding 225 mm	8.43	5.18	93.24	-	11.31	m²	104.55
F31 PRECAST CONCRETE SILLS/LINTELS/COPINGS/FEATURES							
Mix 21.00 N/mm2 - 20 aggregate (1:2:4)							
Copings; once weathered; twice grooved							
152 x 75 mm	4.68	0.40	7.20	-	4.95	m	12.15
178 x 65 mm	6.85	0.40	7.20	-	7.09	m	14.29
305 x 75 mm	9.78	0.50	9.00	-	10.16	m	19.16
Pier caps; four sides weathered							
305 x 305 mm	5.10	1.00	18.00	-	5.22	nr	23.22
381 x 381 mm	5.35	1.00	18.00	-	5.47	nr	23.47
533 x 533 mm	5.90	1.20	21.60	-	6.02	nr	27.62
Copings; Milner Delvaux; "Blanc de Bierges"							
200 x 50 mm	8.20	0.50	9.00	-	8.73	m	17.73
300 x 50 mm	11.77	0.50	9.00	-	12.41	m	21.41
400 x 50 mm	12.91	0.50	9.00	-	13.59	m	22.59
300 x 80 mm	17.77	0.50	9.00	-	18.59	m	27.59
400 x 80 mm	20.87	0.50	9.00	-	21.93	m	30.93

Prices for Measured Works
G STRUCTURAL/ CARCASSING METAL/TIMBER

Item Excluding site overheads and profit	PC £	Labour hours	Labour £	Plant £	Material £	Unit	Total rate £
G31 PREFABRICATED TIMBER UNIT DECKING							
Timber Decking; Wyckham - Blackwell Ltd							
Supports for timber decking; treated softwood joists to receive decking boards; joists at 400 mm centres;							
50 x 150 mm	4.55	1.00	18.00	-	8.48	m²	26.48
50 x 200 mm	7.00	1.00	18.00	-	11.17	m²	29.17
50 x 250 mm;	13.75	1.00	18.00	-	18.60	m²	36.60
Hardwood Decking Yellow Balau or Tatjuba; grooved or smooth; 6mm joints							
Deck boards; 90 mm wide x 19 mm thick	-	1.00	18.00	-	3.64	m²	21.64
Deck boards; 145 mm wide x 21 mm thick	17.32	1.00	18.00	-	22.69	m²	40.69
Deck boards; 145 mm wide x 28 mm thick	23.17	1.00	18.00	-	29.13	m²	47.13
Western Red Cedar; 6mm joints							
Prime Deck Grade; 90 mm wide x 40 mm thick	-	1.00	18.00	-	22.63	m²	40.63
Prime Deck Grade; 142 mm wide x 40 mm thick	19.31	1.00	18.00	-	24.81	m²	42.81
Patio Grade; 90 mm wide x 35 mm thick	23.40	1.00	18.00	-	29.30	m²	47.30
Redwood; Kiln dries PAR Tanalised; 6mm joints							
Deck boards 120 mm wide x 28 mm thick; grooved	5.50	0.75	13.50	-	9.69	m²	23.19
Handrails and base rail; fixed to posts at 2.00 m centres							
Posts;100 x 100 x 1370 high	48.31	1.00	18.00	-	51.48	m	69.48
Posts turned 1220 high	40.67	1.00	18.00	-	43.83	m	61.83
Hand rails; balusters							
Square balusters at 100 mm centres	33.60	0.50	9.00	-	34.29	m	43.29
Square balusters at 300 mm centres	11.19	0.33	5.99	-	11.74	m	17.74
Turned balusters at 100 mm centres	52.80	0.50	9.00	-	53.49	m	62.49
Turned balusters at 300 mm centres	17.58	0.33	5.94	-	18.14	m	24.08

Prices for Measured Works
H CLADDING/COVERING

Item Excluding site overheads and profit	PC £	Labour hours	Labour £	Plant £	Material £	Unit	Total rate £
H51 NATURAL STONE/SLAB CLADDING FEATURES							
Sawn Yorkstone cladding; Johnsons Stone Ltd; Six sides sawn stone; rubbed face ; sawn and jointed edges; fixed to blockwork (not included) with stainless steel fixings "Ancon Ltd" grade 304 stainless steel frame cramp and dowel 7mm; cladding units drilled 4 x to receive dowels 440 x 200 x 50 mm thick	78.70	1.87	33.66	-	80.86	m²	**114.52**
H52 CAST STONE/SLAB CLADDING FEATURES							
Cast Stone cladding; Haddonstone Ltd; Reconstituted stone in Portland Bath or Terracotta; fixed to blockwork or concrete (not included) with stainless steel fixings "Ancon Ltd" grade 304 stainless steel frame cramp and dowel M6 mm; cladding units drilled 4 x to receive dowels 440 x 200 x 50 mm thick	90.14	1.87	33.66	-	92.30	m²	**125.96**

Prices for Measured Works
J WATERPROOFING

Item Excluding site overheads and profit	PC £	Labour hours	Labour £	Plant £	Material £	Unit	Total rate £
J10 SPECIALIST WATERPROOF RENDERING							
"Sika" waterproof rendering; steel trowelled							
Walls; 20 thick; three coats; to concrete base							
width exceeding 300 mm	-	-	-	-	-	m²	46.51
width not exceeding 300 mm	-	-	-	-	-	m²	72.89
Walls; 25 thick; three coats; to concrete base							
width exceeding 300 mm	-	-	-	-	-	m²	52.74
width not exceeding 300 mm	-	-	-	-	-	m²	83.74
J20 MASTIC ASPHALT TANKING/DAMP PROOFING							
Tanking and damp proofing; mastic asphalt; to BS 6925; type T 1097; Coverite Ltd							
13 mm thick; one coat covering; to concrete base; flat; work subsequently covered							
width exceeding 300 mm	-	-	-	-	-	m²	9.83
20 mm thick; two coat coverings; to concrete base; flat; work subsequently covered							
width exceeding 300 mm	-	-	-	-	-	m²	12.31
30 mm thick; three coat coverings; to concrete base; flat; work subsequently covered							
width exceeding 300 mm	-	-	-	-	-	m²	16.72
13 mm thick; two coat coverings; to brickwork base; vertical; work subsequently covered							
width exceeding 300 mm	-	-	-	-	-	m²	32.64
20 mm thick; three coat coverings; to brickwork base; vertical; work subsequently covered							
width exceeding 300 mm	-	-	-	-	-	m²	44.56
Internal angle fillets; work subsequently covered	-	-	-	-	-	m	3.47
Turning asphalt nibs into grooves; 20 mm deep	-	-	-	-	-	m	2.21
J30 LIQUID APPLIED TANKING/DAMP PROOFING							
Tanking and damp proofing; Ruberoid Building Products, "Synthaprufe" cold applied bituminous emulsion waterproof coating							
"Synthaprufe"; to smooth finished concrete or screeded slabs; flat; blinding with sand							
two coats	2.26	0.22	4.00	-	2.60	m²	6.60
three coats	3.39	0.31	5.58	-	3.84	m²	9.42
"Synthaprufe"; to fair faced brickwork with flush joints, rendered brickwork, or smooth finished concrete walls; vertical							
two coats	2.54	0.29	5.14	-	2.91	m²	8.05
three coats	3.73	0.40	7.20	-	4.21	m²	11.41

Prices for Measured Works
J WATERPROOFING

Item Excluding site overheads and profit	PC £	Labour hours	Labour £	Plant £	Material £	Unit	Total rate £
Tanking and damp proofing; RIW Ltd							
Liquid asphaltic composition; to smooth finished concrete screeded slabs or screeded slabs; flat							
two coats	6.93	0.33	6.00	-	6.93	m²	**12.93**
Liquid asphaltic composition; fair faced brickwork with flush joints, rendered brickwork, or smooth finished concrete walls; vertical							
two coats	6.93	0.50	9.00	-	6.93	m²	**15.93**
"Heviseal"; to smooth finished concrete or screeded slabs; to surfaces of ponds, tanks, planters; flat							
two coats	3.21	0.33	6.00	-	3.53	m²	**9.53**
"Heviseal"; to fair faced brickwork with flush joints, rendered brickwork, or smooth finished concrete walls; to surfaces of retaining walls, ponds, tanks, planters; vertical							
two coats	3.21	0.50	9.00	-	3.53	m²	**12.53**
J40 FLEXIBLE SHEET TANKING/DAMP PROOFING							
Tanking and damp proofing; Servicised Ltd							
"Bitu-thene 500x"; 1.50 mm thick; overlapping and bonding; including sealing all edges							
to concrete slabs; flat	3.69	0.25	4.50	-	4.06	m²	**8.56**
to brick/concrete walls; vertical	3.69	0.40	7.20	-	4.32	m²	**11.52**
J50 GREEN ROOF SYSTEMS							
Preamble - A variety of systems are available which address all the varied requirements for a successful Green Roof. For installation by approved contractors only the prices shown are for budgeting purposes only as each installation is site specific and may incorporate some or all of the resources shown. Specifiers should verify that the systems specified include for design liability and inspections by the suppliers. The systems below assume commercial insulation levels are required to the space below the proposed Green Roof. Extensive Green roofs are those of generally lightweight construction with low maintenance planting and shallow soil designed for aesthetics only; Intensive Green roofs are designed to allow use for recreation and trafficking. They require more maintenance and allow a greater variety of surfaces and plant types.							

Item Excluding site overheads and profit	PC £	Labour hours	Labour £	Plant £	Material £	Unit	Total rate £
J50 GREEN ROOF SYSTEMS – cont'd							
Intensive Green Roof; Erisco Bauder Ltd; soil based systems able to provide a variety of hard and soft landscaping; laid to the surface of an unprepared roof deck.							
Vapour barrier laid to prevent intersticial condensation from spaces below the roof applied by torching to the roof deck							
VB4-Expal aluminium lined	-	-	-	-	-	m²	7.91
Insulation laid and hot bitumen bonded to vapour barrier							
PUR-V Polyurethane	-	-	-	-	-	m²	15.51
Underlayer to receive rootbarrier partially bonded to insulation by torching							
G4E	-	-	-	-	-	m²	8.82
Root barrier							
"Plant E"; chemically treated root resistant capping sheet fully bonded to G4E underlayer by torching	-	-	-	-	-	m²	10.10
Slip layers to absorb differential movement							
PE Foil 2 layers laid to root barriers	-	-	-	-	-	m²	2.29
Optional protection layer to prevent mechanical damage							
"Protection mat" 6 mm thick rubber matting loose laid	-	-	-	-	-	m²	8.70
Drainage medium laid to root barrier							
"Drainage board" free draining EPS 50 mm thick	-	-	-	-	-	m²	5.42
"Reservoir board" up to 21.5 litre water storage capacity EPS 75 mm thick	-	-	-	-	-	m²	13.61
Filtration to prevent soil migration to drainage system							
"Filter fleece" 3 mm thick polyester geotextile loose laid over drainage/reservoir layer	-	-	-	-	-	m²	2.49
Extensive Green Roof System; Erisco Bauder Ltd; low maintenance soil free system incorporating single layer growing and planting medium.							
Vapour barrier laid to prevent intersticial condensation applied by torching to the roof deck							
VB4-Expal aluminium lined	-	-	-	-	-	m²	7.91
Insulation laid and hot bitumen bonded to vapour barrier							
PUR-V Polyurethane	-	-	-	-	-	m²	15.51
Underlayer to receive rootbarrier partially bonded to insulation by torching							
G4E	-	-	-	-	-	m²	8.82
Root barrier							
"Plant E"; chemically treated root resistant capping sheet fully bonded to G4E underlayer by torching	-	-	-	-	-	m²	10.10

Prices for Measured Works
J WATERPROOFING

Item Excluding site overheads and profit	PC £	Labour hours	Labour £	Plant £	Material £	Unit	Total rate £
Landscape Options Hydroplanting system; Erisco Bauder Ltd; to Extensive Green Roof as detailed above Ecomat 6mm thick geotextile loose laid Hydroplanting with sedum and succulent coagulant directly onto plant substrate to 60 mm deep	-	-	-	-	-	m²	10.08
Xeroflor vegetation blanket; Erisco Bauder Ltd; to Extensive Green Roof as detailed above Xeroflor Xf 301 pre cultivated sedum blanket incorporating 800 gram recycled fibre water retention layer laid loose	-	-	-	-	-	m²	32.10
Inverted waterproofing systems; **Alumasc Exterior Building Products Ltd (Euroroof); to roof surfaces to receive Green Roof systems.** "Hydrotech 6125"; monolithic hot melt rubberised bitumen; Applied in two 3 mm layers incorporating a polyester reinforcing sheet with 4 mm thick protection sheet and chemically impregnated root barrier; fully bonded into the Hydrotech; applied to plywood or suitably prepared wood float finish and primed concrete deck or screeds. 10 mm thick	-	-	-	-	-	m²	22.00
"Topfoam 350"; extruded polystyrene insulation; Optional system; thickness to suit required U value; calculated at design stage; indicative thicknesses; laid to Hydrotech 6125							
0.25 U value; average requirement 125 mm	-	-	-	-	-	m²	16.00
0.35 U value; average requirement 80 mm thick	-	-	-	-	-	m²	10.00
0.45 U value average requirement 60 mm	-	-	-	-	-	m²	7.00
Warm Roof waterproofing systems; **Alumasc Exterior Building Products Ltd (Euroroof); to roof surfaces to receive Green Roof systems.** "Derbigum" system "Nilperm" aluminium lined vapour barrier; 2 mm thick bonded in hot bitumen to the roof deck	-	-	-	-	-	m²	6.00
"Korklite" insulation bonded to the vapour barrier in hot bitumen; U value dependent; 80 mm thick	-	-	-	-	-	m²	14.00
"Hi-Ten Bond" 2 mm thick underlayer; fully bonded to the insulation	-	-	-	-	-	m²	4.00
"Derbigum SP" cap sheet impregnated with root resisting chemical bonded to the underlayer	-	-	-	-	-	m²	15.00

Prices for Measured Works
J WATERPROOFING

Item Excluding site overheads and profit	PC £	Labour hours	Labour £	Plant £	Material £	Unit	Total rate £
J50 GREEN ROOF SYSTEMS – cont'd							
Intensive Green Roof Systems; Alumasc Exterior Building Products Ltd (Euroroof); components laid to the insulation over the Hydrotech or Derbigum waterproofing above.							
Optional inclusion; moisture retention layer;							
SSM-45; moisture mat	-	-	-	-	-	m²	4.00
Drainage layer; "Floradrain" recycled polypropylene; providing water reservoir, multi-directional drainage and mechanical damage protection;							
FD25; 25 mm deep; rolls inclusive of filter sheet SF	-	-	-	-	-	m²	10.00
FD40; 40 mm deep; rolls inclusive of filter sheet SF	-	-	-	-	-	m²	15.00
FD25; 25 mm deep; sheets excluding filter sheet	-	-	-	-	-	m²	10.00
FD40; 40 mm deep; sheets excluding filter sheet	-	-	-	-	-	m²	14.00
FD60; 60 mm deep; sheets excluding filter sheet	-	-	-	-	-	m²	19.00
Drainage layer; "Elastodrain" recycled rubber mat; providing multi-directional drainage and mechanical damage protection;							
EL200; 20 mm deep	-	-	-	-	-	m²	10.00
"Zincolit" recycled crushed brick; Optional drainage infil to Floradrain layers;							
FD25; 10 litres per m²	-	0.02	0.36	-	42.00	m²	42.36
FD40; 17 litres per m²	-	0.03	0.61	-	42.00	m²	42.61
FD60; 27 litres per m²	-	0.05	0.97	-	42.00	m²	42.97
Filter sheet; rolled out onto drainage layer							
"Filter sheet TG" for Elastodrain EL200	-	-	-	-	-	m²	2.00
"Filter sheet SF" for Floradrain range	-	-	-	-	-	m²	1.50
Intensive soil; lightweight growing medium laid to filter sheet	-	-	-	-	-	m³	95.00
Extensive Green Roof Systems; Alumasc Exterior Building Products Ltd (Euroroof); components laid to the Hydrotech or insulation layers above.							
Moisture retention layer;							
SSM-45; moisture mat	-	-	-	-	-	m²	4.00
Drainage layer for flat roofs; "Floradrain"; recycled polypropylene; providing water reservoir, multi-directional drainage and mechanical damage protection; supplied in sheet or roll form							
FD25; 25 mm deep; rolls inclusive of filter sheet SF	-	-	-	-	-	m²	10.00
FD40; 40 mm deep; rolls inclusive of filter sheet SF	-	-	-	-	-	m²	15.00
FD25; 25 mm deep; sheets excluding filter sheet	-	-	-	-	-	m²	10.00
FD40; 40 mm deep; sheets excluding filter sheet	-	-	-	-	-	m²	14.00
FD60; 60 mm deep; sheets excluding filter sheet	-	-	-	-	-	m²	19.00
Drainage layer for pitched roofs; "Floratec"; recycled polystyrene; providing water reservoir and multi-directional drainage; laid to moisture mat or to the waterproofing.							
FS50	-	-	-	-	-	m²	10.00
FS75	-	-	-	-	-	m²	14.00
FS100	-	-	-	-	-	m²	15.00

Prices for Measured Works
J WATERPROOFING

Item Excluding site overheads and profit	PC £	Labour hours	Labour £	Plant £	Material £	Unit	Total rate £
Filter sheet; only required at perimeters and protrusions	-	-	-	-	-	m	1.50
Extensive soil; lightweight growing medium laid to filter sheet; choice of soil types is dependent on planting design requirements							
"E110" 10% humus 90% Zincolit	-	-	-	-	-	m²	95.00
"E120" 20% humus 80% Zincolit	-	-	-	-	-	m²	95.00
"E130" 30% humus 70% Zincolit	-	-	-	-	-	m²	95.00
Landscape Options							
Sedum Mat vegetation layer; Alumasc Exterior Building Products Ltd (Euroroof); to Extensive Green Roof as detailed above	-	-	-	-	-	m²	23.00
Green Roof Components; Alumasc Exterior Building Products Ltd (Euroroof)							
Outlet inspection chambers							
"KS 10" 100 mm deep	-	-	-	-	-	nr	52.00
"KS 15" 150 mm deep	-	-	-	-	-	nr	62.00
Height extension piece 100 mm	-	-	-	-	-	nr	21.00
Height extension piece 200 mm	-	-	-	-	-	nr	31.00
Outlet and Irrigation control chambers							
"B32" 300 x 300 x 300 mm high	-	-	-	-	-	nr	262.00
"B52" 400 x 500 x 500 mm high	-	-	-	-	-	nr	315.00
Outlet damming piece for water retention	-	-	-	-	-	nr	31.00
Linear drainage channel; collects surface water from adjacent hard surfaces or down pipes for distribution to the drainage layer.	-	-	-	-	-	m	52.00

M SURFACE FINISHES

Item Excluding site overheads and profit	PC £	Labour hours	Labour £	Plant £	Material £	Unit	Total rate £
M10 CEMENT: SAND/CONCRETE SCREEDS/TOPPINGS							
Granolithic paving; cement and granite chippings 5 mm down (1:2:5); steel trowelled; Jonathan James Ltd							
Floors; one coat; laid on concrete while green; width exceeding 300 mm							
20 mm thick	-	-	-	-	-	m²	14.84
25 mm thick	-	-	-	-	-	m²	16.25
Floors; two coats; laid on hacked concrete with slurry; width exceeding 300 mm							
38 mm thick	-	-	-	-	-	m²	19.91
50 mm thick	-	-	-	-	-	m²	23.30
75 mm thick	-	-	-	-	-	m²	30.34
M20 PLASTERED/RENDERED/ROUGHCAST COATING							
Cement:lime:sand (1:1:6); 19 mm thick; two coats; wood floated finish							
Walls							
width exceeding 300 mm; to brickwork or blockwork base	-	-	-	-	-	m²	14.43
Extra over cement:sand:lime (1:1:6) coatings for decorative texture finish							
combed or floated finish	-	-	-	-	-	m²	1.90
M40 STONE/CONCRETE/QUARRY/CERAMIC TILING							
Ceramic tiles; unglazed slip resistant; various colours and textures; jointing							
Floors							
level or to falls only not exceeding 15 degrees from horizontal; 150 x 150 x 8 mm thick	16.44	1.00	18.00	-	19.18	m²	37.18
level or to falls only not exceeding 15 degrees from horizontal; 150 x 150 x 12 mm thick	20.03	1.00	18.00	-	22.95	m²	40.95
Clay tiles - General Preamble: Typical Specification - Clay tiles should be to BS 6431 and shall be reasonably true to shape, flat, free from flaws, frost resistant and true to sample approved by the Landscape Architect prior to laying. Quarry Tiles (or semi-vitrified tiles) shall be of external quality, either heather brown or blue, to size specified, laid on 1:2:4 concrete, with 20 maximum aggregate 100 thick, on 100 hardcore. The hardened concrete should be well wetted and the surplus water taken off. Clay tiles shall be thoroughly wetted immediately before laying and then drained and shall be bedded to 19 thick cement: sand (1:3) screed. Joints should be approximately 4mm (or 3mm for vitrified tiles) grouted in cement: sand (1:2) and cleaned off immediately.							

Prices for Measured Works
M SURFACE FINISHES

Item Excluding site overheads and profit	PC £	Labour hours	Labour £	Plant £	Material £	Unit	Total rate £
Quarry tiles; external quality; including bedding; jointing							
Floors							
level or to falls only not exceeding 15 degrees from horizontal; 150 x 150 x 12.5 mm thick; heather brown	20.54	0.80	14.40	-	23.49	m²	**37.89**
level or to falls only not exceeding 15 degrees from horizontal; 225 x 225 x 29 mm thick; heather brown	20.54	0.67	12.00	-	23.49	m²	**35.49**
level or to falls only not exceeding 15 degrees from horizontal; 150 x 150 x 12.5 mm thick; blue/black	13.05	1.00	18.00	-	15.62	m²	**33.62**
level or to falls only not exceeding 15 degrees from horizontal; 194 x 194 x 12.5 mm thick; heather brown	12.02	0.80	14.40	-	14.54	m²	**28.94**
M60 PAINTING/CLEAR FINISHING – EXTERNALLY							
Prepare; touch up primer; two undercoats and one finishing coat of gloss oil paint; on metal surfaces							
General surfaces							
girth exceeding 300 mm	0.75	0.33	6.00	-	0.75	m²	**6.75**
isolated surfaces; girth not exceeding 300 mm	0.21	0.13	2.40	-	0.21	m	**2.61**
isolated areas not exceeding 0.50 m² irrespective of girth	0.40	0.13	2.40	-	0.40	nr	**2.80**
Ornamental railings; each side measured overall							
girth exceeding 300 mm	0.75	0.75	13.50	-	0.75	m²	**14.25**
Prepare; one coat primer; two undercoats and one finishing coat of gloss oil paint; on wood surfaces							
General Surfaces							
girth exceeding 300 mm	1.01	0.40	7.20	-	1.01	m²	**8.21**
isolated areas not exceeding 0.50 m² irrespective of girth	0.39	0.36	6.54	-	0.39	nr	**6.93**
isolated surfaces; girth not exceeding 300 mm	0.33	0.20	3.60	-	0.33	m	**3.93**
Prepare; two coats of creosote; on wood surfaces							
General surfaces							
girth exceeding 300 mm	0.17	0.21	3.74	-	0.17	m²	**3.91**
isolated surfaces; girth not exceeding 300 mm	0.05	0.12	2.08	-	0.05	m	**2.13**
Prepare, proprietary solution primer; two coats of dark stain; on wood surfaces							
General surfaces							
girth exceeding 300 mm	0.11	0.10	1.80	-	0.11	m²	**1.91**
isolated surfaces; girth not exceeding 300 mm	0.01	0.05	0.90	-	0.01	m	**0.91**

Prices for Measured Works
M SURFACE FINISHES

Item Excluding site overheads and profit	PC £	Labour hours	Labour £	Plant £	Material £	Unit	Total rate £
M60 PAINTING/CLEAR FINISHING – EXTERNALLY – cont'd							
Three coats "Dimex Shield"; to clean, dry surfaces; in accordance with manufacturer's instructions							
Brick or block walls							
girth exceeding 300 mm	0.92	0.28	5.04	-	0.92	m²	**5.96**
Cement render or concrete walls							
girth exceeding 300 mm	0.72	0.25	4.50	-	0.72	m²	**5.22**
Two coats resin based paint; "Sandtex Matt"; in accordance with manufacturer's instructions							
Brick or block walls							
girth exceeding 300 mm	6.16	0.20	3.60	-	6.16	m²	**9.76**
Cement render or concrete walls							
girth exceeding 300 mm	4.19	0.17	3.00	-	4.19	m²	**7.19**

Prices for Measured Works
P BUILDING FABRIC SUNDRIES

Item Excluding site overheads and profit	PC £	Labour hours	Labour £	Plant £	Material £	Unit	Total rate £
P30 TRENCHES/PIPEWAYS/PITS FOR BURIED ENGINEERING SERVICES							
Excavating trenches; using 3 tonne tracked excavator; to receive pipes; grading bottoms; earthwork support; filling with excavated material to within 150 mm of finished surfaces and compacting; completing fill with topsoil; disposal of surplus soil							
Services not exceeding 200 mm nominal size							
average depth of run not exceeding 0.50 m	0.78	0.12	2.16	1.14	0.94	m	**4.24**
average depth of run not exceeding 0.75 m	0.78	0.16	2.93	1.57	0.94	m	**5.44**
average depth of run not exceeding 1.00 m	0.78	0.28	5.10	2.74	0.94	m	**8.78**
average depth of run not exceeding 1.25 m	0.78	0.38	6.90	3.69	0.78	m	**11.37**
Excavating trenches; using 3 tonne tracked excavator; to receive pipes; grading bottoms; earthwork support; filling with imported granular material and compacting; disposal of surplus soil							
Services not exceeding 200 mm nominal size							
average depth of run not exceeding 0.50 m	4.37	0.09	1.56	0.80	5.72	m	**8.09**
average depth of run not exceeding 0.75 m	6.56	0.11	1.95	1.01	8.59	m	**11.54**
average depth of run not exceeding 1.00 m	8.75	0.14	2.51	1.31	11.45	m	**15.27**
average depth of run not exceeding 1.25 m	10.94	0.23	4.11	2.21	14.32	m	**20.63**
Excavating trenches, using 3 tonne tracked excavator, to receive pipes; grading bottoms; earthwork support; filling with lean mix concrete; disposal of surplus soil							
Services not exceeding 200 mm nominal size							
average depth of run not exceeding 0.50 m	9.77	0.11	1.92	0.40	11.12	m	**13.44**
average depth of run not exceeding 0.75 m	14.65	0.13	2.34	0.50	16.68	m	**19.52**
average depth of run not exceeding 1.00 m	19.54	0.17	3.00	0.67	22.24	m	**25.91**
average depth of run not exceeding 1.25 m	24.42	0.23	4.05	1.01	27.80	m	**32.86**
Earthwork Support; providing support to opposing faces of excavation; moving along as work proceeds; A Plant Acrow							
Maximum depth not exceeding 2.00 m							
distance between opposing faces not exceeding 2.00 m	-	0.80	14.40	16.11	-	m	**30.51**

Prices for Measured Works
Q PAVING/PLANTING/FENCING/SITE FURNITURE

Item Excluding site overheads and profit	PC £	Labour hours	Labour £	Plant £	Material £	Unit	Total rate £
Q10 STONE/CONCRETE/BRICK KERBS/EDGINGS/CHANNELS							
Foundations to kerbs							
Excavating trenches; 450 x 150 mm; for concrete foundations to kerbs, edgings or channels							
by machine	-	0.14	2.52	0.71	-	m	3.23
by hand	-	0.66	11.88	-	-	m	11.88
Foundations; to kerbs, edgings, or channels; site mixed reinforced in situ concrete; 21 N/mm2 - 20 aggregate (1:2:4) site mixed); one side against earth face, other against formwork (not included)							
150 wide x 100 mm deep	-	0.13	2.40	-	1.21	m	3.61
150 wide x 150 mm deep	-	0.17	3.00	-	1.81	m	4.81
200 wide x 150 mm deep	-	0.20	3.60	-	2.42	m	6.02
300 wide x 150 mm deep	-	0.23	4.20	-	3.63	m	7.83
600 wide x 200 mm deep	-	0.29	5.14	-	9.68	m	14.82
Formwork; sides of foundations (this will usually be required to one side of each kerb foundation adjacent to road sub-bases)							
100 mm deep	-	0.04	0.75	-	0.53	m	1.28
150 mm deep	-	0.04	0.75	-	0.80	m	1.54
Precast concrete kerbs, channels, edgings etc.; to BS 340; Marshalls Mono; bedding, jointing and pointing in cement mortar (1:3); on 150 mm deep concrete foundation; including haunching with in situ concrete; 11.50 N/mm2 - 40 aggregate one side							
Kerbs; straight							
150 x 305 mm; ref HB1	5.86	0.50	9.00	-	8.02	m	17.02
125 x 255 mm; ref HB2; SP	3.05	0.44	8.00	-	5.21	m	13.21
125 x 150 mm; ref BN	2.05	0.40	7.20	-	4.21	m	11.41
Dropper kerbs; left and right handed							
125 x 255 - 150 mm; ref DL1 or DR1	4.26	0.50	9.00	-	6.43	m	15.43
125 x 255 - 150 mm; ref DL2 or DR2	4.26	0.50	9.00	-	6.43	m	15.43
Quadrant kerbs							
305 mm radius	6.60	0.50	9.00	-	8.04	nr	17.04
455 mm radius	7.10	0.50	9.00	-	9.26	nr	18.26
Internal or external angles							
127 x 254 mm section	12.41	0.50	9.00	-	13.85	m	22.85
Straight kerbs or channels; to radius; 125 x 255 mm							
0.90 m radius (2 units per quarter circle)	5.36	0.80	14.40	-	7.60	m	22.00
1.80 m radius (4 units per quarter circle)	5.36	0.73	13.09	-	7.60	m	20.69
2.40 m radius (5 units per quarter circle)	5.36	0.68	12.20	-	7.60	m	19.80
3.00 m radius (5 units per quarter circle)	5.36	0.67	12.00	-	7.60	m	19.59
4.50 m radius (2 units per quarter circle)	5.36	0.60	10.81	-	7.60	m	18.41
6.10 m radius (11 units per quarter circle)	5.36	0.58	10.44	-	7.60	m	18.03
7.60 m radius (14 units per quarter circle)	5.36	0.57	10.29	-	7.60	m	17.88
9.15 m radius (17 units per quarter circle)	5.36	0.56	10.00	-	7.60	m	17.60
10.70 m radius (20 units per quarter circle)	5.36	0.56	10.00	-	7.60	m	17.60
12.20 m radius (22 units per quarter circle)	5.36	0.53	9.48	-	7.60	m	17.07

Prices for Measured Works
Q PAVING/PLANTING/FENCING/SITE FURNITURE

Item Excluding site overheads and profit	PC £	Labour hours	Labour £	Plant £	Material £	Unit	Total rate £
Kerbs; "Conservation Kerb" units; to simulate natural granite kerbs							
255 x 150 x 914 mm; laid flat	14.66	0.57	10.29	-	18.54	m	28.82
150 x 255 x 914 mm; laid vertical	14.66	0.57	10.29	-	17.90	m	28.19
145 x 255 mm; radius internal 3.25 m	18.54	0.83	15.00	-	22.25	m	37.25
150 x 255 mm; radius external 3.40 m	17.71	0.83	15.00	-	21.40	m	36.40
145 x 255 mm; radius internal 6.50 m	16.98	0.67	12.00	-	20.23	m	32.22
145 x 255 mm; radius external 6.70 m	16.47	0.67	12.00	-	20.13	m	32.13
150 x 255 mm; radius internal 9.80 m	16.38	0.67	12.00	-	20.04	m	32.03
150 x 255 mm; radius external 10.00 m	15.92	0.67	12.00	-	19.16	m	31.16
305 x 305 x 255 mm; solid quadrants	20.82	0.57	10.29	-	24.07	nr	34.35
Channels; dished							
125 x 225 x 915 mm long; square; ref CS	3.10	0.40	7.20	-	9.80	m	17.00
125 x 150 x 915 mm long; square; ref CS2	2.15	0.40	7.20	-	7.58	m	14.78
305 x 150 x 915 mm	7.20	0.40	7.20	-	14.51	m	21.71
150 x 125 x 915 mm	2.24	0.40	7.20	-	8.61	m	15.81
150 x 100 x 915 mm	3.07	0.40	7.20	-	7.17	m	14.37
Precast concrete edging units; including haunching with in situ concrete; 11.50 N/mm2 - 40 aggregate both sides							
Edgings; rectangular, bullnosed, or chamfered							
50 x 150 mm	1.60	0.33	6.00	-	5.21	m	11.21
125 x 150 mm bullnosed	2.01	0.33	6.00	-	5.62	m	11.62
50 x 200 mm	1.67	0.33	6.00	-	5.28	m	11.28
50 x 250 mm	1.94	0.33	6.00	-	5.55	m	11.55
50 x 250 mm flat top	2.20	0.33	6.00	-	5.80	m	11.81
Precast concrete kerbs; to BS 340; RMC Concrete Products; on 150 mm deep concrete foundation; including haunching with in situ concrete; 11.50 N/mm2 - 40 aggregate one side							
Kerbs; "Kerb-sett" units (prices shown are for grey units, prices for other colours vary)							
standard blocks; 100 wide x 190 x 160 mm (set as high or low rise)	5.40	0.80	14.40	-	14.15	m	28.55
external angles; high or low profiles	5.24	0.80	14.40	-	13.99	m	28.39
Dropper kerbs (handed pairs); ref H-L	3.21	1.00	18.00	-	5.63	set	23.63
Crossovers; low; ref L-X	3.21	0.20	3.60	-	3.75	set	7.35
Crossovers; high; ref H-X	5.89	0.20	3.60	-	6.90	set	10.50
Curves; internal; low; ref IR/L	28.90	0.80	14.40	-	31.32	m	45.72
Curves; external; high; ref ER/H	28.90	0.80	14.40	-	31.32	m	45.72

Dressed natural stone kerbs - General

Preamble: BS 435 includes the following conditions for dressed natural stone kerbs. The kerbs are to be good, sound, uniform in texture and free from defects; worked straight or to radius, square and out of wind, with the top front and back edges parallel or concentric to the dimensions specified. All drill and pick holes shall be removed from dressed faces. Standard dressings shall be in accordance with one of three illustrations in BS 435; and designated as either fine picked, single axed or nidged, or rough punched.

Prices for Measured Works
Q PAVING/PLANTING/FENCING/SITE FURNITURE

Item Excluding site overheads and profit	PC £	Labour hours	Labour £	Plant £	Material £	Unit	Total rate £
Q10 STONE/CONCRETE/BRICK KERBS/EDGINGS/CHANNELS – cont'd							
Dressed natural stone kerbs; to BS435; CED Ltd; on concrete foundations (not included); including haunching with in situ concrete; 11.50 N/mm2 - 40 aggregate one side							
Granite kerbs; 125 x 250 mm							
rough, straight, random lengths	11.11	0.80	14.40	-	13.63	m	28.03
special quality, straight, random lengths	17.00	0.80	14.40	-	19.52	m	33.92
Granite kerbs; 125 x 250 mm; curved to mean radius 3 m							
special quality, random lengths	23.00	0.91	16.38	-	25.52	m	41.90
rough, random lengths	14.12	0.91	16.38	-	16.64	m	33.02
York stone kerbs; on 150 mm deep concrete foundations; including haunching with in situ concrete; 11.50 N/mm2 - 40 aggregate one side							
Kerbs							
150 x 254 mm; sawn	41.00	0.80	14.40	-	43.52	m	57.92
Kerbs; curved to mean radius 5 m							
150 x 254 mm; fine picked	192.00	0.80	14.40	-	194.52	m	208.92
Second-hand granite setts; 100 x 100 mm; bedding in cement mortar (1:4); on 150 mm deep concrete foundations; including haunching with in situ concrete; 11.50 N/mm2 - 40 aggregate one side							
Edgings							
300 mm wide	9.02	1.33	24.00	-	12.77	m	36.77
Concrete setts; Blanc de Bierges; bedding in cement mortar (1:4); on 150 mm deep concrete foundations, including haunching with in situ concrete; 11.50 N/mm2 - 40 aggregate one side							
Edgings							
70 x 70 x 70 mm; in strips 230 mm wide	5.32	1.33	24.00	-	9.07	m	33.07
140 x 140 x 80 mm; in strips 310 mm wide	6.84	1.33	24.00	-	10.92	m	34.92
210 x 140 x 80 mm; in strips 290 mm wide	6.40	1.33	24.00	-	10.46	m	34.46
"Kidney" flint cobbles; bedding in 50 mm concrete (1:2:4); on concrete foundations (not included)							
Edgings							
200 mm wide	6.20	1.33	24.00	-	7.01	m	31.01

Prices for Measured Works
Q PAVING/PLANTING/FENCING/SITE FURNITURE

Item Excluding site overheads and profit	PC £	Labour hours	Labour £	Plant £	Material £	Unit	Total rate £
Brick or block stretchers; bedding in cement mortar (1:4); on 150 mm deep concrete foundations, including haunching with in situ concrete; 11.50 N/mm2 - 40 aggregate one side							
Single course							
concrete paving blocks; PC £8.85 /m²; 200 x 100 x 60 mm	1.60	0.31	5.54	-	3.82	m	**9.36**
engineering bricks; PC £260.00/1000; 215 x 102.5 x 65 mm	1.16	0.40	7.20	-	3.44	m	**10.64**
paving bricks; PC £450.00/1000; 215 x 102.5 x 65 mm	2.00	0.40	7.20	-	4.22	m	**11.42**
Two courses							
concrete paving blocks; PC £8.85 /m²; 200 x 100 x 60 mm	3.19	0.40	7.20	-	5.91	m	**13.11**
engineering bricks; PC £260.00/1000; 215 x 102.5 x 65 mm	2.31	0.57	10.29	-	5.14	m	**15.43**
paving bricks; PC £450.00/1000; 215 x 102.5 x 65 mm	4.00	0.57	10.29	-	6.72	m	**17.00**
Three courses							
concrete paving blocks; PC £8.85/m²; 200 x 100 x 60 mm	4.79	0.44	8.00	-	7.50	m	**15.50**
engineering bricks; PC £260.00/1000; 215 x 102.5 x 65 mm	3.47	0.67	12.00	-	6.36	m	**18.35**
paving bricks; PC £450.00/1000; 215 x 102.5 x 65 mm	6.00	0.67	12.00	-	8.71	m	**20.71**
Bricks on edge; bedding in cement mortar (1:4); on 150 mm deep concrete foundations; including haunching with in situ concrete; 11.50 N/mm2 - 40 aggregate one side							
One brick wide							
engineering bricks; 215 x 102.5 x 65 mm	3.47	0.57	10.29	-	6.48	m	**16.76**
paving bricks; 215 x 102.5 x 65 mm	6.00	0.57	10.29	-	8.83	m	**19.12**
Two courses; stretchers laid on edge; 225 mm wide							
engineering bricks; 215 x 102.5 x 65 mm	6.93	1.20	21.60	-	11.21	m	**32.81**
paving bricks; 215 x 102.5 x 65 mm	12.00	1.20	21.60	-	15.93	m	**37.53**
Extra over bricks on edge for standard kerbs to one side; haunching in concrete							
125 x 255 mm; ref HB2; SP	3.05	0.44	8.00	-	5.21	m	**13.21**
Channels; bedding in cement mortar (1:3); joints pointed flush; on concrete foundations (not included)							
Three courses stretchers; 350 mm wide; quarter bond to form dished channels							
engineering bricks; PC £260.00/1000 ; 215 x 102.5 x 65 mm	3.47	1.00	18.00	-	4.74	m	**22.74**
paving bricks; PC £450.00/1000; 215 x 102.5 x 65 mm	6.00	1.00	18.00	-	7.10	m	**25.10**
Three courses granite setts; 340 mm wide; to form dished channels							
340 mm wide	9.02	2.00	36.00	-	11.56	m	**47.56**
Precast concrete dished channels; sides haunching up							
255 x 125 mm	2.92	0.50	9.00	-	5.45	m	**14.45**

Prices for Measured Works
Q PAVING/PLANTING/FENCING/SITE FURNITURE

Item Excluding site overheads and profit	PC £	Labour hours	Labour £	Plant £	Material £	Unit	Total rate £
Q10 STONE/CONCRETE/BRICK KERBS/EDGINGS/CHANNELS – cont'd							
Precast concrete channels; Charcon Hard Landscaping; on 150 mm deep concrete foundations; including haunching with in situ concrete; 21.00 N/mm2 - 20 aggregate; both sides							
"Charcon Safeticurb"; slotted safety channels; for pedestrians and light vehicles							
ref DBJ; 305 x 305 mm	55.95	0.67	12.00	-	69.66	m	81.66
ref DBA; 250 x 250 mm	24.82	0.67	12.00	-	38.53	m	50.53
"Charcon Safeticurb"; slotted safety channels; for heavy vehicles							
ref DBM; 248 x 248 mm	44.14	0.80	14.40	-	57.85	m	72.25
ref Clearway; 305 x 305 mm x 400 mm long	116.48	0.80	14.40	-	130.19	m	144.59
Inspection units; cast iron lids; including jointing to drainage channels							
248 x 248 x 914 mm	59.75	1.50	27.00	-	60.56	nr	87.56
Silt box tops; concrete frame; cast iron grid lids; type 1; set over gully							
457 x 610 mm	265.96	2.00	36.00	-	266.77	nr	302.77
Manhole covers; type K; cast iron; providing inspection to blocks and back gullies	278.21	2.00	36.00	-	280.24	nr	316.24
Kerb drainage systems; Marshalls Mono "Beany Block"; on foundations (not included); including haunching with class "E" concrete one side							
base block with top block	39.36	1.00	18.00	-	41.78	m	59.78
base block and cover plate	68.62	2.00	36.00	-	71.04	m	107.04
Channels; Ensor Building Products; on 100 mm deep concrete bed; 100 mm concrete fill both sides							
Polymer concrete drain units; tapered to falls							
channel units; Grade 'A', ref Stora-drain 100; 100 mm wide including galvanised grate and locks	18.50	0.80	14.40	-	23.74	m	38.14
channel units; Grade 'B', ref Stora-drain 100; 100 mm wide including galvanised grate and locks	22.30	0.80	14.40	-	27.54	m	41.94
channel units; Grade 'C', ref Stora-drain 100; 100 mm wide including galvanised grate and locks	23.95	0.80	14.40	-	29.19	m	43.59
channel units; Grade 'C', ref Stora-drain 100; 100 mm wide including anti-heel grate and locks	22.95	0.80	14.40	-	28.19	m	42.59
Stora Drain Sump unit, steel bucket, Class A15 galv.slotted grating	37.50	2.00	36.00	-	42.74	nr	78.74
Stora Drain Sump unit, steel bucket, Class B125 galv.mesh grating	38.50	2.00	36.00	-	43.74	nr	79.74
Stora Drain Sump unit, steel bucket, Class C250 galv.slotted grating	39.50	2.00	36.00	-	44.74	nr	80.74
Stora Drain Sump unit, steel bucket, C250 'Heelsafe' slotted grating	39.00	2.00	36.00	-	44.24	nr	80.24

Prices for Measured Works
Q PAVING/PLANTING/FENCING/SITE FURNITURE

Item Excluding site overheads and profit	PC £	Labour hours	Labour £	Plant £	Material £	Unit	Total rate £
Channels; Hodkin & Jones; on well rammed subsoil base; set to gradient; backfilling with compacted excavated material							
"Decathlon"; precast concrete drainage channels							
150 mm wide channels	42.00	1.00	18.00	-	42.00	m	**60.00**
250 mm wide channels	53.00	1.00	18.00	-	53.00	m	**71.00**
375 mm wide channels	71.40	1.10	19.80	-	71.40	m	**91.20**
525 mm wide channels	84.00	1.20	21.60	0.33	84.00	m	**105.93**
600 mm wide channels	96.60	1.60	28.80	0.42	96.60	m	**125.82**
Q20 BASES AND SUB-BASES TO ROADS/PAVINGS							
Filling to make up levels							
Obtained off site; hardcore; PC £16.64 /m³							
100 mm thick	1.66	0.05	0.90	0.94	1.66	m²	**3.50**
100 mm thick	16.64	0.50	9.00	9.38	16.64	m³	**35.02**
150 mm thick	2.50	0.07	1.20	1.19	2.50	m²	**4.89**
200 mm thick	3.33	0.08	1.44	1.29	3.33	m²	**6.06**
300 mm thick	4.99	0.07	1.20	1.33	4.99	m²	**7.52**
exceeding 300 mm thick	16.64	0.17	3.00	3.12	19.97	m³	**26.09**
Obtained off site; granular fill type 1; PC £14.45 /tonne (£31.80/m³ compacted)							
100 mm thick	3.18	0.03	0.50	0.31	3.18	m²	**3.99**
150 mm thick	4.77	0.03	0.45	0.47	4.77	m²	**5.69**
250 mm thick	7.95	0.02	0.37	0.78	7.95	m²	**9.11**
Difference for each £1.00 increase/decrease in PC price per m³; price will vary with type and source of hardcore							
average 75 mm thick	-	-	-	-	0.08	m²	**0.08**
average 100 mm thick	-	-	-	-	0.10	m²	**0.10**
average 150 mm thick	-	-	-	-	0.16	m²	**0.16**
average 200 mm thick	-	-	-	-	0.21	m²	**0.21**
average 250 mm thick	-	-	-	-	0.26	m²	**0.26**
average 300 mm thick	-	-	-	-	0.31	m²	**0.31**
exceeding 300 mm thick	-	-	-	-	1.14	m³	**1.14**
Surface treatments							
Sand blinding; to hardcore base (not included); 25 mm thick	0.66	0.03	0.60	-	0.66	m²	**1.25**
Sand blinding; to hardcore base (not included); 50 mm thick	1.31	0.05	0.90	-	1.31	m²	**2.21**
Filter fabrics; to hardcore base (not included)	0.43	0.01	0.18	-	0.46	m²	**0.64**
Herbicides; ICI							
"Casoron G" (residual) herbicide; treating substrate before laying base							
at 1kg / 125m2	3.13	0.05	0.90	-	3.44	100 m²	**4.34**

Prices for Measured Works
Q PAVING/PLANTING/FENCING/SITE FURNITURE

Item Excluding site overheads and profit	PC £	Labour hours	Labour £	Plant £	Material £	Unit	Total rate £
Q21 IN SITU CONCRETE ROADS/PAVINGS/BASES							
Unreinforced concrete; on prepared sub-base (not included)							
Roads; 21.00 N/mm2 - 20 aggregate (1:2:4) mechanically mixed on site							
100 mm thick	8.07	0.13	2.25	-	8.47	m²	10.72
150 mm thick	12.10	0.17	3.00	-	12.70	m²	15.70
Reinforced in situ concrete; mechanically mixed on site; normal Portland cement; on hardcore base (not included); reinforcement (not included)							
Roads; 11.50 N/mm2 - 40 aggregate (1:3:6)							
100 mm thick	7.21	0.40	7.20	-	7.39	m²	14.59
150 mm thick	10.82	0.60	10.80	-	11.09	m²	21.89
200 mm thick	14.42	0.80	14.40	-	15.15	m²	29.55
250 mm thick	18.03	1.00	18.00	0.41	18.48	m²	36.89
300 mm thick	21.64	1.20	21.60	0.41	22.18	m²	44.19
Roads; 21.00 N/mm2 - 20 aggregate (1:2:4)							
100 mm thick	8.07	0.40	7.20	-	8.27	m²	15.47
150 mm thick	12.10	0.60	10.80	-	12.41	m²	23.21
200 mm thick	16.13	0.80	14.40	-	16.54	m²	30.94
250 mm thick	20.16	1.00	18.00	0.41	20.67	m²	39.07
300 mm thick	24.20	1.20	21.60	0.41	24.80	m²	46.81
Roads; 25.00 N/mm2 - 20 aggregate GEN 4 ready mixed							
100 mm thick	6.58	0.40	7.20	-	6.74	m²	13.94
150 mm thick	9.87	0.60	10.80	-	10.12	m²	20.92
200 mm thick	13.16	0.80	14.40	-	13.49	m²	27.89
250 mm thick	16.45	1.00	18.00	0.41	16.86	m²	35.27
300 mm thick	19.74	1.20	21.60	0.41	20.23	m²	42.24
Reinforced in situ concrete; ready mixed; discharged directly into location from supply lorry; normal Portland cement; on hardcore base (not included); reinforcement (not included)							
Roads; 11.50 N/mm2 - 40 aggregate (1:3:6)							
100 mm thick	6.51	0.16	2.88	-	6.51	m²	9.39
150 mm thick	9.77	0.24	4.32	-	9.77	m²	14.09
200 mm thick	13.03	0.36	6.48	-	13.03	m²	19.51
250 mm thick	16.28	0.54	9.72	0.41	16.28	m²	26.41
300 mm thick	19.54	0.66	11.88	0.41	19.54	m²	31.83
Roads; 21.00 N/mm2 - 20 aggregate (1:2:4)							
100 mm thick	6.81	0.16	2.88	-	6.81	m²	9.69
150 mm thick	10.21	0.24	4.32	-	10.21	m²	14.53
200 mm thick	13.61	0.36	6.48	-	13.61	m²	20.09
250 mm thick	17.02	0.54	9.72	0.41	17.02	m²	27.14
300 mm thick	20.42	0.66	11.88	0.41	20.42	m²	32.71
Roads; 26.00 N/mm2 - 20 aggregate (1:1:5:3)							
100 mm thick	7.05	0.16	2.88	-	7.05	m²	9.93
150 mm thick	10.57	0.24	4.32	-	10.57	m²	14.89
200 mm thick	13.61	0.36	6.48	-	13.61	m²	20.09
250 mm thick	17.62	0.54	9.72	0.41	17.62	m²	27.75
300 mm thick	21.15	0.66	11.88	0.41	21.15	m²	33.44

Prices for Measured Works
Q PAVING/PLANTING/FENCING/SITE FURNITURE

Item Excluding site overheads and profit	PC £	Labour hours	Labour £	Plant £	Material £	Unit	Total rate £
Roads; PAV1 Concrete - 35 N/mm2; Designated mix							
100 mm thick	7.37	0.16	2.88	-	7.56	m²	10.44
150 mm thick	11.06	0.24	4.32	-	11.34	m²	15.66
200 mm thick	14.75	0.36	6.48	-	15.11	m²	21.59
250 mm thick	18.43	0.54	9.72	0.41	18.89	m²	29.02
300 mm thick	22.12	0.66	11.88	0.41	22.67	m²	34.96
Concrete sundries							
Treating surfaces of unset concrete; grading to cambers, tamping with 75 mm thick steel shod tamper or similar	-	0.13	2.40	-	-	m²	2.40
Expansion joints							
13 mm thick joint filler; formwork							
width or depth not exceeding 150 mm	1.45	0.20	3.60	-	2.63	m	6.23
width or depth 150 - 300 mm	1.05	0.25	4.50	-	3.40	m	7.90
width or depth 300 - 450 mm	0.47	0.30	5.40	-	4.00	m	9.40
25 mm thick joint filler; formwork							
width or depth not exceeding 150 mm	1.90	0.20	3.60	-	3.08	m	6.68
width or depth 150 - 300 mm	1.90	0.25	4.50	-	4.25	m	8.75
width or depth 300 - 450 mm	1.90	0.30	5.40	-	5.42	m	10.83
Sealants; sealing top 25 mm of joint with rubberized bituminous compound	1.01	0.25	4.50	-	1.01	m	5.51
Formwork for in situ concrete							
Sides of foundations							
height exceeding 1.00 m	5.30	2.00	36.00	-	9.07	m²	45.08
height not exceeding 250 mm	1.33	1.00	18.00	-	2.42	m²	20.42
height 250 - 500 mm	2.65	1.00	18.00	-	4.54	m²	22.54
height 500 mm - 1.00 m	5.30	1.50	27.00	-	9.07	m²	36.08
Extra over formwork for curved work 6 m radius	-	0.25	4.50	-	-	m²	4.50
Steel road forms; to edges of beds or faces of foundations							
150 mm wide	-	0.20	3.60	-	0.53	m	4.13
Reinforcement; fabric; BS 4483; side laps 150 mm; head laps 300 mm; mesh 200 x 200 mm; in roads, footpaths or pavings							
Fabric							
ref A142 (2.22 kg/m²)	0.75	0.08	1.49	-	0.83	m²	2.32
ref A193 (3.02 kg/m²)	1.03	0.08	1.49	-	1.13	m²	2.63

Prices for Measured Works
Q PAVING/PLANTING/FENCING/SITE FURNITURE

Item Excluding site overheads and profit	PC £	Labour hours	Labour £	Plant £	Material £	Unit	Total rate £
Q22 COATED MACADAM/ASPHALT ROADS/PAVINGS							
Coated macadam/asphalt roads/pavings - General							
Preamble: The prices for all in situ finishings to roads and footpaths include for work to falls, crossfalls or slopes not exceeding 15 degrees from horizontal; for laying on prepared bases (not included) and for rolling with an appropriate roller.							
Preamble: Users should note the new terminology for the surfaces described below which is to European standard descriptions. The now redundant descriptions for each course are shown in brackets.							
Macadam Surfacing; Spadeoak Construction Co Ltd; Surface (Wearing) Course; 20 mm of 6 mm dense bitumen macadam to BS4987-1 2001 ref 7.5							
Machine lay; areas 1000 m² and over							
limestone aggregate	-	-	-	-	-	m²	4.10
granite aggregate	-	-	-	-	-	m²	4.16
red	-	-	-	-	-	m²	7.53
Hand Lay; areas 400 m² and over							
limestone aggregate	-	-	-	-	-	m²	5.21
granite aggregate	-	-	-	-	-	m²	5.27
red	-	-	-	-	-	m²	9.32
Macadam Surfacing; Spadeoak Construction Co Ltd; Surface (Wearing) Course; 30 mm of 10 mm dense bitumen macadam to BS4987-1 2001 ref 7.4							
Machine lay; areas 1000 m² and over							
limestone aggregate	-	-	-	-	-	m²	4.93
granite aggregate	-	-	-	-	-	m²	5.01
red	-	-	-	-	-	m²	10.55
Hand Lay; areas 400 m² and over							
limestone aggregate	-	-	-	-	-	m²	6.09
granite aggregate	-	-	-	-	-	m²	6.18
red	-	-	-	-	-	m²	12.54
Macadam Surfacing; Spadeoak Construction Co Ltd; Surface (Wearing) Course; 40 mm of 10 mm dense bitumen macadam to BS4987-1 2001 ref 7.4							
Machine lay; areas 1000 m² and over							
limestone aggregate	-	-	-	-	-	m²	6.25
granite aggregate	-	-	-	-	-	m²	6.39
red	-	-	-	-	-	m²	11.97
Hand Lay; areas 400 m² and over							
limestone aggregate	-	-	-	-	-	m²	7.63
granite aggregate	-	-	-	-	-	m²	7.65
red	-	-	-	-	-	m²	14.06

Prices for Measured Works
Q PAVING/PLANTING/FENCING/SITE FURNITURE

Item Excluding site overheads and profit	PC £	Labour hours	Labour £	Plant £	Material £	Unit	Total rate £
Macadam Surfacing; Spadeoak Construction Co Ltd; Binder (Base) Course; 50 mm of 20 mm dense bitumen macadam to BS4987-1 2001 ref 6.5							
Machine lay; areas 1000 m² and over							
limestone aggregate	-	-	-	-	-	m²	6.71
granite aggregate	-	-	-	-	-	m²	6.55
Hand Lay; areas 400 m² and over							
limestone aggregate	-	-	-	-	-	m²	8.00
granite aggregate	-	-	-	-	-	m²	8.14
Macadam Surfacing; Spadeoak Construction Co Ltd; Binder (Base) Course; 60 mm of 20 mm dense bitumen macadam to BS4987-1 2001 ref 6.5							
Machine lay; areas 1000 m² and over							
limestone aggregate	-	-	-	-	-	m²	7.40
granite aggregate	-	-	-	-	-	m²	7.55
Hand Lay; areas 400 m² and over							
limestone aggregate	-	-	-	-	-	m²	8.73
granite aggregate	-	-	-	-	-	m²	8.89
Macadam Surfacing; Spadeoak Construction Co Ltd; Base (Roadbase) Course; 75 mm of 28 mm dense bitumen macadam to BS4987-1 2001 ref 5.2							
Machine lay; areas 1000 m² and over							
limestone aggregate	-	-	-	-	-	m²	8.92
granite aggregate	-	-	-	-	-	m²	9.12
Hand Lay; areas 400 m² and over							
limestone aggregate	-	-	-	-	-	m²	10.35
granite aggregate	-	-	-	-	-	m²	10.56
Macadam Surfacing; Spadeoak Construction Co Ltd; Base (Roadbase) Course; 100 mm of 28 mm dense bitumen macadam to BS4987-1 2001 ref 5.2							
Machine lay; areas 1000 m² and over							
limestone aggregate	-	-	-	-	-	m²	11.29
granite aggregate	-	-	-	-	-	m²	11.55
Hand Lay; areas 400 m² and over							
limestone aggregate	-	-	-	-	-	m²	13.33
granite aggregate	-	-	-	-	-	m²	13.60
Macadam Surfacing; Spadeoak Construction Co Ltd; Base (Roadbase) Course; 150 mm of 28 mm dense bitumen macadam in two layers to BS4987-1 2001 ref 5.2							
Machine lay; areas 1000 m² and over							
limestone aggregate	-	-	-	-	-	m²	17.75
granite aggregate	-	-	-	-	-	m²	18.13
Hand Lay; areas 400 m² and over							
limestone aggregate	-	-	-	-	-	m²	21.50
granite aggregate	-	-	-	-	-	m²	21.91

Prices for Measured Works
Q PAVING/PLANTING/FENCING/SITE FURNITURE

Item Excluding site overheads and profit	PC £	Labour hours	Labour £	Plant £	Material £	Unit	Total rate £
Q22 COATED MACADAM/ASPHALT ROADS/PAVINGS – cont'd							
Base (Roadbase) Course; 200 mm of 28 mm dense bitumen macadam in two layers to BS4987-1 2001 ref 5.2 Machine lay; areas 1000 m² and over							
limestone aggregate	-	-	-	-	-	m²	22.63
granite aggregate	-	-	-	-	-	m²	23.14
Hand Lay; areas 400 m² and over							
limestone aggregate	-	-	-	-	-	m²	26.70
granite aggregate	-	-	-	-	-	m²	27.25
Resin Bound Macadam Pavings; **Machine laid; Colourpave Ltd;** "Naturatex" Clear resin bound macadam to pedestrian or vehicular hard landscape areas; laid to base course (not included)							
25 mm thick to pedestrian areas	-	-	-	-	-	m²	25.00
30 mm thick to vehicular areas	-	-	-	-	-	m²	28.00
"Colourtex" Coloured resin bound macadam to pedestrian or vehicular hard landscape areas							
25 mm thick	-	-	-	-	-	m²	22.00
Road repairs; Colas Ltd "Colas Pre-Formed Surface Dressing." preformed membrane; non-slip surface dressing; on clean smooth surface; concrete, wood, steel etc (not included); treated with "Cariphalte CP Primer"							
"Calcinated Bauxite" finish	-	0.20	3.60	-	12.67	m²	16.27
Marking car parks Car parking space division strips; in accordance with BS 3262; laid hot at 115 degrees C; on bitumen macadam surfacing							
100 mm wide thermoplastic strips	-	-	-	-	-	m	0.95
Stainless metal road studs							
100 x 100 mm	4.90	0.25	4.50	-	4.90	nr	9.40
Q23 GRAVEL/HOGGIN ROADS/PAVINGS							
Excavation and path preparation Excavating; 300 mm deep; to width of path; depositing excavated material at sides of excavation							
width 1.00 m	-	-	-	2.06	-	m²	2.06
width 1.50 m	-	-	-	1.72	-	m²	1.72
width 2.00 m	-	-	-	1.47	-	m²	1.47
width 3.00 m	-	-	-	1.24	-	m²	1.24
Excavating trenches; in centre of pathways; 100 flexible drain pipes; filling with clean broken stone or gravel rejects							
300 x 450 mm deep	3.51	0.10	1.80	1.03	3.58	m	6.41

Prices for Measured Works
Q PAVING/PLANTING/FENCING/SITE FURNITURE

Item Excluding site overheads and profit	PC £	Labour hours	Labour £	Plant £	Material £	Unit	Total rate £
Hand trimming and compacting reduced surface of pathway; by machine							
width 1.00 m	-	0.05	0.90	0.17	-	m	1.07
width 1.50 m	-	0.04	0.80	0.15	-	m	0.95
width 2.00 m	-	0.04	0.72	0.13	-	m	0.85
width 3.00 m	-	0.04	0.72	0.13	-	m	0.85
Permeable membranes; to trimmed and compacted surface of pathway							
"Terram 1000"	0.43	0.02	0.36	-	0.46	m²	0.82
Edging boards							
Boards; 50 x 50 x 750 mm timber pegs at 1000 mm centres (excavations and hardcore under edgings not included)							
38 x 150 mm treated softwood edge boards	2.08	0.10	1.80	-	2.08	m	3.88
50 x 150 mm treated softwood edge boards	2.34	0.10	1.80	-	2.34	m	4.14
38 x 150 mm hardwood (iroko) edge boards	7.08	0.10	1.80	-	7.08	m	8.88
50 x 150 mm hardwood (iroko) edge boards	8.94	0.10	1.80	-	8.94	m	10.74
Filling to make up levels							
Obtained off site; hardcore; PC £16.64 /m³							
150 mm thick	2.66	0.04	0.72	0.44	2.66	m²	3.82
Obtained off site; granular fill type 1; PC £14.45 /tonne (£31.80/m³ compacted)							
100 mm thick	3.18	0.03	0.50	0.31	3.18	m²	3.99
150 mm thick	4.77	0.03	0.45	0.47	4.77	m²	5.69
Surface treatments							
Sand blinding; to hardcore (not included)							
50 mm thick	1.31	0.04	0.72	-	1.44	m²	2.16
Filter fabric; to hardcore (not included)	0.43	0.01	0.18	-	0.46	m²	0.64
Herbicides; ICI							
"Casoron G" (residual) herbicide; treating substrate before laying base							
at 1kg / 125m2	3.13	0.05	0.90	-	3.44	100 m²	4.34
Granular pavings							
Approved washed river or pit gravel; on base layer of coarse aggregate; wearing layer 10 mm gravel graded; watering; rolling; on hardcore base (not included); for pavements; to falls and crossfalls and to slopes not exceeding 15 degrees from horizontal; over 300 mm wide							
50 mm thick	1.53	0.09	1.58	-	1.84	m²	3.41
63 mm thick	2.08	0.13	2.27	-	2.50	m²	4.77
Breedon PLC; "Golden Gravel" or equivalent; rolling wet; on hardcore base (not included); for pavements; to falls and crossfalls and to slopes not exceeding 15 degrees from horizontal; over 300 mm wide							
50 mm thick	7.53	0.10	1.80	0.51	7.53	m²	9.84
75 mm thick	11.30	0.10	1.80	0.60	11.30	m²	13.69
Breedon PLC; 'Wayfarer' specially formulated fine gravel for use on golf course pathways							
50 mm thick	5.97	0.10	1.80	0.51	5.97	m²	8.28
75 mm thick	8.95	0.11	2.00	0.60	8.95	m²	11.54

Prices for Measured Works
Q PAVING/PLANTING/FENCING/SITE FURNITURE

Item Excluding site overheads and profit	PC £	Labour hours	Labour £	Plant £	Material £	Unit	Total rate £
Q23 GRAVEL/HOGGIN ROADS/PAVINGS – cont'd							
Breedon PLC; Breedon Buff decorative limestone chippings							
50 mm thick	-	0.01	0.18	0.10	4.79	m²	5.07
75 mm thick	9.41	0.01	0.23	0.13	9.89	m²	10.24
Hoggin (stabilized);PC £14.50/m³ on hardcore base (not included); to falls and crossfalls and to slopes not exceeding 15 degrees from horizontal; over 300 mm wide							
100 mm thick	1.45	0.13	2.40	1.11	2.17	m²	5.69
150 mm thick	2.17	0.13	2.27	1.05	3.26	m²	6.58
Washed shingle; on prepared base (not included)							
25 - 50 size, 25 mm thick	0.65	0.02	0.32	0.09	0.69	m²	1.09
25 - 50 size, 75 mm thick	1.96	0.05	0.95	0.27	2.06	m²	3.27
50 - 75 size, 25 mm thick	0.65	0.02	0.36	0.10	0.69	m²	1.15
50 - 75 size, 75 mm thick	1.96	0.07	1.20	0.34	2.06	m²	3.59
Pea shingle; on prepared base (not included)							
10 - 15 size, 25 mm thick	0.68	0.02	0.32	0.09	0.71	m²	1.12
5 - 10 size, 75 mm thick	2.04	0.05	0.95	0.27	2.14	m²	3.35
Ballast ; as dug; watering; rolling; on hardcore base (not included)							
100 mm thick	2.81	0.13	2.40	1.11	3.51	m²	7.02
150 mm thick	4.21	0.13	2.27	1.05	5.27	m²	8.58
CED Ltd; Cedec golden gravel; self binding; laid to base measured separately; compacting							
50 mm thick	6.96	0.03	0.51	0.07	6.96	m²	7.54
Grundon Ltd; Coxwell self binding path gravels laid and compacted to excavation or base measured separately							
50 mm thick	1.56	0.03	0.51	0.07	1.56	m²	2.14
Bark surfaces; ICI "Casoron G" (residual) herbicide; treating substrate before laying base							
at 1kg / 125m2	3.13	0.05	0.90	-	3.44	100 m²	4.34
Bark surfaces; Melcourt Industries Bark chips; to surface of pathways by machine; material delivered in 70 m³ loads; levelling and spreading by hand (excavation and preparation not included)							
conifer walk chips; 100 mm thick	1.65	0.03	0.60	0.26	1.73	m²	2.58
wood fibre; 100 mm thick	1.60	0.03	0.60	0.26	1.67	m²	2.53
hardwood chips; 100 mm thick	1.72	0.03	0.60	0.26	1.80	m²	2.66
Bound aggregates; Addagrip Surface Treatments UK Ltd; Natural decorative Resin bonded surface dressing laid to concrete, macadam or to plywood panels priced separately.							
Primer coat to macadam or concrete base	-	-	-	-	-	m²	4.00
Golden pea Gravel 1-3 mm							
Buff adhesive	-	-	-	-	-	m²	19.36
Red adhesive	-	-	-	-	-	m²	19.36
Green adhesive	-	-	-	-	-	m²	19.36
Golden pea Gravel 2-5 mm							
Buff adhesive	-	-	-	-	-	m²	22.99
Chinese bauxite 1-3 mm							
Buff adhesive	-	-	-	-	-	m²	19.36

Prices for Measured Works
Q PAVING/PLANTING/FENCING/SITE FURNITURE

Item Excluding site overheads and profit	PC £	Labour hours	Labour £	Plant £	Material £	Unit	Total rate £
Q24 INTERLOCKING BRICK/BLOCK ROADS/PAVINGS							
Precast concrete block edgings; PC £7.17/m²; 200 x 100 x 60 mm; on prepared base (not included); haunching one side							
Edgings; butt joints							
stretcher course	1.43	0.17	3.00	-	3.44	m	6.44
header course	0.72	0.27	4.80	-	2.86	m	7.66
Precast concrete vehicular paving blocks; Marshalls Mono; on prepared base (not included); on 50 mm compacted sharp sand bed; blocks laid in 7 mm loose sand and vibrated; joints filled with sharp sand and vibrated; level and to falls only							
"Trafica" paving blocks; 450 x 450 x 70 mm							
"Perfecta" finish; colour natural	18.78	0.50	9.00	0.09	20.69	m²	29.78
"Perfecta" finish; colour buff	21.74	0.50	9.00	0.09	23.73	m²	32.81
"Saxon" finish; colour natural	17.16	0.50	9.00	0.09	19.03	m²	28.11
"Saxon" finish; colour buff	20.31	0.50	9.00	0.09	22.26	m²	31.35
Precast concrete vehicular paving blocks; Marshalls Mono; on prepared base (not included); on 50 mm compacted sharp sand bed; blocks laid in 7 mm loose sand and vibrated; joints filled with sharp sand and vibrated; level and to falls only							
"Keyblock" concrete blocks							
200 x 100 x 60 mm; natural grey	7.17	0.40	7.20	0.09	9.14	m²	16.42
200 x 100 x 60 mm; colours	7.98	0.40	7.20	0.09	9.99	m²	17.27
200 x 100 x 80 mm; natural grey	8.00	0.40	7.20	0.09	10.01	m²	17.29
200 x 100 x 80 mm; colours	9.22	0.40	7.20	0.09	11.29	m²	18.58
Precast concrete vehicular paving blocks; Charcon Hard Landscaping; on prepared base (not included); on 50 mm compacted sharp sand bed; blocks laid in 7 mm loose sand and vibrated; joints filled with sharp sand and vibrated; level and to falls only							
"Europa" concrete blocks							
200 x 100 x 60 mm; natural grey	7.73	0.40	7.20	0.09	9.72	m²	17.01
200 x 100 x 60 mm; colours	8.28	0.40	7.20	0.09	10.30	m²	17.59
"Parliament" concrete blocks							
200 x 100 x 65 mm; natural grey	18.69	0.40	7.20	0.09	20.77	m²	28.05
200 x 100 x 65 mm; colours	18.69	0.40	7.20	0.09	20.77	m²	28.05

Prices for Measured Works
Q PAVING/PLANTING/FENCING/SITE FURNITURE

Item Excluding site overheads and profit	PC £	Labour hours	Labour £	Plant £	Material £	Unit	Total rate £
Q24 INTERLOCKING BRICK/BLOCK ROADS/PAVINGS – cont'd							
Recycled polyethylene grassblocks; Netlon Ltd; interlocking units laid to prepared base or rootzone (not included);							
"Netpave 50"; load bearing 50 tonnes per m²; 500 x 500 x 50 mm deep							
minimum area 40 m²	-	0.20	3.60	0.20	17.11	m²	20.91
40 -1039 m²	-	0.20	3.60	0.20	12.51	m²	16.31
over 1039 m²	-	0.20	3.60	0.20	11.61	m²	15.41
"Netpave 25"; load bearing :light vehicles and pedestrians; 500 x 500 x 25 mm deep; laid onto established grass surface							
minimum area 40 m²	14.00	0.10	1.80	-	14.00	m²	15.80
40 -1039 m²	9.40	0.10	1.80	-	9.40	m²	11.20
over 1039 m²	8.50	0.10	1.80	-	8.50	m²	10.30
"Grassroad"; Cooper Clarke Group; heavy duty for car parking and fire paths verge hardening and shallow embankments							
Honeycomb cellular polyproylene interconnecting paviors with integral downstead anti-shear cleats including topsoil but excluding edge restraints; to granular sub-base (not included)							
635 x 330 x 42 overall laid to a module of 622 x 311 x 32	-	-	-	-	-	m²	21.23
Extra over for green colour	-	-	-	-	-	m²	0.26
Q25 SLAB/BRICK/SETT/COBBLE PAVINGS							
Bricks - General							
Preamble: BS 3921 includes the following specification for bricks for paving: bricks shall be hard, well burnt, non-dusting, resistant to frost and sulphate attack and true to shape, size and sample.							
Movement of materials							
Mechanically offloading bricks; loading wheel barrows; transporting maximum 25 m distance	-	0.20	3.60	-	-	m²	3.60
Edge restraints; to brick paving; on prepared base (not included); 65 mm thick bricks; PC £300.00 /1000; haunching one side							
Header Course							
200 x 100 mm; butt joints	3.00	0.27	4.80	-	5.14	m	9.94
210 x 105 mm; mortar joints	2.67	0.50	9.00	-	5.13	m	14.13
Stretcher course							
200 x 100 mm; butt joints	1.50	0.17	3.00	-	3.51	m	6.51
210 x 105 mm; mortar joints	1.36	0.33	6.00	-	3.52	m	9.52

Prices for Measured Works
Q PAVING/PLANTING/FENCING/SITE FURNITURE

Item Excluding site overheads and profit	PC £	Labour hours	Labour £	Plant £	Material £	Unit	Total rate £
Variation in brick prices; Add or subtract the following amounts for every £1.00/1000 difference in the PC price							
Edgings							
100 wide	-	-	-	-	0.05	10 m	0.05
200 wide	-	-	-	-	0.11	10 m	0.11
102.5 wide	-	-	-	-	0.05	10 m	0.05
215 wide	-	-	-	-	0.09	10 m	0.09
Clay brick pavings; on prepared base (not included); bedding on 50 mm sharp sand; kiln dried sand joints							
Pavings; Marshalls Clay Products; "Nori"; 200 x 100 x 65 mm wirecut chamfered paviors							
"Sherbourne Red"; PC £16.45/m²	16.05	1.06	19.05	0.25	18.21	m²	37.51
"Allandale Gold"; PC £19.88/m²	19.88	1.06	19.05	0.25	21.63	m²	40.94
Paving bricks; PC £300.00 per 1000							
215 x 102.5 x 65 mm	12.40	1.00	18.00	-	15.74	m²	33.74
200 x 100.5 x 50 mm	15.00	1.10	19.80	-	18.34	m²	38.14
Reduction for bedding on 50 mm compacted sharp sand; setting; vibrating; sand joints							
Clay brick pavings; on prepared base (not included); bedding on cement:sand (1:4) pointing mortar as work proceeds							
Pavings; Hanson Brick							
Buff Multi Chamfered Paver; 200 x 100 x 50 mm; PC £324.24/1000	13.39	2.20	39.60	-	16.79	m²	56.39
Antique Chamfered Paver; 200 x 100 x 50 mm; PC £280.98/1000	11.60	2.20	39.60	-	14.96	m²	54.56
Blue Brown Chamfered Cobble Paver; 200 x 100 x 50 mm; PC £324.24/1000	13.39	2.20	39.60	-	16.79	m²	56.39
Paving bricks; PC £300.00 per 1000							
215 x 102.5 x 65 mm	11.85	2.22	40.00	-	15.19	m²	55.19
200 x 100 x 50 mm	12.99	2.50	45.00	-	16.33	m²	61.33
Extra over clay brick pavings for labours							
basketweave bond	-	0.20	3.60	-	-	m²	3.60
herringbone; laid on edge	-	2.00	36.00	-	-	m²	36.00
Cutting							
raking cutting	-	0.33	6.00	4.22	-	m	10.22
curved cutting	-	0.44	8.00	5.29	-	m	13.29
Add or subtract the following amounts for every £10.00 /1000 difference in the prime cost of bricks							
Butt joints							
200 x 100	-	-	-	-	0.50	m²	0.50
215 x 102.5	-	-	-	-	0.45	m²	0.45
10 mm mortar joints							
200 x 100	-	-	-	-	0.43	m²	0.43
215 x 102.5	-	-	-	-	0.40	m²	0.40

Prices for Measured Works
Q PAVING/PLANTING/FENCING/SITE FURNITURE

Item Excluding site overheads and profit	PC £	Labour hours	Labour £	Plant £	Material £	Unit	Total rate £
Q25 SLAB/BRICK/SETT/COBBLE PAVINGS - cont'd							
Precast concrete pavings; Charcon Hard Landscaping; to BS 7263; on prepared sub-base (not included); bedding on 25 mm thick cement:sand mortar (1:4); butt joints; straight both ways; jointing in cement:sand (1:3) brushed in; on 50 mm thick sharp sand base							
Pavings; natural grey							
450 x 450 x 70 mm chamfered	10.86	0.44	8.00	-	14.90	m²	**22.90**
450 x 450 x 50 mm chamfered	9.38	0.44	8.00	-	13.42	m²	**21.42**
600 x 300 x 50 mm	8.44	0.44	8.00	-	12.48	m²	**20.48**
400 x 400 x 65 mm chamfered	17.44	0.40	7.20	-	21.48	m²	**28.68**
450 x 600 x 50 mm	7.52	0.44	8.00	-	11.56	m²	**19.56**
600 x 600 x 50 mm	6.06	0.40	7.20	-	10.10	m²	**17.30**
750 x 600 x 50 mm	5.71	0.40	7.20	-	9.75	m²	**16.95**
900 x 600 x 50 mm	5.22	0.40	7.20	-	9.26	m²	**16.46**
Pavings; coloured							
450 x 450 x 70 mm chamfered	12.69	0.44	8.00	-	16.73	m²	**24.73**
450 x 600 x 50 mm	11.33	0.44	8.00	-	15.37	m²	**23.37**
400 x 400 x 65 mm chamfered	17.44	0.40	7.20	-	21.48	m²	**28.68**
600 x 600 x 50 mm	9.47	0.40	7.20	-	13.51	m²	**20.71**
750 x 600 x 50 mm	8.73	0.40	7.20	-	12.77	m²	**19.97**
900 x 600 x 50 mm	7.75	0.40	7.20	-	11.79	m²	**18.99**
Precast concrete pavings; Charcon Hard Landscaping; to BS 7263; on prepared sub-base (not included); bedding on 25 mm thick cement:sand mortar (1:4); butt joints; straight both ways; jointing in cement:sand (1:3) brushed in; on 50 mm thick sharp sand base "Appalacian" rough textured exposed aggregate pebble paving							
600 mm x 600 mm x 65 mm	22.50	0.50	9.00	-	25.79	m²	**34.79**
Pavings; Atlas Stone Products; on prepared base (not included); "French Limestone" precast slab in random sizes laid to full mortar bed; butt jointed							
340 x 340 x 25 mm	16.61	0.44	8.00	-	19.34	m²	**27.34**
685 x 340 x 25 mm	16.82	0.44	8.00	-	19.97	m²	**27.97**
455 x 340 x 25 mm	16.86	0.57	10.29	-	20.01	m²	**30.30**
"Broadway paving" in Cotswold; spot bedded on 5 pads of sand:cement mortar (1:4) on sharp sand							
600 x 600 mm	7.64	0.40	7.20	-	9.19	m²	**16.39**
450 x 450 mm	6.57	0.50	9.00	-	8.10	m²	**17.10**
600 x 300 mm	9.44	0.50	9.00	-	11.05	m²	**20.05**
Pavings; Marshalls Mono; spot bedding on 5 nr pads of cement:sand mortar (1:4); on sharp sand "Blister Tactile" pavings; specially textured slabs for guidance of blind pedestrians; red or buff							

Prices for Measured Works
Q PAVING/PLANTING/FENCING/SITE FURNITURE

Item Excluding site overheads and profit	PC £	Labour hours	Labour £	Plant £	Material £	Unit	Total rate £
400 x 400 x 50 mm	19.88	0.50	9.00	-	21.74	m²	30.74
450 x 450 x 50 mm	17.48	0.50	9.00	-	19.28	m²	28.28
"Metric Four Square" pavings							
496 x 496 x 50 mm; exposed river gravel aggregate	50.87	0.50	9.00	-	52.24	m²	61.24
"Metric Four Square" cycle blocks							
496 x 496 x 50 mm; exposed aggregate	50.87	0.25	4.50	-	52.87	m²	57.37
Precast concrete pavings; Marshalls; "Heritage" imitation riven yorkstone paving; on prepared sub-base measured separately; bedding on 25 mm thick cement:sand mortar (1:4); pointed straight both ways cement:sand (1:3)							
Square and rectangular paving							
450 x 300 x 38 mm	26.42	1.00	18.00	-	29.14	m²	47.14
450 x 450 x 38 mm	15.93	0.75	13.50	-	18.65	m²	32.15
600 x 300 x 38 mm	17.44	0.80	14.40	-	20.16	m²	34.56
600 x 450 x 38 mm	17.69	0.75	13.50	-	20.41	m²	33.91
600 x 600 x 38 mm	17.11	0.50	9.00	-	19.88	m²	28.88
Extra labours for laying the a selection of the above sizes to random rectangular pattern	-	0.33	6.00	-	-	m²	6.00
Radial paving for circles							
circle with centre stone and first ring (8 slabs), 450 x 230/560 x 38 mm; diameter 1.54 m (total area 1.86 m²)	46.11	1.50	27.00	-	51.15	nr	78.15
circle with second ring (16 slabs), 450 x 300/460 x 38 mm; diameter 2.48 m (total area 4.83 m²)	46.11	4.00	72.00	-	131.71	nr	203.71
circle with third ring (16 slabs), 450 x 470/625 x 38 mm; diameter 3.42 m (total area 9.18 m²)	46.11	8.00	144.00	-	240.78	nr	384.78
Stepping stones							
380 dia x 38 mm	3.29	0.20	3.60	-	8.75	nr	12.35
Asymmetrical 560 x 420 x 38 mm	4.90	0.20	3.60	-	10.36	nr	13.96
Precast concrete pavings; Marshalls; "Chancery" imitation reclaimed riven yorkstone paving; on prepared sub-base measured separately; bedding on 25 mm thick cement:sand mortar (1:4); pointed straight both ways cement:sand (1:3)							
Square and rectangular paving							
300 x 300 x 45 mm	21.84	1.00	18.00	-	24.57	m²	42.57
450 x 300 x 45 mm	20.52	0.90	16.20	-	23.25	m²	39.45
600 x 300 x 45 mm	20.09	0.80	14.40	-	22.81	m²	37.21
600 x 450 x 45 mm	20.05	0.75	13.50	-	22.78	m²	36.28
450 x 450 x 45 mm	18.13	0.75	13.50	-	20.86	m²	34.36
600 x 600 x 45 mm	20.52	0.50	9.00	-	23.25	m²	32.25
Extra labours for laying the a selection of the above sizes to random rectangular pattern	-	0.33	6.00	-	-	m²	6.00
Radial paving for circles							
circle with centre stone and first ring (8 slabs), 450 x 230/560 x 38 mm; diameter 1.54 m (total area 1.86 m²)	55.33	1.50	27.00	-	60.37	nr	87.37
circle with second ring (16 slabs), 450 x 300/460 x 38 mm; diameter 2.48 m (total area 4.83 m²)	142.69	4.00	72.00	-	155.65	nr	227.65
circle with third ring (16 slabs), 450 x 470/625 x 38 mm; diameter 3.42 m (total area 9.18 m²)	259.65	8.00	144.00	-	284.08	nr	428.08
Squaring off set for 2 ring circle							
16 slabs; 2.72 m²	79.79	1.00	18.00	-	79.79	nr	97.79

Prices for Measured Works
Q PAVING/PLANTING/FENCING/SITE FURNITURE

Item Excluding site overheads and profit	PC £	Labour hours	Labour £	Plant £	Material £	Unit	Total rate £
Q25 SLAB/BRICK/SETT/COBBLE PAVINGS – cont'd							
Pavings; Milner Delvaux Ltd; bedding on 50 mm cement:sand mortar (1:4) "Blanc de Bierges" precast concrete pavings							
200 x 200 mm	20.32	0.80	14.40	-	26.28	m²	40.68
400 x 200 mm	19.32	0.74	13.33	-	25.26	m²	38.60
400 x 300 mm	19.48	0.71	12.86	-	25.43	m²	38.28
400 x 400 mm	19.32	0.71	12.86	-	25.26	m²	38.12
600 x 300 mm	18.26	0.61	10.91	-	24.17	m²	35.08
400 x 600 mm	18.26	0.50	9.05	-	24.17	m²	33.22
600 x 600 mm	18.26	0.44	7.96	-	24.17	m²	32.14
300 mm diameter	2.46	0.40	7.20	-	2.84	nr	10.04
Marshalls; La Linia Pavings; Fine textured exposed aggregate 80 mm thick pavings in various sizes to designed laying patterns; laid to 50mm sharp sand bed on Type 1 base all priced separately Bonded laying patterns 300 x 300							
Light Granite / Anthracite basalt	23.82	2.00	36.00	-	24.42	m²	60.42
Indian granite / Yellow	-	2.00	36.00	-	26.55	m²	62.55
Random Scatter pattern incorporating 100 x 200, 200 x 200 and 300 x 200 units							
Light Granite / Anthracite basalt	23.82	3.00	54.05	-	23.82	m²	77.87
Indian granite / Yellow	-	3.00	54.05	-	25.90	m²	79.95
Marshalls; "La Linia Grande Paving"; as above but 140 mm thick Bonded laying patterns 300 x 300							
All colours	46.54	3.00	54.05	-	46.54	m²	100.59
Random Scatter pattern incorporating 100 x 200, 200 x 200 and 300 x 200 units							
Light Granite / Anthracite basalt	46.54	4.00	72.00	-	46.54	m²	118.54
Extra over to the above for incorporating inlay stones to patterns; blue, light granite or anthracite basalt							
to prescribed patterns	100.89	1.50	27.00	-	100.89	m²	127.89
as individual units; triangular 200 x 200 x 282 mm	0.57	0.05	0.90	-	0.57	nr	1.47
Pedestrian deterrent pavings; Marshalls Mono; on prepared base (not included); bedding on 25 mm cement:sand (1:3); cement:sand (1:3) joints "Lambeth" pyramidal pavings							
600 x 600 x 75 mm	34.40	0.25	4.50	-	37.13	m²	41.63
"Thaxted" pavings; granite sett appearance							
600 x 600 x 75 mm	39.88	0.33	6.00	-	42.61	m²	48.61

Prices for Measured Works
Q PAVING/PLANTING/FENCING/SITE FURNITURE

Item Excluding site overheads and profit	PC £	Labour hours	Labour £	Plant £	Material £	Unit	Total rate £
Pedestrian deterrent pavings; **Townscape Products; on prepared base** **(not included); bedding on 25 mm** **cement:sand (1:3); cement:sand (1:3)** **joints**							
"Strata" striated textured slab pavings; giving bonded appearance; grey							
600 x 600 x 60 mm	23.19	0.50	9.00	-	25.92	m²	**34.92**
"Geoset" raised chamfered studs pavings; grey							
600 x 600 x 60 mm	23.19	0.50	9.00	-	25.92	m²	**34.92**
"Abbey" square cobble pattern pavings; reinforced							
600 x 600 x 65 mm	21.92	0.80	14.40	-	24.65	m²	**39.05**
Surface treatments; ICI							
Treating substrate before laying base							
"Casoron G" (residual); at 1kg / 125m2	3.13	0.05	0.90	-	3.44	100 m²	**4.34**
Slip-resistant surfaces							
Resin-base coat 1 - 2 kg/m² with applied grit or aggregate finish; 0.50 - 0.60 kg/m²; on screeded sub-base	-	-	-	-	-	m²	**10.34**
Edge restraints; to block paving; on **prepared base (not included); 200 x 100** **x 80 mm; PC £80.00/m²); haunching** **one side**							
Header course							
200 x 100 mm; butt joints	80.00	0.27	4.80	-	82.14	m	**86.94**
Stretcher course							
200 x 100 mm; butt joints	40.00	0.17	3.00	-	42.01	m	**45.01**
Concrete paviors; Marshalls Mono; on **prepared base (not included); bedding** **on 50 mm sand; kiln dried sand joints** **swept in**							
"Keyblock" paviors							
200 x 100 x 60 mm; grey	7.17	0.40	7.20	0.09	8.78	m²	**16.06**
200 x 100 x 60 mm; colours	7.98	0.40	7.20	0.09	9.59	m²	**16.87**
200 x 100 x 80 mm; grey	8.00	0.44	8.00	0.09	9.61	m²	**17.69**
200 x 100 x 80 mm; colours	9.22	0.44	8.00	0.09	10.83	m²	**18.91**
Concrete cobble paviors; Charcon Hard **Landscaping; Concrete Products; on** **prepared base (not included); bedding** **on 50 mm sand; kiln dried sand joints** **swept in**							
Paviors							
"Woburn" blocks; 100 - 201 x 134 x 80 mm; random sizes	19.65	0.67	12.00	0.09	21.75	m²	**33.84**
"Woburn" blocks; 100 - 201 x 134 x 80 mm; single size	19.65	0.50	9.00	0.09	21.75	m²	**30.84**
"Woburn" blocks; 100 - 201 x 134 x 60 mm; random sizes	15.90	0.67	12.00	0.09	17.91	m²	**29.99**
"Woburn" blocks; 100 - 201 x 134 x 60 mm; single size	15.90	0.50	9.00	0.09	17.89	m²	**26.98**

Prices for Measured Works
Q PAVING/PLANTING/FENCING/SITE FURNITURE

Item Excluding site overheads and profit	PC £	Labour hours	Labour £	Plant £	Material £	Unit	Total rate £
Q25 SLAB/BRICK/SETT/COBBLE PAVINGS - cont'd							
Concrete setts; on 25 mm sand; compacted; vibrated; joints filled with sand; natural or coloured; well rammed hardcore base (not included) Marshalls Mono; Tegula Paving							
60 mm thick; random sizes	17.45	0.57	10.29	0.09	20.06	m²	30.43
60 mm thick; single size	17.45	0.45	8.18	0.09	20.06	m²	28.33
80 mm thick; random sizes	19.66	0.57	10.29	0.09	22.38	m²	32.75
80 mm thick; single size	19.66	0.45	8.18	0.09	22.38	m²	30.65
Cobbles 80 x 80 x 60 mm thick; traditional	-	0.56	10.00	0.09	28.32	m²	38.40
Cobbles Charcon Hard Landscaping; Country setts							
100 mm thick; random sizes	24.25	1.00	18.00	0.09	25.99	m²	44.08
100 mm thick; single size	24.25	0.67	12.00	0.09	26.60	m²	38.68
Concrete setts; Milner Delvaux Ltd; on prepared base (not included); bedding on 30 mm cement:sand (1:3); cement:sand (1:3) joints "Blanc de Bierges" concrete setts							
140 x 140 x 80 mm	22.06	2.50	45.00	-	26.19	m²	71.19
210 x 140 x 80 mm	22.06	2.30	41.40	-	26.19	m²	67.59
70 x 70 x 70 mm	23.14	2.75	49.50	-	27.32	m²	76.82
Natural stone, slab or granite paving - General Preamble: provide paving slabs of the specified thickness in random sizes but not less than 25 slabs per 10 m² of surface area, to be laid in parallel courses with joints alternately broken and laid to falls.							
Reconstituted Yorkstone aggregate pavings; Marshalls; "Saxon" on prepared sub-base measured separately; bedding on 25 mm thick cement:sand mortar (1:4); on 50 mm thick sharp sand base square and rectangular paving in various colours; butt joints straight both ways							
300 x 300 x 35 mm	24.31	0.88	15.84	-	30.27	m²	46.11
600 x 300 x 35 mm	15.96	0.72	12.87	-	21.92	m²	34.79
600 x 600 x 35 mm	11.96	0.55	9.90	-	17.91	m²	27.81
450 x 450 x 50 mm	18.23	0.77	13.86	-	24.19	m²	38.05
600 x 600 x 50 mm	15.96	0.61	10.89	-	21.92	m²	32.81
square and rectangular paving in natural; butt joints straight both ways							
300 x 300 x 35 mm	20.20	0.88	15.84	-	26.16	m²	42.00
600 x 300 x 35 mm	14.06	0.83	14.85	-	20.01	m²	34.86
600 x 600 x 35 mm	10.30	0.55	9.90	-	16.26	m²	26.16
450 x 450 x 50 mm	15.88	0.77	13.86	-	21.83	m²	35.69
600 x 600 x 50 mm	13.52	0.66	11.88	-	19.48	m²	31.36

Prices for Measured Works
Q PAVING/PLANTING/FENCING/SITE FURNITURE

Item Excluding site overheads and profit	PC £	Labour hours	Labour £	Plant £	Material £	Unit	Total rate £
Radial paving for circles; 20 mm joints							
circle with centre stone and first ring (8 slabs), 450 x 230/560 x 35 mm; diameter 1.54 m (total area 1.86 m²)	47.63	1.50	27.00	-	52.96	nr	79.96
circle with second ring (16 slabs), 450 x 300/460 x 35 mm; diameter 2.48 m (total area 4.83 m²)	115.95	4.00	72.00	-	129.71	nr	201.71
circle with third ring (24 slabs), 450 x 310/430 x 35 mm; diameter 3.42 m (total area 9.18 m²)	218.43	8.00	144.00	-	244.74	nr	388.74
Granite setts; bedding on 25 mm cement:sand (1:3)							
Natural granite setts; 100 x 100 mm to 125 x 150 mm; x 150 to 250 mm length; riven surface; silver grey							
new	22.85	2.50	45.00	-	30.06	m²	75.06
reclaimed; cleaned	30.00	2.50	45.00	-	37.81	m²	82.81
Natural stone, slate or granite flag pavings; CED Limited; on prepared base (not included); bedding on 25 mm cement:sand (1:3); cement:sand (1:3) joints							
York stone; riven laid random rectangular							
new slabs; 40 - 60 mm thick	57.90	1.71	30.86	-	66.21	m²	97.07
reclaimed slabs, Cathedral grade; 50 - 75 mm thick	54.00	2.80	50.40	-	61.92	m²	112.32
Donegal quartzite slabs; standard tiles 595 x 595 x 7 - 13 mm	50.40	2.33	42.00	-	55.44	m²	97.45
Natural Yorkstone, pavings or edgings; Johnsons Wellfield Quarries; sawn 6 sides; 50 mm thick; on prepared base measured separately; bedding on 25 mm cement:sand (1:3); cement:sand (1:3) joints							
Paving							
laid to random rectangular pattern	47.25	1.71	30.85	-	52.48	m²	83.34
laid to coursed laying pattern; 3 sizes	51.45	1.72	31.03	-	56.42	m²	87.46
Paving; single size							
600 x 600 mm	57.23	0.85	15.30	-	62.49	m²	77.79
600 x 400 mm	57.23	1.00	18.00	-	62.49	m²	80.49
300 x 200 mm	65.63	2.00	36.00	-	71.31	m²	107.31
215 x 102.5 mm	68.78	2.50	45.00	-	74.62	m²	119.62
Paving; cut to template off site; 600 x 600; radius:							
1.00 m	168.00	3.33	60.00	-	170.40	m²	230.40
2.50 m	162.75	2.00	36.00	-	165.15	m²	201.15
5.00 m	157.50	2.00	36.00	-	159.90	m²	195.90
Edgings							
100 mm wide x random lengths	6.56	0.50	9.00	-	7.13	m	16.13
100 mm x 100 mm	6.56	0.50	9.00	-	9.29	m	18.29
100 mm x 200 mm	13.13	0.50	9.00	-	16.18	m	25.18
250 mm wide x random lengths	12.86	0.40	7.20	-	15.90	m	23.10
500 mm wide x random lengths	25.73	0.33	6.00	-	29.42	m	35.42
York stone edgings; 600 mm long x 250 mm wide; cut to radius							
1.00 m to 3.00	60.38	0.50	9.00	-	63.63	m	72.63
3.00 m to 5.00 m	59.06	0.44	8.00	-	62.26	m	70.25
exceeding 5.00 m	57.75	0.40	7.20	-	60.88	m	68.08

Prices for Measured Works
Q PAVING/PLANTING/FENCING/SITE FURNITURE

Item Excluding site overheads and profit	PC £	Labour hours	Labour £	Plant £	Material £	Unit	Total rate £
Q25 SLAB/BRICK/SETT/COBBLE PAVINGS - cont'd							
Natural Yorkstone, pavings or edgings; Johnsons Wellfield Quarries; sawn 6 sides; 75 mm thick; on prepared base measured separately; bedding on 25 mm cement:sand (1:3); cement:sand (1:3) joints							
Paving							
laid to random rectangular pattern	55.13	0.95	17.10	-	60.29	m²	77.39
laid to coursed laying pattern; 3 sizes	61.95	0.95	17.10	-	67.45	m²	84.55
Paving; single size							
600 x 600 mm	69.04	0.95	17.10	-	74.89	m²	91.99
600 x 400 mm	69.04	0.95	17.10	-	74.89	m²	91.99
300 x 200 mm	79.28	0.75	13.50	-	85.64	m²	99.14
215 x 102.5 mm	83.21	2.50	45.00	-	89.77	m²	134.77
Paving; cut to template off site; 600 x 600; radius:							
1.00 m	199.50	4.00	72.00	-	201.90	m²	273.90
2.50 m	194.25	2.50	45.00	-	196.65	m²	241.65
5.00 m	189.00	2.50	45.00	-	191.40	m²	236.40
Edgings							
100 mm wide x random lengths	8.40	0.60	10.80	-	9.06	m	19.86
100 mm x 100 mm	8.32	0.60	10.80	-	11.14	m	21.94
100 mm x 200 mm	16.64	0.50	9.00	-	19.87	m	28.87
250 mm wide x random lengths	15.49	0.50	9.00	-	18.66	m	27.66
500 mm wide x random lengths	30.98	0.40	7.20	-	34.93	m	42.13
Edgings; 600 mm long x 250 mm wide; cut to radius							
1.00 m to 3.00	73.50	0.60	10.80	-	77.42	m	88.21
3.00 m to 5.00 m	72.19	0.50	9.00	-	76.04	m	85.04
exceeding 5.00 m	70.88	0.44	8.00	-	74.66	m	82.66
Cobble pavings - General Preamble: Excavate for and supply and lay hard coarse clinker, consolidated to a thickness of 75 mm. Lay a 50 mm bed of concrete (1:2:4) with small aggregate. Cobbles should be embedded by hand, tight-butted, endwise to a depth of 60% of their length. The tops of cobbles generally shall be about 25 mm above the level of the adjoining paving, except where cobbled areas abut flag paving when the last three rows of cobbles shall be graded down so that the tops of the cobbles shall be level with the flags. Each flint shall be consolidated, using a wooden mallet. Preamble: Upon completion of each area, a dry grout of rapid-hardening cement:sand (1:2) shall be brushed over the cobbles until the interstices are filled to the level of the adjoining paving. Surplus grout shall then be brushed off and a light, fine spray of water applied over the area. The area shall again be brushed after 24 hours to remove any free grout adhering to exposed faces of cobbles. Cement for grouting shall be							

Prices for Measured Works
Q PAVING/PLANTING/FENCING/SITE FURNITURE

Item Excluding site overheads and profit	PC £	Labour hours	Labour £	Plant £	Material £	Unit	Total rate £
rapid-hardening Portland Cement conforming to BS12:1978, and shall be stored on site in a proper manner to avoid deterioration (manufacturer's instructions to be carefully followed).							
Cobble pavings							
Cobbles; to present a uniform colour in panels; or varied in colour as required							
Scottish Beach Cobbles; 200 - 100 mm	25.73	2.00	36.00	-	31.46	m²	67.46
Scottish Beach Cobbles; 100 - 75 mm	23.30	2.50	45.00	-	29.03	m²	74.03
Scottish Beach Cobbles; 75 - 50 mm	22.25	3.33	60.00	-	27.98	m²	87.98
Single layer of kidney flint cobbles; on prepared base (not included); bedding in sand; on 50 mm compacted sand							
150 - 100 mm average size	22.00	3.00	54.00	-	26.03	m²	80.03
100 - 75 mm average size	22.00	3.00	54.00	-	26.03	m²	80.03
75 - 50 mm average size	22.75	2.50	45.00	-	26.78	m²	71.78
Concrete cycle blocks; Marshalls Mono; on 100 mm concrete (1:2:4); on 150 mm hardcore; bedding in cement:sand (1:4)							
"Metric 4 Square" cycle stand blocks, smooth grey concrete							
496 x 496 x 100 mm	22.91	0.25	4.50	-	25.89	nr	30.39
Concrete cycle blocks; Townscape Products; on 100 mm concrete (1:2:4); on 150 mm hardcore; bedding in cement:sand (1:4)							
Cycle Blocks							
"Cycle Bloc"; in white concrete	16.94	0.25	4.50	-	22.27	nr	26.77
"Mountain Cycle Bloc"; in white concrete	23.44	0.25	4.50	-	28.77	nr	33.27
Grass concrete - General							
Preamble: Grass seed should be a perennial ryegrass mixture, with the proportion depending on expected traffic. Hardwearing winter sportsground mixtures are suitable for public areas. Loose gravel, shingle or sand is liable to be kicked out of the blocks; rammed hoggin or other stabilized material should be specified.							
Grass concrete; Grass Concrete Ltd; on blinded Granular Type 1 sub-base (not included)							
"Grasscrete" in situ concrete continuously reinforced surfacing; including expansion joints at 10 m centres; soiling; seeding							
ref GC2; 150 mm thick; traffic up to 40.00 tonnes	-	-	-	-	-	m²	28.81
ref GC1; 100 mm thick; traffic up to 13.30 tonnes	-	-	-	-	-	m²	24.07
ref GC3; 76 mm thick; traffic up to 4.30 tonnes	-	-	-	-	-	m²	20.63

Prices for Measured Works
Q PAVING/PLANTING/FENCING/SITE FURNITURE

Item Excluding site overheads and profit	PC £	Labour hours	Labour £	Plant £	Material £	Unit	Total rate £
Q25 SLAB/BRICK/SETT/COBBLE PAVINGS – cont'd							
Grass concrete; Grass Concrete Ltd; 406 x 406 mm blocks; on 20 mm sharp sand; on blinded MOT type 1 sub-base (not included); level and to falls only; including filling with topsoil; seeding with dwarf rye grass at £2.40/kg Pavings							
ref GB103; 103 mm thick	12.56	0.40	7.20	0.20	14.29	m²	21.69
ref GB83; 83 mm thick	11.27	0.36	6.54	0.20	13.00	m²	19.75
Grass concrete; Charcon Hard Landscaping; on 25 mm sharp sand; including filling with topsoil; seeding with dwarf rye grass at £2.40/kg "Grassgrid" grass/concrete paving blocks							
366 x 274 x 100 mm thick	16.07	0.38	6.75	0.20	17.80	m²	24.75
Grass concrete; Marshalls Mono; on 25 mm sharp sand; including filling with topsoil; seeding with dwarf rye grass at £2.40/kg Concrete grass pavings							
"Grassguard 130"; for light duty applications; (80 mm prepared base not included)	12.52	0.38	6.75	0.20	14.71	m²	21.66
"Grassguard 160"; for medium duty applications; (8C - 150 mm prepared base not included)	14.75	0.46	8.31	0.20	17.00	m²	25.50
"Grassguard 180"; for heavy duty applications; (150 mm prepared base not included)	16.52	0.60	10.80	0.20	18.81	m²	29.81
Precast concrete units; Marshalls Mono; including filling with topsoil; seeding with dwarf rye grass at £2.40/kg "Mono Hexpot" paving; 100 mm thick							
standard; 292 x 339 mm	23.60	0.33	6.00	0.20	27.63	m²	33.84
large; 360 x 415 mm	20.68	0.33	6.00	0.20	24.68	m²	30.88
Full mortar bedding Extra over pavings for bedding on 25 mm cement:sand (1:4); in lieu of spot bedding on sharp sand	-	0.03	0.45	-	1.53	m²	1.98

Prices for Measured Works
Q PAVING/PLANTING/FENCING/SITE FURNITURE

Item Excluding site overheads and profit	PC £	Labour hours	Labour £	Plant £	Material £	Unit	Total rate £
Q26 SPECIAL SURFACINGS FOR SPORT							
Market Prices of Surfacing Materials							
Surfacings; Melcourt Industries Ltd							
Playbark 10/50®; per 25m3 load	-	-	-	-	46.00	m³	46.00
Playbark 10/50®; per 70m3 load	-	-	-	-	33.60	m³	33.60
Playbark 8/25®; per 25m3 load	-	-	-	-	43.50	m³	43.50
Playbark 8/25®; per 70m3 load	-	-	-	-	31.10	m³	31.10
Adventure Bark 30; per 25m3 load	-	-	-	-	44.90	m³	44.90
Adventure Bark 30; per 70m3 load	-	-	-	-	32.50	m³	32.50
Conifer Playchips®; per 25m3 load	-	-	-	-	31.40	m³	31.40
Conifer Playchips®; per 70m3 load	-	-	-	-	19.00	m³	19.00
Kushyfall; per 25m3 load	-	-	-	-	30.90	m³	30.90
Kushyfall; per 70m3 load	-	-	-	-	18.50	m³	18.50
Softfall; per 25m3 load	-	-	-	-	24.60	m³	24.60
Softfall; per 70m3 load	-	-	-	-	12.20	m³	12.20
Playsand; per 10t load	-	-	-	-	78.30	m³	78.30
Playsand; per 20t load	-	-	-	-	66.06	m³	66.06
Hardwood walk chips; per 25m3 load	-	-	-	-	29.85	m³	29.85
Hardwood walk chips; per 70m3 load	-	-	-	-	17.15	m³	17.15
Conifer walk chips; per 25m3 load	-	-	-	-	28.85	m³	28.85
Conifer walk chips; per 70m3 load	-	-	-	-	16.45	m³	16.45
Woodfibre; per 25m3 load	-	-	-	-	28.35	m³	28.35
Woodfibre; per 70m3 load	-	-	-	-	15.95	m³	15.95
Artificial surfaces and finishes - General							
Preamble: Manufacturer's should be consulted before specifying artificial playing surfaces as usage, local site conditions and maintenance will affect the choice of material. Advice should also be sought from the Technical Unit for Sport, the appropriate regional office of the Sports Council or the National Playing Fields Association as many surfaces have not yet been fully tested in use so that final opinions on durability, value for money or permanence cannot always be formed. Note that finishes which are inexpensive to lay are often expensive to maintain. Prices for the construction of average facilities are given in the Approximate Estimates section (book only). Some of the following prices include base work whereas others are for a specialist surface only on to a base prepared and costed separately.							
Sports areas; Baylis Landscapes							
Sports tracks; polyurethane rubber surfacing; on bitumen-macadam (not included); (prices for 5500 m² minimum)							
"International"	-	-	-	-	-	m²	43.05
"Club Grade"	-	-	-	-	-	m²	25.63
Sports areas; multi-component polyurethane-rubber surfacing; on bitumen-macadam; finished with polyurethane or acrylic coat; green or red							
"Permaprene"	-	-	-	-	-	m²	60.27

Prices for Measured Works
Q PAVING/PLANTING/FENCING/SITE FURNITURE

Item Excluding site overheads and profit	PC £	Labour hours	Labour £	Plant £	Material £	Unit	Total rate £
Q26 SPECIAL SURFACINGS FOR SPORT – cont'd							
Sports areas; Exclusive Leisure "Tom Graveney Cricketweave" artificial wicket system; including proprietary sub-base (all by specialist sub-contractor)							
28 x 2.74 m	-	-	-	-	-	nr	5500.00
"Tom Graveney Cricketweave" practice batting end							
14 x 2 m	-	-	-	-	-	each	2800.00
"Tom Graveney Cricketweave" practice bowling end							
7 x 2 m	-	-	-	-	-	each	1700.00
Supply and erection of single bay Cricket cage							
18.3 x 3.65 x 3.2 m high	-	-	-	-	-	each	1700.00
Tennis courts; Duracourt Hard playing surfaces to SAPCA Code of Practice minimum requirements; laid on 65 mm thick macadam base on 150 mm thick stone foundation, to include lines, nets, posts, brick edging, 2.75 m high fence and perimeter drainage (excluding excavation, levelling or additional foundation). Based on court size 36.6 m x 18.3 m - 670 m²							
"Durapore"; 'All Weather' porous macadam and acrylic colour coating	-	-	-	-	-	m²	32.70
"Tiger Turf TT"; sand filled artificial grass	-	-	-	-	-	m²	45.10
"Tiger Turf Grand Prix"; short pile sand filled artificial grass	-	-	-	-	-	m²	47.50
"DecoColour"; impervious acrylic hardcourt (200 mm foundation)	-	-	-	-	-	m²	43.60
"DecoTurf"; cushioned impervious acrylic tournament surface (200 mm foundation)	-	-	-	-	-	m²	51.10
"Porous Kushion Kourt"; porous cushioned acrylic surface	-	-	-	-	-	m²	51.60
"Porous Klay Kourt"; synthetic clay system	-	-	-	-	-	m²	53.40
"Canada Tenn"; American green clay, fast dry surface (no macadam but including irrigation)	-	-	-	-	-	m²	53.40
Playgrounds; Baylis Landscapes; "Rubaflex" in-situ playground surfacing, porous; on prepared stone/granular base Type 1 (not included) and macadam base course (not included) Black							
15 mm thick (0.50 m critical fall height)	-	-	-	-	-	m²	24.60
35 mm thick (1.00 m critical fall height)	-	-	-	-	-	m²	37.92
60 mm thick (1.50 m critical fall height)	-	-	-	-	-	m²	50.23

Prices for Measured Works
Q PAVING/PLANTING/FENCING/SITE FURNITURE

Item Excluding site overheads and profit	PC £	Labour hours	Labour £	Plant £	Material £	Unit	Total rate £
Playgrounds; Baylis Landscapes; **"Rubaflex" in-situ playground surfacing,** **porous; on prepared stone/granular base** **Type 1 (not included) and macadam** **base course (not included)** Coloured							
15 mm thick (0.50 m critical fall height)	-	-	-	-	-	m²	58.42
35 mm thick (1.00 m critical fall height)	-	-	-	-	-	m²	63.55
60 mm thick (1.50 m critical fall height)	-	-	-	-	-	m²	71.75
Playgrounds; Wicksteed Leisure Safety tiles; on prepared base (not included)							
1000 x 1000 x 60 mm; red or green	-	-	-	-	41.00	m²	41.00
1000 x 1000 x 60 mm; black	-	-	-	-	38.00	m²	38.00
1000 x 1000 x 43 mm; red or green	-	-	-	-	38.00	m²	38.00
1000 x 1000 x 43 mm; black	-	-	-	-	35.00	m²	35.00
Playgrounds; SMP Tiles; on prepared base (not included)							
Premier 25; 1000 x 1000 x 25 mm; black; for general use	22.00	0.20	3.60	-	27.00	m²	30.60
Premier 70; 1000 x 1000 x 70 mm; black; for higher equipment	42.40	0.20	3.60	-	47.40	m²	51.00
Playgrounds; Melcourt Industries "Playbark"; on drainage layer (not included); to BS5696; minimum 300 mm settled depth							
"Playbark 8/25", 8 - 25 mm particles; red/brown	9.33	0.35	6.31	-	10.26	m²	16.58
"Playbark 10/50"; 10 - 50 mm particles; red/brown	11.20	0.35	6.31	-	12.32	m²	18.63
Playgrounds; timber edgings Timber edging boards; 50 x 50 x.750 mm timber pegs at 1000 mm centres; excavations and hardcore under edgings (not included)							
50 x 150 mm; hardwood (iroko) edge boards	8.94	0.10	1.80	-	8.94	m	10.74
38 x 150 mm; hardwood (iroko) edge boards	7.08	0.10	1.80	-	7.08	m	8.88
50 x 150 mm; treated softwood edge boards	2.34	0.10	1.80	-	2.34	m	4.14
38 x 150 mm; treated softwood edge boards	2.08	0.10	1.80	-	2.08	m	3.88

Prices for Measured Works
Q PAVING/PLANTING/FENCING/SITE FURNITURE

Item Excluding site overheads and profit	PC £	Labour hours	Labour £	Plant £	Material £	Unit	Total rate £
Q30 SEEDING/TURFING							
Seeding/turfing - General							
Preamble: The following market prices generally reflect the manufacturer's recommended retail prices. Trade and bulk discounts are often available on the prices shown. The manufacturer's of these products generally recommend application rates. Note the following rates reflect the average rate for each product.							
Market Prices of pre-seeding materials							
Scotts UK Professional							
herbicide; "Casoron G"; minimum application rate	-	-	-	-	2.19	100 m²	2.19
"Casoron G" as a selective herbicide	-	-	-	-	4.89	100 m²	4.89
herbicide; "Casoron G"; maximum application rate	-	-	-	-	8.81	100 m²	8.81
"Dextrone X"	-	-	-	-	0.49	100 m²	0.49
"Intrepid"	-	-	-	-	0.54	100 m²	0.54
"Speedway"	-	-	-	-	0.96	100 m²	0.96
turf herbicide; "Tritox", 5 litre	-	-	-	-	2.55	100 m²	2.55
turf herbicide; "Tritox", 20 litre	-	-	-	-	2.38	100 m²	2.38
Rigby Taylor Ltd							
outfield fertilizer; "Taylor's Pre-Seeding"	-	-	-	-	1.43	100 m²	1.43
Boughton Loam							
Screened topsoil 100 mm	-	-	-	-	48.51	m³	48.51
Screened Kettering loam 3 mm	-	-	-	-	68.31	m³	68.31
Top dressing; sand soil mixtures; 90/10 to 50/50	-	-	-	-	54.45	m³	54.45
Market Prices of grass maintenance materials; recommended application rates for most products vary. The rates below reflect an application rate for "normal" application							
Scotts UK Professional; Fungicides							
fungicide; "Daconil"; turf	-	-	-	-	6.72	100 m²	6.72
fungicide; "Nimrod-T"	-	-	-	-	19.56	litre	19.56
turf fungicide and wormcast control; "Turfclear"; 5 litre	-	-	-	-	1.05	100 m²	1.05
turf fungicide and wormcast control; "Turfclear"; 800 ml	-	-	-	-	24.72	800 ml	24.72
Scotts UK Professional; Turf chemicals							
growth regulator; "Shortcut"	-	-	-	-	1.59	100 m²	1.59
spray indicator; "Turf Mark"; per litre of water	-	-	-	-	0.36	litre	0.36
wetting and penetrating agent; "Aquamaster"; applied quarterly	-	-	-	-	2.33	100 m²	2.33
Scotts UK Professional; Moss killers							
mosskiller; "Enforcer"	-	-	-	-	5.57	100 m²	5.57
Scotts UK Professional; Turf fertilizers							
grass fertilizer; "Longlife Fine Turf - Spring & Summer"	-	-	-	-	4.34	100 m²	4.34
grass fertilizer; "Longlife Fine Turf - Autumn Feed"	-	-	-	-	2.52	100 m²	2.52
grass fertilizer; "Longlife Fine Turf - Finegreen NK"	-	-	-	-	4.34	100 m²	4.34
grass fertilizer; "Greenmaster - Invigorator"	-	-	-	-	2.32	100 m²	2.32
grass fertilizer; "Longlife - Renovator"	-	-	-	-	5.28	100 m²	5.28

Prices for Measured Works
Q PAVING/PLANTING/FENCING/SITE FURNITURE

Item Excluding site overheads and profit	PC £	Labour hours	Labour £	Plant £	Material £	Unit	Total rate £
grass fertilizer; slow release; "Sierraform", 18+9+18+Fe+Mn	-	-	-	-	3.81	100 m²	3.81
grass fertilizer; slow release; "Sierraform", 15+0+26	-	-	-	-	4.58	100 m²	4.58
grass fertilizer; slow release; "Sierraform", 22+5+10	-	-	-	-	3.05	100 m²	3.05
grass fertilizer; slow release; "Sierraform", 16+0+15+Fe+Mn	-	-	-	-	3.81	100 m²	3.81
grass fertilizer; water soluble; "Sierrasol", 28+5+18+TE	-	-	-	-	6.30	100 m²	6.30
grass fertilizer; water soluble; "Sierrasol", 20+5+30+TE	-	-	-	-	4.90	100 m²	4.90
grass fertilizer; controlled release; "Sierrablen", 28+5+5+Fe	-	-	-	-	4.24	100 m²	4.24
grass fertilizer; controlled release; "Sierrablen", 27+5+5+Fe	-	-	-	-	4.44	100 m²	4.44
grass fertilizer; controlled release; "Sierrablen", 15+0+22+Fe	-	-	-	-	4.24	100 m²	4.24
grass fertilizer; controlled release; "Sierrablen Fine", 38+0+0	-	-	-	-	1.61	100 m²	1.61
grass fertilizer; controlled release; "Sierrablen Fine", 25+5+12	-	-	-	-	2.68	100 m²	2.68
grass fertilizer; controlled release; "Sierrablen Fine", 21+0+20	-	-	-	-	2.68	100 m²	2.68
grass fertilizer; controlled release; "Sierrablen Fine", 15+0+29	-	-	-	-	2.68	100 m²	2.68
grass fertilizer; controlled release; "Sierrablen Mini", 22+5+10	-	-	-	-	8.44	100 m²	8.44
grass fertilizer; controlled release; "Sierrablen Mini", 0+0+45	-	-	-	-	4.21	100 m²	4.21
grass fertilizer; "Greenmaster Turf Tonic"	-	-	-	-	2.15	100 m²	2.15
grass fertilizer; "Greenmaster Spring & Summer"	-	-	-	-	3.08	100 m²	3.08
grass fertilizer; "Greenmaster Zero Phosphate"	-	-	-	-	3.08	100 m²	3.08
grass fertilizer; "Greenmaster Mosskiller"	-	-	-	-	3.42	100 m²	3.42
grass fertilizer; "Greenmaster Extra"	-	-	-	-	3.89	100 m²	3.89
grass fertilizer; "Greenmaster Autumn"	-	-	-	-	3.31	100 m²	3.31
grass fertilizer; "Greenmaster Double K"	-	-	-	-	3.31	100 m²	3.31
grass fertilizer; "Greenmaster NK"	-	-	-	-	3.31	100 m²	3.31
grass & soil fertilizer; "Greenmaster Liquid Seafeed"	-	-	-	-	0.68	100 m²	0.68
outfield turf fertilizer; "Sportsmaster PS3"	-	-	-	-	1.39	100 m²	1.39
outfield turf fertilizer; "Sportsmaster PS4"	-	-	-	-	1.39	100 m²	1.39
outfield turf fertilizer; "Sportsmaster PS5"	-	-	-	-	1.78	100 m²	1.78
outfield turf fertilizer; "Sportsmaster Fairway"	-	-	-	-	3.17	100 m²	3.17
outfield turf fertilizer; "Sportsmaster Municipality"	-	-	-	-	4.14	100 m²	4.14
"TPMC"	-	-	-	-	4.28	80 lt	4.28

Prices for Measured Works
Q PAVING/PLANTING/FENCING/SITE FURNITURE

Item Excluding site overheads and profit	PC £	Labour hours	Labour £	Plant £	Material £	Unit	Total rate £
Q30 SEEDING/TURFING – cont'd							
Rigby Taylor Ltd; 35 g/m²							
grass fertilizer; "Mascot Microfine 20-0-15" + 2% Mg	-	-	-	-	7.00	100 m²	7.00
grass fertilizer; "Mascot Microfine 15-5-15" +2% Mg	-	-	-	-	5.02	100 m²	5.02
grass fertilizer; "Mascot Microfine 18-0-0" + 6% Fe	-	-	-	-	6.11	100 m²	6.11
grass fertilizer; "Mascot Microfine 14-4-7" + 2% Mg	-	-	-	-	4.47	100 m²	4.47
grass fertilizer; "Mascot Microfine 8-0-0" + 4% Fe + 2% Mg	-	-	-	-	3.21	100 m²	3.21
grass fertilizer; "Mascot Microfine 4-0-8" + 6% Fe + 2% Mg	-	-	-	-	4.97	100 m²	4.97
grass fertilizer; "Taylor's Organic SS2"	-	-	-	-	2.88	100 m²	2.88
grass fertilizer; "Taylor's Mini-Granular 11-2-7"	-	-	-	-	3.05	100 m²	3.05
grass fertilizer/weedkiller; "Taylor's Weed & Feed"	-	-	-	-	2.63	100 m²	2.63
grass fertilizer/mosskiller; "Taylor's Lawn Sand"	-	-	-	-	1.43	100 m²	1.43
outfield fertilizer; "Taylor's Spring/Summer Outfield 9-7-7"	-	-	-	-	2.09	100 m²	2.09
outfield fertilizer; "Taylor's Spring/Summer Outfield 7-7-7"	-	-	-	-	1.73	100 m²	1.73
outfield fertilizer; "Taylor's Spring/Summer Outfield 20-10-10"	-	-	-	-	2.44	100 m²	2.44
outfield fertilizer; "Taylor's Autumn/Winter Outfield"	-	-	-	-	2.03	100 m²	2.03
outfield fertilizer; "Taylor's Spring/Summer Sportsfield"	-	-	-	-	2.44	100 m²	2.44
liquid fertilizer; "Vitax 50/50 Fine Turf"; 56 ml/100m2	-	-	-	-	13.99	100 m²	13.99
liquid fertilizer; "Vitax 50/50 Fine Turf Special"; 56 ml/100m2	-	-	-	-	11.97	100 m²	11.97
liquid fertilizer; "Vitax 50/50 Autumn & Winter"; 56 ml/100m2	-	-	-	-	12.84	100 m²	12.84
Cultivation							
Treating soil with "Paraquat-Diquat" weedkiller at rate of 5 litre/ha; PC £22.20 per litre; in accordance with manufacturer's instructions; including all safety precautions							
by machine	1.11	-	-	0.24	1.11	100 m²	1.35
by hand	1.11	0.13	2.40	-	1.11	100 m²	3.51
Ripping up subsoil; using approved subsoiling machine; minimum depth 250 mm below topsoil; at 1.20 m centres; in							
gravel or sandy clay	-	-	-	2.31	-	100m2	2.31
soil compacted by machines	-	-	-	2.70	-	100m2	2.70
clay	-	-	-	2.89	-	100m2	2.89
chalk or other soft rock	-	-	-	5.78	-	100m2	5.78
Extra for subsoiling at 1 m centres	-	-	-	0.58	-	100m2	0.58
Breaking up existing ground; using pedestrian operated tine cultivator or rotavator							
100 mm deep	-	0.22	3.96	2.24	-	100m2	6.20
150 mm deep	-	0.28	4.95	2.80	-	100m2	7.75
200 mm deep	-	0.37	6.60	3.73	-	100m2	10.33

Prices for Measured Works
Q PAVING/PLANTING/FENCING/SITE FURNITURE

Item Excluding site overheads and profit	PC £	Labour hours	Labour £	Plant £	Material £	Unit	Total rate £
As above but in heavy clay or wet soils							
100 mm deep	-	0.44	7.92	4.48	-	100m2	12.40
150 mm deep	-	0.66	11.88	6.72	-	100m2	18.60
200 mm deep	-	0.83	14.85	8.40	-	100m2	23.25
Breaking up existing ground; using tractor drawn tine cultivator or rotavator single pass							
100 mm deep	-	-	-	0.52	-	100 m²	0.52
150 mm deep	-	-	-	0.65	-	100 m²	0.65
200 mm deep	-	-	-	0.86	-	100 m²	0.86
600 mm deep	-	-	-	2.59	-	100 m²	2.59
Cultivating ploughed ground; using disc, drag, or chain harrow							
4 passes	-	-	-	3.10	-	100 m²	3.10
Rolling cultivated ground lightly; using self-propelled agricultural roller	-	0.06	1.00	0.58	-	100 m²	1.58
Importing and storing selected and approved topsoil; to BS 3882; from source not exceeding 13 km from site; inclusive of settlement							
small quantities, less than 15 m³	26.00	-	-	-	31.20	m³	31.20
over 15 m³	17.32	-	-	-	20.78	m³	20.78
Spreading and lightly consolidating approved topsoil (imported or from spoil heaps); in layers not exceeding 150 mm; travel distance from spoil heaps not exceeding 100 m; by machine (imported topsoil not included)							
minimum depth 100 mm	-	1.55	27.90	35.54	-	100 m²	63.44
minimum depth 150 mm	-	2.33	42.00	53.49	-	100 m²	95.49
minimum depth 300 mm	-	4.67	84.00	106.97	-	100 m²	190.97
minimum depth 450 mm	-	6.99	125.82	160.29	-	100 m²	286.11
Spreading and lightly consolidating approved topsoil (imported or from spoil heaps); in layers not exceeding 150 mm; travel distance from spoil heaps not exceeding 100 m; by hand (imported topsoil not included)							
minimum depth 100 mm	-	20.00	360.07	-	-	100 m²	360.07
minimum depth 150 mm	-	30.01	540.11	-	-	100 m²	540.11
minimum depth 300 mm	-	60.01	1080.22	-	-	100 m²	1080.22
minimum depth 450 mm	-	90.02	1620.32	-	-	100 m²	1620.32
Extra over for spreading topsoil to slopes 15 - 30 degrees by machine or hand	-	-	-	-	-	10%	-
Extra over for spreading topsoil to slopes over 30 degrees by machine or hand	-	-	-	-	-	25%	-
Extra over for spreading topsoil from spoil heaps travel exceeding 100 m; by machine							
100 - 150 m	-	0.01	0.22	0.09	-	m³	0.31
150 - 200 m	-	0.02	0.33	0.13	-	m³	0.46
200 - 300 m	-	0.03	0.50	0.19	-	m³	0.69
Extra over spreading topsoil for travel exceeding 100 m; by hand							
100 m	-	0.83	15.00	-	-	m³	15.00
200 m	-	1.67	30.00	-	-	m³	30.00
300 m	-	2.50	45.00	-	-	m³	45.00
Evenly grading; to general surfaces to bring to finished levels							
by machine (tractor mounted rotavator)	-	-	-	0.03	-	m²	0.03
by pedestrian operated rotavator	-	-	0.07	0.04	-	m²	0.12
by hand	-	0.01	0.18	-	-	m²	0.18

Prices for Measured Works
Q PAVING/PLANTING/FENCING/SITE FURNITURE

Item Excluding site overheads and profit	PC £	Labour hours	Labour £	Plant £	Material £	Unit	Total rate £
Q30 SEEDING/TURFING – cont'd							
Extra over grading for slopes 15 - 30 degrees by machine or hand	-	-	-	-	-	10%	-
Extra over grading for slopes over 30 degrees by machine or hand	-	-	-	-	-	25%	-
Apply screened topdressing to grass surfaces. Spread using Tru-Lute							
Sand soil mixes 90/10 to 50/50	0.08	-	0.04	0.03	0.08	m²	**0.15**
Spread only existing cultivated soil to final levels using Tru-Lute							
Cultivated soil	-	-	0.04	0.03	-	m²	**0.06**
Clearing stones; disposing off site; to distance not exceeding 13 km							
by hand; stones not exceeding 50 mm in any direction; loading to skip 5.35 m³	-	0.01	0.18	0.03	-	m²	**0.21**
by mechanical stone rake; stones not exceeding 50 mm in any direction; loading to 15 m³ truck by mechanical loader	-	-	0.04	0.08	-	m²	**0.11**
Lightly cultivating; weeding; to fallow areas; disposing debris off site; to distance not exceeding 13 km							
by hand	-	0.01	0.26	-	0.06	m²	**0.32**
Surface applications; soil additives; pre-seeding; material delivered to a maximum of 25 m from area of application; applied; by machine							
Soil conditioners; to cultivated ground; ground limestone; PC £9.30/tonne; including turning in							
0.25 kg/m² = 2.50 tonnes/ha	0.23	-	-	4.03	0.23	100 m²	**4.26**
0.50 kg/m² = 5.00 tonnes/ha	0.47	-	-	4.03	0.47	100 m²	**4.49**
0.75 kg/m² = 7.50 tonnes/ha	0.70	-	-	4.03	0.70	100 m²	**4.73**
1.00 kg/m² = 10.00 tonnes/ha	0.93	-	-	4.03	0.93	100 m²	**4.96**
Soil conditioners; to cultivated ground; medium bark; based on deliveries of 15 m³ loads; PC £32.00/m³; including turning in							
1 m³ per 40 m² = 25 mm thick	0.80	-	-	0.09	0.80	m²	**0.89**
1 m³ per 20 m² = 50 mm thick	-	-	-	0.14	1.60	m²	**1.74**
1 m³ per 13.33 m² = 75 mm thick	-	-	-	0.19	2.40	m²	**2.59**
1 m³ per 10 m² = 100 mm thick	3.20	-	-	0.25	3.20	m²	**3.45**
Soil conditioners; to cultivated ground; peat loose 55 m³ loads; PC £28.00/m³; including turning in							
1 m³ per 40 m² = 25 mm thick	0.70	-	-	0.09	0.70	m²	**0.79**
1 m³ per 20 m² = 50 mm thick	1.40	-	-	0.14	1.40	m²	**1.54**
1 m³ per 13.33 m² = 75 mm thick	2.10	-	-	0.19	2.10	m²	**2.29**
1 m³ per 10 m² = 100 mm thick	2.80	-	-	0.25	2.80	m²	**3.05**
Soil conditioners; to cultivated ground; mushroom compost; delivered in 20 m³ loads; PC £8.00/m³; including turning in							
1 m³ per 40 m² = 25 mm thick	0.45	0.02	0.30	-	0.45	m²	**0.75**
1 m³ per 40 m² = 25 mm thick	0.45	0.02	0.30	-	0.45	m²	**0.75**
1 m³ per 20 m² = 50 mm thick	0.90	0.03	0.55	-	0.90	m²	**1.45**
1 m³ per 13.33 m² = 75 mm thick	1.34	0.04	0.72	-	1.34	m²	**2.06**
1 m³ per 10 m² = 100 mm thick	1.79	0.05	0.90	-	1.79	m²	**2.69**
Soil conditioners; to cultivated ground; mushroom compost; delivered in 55 m³ loads; PC £4.00/m³; including turning in							

Prices for Measured Works
Q PAVING/PLANTING/FENCING/SITE FURNITURE

Item Excluding site overheads and profit	PC £	Labour hours	Labour £	Plant £	Material £	Unit	Total rate £
1 m³ per 40 m² = 25 mm thick	0.10	0.02	0.30	-	0.10	m²	0.40
1 m³ per 20 m² = 50 mm thick	0.20	0.03	0.55	-	0.20	m²	0.75
1 m³ per 13.33 m² = 75 mm thick	0.30	0.04	0.72	-	0.30	m²	1.02
1 m³ per 10 m² = 100 mm thick	0.40	0.05	0.90	-	0.40	m²	1.30
Surface applications and soil additives; pre-seeding; material delivered to a maximum of 25 m from area of application; applied; by hand							
Soil conditioners; to cultivated ground; ground limestone; PC £9.30/tonne; including turning in							
0.25 kg/m² = 2.50 tonnes/ha	0.23	1.20	21.60	-	0.23	100 m²	21.83
0.50 kg/m² = 5.00 tonnes/ha	0.47	1.33	24.00	-	0.47	100 m²	24.46
0.75 kg/m² = 7.50 tonnes/ha	0.70	1.50	27.00	-	0.70	100 m²	27.70
1.00 kg/m² = 10.00 tonnes/ha	0.93	1.71	30.86	-	0.93	100 m²	31.79
Soil conditioners; to cultivated ground; medium bark; based on deliveries of 15 m³ loads; PC £32.00/ m³; including turning in							
1 m³ per 40 m² = 25 mm thick	0.80	0.02	0.40	-	0.80	m²	1.20
1 m³ per 20 m² = 50 mm thick	1.60	0.04	0.80	-	1.60	m²	2.40
1 m³ per 13.33 m² = 75 mm thick	2.40	0.07	1.20	-	2.40	m²	3.60
1 m³ per 10 m² = 100 mm thick	3.20	0.08	1.44	-	3.20	m²	4.64
Soil conditioners; to cultivated ground; peat loose 55 m³ loads; PC £28.00 /m³; including turning in							
1 m³ per 40 m² = 25 mm thick	0.70	0.02	0.40	-	0.70	m²	1.10
1 m³ per 20 m² = 50 mm thick	1.40	0.04	0.80	-	1.40	m²	2.20
1 m³ per 13.33 m² = 75 mm thick	2.10	0.07	1.20	-	2.10	m²	3.30
1 m³ per 10 m² = 100 mm thick	2.80	0.08	1.44	-	2.80	m²	4.24
Soil conditioners; to cultivated ground; mushroom compost; delivered in 25 m³ loads; PC £17.92/m³; including turning in							
1 m³ per 40 m² = 25 mm thick	0.45	0.02	0.40	-	0.45	m²	0.85
1 m³ per 20 m² = 50 mm thick	0.90	0.04	0.80	-	0.90	m²	1.69
1 m³ per 13.33 m² = 75 mm thick	1.34	0.07	1.20	-	1.34	m²	2.54
1 m³ per 10 m² = 100 mm thick	1.79	0.08	1.44	-	1.79	m²	3.23
Soil conditioners; to cultivated ground; mushroom compost; delivered in 55 m³ loads; PC £4.00/m³; including turning in							
1 m³ per 40 m² = 25 mm thick	0.10	0.02	0.40	-	0.10	m²	0.50
1 m³ per 20 m² = 50 mm thick	0.20	0.04	0.80	-	0.20	m²	1.00
1 m³ per 13.33 m² = 75 mm thick	0.30	0.07	1.20	-	0.30	m²	1.50
1 m³ per 10 m² = 100 mm thick	0.40	0.08	1.44	-	0.40	m²	1.84
Preparation of seedbeds - General							
Preamble: For preliminary operations see "Cultivation" section.							
Preparation of seedbeds; soil preparation							
Lifting selected and approved topsoil from spoil heaps; passing through 6 mm screen; removing debris	-	0.08	1.50	3.17	0.01	m³	4.68
Topsoil; supply only; PC £17.32 /m³; allowing for 20% settlement							
25 mm	0.43	-	-	-	0.52	m²	0.52
50 mm	0.87	-	-	-	1.04	m²	1.04
100 mm	1.73	-	-	-	2.08	m²	2.08
150 mm	2.60	-	-	-	3.12	m²	3.12

Prices for Measured Works
Q PAVING/PLANTING/FENCING/SITE FURNITURE

Item Excluding site overheads and profit	PC £	Labour hours	Labour £	Plant £	Material £	Unit	Total rate £
Q30 SEEDING/TURFING – cont'd							
Preparation of seedbeds; soil Preparation – cont'd							
200 mm	3.46	-	-	-	4.16	m²	**4.16**
250 mm	4.33	-	-	-	5.20	m²	**5.20**
300 mm	5.20	-	-	-	6.24	m²	**6.24**
400 mm	6.93	-	-	-	8.31	m²	**8.31**
450 mm	7.79	-	-	-	9.35	m²	**9.35**
Spreading topsoil to form seedbeds (topsoil not included); by machine							
25 mm deep	-	-	0.05	0.10	-	m²	**0.14**
50 mm deep	-	-	0.06	0.12	-	m²	**0.18**
75 mm deep	-	-	0.07	0.14	-	m²	**0.21**
100 mm deep	-	0.01	0.09	0.19	-	m²	**0.28**
150 mm deep	-	0.01	0.14	0.29	-	m²	**0.42**
Spreading only topsoil to form seedbeds (topsoil not included); by hand							
25 mm deep	-	0.03	0.45	-	-	m²	**0.45**
50 mm deep	-	0.03	0.60	-	-	m²	**0.60**
75 mm deep	-	0.04	0.77	-	-	m²	**0.77**
100 mm deep	-	0.05	0.90	-	-	m²	**0.90**
150 mm deep	-	0.08	1.35	-	-	m²	**1.35**
Bringing existing topsoil to a fine tilth for seeding; by raking or harrowing; stones not to exceed 6 mm; by machine	-	-	0.07	0.04	-	m²	**0.12**
Bringing existing topsoil to a fine tilth for seeding; by raking or harrowing; stones not to exceed 6 mm; by hand	-	0.01	0.16	-	-	m²	**0.16**
Preparation of seedbeds; soil treatments							
For the following topsoil improvement and seeding operations add or subtract the following amounts for every £0.10 difference in the material cost price							
35 g/m²	-	-	-	-	0.35	100 m²	**0.35**
50 g/m²	-	-	-	-	0.50	100 m²	**0.50**
70 g/m²	-	-	-	-	0.70	100 m²	**0.70**
100 g/m²	-	-	-	-	1.00	100 m²	**1.00**
125 kg/ ha	-	-	-	-	12.50	ha	**12.50**
150 kg/ ha	-	-	-	-	15.00	ha	**15.00**
175 kg/ ha	-	-	-	-	17.50	ha	**17.50**
200 kg/ ha	-	-	-	-	20.00	ha	**20.00**
225 kg/ ha	-	-	-	-	22.50	ha	**22.50**
250 kg/ ha	-	-	-	-	25.00	ha	**25.00**
300 kg/ ha	-	-	-	-	30.00	ha	**30.00**
350 kg/ ha	-	-	-	-	35.00	ha	**35.00**
400 kg/ ha	-	-	-	-	40.00	ha	**40.00**
500 kg/ ha	-	-	-	-	50.00	ha	**50.00**
700 kg/ ha	-	-	-	-	70.00	ha	**70.00**
1000 kg/ ha	-	-	-	-	100.00	ha	**100.00**
1250 kg/ ha	-	-	-	-	125.00	ha	**125.00**

Prices for Measured Works
Q PAVING/PLANTING/FENCING/SITE FURNITURE

Item Excluding site overheads and profit	PC £	Labour hours	Labour £	Plant £	Material £	Unit	Total rate £
Pre-seeding fertilizers (6:9:6); PC £0.34/kg; to seedbeds; by machine							
35 g/m²	1.19	-	-	0.18	1.19	100 m²	**1.37**
50 g/m²	1.70	-	-	0.18	1.70	100 m²	**1.88**
70 g/m²	2.37	-	-	0.18	2.37	100 m²	**2.56**
100 g/m²	3.39	-	-	0.18	3.39	100 m²	**3.57**
125 g/m²	4.24	-	-	0.18	4.24	100 m²	**4.42**
125 kg/ha	42.40	-	-	18.05	42.40	ha	**60.45**
250 kg/ha	84.80	-	-	18.05	84.80	ha	**102.85**
300 kg/ha	101.76	-	-	18.05	101.76	ha	**119.81**
350 kg/ha	118.72	-	-	18.05	118.72	ha	**136.77**
400 kg/ha	135.68	-	-	18.05	135.68	ha	**153.73**
500 kg/ha	169.60	-	-	28.89	169.60	ha	**198.49**
700 kg/ha	237.44	-	-	28.89	237.44	ha	**266.33**
1250 kg/ha	424.00	-	-	48.14	424.00	ha	**472.14**
Pre-seeding fertilizers (6:9:6); PC £0.34/kg; to seedbeds; by hand							
35 g/m²	1.19	0.17	3.00	-	1.19	100 m²	**4.19**
50 g/m²	1.70	0.17	3.00	-	1.70	100 m²	**4.70**
70 g/m²	2.37	0.17	3.00	-	2.37	100 m²	**5.38**
100 g/m²	3.39	0.20	3.60	-	3.39	100 m²	**6.99**
125 g/m²	4.24	0.20	3.60	-	4.24	100 m²	**7.84**
Pre-seeding pesticides; to seedbeds; in accordance with manufacturer's instructions; by machine							
35 g/m²	16.45	-	-	0.18	16.45	100 m²	**16.63**
50 g/m²	23.50	-	-	0.18	23.50	100 m²	**23.68**
70 g/m²	32.90	-	-	0.18	32.90	100 m²	**33.08**
100 g/m²	47.00	-	-	0.18	47.00	100 m²	**47.18**
125 g/m²	58.75	-	-	0.18	58.75	100 m²	**58.93**
125 kg/ha	587.50	-	-	18.05	587.50	ha	**605.55**
250 kg/ha	1175.00	-	-	18.05	1175.00	ha	**1193.05**
300 kg/ha	1410.00	-	-	18.05	1410.00	ha	**1428.05**
350 kg/ha	1645.00	-	-	18.05	1645.00	ha	**1663.05**
400 kg/ha	1880.00	-	-	18.05	1880.00	ha	**1898.05**
500 kg/ha	2350.00	-	-	28.89	2350.00	ha	**2378.89**
700 kg/ha	3290.00	-	-	28.89	3290.00	ha	**3318.89**
1250 kg/ha	5875.00	-	-	18.05	5875.00	ha	**5893.05**
Pre-seeding pesticides; to seedbeds; in accordance with manufacturer's instructions; by hand							
35 g/m²	16.45	0.17	3.00	-	16.45	100 m²	**19.45**
50 g/m²	23.50	0.17	3.00	-	23.50	100 m²	**26.50**
70 g/m²	32.90	0.17	3.00	-	32.90	100 m²	**35.90**
100 g/m²	47.00	0.20	3.60	-	47.00	100 m²	**50.60**
125 g/m²	58.75	0.20	3.60	-	58.75	100 m²	**62.35**
Pre-emergent weedkillers; in accordance with manufacturer's instructions; including all safety precautions; by machine (coverage 0.80 ha /hr)							
15 ml	1.30	-	-	0.18	1.30	100 m²	**1.48**
20 ml	1.74	-	-	0.18	1.74	100 m²	**1.92**
25 ml	2.17	-	-	0.18	2.17	100 m²	**2.35**
50 ml	4.34	-	-	0.18	4.34	100 m²	**4.52**

Prices for Measured Works
Q PAVING/PLANTING/FENCING/SITE FURNITURE

Item Excluding site overheads and profit	PC £	Labour hours	Labour £	Plant £	Material £	Unit	Total rate £
Q30 SEEDING/TURFING – cont'd							
Preparation of seedbeds; soil treatments - cont'd Pre-emergent liquid applied weedkillers; in accordance with manufacturer's instructions; including all safety precautions; by hand							
10 ml	0.04	0.50	9.00	-	0.04	100 m²	9.04
12 ml	0.05	0.50	9.00	-	0.05	100 m²	9.05
15 ml	0.06	0.50	9.00	-	0.06	100 m²	9.06
20 ml	0.08	0.50	9.00	-	0.08	100 m²	9.08
25 ml	0.10	0.50	9.00	-	0.10	100 m²	9.10
30 ml	0.12	0.50	9.00	-	0.12	100 m²	9.12
35 ml	0.14	0.50	9.00	-	0.14	100 m²	9.14
40 ml	0.16	0.50	9.00	-	0.16	100 m²	9.16
45 ml	0.18	0.50	9.00	-	0.18	100 m²	9.18
50 ml	0.20	0.50	9.00	-	0.20	100 m²	9.20
Lime granules; on seedbeds; by machine							
35 g/m²	0.74	-	-	0.18	0.74	100 m²	0.92
50 g/m²	1.05	-	-	0.18	1.05	100 m²	1.23
70 g/m²	1.47	-	-	0.18	1.47	100 m²	1.65
100 g/m²	2.10	-	-	0.18	2.10	100 m²	2.28
125 g/m²	2.63	-	-	0.18	2.63	100 m²	2.81
125 kg/ha	26.25	-	-	18.05	26.25	ha	44.30
250 kg/ha	52.50	-	-	18.05	52.50	ha	70.55
300 kg/ha	63.00	-	-	18.05	63.00	ha	81.05
350 kg/ha	73.50	-	-	18.05	73.50	ha	91.55
400 kg/ha	84.00	-	-	18.05	84.00	ha	102.05
500 kg/ha	105.00	-	-	28.89	105.00	ha	133.89
700 kg/ha	147.00	-	-	72.22	147.00	ha	219.21
1250 kg/ha	262.50	-	-	72.22	262.50	ha	334.72
Lime granules; on seedbeds; by hand							
35 g/m²	0.74	0.17	3.00	-	0.74	100 m²	3.74
50 g/m²	1.05	0.17	3.00	-	1.05	100 m²	4.05
70 g/m²	1.47	0.17	3.00	-	1.47	100 m²	4.47
100 g/m²	2.10	0.17	3.00	-	2.10	100 m²	5.10
125 g/m²	2.63	0.17	3.00	-	2.63	100 m²	5.63

Seeding grass areas - General
Preamble: The British Standard recommendations for seed and seeding of grass areas are contained in BS 4428: 1989. The prices given in this section are based on compliance with the standard.

Market Prices of grass seed
Preamble: The prices shown are for supply only at one number 25 kg bag purchase price unless otherwise stated. Rates shown are based on the manufacturer's maximum recommendation for each seed type. Trade and bulk discounts are often available on the prices shown for quantities of more than one bag.

Prices for Measured Works
Q PAVING/PLANTING/FENCING/SITE FURNITURE

Item Excluding site overheads and profit	PC £	Labour hours	Labour £	Plant £	Material £	Unit	Total rate £
Bowling greens; fine lawns; ornamental turf; croquet lawns							
"British Seed Houses"; ref A1Greens; 35g/m² (New Formulation)	-	-	-	-	17.19	100m2	17.19
"Johnsons Seeds"; ref JL 21; 34 - 50g/m²	-	-	-	-	10.00	100m2	10.00
"Johnsons Seeds"; ref J1; 34 - 50g/m²	-	-	-	-	16.60	100m2	16.60
"Johnsons Seeds"; ref J2; 34 - 50g/m²	-	-	-	-	20.40	100m2	20.40
"Perryfields"; ref Pro 20; 35 - 50g/m²	-	-	-	-	14.80	100m2	14.80
"Perryfields"; ref Pro 10; 35 - 50g/m²	-	-	-	-	19.40	100m2	19.40
Tennis courts; cricket squares							
"British Seed Houses"; ref A2 Lawn Tennis; 35g/m²	-	-	-	-	12.32	100m2	12.32
"British Seed Houses"; ref A5 Cricket Square; 35g/m²	-	-	-	-	15.30	100m2	15.30
"Johnsons Seeds"; ref JL 21; 34 - 50g/m²	-	-	-	-	10.00	100m2	10.00
"Johnsons Seeds"; ref J2; 34 - 50g/m²	-	-	-	-	20.40	100m2	20.40
"Johnsons Seeds"; ref Taskmaster; 18 - 25g/m²	-	-	-	-	5.33	100m2	5.33
"Perryfields"; ref Pro 35; 35g/m²	-	-	-	-	10.08	100m2	10.08
Amenity grassed areas; general purpose lawns							
"British Seed Houses"; ref A3 Landscape; 25 - 50g/m²	-	-	-	-	13.70	100m2	13.70
"Johnsons Seeds"; ref J5; 18 - 25g/m²	-	-	-	-	7.20	100m2	7.20
"Johnsons Seeds"; ref Taskmaster; 18 - 25gm2	-	-	-	-	5.33	100m2	5.33
"Perryfields"; ref Pro 50; 25 - 35g/m²	-	-	-	-	11.34	100m2	11.34
"Perryfields"; ref Pro 120; 25 - 35g/m²	-	-	-	-	13.02	100m2	13.02
Conservation; country parks; slopes and banks							
"British Seed Houses"; ref A4 Parkland; 17 - 35g/m²	-	-	-	-	10.01	100m2	10.01
"British Seed Houses"; ref A16 Country Park; 8 - 19g/m²	-	-	-	-	6.76	100m2	6.76
"British Seed Houses"; ref A17 (legume and clover); 2g/m²	-	-	-	-	13.00	100m2	13.00
"Johnsons Seeds"; ref J8 Slowgrow; 8 - 30g/m²	-	-	-	-	8.04	100m2	8.04
"Perryfields"; ref Pro 30; 25 - 35g/m²; up to 50g/m² for banks	-	-	-	-	9.24	100m2	9.24
"Perryfields"; ref Pro 30; up to 50g/m² for banks	-	-	-	-	13.20	100m2	13.20
Shaded areas							
"British Seed Houses"; ref A6 Supra Shade; 50g/m²	-	-	-	-	21.10	100m2	21.10
"Johnsons Seeds"; ref JL 21; 34 - 50g/m²	-	-	-	-	10.00	100m2	10.00
"Johnsons Seeds"; ref J1; 34 - 50g/m²	-	-	-	-	16.60	100m2	16.60
"Perryfields"; ref Pro 60; 35 - 50g/m²	-	-	-	-	16.80	100m2	16.80
Sports pitches; rugby; soccer pitches							
"British Seed Houses"; ref A7 Sports Ground; 20g/m²	-	-	-	-	5.76	100m2	5.76
"Johnson Seeds"; ref JL 24; 18 - 30g/m²	-	-	-	-	5.10	100m2	5.10
"Johnson Seeds"; ref J7; 18 - 25g/m²	-	-	-	-	4.88	100m2	4.88
"Johnson Seeds"; ref Sportsmaster (winter wearing); 18 - 30g/m²	-	-	-	-	6.60	100m2	6.60
"Perryfields"; ref Pro 70; 15 - 35g/m²	-	-	-	-	9.66	100m2	9.66
"Perryfields"; ref Pro 75; 15 - 35g/m²	-	-	-	-	13.30	100m2	13.30
"Perryfields"; ref Pro 80; 17 - 35g/m²	-	-	-	-	10.36	100m2	10.36
Outfields							
"British Seed Houses"; ref A7 Sports Ground; 20g/m²	-	-	-	-	5.76	100m2	5.76
"British Seed Houses"; ref A9 Universal; 17 - 35g/m²	-	-	-	-	9.10	100m2	9.10

Prices for Measured Works
Q PAVING/PLANTING/FENCING/SITE FURNITURE

Item Excluding site overheads and profit	PC £	Labour hours	Labour £	Plant £	Material £	Unit	Total rate £
Q30 SEEDING/TURFING – cont'd							
Market Prices of grass seed – cont'd							
"Johnsons Seeds"; ref J4; 12 - 25g/m²	-	-	-	-	5.15	100m2	5.15
"Perryfields"; ref Pro 40; 35g/m²	-	-	-	-	9.94	100m2	9.94
"Perryfields"; ref Pro 70; 15 - 35g/m²	-	-	-	-	9.66	100m2	9.66
Hockey pitches							
"Johnsons Seeds"; ref J4; 12 - 25g/m²	-	-	-	-	5.15	100m2	5.15
"Perryfields"; ref Pro 70; 15 - 35g/m²	-	-	-	-	9.66	100m2	9.66
Parks							
"British Seed Houses"; ref A7; Sportsground 20g/m²	-	-	-	-	5.76	100m2	5.76
"British Seed Houses"; ref A9 Universal; 17 - 35g/m²	-	-	-	-	9.10	100m2	9.10
"Johnson Seeds"; ref JL 24; 18 - 30g/m²	-	-	-	-	5.10	100m2	5.10
"Perryfields"; ref Pro 120; 25 - 35g/m²	-	-	-	-	13.02	100m2	13.02
Informal playing fields							
"Johnsons seeds"; ref J5; 18 - 25g/m²	-	-	-	-	7.20	100m2	7.20
"Perryfields"; ref Pro 45; 35g/m²	-	-	-	-	10.36	100m2	10.36
Caravan sites							
"British Seed Houses"; ref A9 Universal; 17 - 35g/m²	-	-	-	-	9.10	100m2	9.10
"Johnson Seeds"; ref JL 22; 18 - 30g/m²	-	-	-	-	7.02	100m2	7.02
Sports pitch re-seeding and repair							
"British Seed Houses"; ref A8 Supra Sport; 20 - 35g/m²	-	-	-	-	13.79	100m2	13.79
"British Seed Houses"; ref A20 Ryesport; 20 - 35g/m²	-	-	-	-	8.61	100m2	8.61
"Johnson Seeds"; ref JL 24; 18 - 30g/m²	-	-	-	-	5.10	100m2	5.10
"Johnson Seeds"; ref Sportsmaster (winter wearing); 18 - 30g/m²	-	-	-	-	6.60	100m2	6.60
"Perryfields"; ref Pro 80; 17 - 35g/m²	-	-	-	-	10.36	100m2	10.36
"Perryfields"; ref Pro 81; 17 - 35g/m²	-	-	-	-	11.34	100m2	11.34
Racecourses; gallops; polo grounds; horse rides							
"British Seed Houses"; ref A14; 25 - 30g/m²	-	-	-	-	7.80	100m2	7.80
"Johnsons Seeds"; ref Taskmaster; 18 - 25g/m²	-	-	-	-	5.33	100m2	5.33
"Perryfields"; ref Pro 65; 17 - 35g/m²	-	-	-	-	13.44	100m2	13.44
Motorway and road verges							
"British Seed Houses"; ref A18; 6 - 15g/m²	-	-	-	-	4.50	100m2	4.50
"Johnsons Seeds"; ref J8; 18 - 30g/m²	-	-	-	-	8.04	100m2	8.04
"Perryfields"; ref Pro 120; 25 - 35g/m²	-	-	-	-	13.02	100m2	13.02
"Perryfields"; ref Pro 85 (DOT Official Mix); 10g/m²	-	-	-	-	2.84	100m2	2.84
Golf courses; tees							
"British Seed Houses"; ref A10, 35 - 50g/m²	-	-	-	-	17.70	100m2	17.70
"Johnsons Seeds"; ref J3; 18 - 30g/m²	-	-	-	-	8.64	100m2	8.64
"Johnsons Seeds"; ref Taskmaster; 18 - 25g/m²	-	-	-	-	5.33	100m2	5.33
"Perryfields"; ref Pro 40; 35g/m²	-	-	-	-	9.94	100m2	9.94
"Perryfields"; ref Pro 45; 35g/m²	-	-	-	-	10.36	100m2	10.36
Golf courses; greens							
"British Seed Houses"; ref A11; 35g/m²	-	-	-	-	17.99	100m2	17.99
"British Seed Houses"; ref A13 (for overseeding); 8g/m²	-	-	-	-	11.80	100m2	11.80
"Johnsons Seeds"; ref J1; 34 - 50g/m²	-	-	-	-	16.60	100m2	16.60
"Johnsons Seeds"; ref Greensmaster (for overseeding); 8g/m²	-	-	-	-	6.36	100m2	6.36
"Perryfields"; ref Pro 5; 35 - 50g/m²	-	-	-	-	14.00	100m2	14.00

Prices for Measured Works
Q PAVING/PLANTING/FENCING/SITE FURNITURE

Item Excluding site overheads and profit	PC £	Labour hours	Labour £	Plant £	Material £	Unit	Total rate £
Golf courses; fairways							
"British Seed Houses"; ref A12; 15 - 25g/m²	-	-	-	-	7.55	100m2	7.55
"Johnsons Seeds"; ref J3; 18 - 30g/m²	-	-	-	-	8.64	100m2	8.64
"Johnsons Seeds"; ref J5; 18 - 30g/m²	-	-	-	-	8.64	100m2	8.64
"Perryfields"; ref Pro 40; 35g/m²	-	-	-	-	9.94	100m2	9.94
"Perryfields"; ref Pro 45; 35g/m²	-	-	-	-	10.36	100m2	10.36
Golf courses; roughs							
"Johnsons Seeds"; ref J8; 18 - 30g/m²	-	-	-	-	8.04	100m2	8.04
"Perryfields"; ref Pro 25; 17 - 35g/m²	-	-	-	-	13.44	100m2	13.44
Waste land; spoil heaps; quarries							
"British Seed Houses"; ref A15 (ley type; can be grazed); 15 - 20g/m²	-	-	-	-	6.60	100m2	6.60
"Perryfields"; ref Pro 95; 12 - 35g/m²	-	-	-	-	16.94	100m2	16.94
"Perryfields"; ref Pro 105 Pioneer Mix (fertility builders); 5g/m²	-	-	-	-	2.48	100m2	2.48
Low maintenance; housing estates; amenity grassed areas							
"British Seed Houses"; ref A19; 25 - 35g/m²	-	-	-	-	8.75	100m2	8.75
"British Seed Houses"; ref A22; 25 - 35g/m²	-	-	-	-	13.93	100m2	13.93
"Perryfields"; ref Pro 120; 25 - 35g/m²	-	-	-	-	13.02	100m2	13.02
Saline coastal; road side areas							
"British Seed Houses"; ref A21; 15 - 20g/m²	-	-	-	-	7.48	100m2	7.48
"Perryfields"; ref Pro 90; 15 - 35g/m²	-	-	-	-	13.30	100m2	13.30
Turf production							
"British Seed Houses"; ref A23 (fine turf); 160 kg/ha	-	-	-	-	463.68	ha	463.68
"British Seed Houses"; ref A24 (hardwearing turf); 185 kg/ha	-	-	-	-	522.81	ha	522.81
Market Prices of wild flora seed mixtures							
Acid soils							
"British Seed Houses"; ref WF1 (annual flowering); 1.00 - 2.00 g/m²	-	-	-	-	16.43	100 m²	16.43
Neutral soils							
"British Seed Houses"; ref WF3; 0.50 - 1.00 g/m²	-	-	-	-	8.21	100 m²	8.21
Market Prices of wild flora and grass seed mixtures							
General purpose							
"Perryfields"; ref Pro Flora 8 Old English Country Meadow Mix; 5 g/m²	-	-	-	-	15.00	100 m²	15.00
"Perryfields"; ref Pro Flora 9; 5 g/m²	-	-	-	-	15.00	100 m²	15.00
Acid soils							
"British Seed Houses"; ref WFG2 (meadow mixture); 5 g/m²	-	-	-	-	22.25	100 m²	22.25
"Perryfields"; ref Pro Flora 2; 5 g/m²	-	-	-	-	11.25	100 m²	11.25
Neutral soils							
"British Seed Houses"; ref WFG4; 5 g/m²	-	-	-	-	24.45	100 m²	24.45
"British Seed Houses"; ref WFG13 (Scotland); 5 g/m²	-	-	-	-	21.55	100 m²	21.55
"Perryfields"; ref Pro Flora 3; 5 g/m²	-	-	-	-	15.75	100 m²	15.75
Calcareous soils							
"British Seed Houses"; ref WFG5; 5 g/m²	-	-	-	-	22.80	100 m²	22.80
"Perryfields"; ref Pro Flora 4; 5 g/m²	-	-	-	-	13.50	100 m²	13.50

Prices for Measured Works
Q PAVING/PLANTING/FENCING/SITE FURNITURE

Item Excluding site overheads and profit	PC £	Labour hours	Labour £	Plant £	Material £	Unit	Total rate £
Q30 SEEDING/TURFING – cont'd							
Market Prices of wild flora and grass seed mixtures – cont'd							
Heavy clay soils							
"British Seed Houses"; ref WFG6; 5 g/m²	-	-	-	-	27.50	100 m²	27.50
"British Seed Houses"; ref WFG12 (Ireland); 5 g/m²	-	-	-	-	21.63	100 m²	21.63
"Perryfields"; ref Pro Flora 5; 5 g/m²	-	-	-	-	24.75	100 m²	24.75
Sandy soils							
"British Seed Houses"; ref WFG7; 5 g/m²	-	-	-	-	32.25	100 m²	32.25
"British Seed Houses"; ref WFG11 (Ireland); 5 g/m²	-	-	-	-	32.75	100 m²	32.75
"British Seed Houses"; ref WFG14 (Scotland); 5 g/m²	-	-	-	-	25.00	100 m²	25.00
"Perryfields"; ref Pro Flora 6; 5 g/m²	-	-	-	-	29.25	100 m²	29.25
Shaded areas							
"British Seed Houses"; ref WFG8; 5 g/m²	-	-	-	-	24.00	100 m²	24.00
"Perryfields"; ref Pro Flora 7; 5 g/m²	-	-	-	-	33.75	100 m²	33.75
Educational							
"British Seed Houses"; ref WFG15; 5 g/m²	-	-	-	-	33.75	100 m²	33.75
Wetlands							
"British Seed Houses"; ref WFG9; 5 g/m²	-	-	-	-	29.75	100 m²	29.75
"Perryfields"; ref Pro Flora 5; 5 g/m²	-	-	-	-	24.75	100 m²	24.75
Scrub and moorland							
"British Seed Houses"; ref WFG10; 5 g/m²	-	-	-	-	32.80	100 m²	32.80
Hedgerow							
"Perryfields"; ref Pro Flora 7; 5 g/m²	-	-	-	-	33.75	100 m²	33.75
Vacant sites							
"Perryfields"; ref Pro Flora 1; 5 g/m²	-	-	-	-	30.00	100 m²	30.00
Seeding							
Seeding labours only in two operations; by machine (for seed prices see above)							
35 g/m²	-	-	-	0.49	-	100 m²	0.49
Grass seed; spreading in two operations; PC £2.46/kg; (for changes in material prices please refer to table above); by machine							
35 g/m²	-	-	-	0.49	8.61	100 m²	9.10
50 g/m²	-	-	-	0.49	12.30	100 m²	12.79
70 g/m²	-	-	-	0.49	17.22	100 m²	17.71
100 g/m²	-	-	-	0.49	24.60	100 m²	25.09
125 kg/ha	-	-	-	49.03	307.50	ha	356.53
150 kg/ha	-	-	-	49.03	369.00	ha	418.03
200 kg/ha	-	-	-	49.03	492.00	ha	541.03
250 kg/ha	-	-	-	49.03	615.00	ha	664.03
300 kg/ha	-	-	-	49.03	738.00	ha	787.03
350 kg/ha	-	-	-	49.03	861.00	ha	910.03
400 kg/ha	-	-	-	49.03	984.00	ha	1033.03
500 kg/ha	-	-	-	49.03	1230.00	ha	1279.03
700 kg/ha	-	-	-	49.03	1722.00	ha	1771.03
1400 kg/ha	-	-	-	49.03	3444.00	ha	3493.03

Prices for Measured Works
Q PAVING/PLANTING/FENCING/SITE FURNITURE

Item Excluding site overheads and profit	PC £	Labour hours	Labour £	Plant £	Material £	Unit	Total rate £
Extra over seeding by machine for slopes over 30 degrees (allowing for the actual area but measured in plan)							
35 g/m²	-	-	-	0.07	1.29	100 m²	1.37
50 g/m²	-	-	-	0.07	1.85	100 m²	1.92
70 g/m²	-	-	-	0.07	2.58	100 m²	2.66
100 g/m²	-	-	-	0.07	3.69	100 m²	3.76
125 kg/ha	-	-	-	7.36	46.13	ha	53.48
150 kg/ha	-	-	-	7.36	55.35	ha	62.71
200 kg/ha	-	-	-	7.36	73.80	ha	81.16
250 kg/ha	-	-	-	7.36	92.25	ha	99.61
300 kg/ha	-	-	-	7.36	110.70	ha	118.06
350 kg/ha	-	-	-	7.36	129.15	ha	136.51
400 kg/ha	-	-	-	7.36	147.60	ha	154.96
500 kg/ha	-	-	-	7.36	184.50	ha	191.86
700 kg/ha	-	-	-	7.36	258.30	ha	265.66
1400 kg/ha	-	-	-	7.36	516.60	ha	523.96
Seeding labours only in two operations; by machine (for seed prices see above)							
35 g/m²	-	0.17	3.00	-	-	100 m²	3.00
Grass seed; spreading in two operations; PC £2.46/kg; (for changes in material prices please refer to table above); by hand							
35 g/m²	-	0.17	3.00	-	8.61	100 m²	11.61
50 g/m²	-	0.17	3.00	-	12.30	100 m²	15.30
70 g/m²	-	0.17	3.00	-	17.22	100 m²	20.22
100 g/m²	-	0.20	3.60	-	24.60	100 m²	28.20
125 g/m²	-	0.20	3.60	-	30.75	100 m²	34.35
Extra over seeding by hand for slopes over 30 degrees (allowing for the actual area but measured in plan)							
35 g/m²	1.28	-	0.07	-	1.28	100 m²	1.35
50 g/m²	1.85	-	0.07	-	1.85	100 m²	1.91
70 g/m²	2.58	-	0.07	-	2.58	100 m²	2.65
100 g/m²	3.69	-	0.08	-	3.69	100 m²	3.77
125 g/m²	4.60	-	0.08	-	4.60	100 m²	4.68
Harrowing seeded areas; light chain harrow	-	-	-	0.09	-	100 m²	0.09
Raking over seeded areas							
by mechanical stone rake	-	-	-	2.10	-	100 m²	2.10
by hand	-	0.80	14.40	-	-	100 m²	14.40
Rolling seeded areas; light roller							
by tractor drawn roller	-	-	-	0.52	-	100 m²	0.52
by pedestrian operated mechanical roller	-	0.08	1.50	0.55	-	100 m²	2.05
by hand drawn roller	-	0.17	3.00	-	-	100 m²	3.00
Extra over harrowing, raking or rolling seeded areas for slopes over 30 degrees; by machine or hand	-	-	-	-	-	25%	-
Turf edging; to seeded areas; 300 mm wide	1.25	0.05	0.86	-	1.25	m²	2.11

Q PAVING/PLANTING/FENCING/SITE FURNITURE

Item Excluding site overheads and profit	PC £	Labour hours	Labour £	Plant £	Material £	Unit	Total rate £
Q30 SEEDING/TURFING – cont'd							
Hydroseeding - General Preamble: Hydroseeding may be specified where conventional seeding is difficult or impossible, ie. where cultivation is hard on certain subsoils, on steep banks, in dangerous situations, where soil erosion by rain or wind is likely. Hydroseeding consists of high pressure spraying a mixture of water, seed, fertilizer, mulches and binders from a special machine which can spray up to 60 m horizontally and 25 m vertically. Hose extensions allow seeding to be carried out up to 200 m depending on how thick a mix is used. Preamble: Seed can be grass, wild flowers, certain shrubs and trees. Seed mixes should be site specific as all ecological requirements for selected species must be present on site if hydroseeding is to be successful. Special soil conditioners, mulches, fertilizers and binders can be used in the hydromix and each contractor has a preference to use his own. Specification and prices vary widely according to the location, hydromix required, and size of contract. The following are price ranges.							
Hydroseeding; Keller Comtec "Terra-soil" process (used where grass is to be established on topsoil or easy grow sites); typical solids application rate 250 g/m²							
from:	-	-	-	-	-	m²	0.15
to:	-	-	-	-	-	m²	0.25
"Terraseal" process (used on most subsoils including sand, clays, silts, shale and rock; the hydramulch includes materials which will kick start the generation of topsoil); typical solids' application rate 400 - 500 g/m²							
from:	-	-	-	-	-	m²	0.30
to:	-	-	-	-	-	m²	0.70
"Terrablanket" process (used on steep slopes to assist in soil stability and erosion control; can be used in place of straw or erosion geotextiles and is more effective in establishing grass cover); typical solids application rate 600 - 700 g/m²							
from:	-	-	-	-	-	m²	0.90
to:	-	-	-	-	-	m²	1.50
"Terraseal Runway" process (used on airport runway aprons to stabilise soil against jet blast); typical solids application rate 300 g/m²							
from:	-	-	-	-	-	m²	0.30
to:	-	-	-	-	-	m²	0.40
Liquid Sod; Turf Management Spray on grass system of grass plantlets fertilizer, bio-degradable mulch carrier, root enhancer and water							
to prepared ground	-	-	-	-	-	m²	1.00

Prices for Measured Works
Q PAVING/PLANTING/FENCING/SITE FURNITURE

Item Excluding site overheads and profit	PC £	Labour hours	Labour £	Plant £	Material £	Unit	Total rate £
Preparation of turf beds							
Rolling turf to be lifted; lifting by hand or mechanical turf stripper; stacks to be not more than 1 m high							
cutting only preparing to lift; pedestrian turf cutter	-	0.75	13.50	9.71	-	100 m²	23.21
lifting and stacking; by hand	-	8.33	150.00	-	-	100 m²	150.00
Rolling up; moving to stacks							
distance not exceeding 100 m	-	2.50	45.00	-	-	100 m²	45.00
extra over rolling and moving turf to stacks to transport per additional 100 m	-	0.83	15.00	-	-	100 m²	15.00
Lifting selected and approved topsoil from spoil heaps							
passing through 6 mm screen; removing debris	-	0.17	3.00	6.35	-	m³	9.35
Extra over lifting topsoil and passing through screen for imported topsoil; plus 20% allowance for settlement	17.32	-	-	-	20.78	m³	20.78
Topsoil; PC £17.32/m³; plus 20% allowance for settlement							
25 mm deep	-	-	-	-	0.52	m²	0.52
50 mm deep	-	-	-	-	1.04	m²	1.04
100 mm deep	-	-	-	-	2.08	m²	2.08
150 mm deep	-	-	-	-	3.12	m²	3.12
200 mm deep	-	-	-	-	4.16	m²	4.16
250 mm deep	-	-	-	-	5.20	m²	5.20
300 mm deep	-	-	-	-	6.24	m²	6.24
400 mm deep	-	-	-	-	8.31	m²	8.31
450 mm deep	-	-	-	-	9.35	m²	9.35
Spreading topsoil to form turfbeds (topsoil not included); by machine							
25 mm deep	-	-	0.05	0.10	-	m²	0.14
50 mm deep	-	-	0.06	0.12	-	m²	0.18
75 mm deep	-	-	0.07	0.14	-	m²	0.21
100 mm deep	-	0.01	0.09	0.19	-	m²	0.28
150 mm deep	-	0.01	0.14	0.29	-	m²	0.42
Spreading topsoil to form turfbeds (topsoil not included); by hand							
25 mm deep	-	0.03	0.45	-	-	m²	0.45
50 mm deep	-	0.03	0.60	-	-	m²	0.60
75 mm deep	-	0.04	0.77	-	-	m²	0.77
100 mm deep	-	0.05	0.90	-	-	m²	0.90
150 mm deep	-	0.08	1.35	-	-	m²	1.35
Bringing existing topsoil to a fine tilth for turfing by raking or harrowing; stones not to exceed 6 mm; by machine	-	-	0.07	0.04	-	m²	0.12
Bringing existing topsoil to a fine tilth for turfing by raking or harrowing; stones not to exceed 6 mm; by hand	-	0.01	0.16	-	-	m²	0.16
Preparation of cricket squares							
Excavating cricket square size 20 x 22 m to a depth of 150 mm; screening and returning topsoil; mixing with imported marl or clay loam at the rate of 1 kg/m²; bringing to accurate levels; hand	-	-	-	-	-	nr	4514.37

Prices for Measured Works
Q PAVING/PLANTING/FENCING/SITE FURNITURE

Item Excluding site overheads and profit	PC £	Labour hours	Labour £	Plant £	Material £	Unit	Total rate £
Q30 SEEDING/TURFING – cont'd							
Turfing							
Turfing; laying only; to stretcher bond; butt joints; including providing and working from barrow plank runs where necessary to surfaces not exceeding 30 degrees from horizontal; average size of turves 0.50 yd2 (0.418 m²); 2.39 turves per m²							
specially selected lawn turves from previously lifted stockpile	-	0.08	1.35	-	-	m²	1.35
cultivated lawn turves; to large open areas	-	0.06	1.05	-	-	m²	1.05
cultivated lawn turves; to domestic or garden areas	-	0.08	1.40	-	-	m²	1.40
meadow turf; to amenity areas	-	0.05	0.90	-	-	m²	0.90
road verge quality turf	-	0.04	0.72	-	-	m²	0.72
Industrially grown turf; PC prices listed represent the general range of industrial turf prices for sportsfields and amenity purposes. Prices will vary with quantity and site location							
"Rolawn"							
ref Standard; golf tees and tennis courts, fine domestic lawns, ornamental landscaping	2.93	0.08	1.35	-	2.93	m²	4.28
ref RB Medallion; sports fields, domestic lawns, general landscape	2.10	0.07	1.26	-	2.10	m²	3.36
Greenkeeper Ltd;							
"Turfliner" light reinforced turf for low maintenance, low fertility areas to steep slopes	2.30	0.10	1.72	-	2.30	m²	4.02
"Turf Carpet" grown on sterile soil on polythene	2.25	0.10	1.72	-	2.25	m²	3.97
"Tensar Turf Mat" reinforced turf for embankments; laid to embankments 3.3 m² turves	5.95	0.10	1.80	-	5.95	m²	7.75
"Tensar Turf Mat" reinforced turf for embankments; laid to embankments 16.5 m² rolls	5.95	0.04	0.63	1.08	5.95	m²	7.67
"Inturf"							
ref SS5; fine texture, special golf and bowling greens	2.35	0.10	1.72	-	2.35	m²	4.07
ref SS1; fine lawns, golf greens, bowling greens	2.35	0.10	1.72	-	2.35	m²	4.07
ref SS3; hockey grounds, polo, medium wearing areas	1.80	0.05	0.98	-	1.80	m²	2.78
ref SS4; hardwearing fine turf, low maintenance areas	1.85	0.04	0.76	-	1.85	m²	2.61
ref SS2; football grounds, parks, hardwearing areas	1.60	0.05	0.95	-	1.60	m²	2.55
Millennium Turf; for difficult site conditions	1.80	0.08	1.44	-	1.80	m²	3.24
Custom Grown Turf; specific seed mixtures to suit soil or site conditions	5.50	0.08	1.44	-	5.50	m²	6.94
Turf Tiles; instant repairs to goal mouths and playing fields providing instant stable surface	-	-	-	-	-	m²	28.62
Turf Modules; for trade shows exhibitions and events	-	-	-	-	-	m²	129.60

Prices for Measured Works
Q PAVING/PLANTING/FENCING/SITE FURNITURE

Item Excluding site overheads and profit	PC £	Labour hours	Labour £	Plant £	Material £	Unit	Total rate £
Reinforced turf; Netlon Advanced Turf; Netlon Ltd; Blended mesh fibre elements incorporated into root zone; Rootzone spread and levelled over cultivated, prepared and reduced and levelled ground (not included); compacted with light roller.							
"ATS 300" with "R300" topping, seed and fertilizer; 100 thick							
100 -199 m²	16.15	0.03	0.45	0.59	16.15	m²	**17.19**
200 - 299 m²	14.65	0.03	0.45	0.59	14.65	m²	**15.69**
300 - 499 m²	14.10	0.03	0.45	0.59	14.10	m²	**15.14**
over 500 m²	13.70	0.03	0.45	0.59	13.70	m²	**14.74**
"ATS 300" with "R300" topping, seed and fertilizer; 150 thick							
100 -199 m²	21.10	0.03	0.60	0.78	21.10	m²	**22.48**
200 - 299 m²	19.60	0.03	0.60	0.78	19.60	m²	**20.98**
300 - 499 m²	18.95	0.03	0.60	0.78	18.95	m²	**20.33**
over 500 m²	18.50	0.03	0.60	0.78	18.50	m²	**19.88**
"ATS 300" with washed turf and fertilizer ("R300" topping not included) 100 thick							
100 -199 m²	19.70	0.03	0.60	0.78	19.70	m²	**21.08**
200 - 499 m²	18.10	0.03	0.60	0.78	18.10	m²	**19.48**
over 500 m²	17.15	0.03	0.60	0.78	17.15	m²	**18.53**
"ATS 300" with washed turf and fertilizer ("R300" topping not included) 150 thick							
100 -199 m²	-	0.04	0.72	0.87	24.55	m²	**26.14**
200 - 499 m²	22.95	0.04	0.72	0.87	22.95	m²	**24.54**
over 500 m²	21.85	0.04	0.72	0.87	21.85	m²	**23.44**
Firming turves with wooden beater	-	0.01	0.18	-	-	m²	**0.18**
Rolling turfed areas; light roller							
by tractor with turf tyres and roller	-	-	-	0.52	-	100 m²	**0.52**
by pedestrian operated mechanical roller	-	0.08	1.50	0.55	-	100 m²	**2.05**
by hand drawn roller	-	0.17	3.00	-	-	100 m²	**3.00**
Dressing with finely sifted topsoil; brushing into joints	0.01	0.05	0.90	-	0.01	m²	**0.91**
Turfing; laying only							
to slopes over 30 degrees; to diagonal bond (measured as plan area - add 15% to these rates for the incline area of 30 degree slopes)	-	0.12	2.16	-	-	m²	**2.16**
Extra over laying turfing for pegging down turves wooden or galvanized wire pegs; 200 mm long; 2 pegs per 0.50 yd2	0.19	0.01	0.24	-	0.19	m²	**0.43**
Initial cutting; to turfed areas							
20 mm high; using pedestrian guided power driven cylinder mower; including boxing off cuttings (stone picking and rolling not included)	-	0.18	3.24	0.25	-	100 m²	**3.49**

Prices for Measured Works
Q PAVING/PLANTING/FENCING/SITE FURNITURE

Item Excluding site overheads and profit	PC £	Labour hours	Labour £	Plant £	Material £	Unit	Total rate £
Q30 SEEDING/TURFING – cont'd							
Maintenance operations (Note: the following rates apply to aftercare maintenance executed as part of a landscaping contract only)							
Repairing damaged grass areas							
scraping out; removing slurry; from ruts and holes; average 100 mm deep	-	0.13	2.40	-	-	m²	2.40
100 mm topsoil	-	0.13	2.40	-	2.08	m²	4.48
Repairing damaged grass areas; sowing grass seed to match existing or as specified; to individually prepared worn patches							
35 g/m²	0.09	0.01	0.18	-	0.10	m²	0.28
50 g/m²	0.12	0.01	0.18	-	0.14	m²	0.32
Sweeping leaves; disposing off site; motorized vacuum sweeper or rotary brush sweeper							
areas of maximum 2500 m² with occasional large tree and established boundary planting; 5.35 m³ (1 skip of material to be removed)	-	0.40	7.20	2.73	-	100 m²	9.93
Leaf Clearance; clearing grassed area of leaves and other extraneous debris							
using equipment towed by tractor							
large grassed areas with perimeters of mature trees such as sports fields and amenity areas	-	0.01	0.23	0.29	-	100 m²	0.51
large grassed areas containing ornamental trees and shrub beds	-	0.03	0.45	0.40	-	100 m²	0.85
using pedestrian operated mechanical equipment and blowers							
grassed areas with perimeters of mature trees such as sports fields and amenity areas	-	0.04	0.72	0.10	-	100 m²	0.82
grassed areas containing ornamental trees and shrub beds	-	0.10	1.80	0.25	-	100 m²	2.05
verges	-	0.07	1.20	0.17	-	100 m²	1.37
by hand							
grassed areas with perimeters of mature trees such as sports fields and amenity areas	-	0.05	0.90	0.09	-	100 m²	0.99
grassed areas containing ornamental trees and shrub beds	-	0.08	1.50	0.15	-	100 m²	1.65
verges	-	1.00	18.00	1.84	-	100 m²	19.84
removal of arisings							
areas with perimeters of mature trees	-	0.01	0.09	0.08	0.64	100 m²	0.82
areas containing ornamental trees and shrub beds	-	0.02	0.28	0.31	1.60	100 m²	2.19
Cutting grass to specified height; per cut							
multi unit gang mower	-	0.59	10.59	15.89	-	ha	26.47
ride-on triple cylinder mower	-	0.01	0.25	0.12	-	100 m²	0.37
ride-on triple rotary mower	-	0.01	0.25	-	-	100 m²	0.25
pedestrian mower	-	0.18	3.24	0.59	-	100 m²	3.83
Cutting grass to banks; per cut							
side arm cutter bar mower	-	0.02	0.42	0.29	-	100 m²	0.71
Cutting rough grass; per cut							
power flail or scythe cutter	-	0.04	0.63	-	-	100 m²	0.63
Extra over cutting grass for slopes not exceeding 30 degrees	-	-	-	-	-	10%	-
Extra over cutting grass for slopes exceeding 30 degrees	-	-	-	-	-	40%	-

Prices for Measured Works
Q PAVING/PLANTING/FENCING/SITE FURNITURE

Item Excluding site overheads and profit	PC £	Labour hours	Labour £	Plant £	Material £	Unit	Total rate £
Cutting fine sward							
pedestrian operated seven-blade cylinder lawn mower	-	0.14	2.52	0.19	-	100 m²	2.71
Extra over cutting fine sward for boxing off cuttings							
pedestrian mower	-	0.03	0.50	0.04	-	100 m²	0.54
Cutting areas of rough grass							
scythe	-	1.00	18.00	-	-	100 m²	18.00
sickle	-	2.00	36.00	-	-	100 m²	36.00
petrol operated strimmer	-	0.30	5.41	0.41	-	100 m²	5.81
Cutting areas of rough grass which contain trees or whips							
petrol operated strimmer	-	0.40	7.20	0.54	-	100 m²	7.74
Extra over cutting rough grass for on site raking up and dumping	-	0.33	6.00	-	-	100 m²	6.00
Trimming edge of grass areas; edging tool							
with petrol powered strimmer	-	0.13	2.40	0.18	-	100 m	2.58
by hand	-	0.67	12.00	-	-	100 m	12.00
Marking out pitches using approved line marking compound; including initial setting out and marking							
discus, hammer, javelin or shot putt area	2.79	2.00	36.00	-	2.79	nr	38.79
cricket square	1.86	2.00	36.00	-	1.86	nr	37.86
cricket boundary	6.51	8.00	144.00	-	6.51	nr	150.51
grass tennis court	2.79	4.00	72.00	-	2.79	nr	74.79
hockey pitch	9.30	8.00	144.00	-	9.30	nr	153.30
football pitch	9.30	8.00	144.00	-	9.30	nr	153.30
rugby pitch	9.30	8.00	144.00	-	9.30	nr	153.30
eight lane running track; 400 m	18.60	16.00	288.00	-	18.60	nr	306.60
Re-marking out pitches using approved line marking compound							
discus, hammer, javelin or shot putt area	1.86	0.50	9.00	-	1.86	nr	10.86
cricket square	1.86	0.50	9.00	-	1.86	nr	10.86
grass tennis court	6.51	1.00	18.00	-	6.51	nr	24.51
hockey pitch	6.51	1.00	18.00	-	6.51	nr	24.51
football pitch	6.51	1.00	18.00	-	6.51	nr	24.51
rugby pitch	6.51	1.00	18.00	-	6.51	nr	24.51
eight lane running track; 400 m	18.60	2.50	45.00	-	18.60	nr	63.60
Rolling grass areas; light roller							
by tractor drawn roller	-	-	-	0.52	-	100 m²	0.52
by pedestrian operated mechanical roller	-	0.08	1.50	0.55	-	100 m²	2.05
by hand drawn roller	-	0.17	3.00	-	-	100 m²	3.00
Aerating grass areas; to a depth of 100 mm							
using tractor-drawn aerator	-	0.06	1.05	2.18	-	100 m²	3.23
using pedestrian-guided motor powered solid or slitting tine turf aerator	-	0.18	3.15	2.89	-	100 m²	6.04
using hollow tine aerator; including sweeping up and dumping corings	-	0.50	9.00	5.78	-	100 m²	14.78
using hand aerator or fork	-	1.67	30.00	-	-	100 m²	30.00
Extra over aerating grass areas for on site sweeping up and dumping corings	-	0.17	3.00	-	-	100 m²	3.00
Switching off dew; from fine turf areas	-	0.20	3.60	-	-	100 m²	3.60
Scarifying grass areas to break up thatch; removing dead grass							
using tractor-drawn scarifier	-	0.07	1.26	1.52	-	100 m²	2.78
using self-propelled scarifier; including removing and disposing of grass on site	-	0.33	6.00	0.46	-	100 m²	6.46

Prices for Measured Works
Q PAVING/PLANTING/FENCING/SITE FURNITURE

Item Excluding site overheads and profit	PC £	Labour hours	Labour £	Plant £	Material £	Unit	Total rate £
Q30 SEEDING/TURFING – cont'd							
Maintenance operations – cont'd							
Harrowing grass areas							
using drag mat	-	0.03	0.50	0.29	-	100 m²	0.80
using chain harrow	-	0.04	0.63	0.36	-	100 m²	0.99
using drag mat	-	2.80	50.41	29.11	-	ha	79.51
using chain harrow	-	3.50	63.00	36.38	-	ha	99.39
Extra for scarifying and harrowing grass areas for disposing excavated material off site; to tip not exceeding 13 km; loading by machine							
slightly contaminated	-	-	-	1.55	12.50	m³	14.05
rubbish	-	-	-	1.55	12.50	m³	14.05
inert material	-	-	-	1.03	7.50	m³	8.53
For the following topsoil improvement and seeding operations add or subtract the following amounts for every £0.10 difference in the material cost price							
35 g/m²	-	-	-	-	0.35	100 m²	0.35
50 g/m²	-	-	-	-	0.50	100 m²	0.50
70 g/m²	-	-	-	-	0.70	100 m²	0.70
100 g/m²	-	-	-	-	1.00	100 m²	1.00
125 kg/ ha	-	-	-	-	12.50	ha	12.50
150 kg/ ha	-	-	-	-	15.00	ha	15.00
175 kg/ ha	-	-	-	-	17.50	ha	17.50
200 kg/ ha	-	-	-	-	20.00	ha	20.00
225 kg/ ha	-	-	-	-	22.50	ha	22.50
250 kg/ ha	-	-	-	-	25.00	ha	25.00
300 kg/ ha	-	-	-	-	30.00	ha	30.00
350 kg/ ha	-	-	-	-	35.00	ha	35.00
400 kg/ ha	-	-	-	-	40.00	ha	40.00
500 kg/ ha	-	-	-	-	50.00	ha	50.00
700 kg/ ha	-	-	-	-	70.00	ha	70.00
1000 kg/ ha	-	-	-	-	100.00	ha	100.00
1250 kg/ ha	-	-	-	-	125.00	ha	125.00
Selective residual pre-emergent weedkiller; Rigby Taylor Ltd; "Flexidor 125"; in accordance with manufacturer's instructions; PC £86.83 per litre; application rate 1.20 - 2 litre/hectare							
1.20 l /ha	-	0.28	5.04	-	1.04	100 m²	6.08
2.00 l /ha	-	0.28	5.04	-	1.74	100 m²	6.78
Top dressing fertilizers (7:7:7); PC £0.49/kg; to seedbeds; by machine							
35 g/m²	1.73	-	-	0.18	1.73	100 m²	1.91
50 g/m²	2.47	-	-	0.18	2.47	100 m²	2.65
300 kg/ha	148.32	-	-	18.05	148.32	ha	166.37
350 kg/ha	173.04	-	-	18.05	173.04	ha	191.09
400 kg/ha	197.76	-	-	18.05	197.76	ha	215.81
500 kg/ha	247.20	-	-	28.89	247.20	ha	276.09
Top dressing fertilizers (7:7:7); PC £0.49/kg; to seedbeds; by hand							
35 g/m²	1.73	0.17	3.00	-	1.73	100 m²	4.73
50 g/m²	2.47	0.17	3.00	-	2.47	100 m²	5.47
70 g/m²	3.46	0.17	3.00	-	3.46	100 m²	6.46
Watering turf; evenly; at a rate of 5 litre/m²							
using movable spray lines powering 3 nr sprinkler heads with a radius of 15 m and allowing for 60% overlap (irrigation machinery costs not included)	-	0.02	0.28	-	-	100 m²	0.28
using sprinkler equipment and with sufficient water pressure to run 1 nr 15 m radius sprinkler	-	0.02	0.36	-	-	100 m²	0.36
using hand-held watering equipment	-	0.25	4.50	-	-	100 m²	4.50

Prices for Measured Works
Q PAVING/PLANTING/FENCING/SITE FURNITURE

Item Excluding site overheads and profit	PC £	Labour hours	Labour £	Plant £	Material £	Unit	Total rate £
Q31 PLANTING							
Market prices of mulching materials							
Melcourt Industries Ltd; 25m3 loads							
"Ornamental Bark Mulch"	-	-	-	-	42.50	m³	42.50
"Bark Nuggets"	-	-	-	-	41.40	m³	41.40
"Graded Bark Flakes"	-	-	-	-	44.00	m³	44.00
"Amenity Bark Mulch"	-	-	-	-	30.15	m³	30.15
"Contract Bark Mulch"	-	-	-	-	27.80	m³	27.80
"Spruce Ornamental"	-	-	-	-	31.55	m³	31.55
"Spruce Flakes"	-	-	-	-	33.85	m³	33.85
"Decorative Biomulch"®	-	-	-	-	28.80	m³	28.80
"Rustic Biomulch"®	-	-	-	-	33.55	m³	33.55
"Mulchip"®	-	-	-	-	29.30	m³	29.30
"Mulch 2000"	-	-	-	-	24.80	m³	24.80
"Mulch 3000"	-	-	-	-	21.80	m³	21.80
"Forest BioMulch"®	-	-	-	-	26.30	m³	26.30
Melcourt Industries Ltd; 50m3 loads							
"Mulch 2000"	-	-	-	-	14.30	m³	14.30
"Mulch 3000"	-	-	-	-	11.30	m³	11.30
Melcourt Industries Ltd; 60m3 loads							
"Contract Bark Mulch"	-	-	-	-	16.10	m³	16.10
Melcourt Industries Ltd; 70m3 loads							
"Ornamental Bark Mulch"	-	-	-	-	30.10	m³	30.10
"Bark Nuggets"	-	-	-	-	29.00	m³	29.00
"Graded Bark Flakes"	-	-	-	-	31.60	m³	31.60
"Amenity Bark Mulch"	-	-	-	-	17.75	m³	17.75
"Spruce Ornamental"	-	-	-	-	19.15	m³	19.15
"Spruce Flakes"	-	-	-	-	21.45	m³	21.45
"Decorative Biomulch"®	-	-	-	-	16.40	m³	16.40
"Rustic Biomulch"®	-	-	-	-	21.15	m³	21.15
"Mulchip"®	-	-	-	-	16.90	m³	16.90
"Forest BioMulch"®	-	-	-	-	13.90	m³	13.90
Market prices of planting materials							
(Note: The rates shown generally reflect the manufacturer's recommended retail prices. Trade and bulk discounts are often available on the prices shown)							
Topsoil; Boughton Loam Ltd;							
50/50; as dug / 10 mm screened	-	-	-	-	38.61	m³	38.61
Mulch; Melcourt Industries Ltd; 25 m³ loads							
"Landscape Bark"	-	-	-	-	31.00	m³	31.00
"Humus 2000"	-	-	-	-	23.05	m³	23.05
"Spent Mushroom Compost"	-	-	-	-	17.90	m³	17.90
"Topgrow"	-	-	-	-	27.00	m³	27.00
Mulch; Melcourt Industries Ltd; 50 m³ loads							
"Humus 2000"	-	-	-	-	12.55	m³	12.55
Mulch; Melcourt Industries Ltd; 60 m³ loads							
"Super Humus"	-	-	-	-	14.30	m³	14.30
"Landscape Bark"	-	-	-	-	19.30	m³	19.30
"Spent Mushroom Compost"	-	-	-	-	6.20	m³	6.20
"Topgrow"	-	-	-	-	15.30	m³	15.30
Fertilizers; Scotts UK Professional							
"Enmag"; 70 g/m²	-	-	-	-	8.43	100 m²	8.43
fertilizer; controlled release; "Osmocote Exact"; 15+10+12+2MgO+TE, tablet; 5gr each	-	-	-	-	2.82	100 nr	2.82

Prices for Measured Works
Q PAVING/PLANTING/FENCING/SITE FURNITURE

Item Excluding site overheads and profit	PC £	Labour hours	Labour £	Plant £	Material £	Unit	Total rate £
Q31 PLANTING – cont'd							
Market prices of planting materials – cont'd							
Fertilizers; Scotts UK Professional; granular "Sierrablen Flora", 15+9+9+3 MgO; controlled release fertilizer							
transplant	-	-	-	-	0.09	nr	0.09
whip	-	-	-	-	0.13	nr	0.13
feathered	-	-	-	-	0.17	nr	0.17
light standard	-	-	-	-	0.17	nr	0.17
standard	-	-	-	-	0.22	nr	0.22
selected standard	-	-	-	-	0.30	nr	0.30
heavy standard	-	-	-	-	0.35	nr	0.35
extra heavy standard	-	-	-	-	0.43	nr	0.43
16 -18 cm girth	-	-	-	-	0.48	nr	0.48
18 - 20 cm girth	-	-	-	-	0.52	nr	0.52
20 - 22 cm girth	-	-	-	-	0.61	nr	0.61
22 - 24 cm girth	-	-	-	-	0.65	nr	0.65
24 - 26 cm girth	-	-	-	-	0.69	nr	0.69
Fertilizers; Farmura Products Ltd; Seanure Root Dip							
to transplants	-	-	-	-	0.22	10 nr	0.22
medium whips	-	-	-	-	0.06	each	0.06
standard trees	-	-	-	-	0.37	each	0.37
Fertilizers; Farmura Products Ltd; Seanure Soilbuilder							
soil amelioration; 70g/m²	-	-	-	-	8.07	100 m²	8.07
to plant pits; 300 x 300 x 300	-	-	-	-	0.03	each	0.03
to plant pits; 600 x 600 x 600	-	-	-	-	0.37	each	0.37
to tree pits; 1.00 x 1.00 x 1.00	-	-	-	-	1.73	each	1.73
Fertilizers; Scotts UK Professional							
grass and soil fertilizer; "Greenmaster Liquid Seafeed"	-	-	-	-	3.40	litre	3.40
"TPMC" tree and shrub planting compost	-	-	-	-	4.28	bag	4.28
Fertilizers and anti-desiccants; Rigby Taylor Ltd; fertilizer application rates 35 g/m² unless otherwise shown							
straight fertilizer; "Bone Meal"; at 70 g/m²	-	-	-	-	6.20	100 m²	6.20
straight fertilizer; "Sulphate of Ammonia"	-	-	-	-	1.78	100 m²	1.78
straight fertilizer; "Sulphate of Iron"	-	-	-	-	1.66	100 m²	1.66
straight fertilizer; "Sulphate of Potash"	-	-	-	-	2.20	100 m²	2.20
straight fertilizer; "Super Phosphate Powder"	-	-	-	-	0.46	100 m²	0.46
liquid fertilizer; "Vitax 50/50 Soluble Iron"; 56 ml/100m2	-	-	-	-	6.79	100 m²	6.79
liquid fertilizer; "Vitax 50/50 Standard"; 56 ml/100m2	-	-	-	-	11.10	100 m²	11.10
liquid fertilizer; "Vitax 50/50 Extra"; 56 ml/100m2	-	-	-	-	13.41	100 m²	13.41
wetting agent; "Breaker liquid"; 10 lt	-	-	-	-	1.24	100 m²	1.24
wetting agent; "Breaker granules"; 25 Kg	-	-	-	-	7.80	100 m²	7.80
Herbicides; Scotts UK Professional; application rates used to produce these rates are the maximum rate recommended by the manufacturer in each case (Note: lower rates may often be appropriate in many cases)							
"Dextrone X"	-	-	-	-	0.49	100 m²	0.49
"Intrepid"	-	-	-	-	0.54	100 m²	0.54
"Speedway"	-	-	-	-	0.96	100 m²	0.96

Prices for Measured Works
Q PAVING/PLANTING/FENCING/SITE FURNITURE

Item Excluding site overheads and profit	PC £	Labour hours	Labour £	Plant £	Material £	Unit	Total rate £
Herbicides; Rigby Taylor Ltd; application rates used to produce these are the maximum rate recommended by the manufacturer in each case (Note: these application rates will vary dependent on season)							
"Roundup Pro Biactive"; 5 lt/ha	-	-	-	-	1.06	100 m²	1.06
"Flexidor 125 (Isoxaben)"; 2lt/ha	-	-	-	-	1.74	100 m²	1.74
"Kerb Flowable (Propyzamide)"	-	-	-	-	2.22	100 m²	2.22
"Kerb Granules" (15 x 120 tree pack)	-	-	-	-	0.31	100 m²	0.31
"Timbrel (Triclopyr)"	-	-	-	-	4.87	100 m²	4.87
"Casoron G" as a selective herbicide	-	-	-	-	4.89	100 m²	4.89
"Casoron G" as a residual herbicide	-	-	-	-	8.81	100 m²	8.81
Market prices of trees, shrubs and plants: Notcutts Nurseries Ltd							
Shrubs							
Acer palmatum Dissectum Atropurpureum; 30-45 3 L	-	-	-	-	15.50	nr	15.50
Arbutus unedo rubra; 30-45 3 L	-	-	-	-	7.00	nr	7.00
Aucuba japonica Variegata; 30-45 3 L	-	-	-	-	2.95	nr	2.95
Berberis candidula; 20-30 2 L	-	-	-	-	1.85	nr	1.85
Berberis thunbergii; 30-45 3 L	-	-	-	-	1.40	nr	1.40
Berberis thunbergii Rose Glow; 30-45 2 L	-	-	-	-	2.80	nr	2.80
Buddleja davidii Black Knight; 45-60 3 L	-	-	-	-	1.80	nr	1.80
Buddleja globosa; 45-60 3 L	-	-	-	-	2.90	nr	2.90
Buxus sempervirens; 30-45 3 L	-	-	-	-	2.90	nr	2.90
Caryopteris clandonensis; 30-45 3 L	-	-	-	-	2.45	nr	2.45
Ceanothus dentatus; 60-90 3 L	-	-	-	-	3.20	nr	3.20
Ceanothus thyrsiflorus Repens; 30-45 2 L	-	-	-	-	2.20	nr	2.20
Chaenomeles superba Jet Trail; 45-60 3 L	-	-	-	-	2.20	nr	2.20
Choisya ternata; 30-45 3 L	-	-	-	-	2.45	nr	2.45
Cornus alba; 60-90 3 L	-	-	-	-	1.40	nr	1.40
Cornus alba Elegantissima; 60-90 3 L	-	-	-	-	2.40	nr	2.40
Cornus stolonifera Flayiramea; 60-90 3 L	-	-	-	-	1.80	nr	1.80
Corylus avellana; 45-60 3 L	-	-	-	-	1.40	nr	1.40
Cotinus coggygria Royal Purple; 30-45 3 L	-	-	-	-	3.30	nr	3.30
Cotoneaster salicifolius Exburyensis; 45-50 3 L	-	-	-	-	2.40	nr	2.40
Cotoneaster suecicus watereri; 60-90 3 L	-	-	-	-	1.80	nr	1.80
Elaeagnus pungens maculata; 45-60 3 L	-	-	-	-	3.20	nr	3.20
Elaeagnus x ebbingei; 45-60 3 L	-	-	-	-	2.20	nr	2.20
Elaeagnus x ebbingei Limelight; 45-60 3 L	-	-	-	-	3.20	nr	3.20
Escallonia Edinensis; 45-60 3 L	-	-	-	-	2.10	nr	2.10
Euonymus fortunei Blondy; 20-30 2 L	-	-	-	-	2.80	nr	2.80
Euonymus fortunei Darts Blanket; 30-45 2 L	-	-	-	-	1.40	nr	1.40
Forsythia intermedia Lynwood; 60-90 3 L	-	-	-	-	1.80	nr	1.80
Genista hispanica; 20-30 2 L	-	-	-	-	1.80	nr	1.80
Griselinia littoralis; 45-60 3 L	-	-	-	-	2.95	nr	2.95
Hebe albicans; 20-30 2 L	-	-	-	-	1.80	nr	1.80
Hebe Great Orme; 20-30 3 L	-	-	-	-	2.60	nr	2.60
Hebe pinguifolia Pagei; 15-20 2 L	-	-	-	-	1.50	nr	1.50
Hebe rakaiensis; 20-30 2 L	-	-	-	-	1.50	nr	1.50
Hedera helix Glacier; 45-60 1.5 L	-	-	-	-	2.10	nr	2.10
Hippophae rhamnoides; 45-60 3 L	-	-	-	-	1.60	nr	1.60
Hydrangea macrophylla Blue Wave; 45-60 4 L	-	-	-	-	4.20	nr	4.20
Hypericum Hidcote; 30-45 3 L	-	-	-	-	1.80	nr	1.80
Ilex aquifolium Pyramidalis; 60-90 3 L	-	-	-	-	3.90	nr	3.90

Prices for Measured Works
Q PAVING/PLANTING/FENCING/SITE FURNITURE

Item Excluding site overheads and profit	PC £	Labour hours	Labour £	Plant £	Material £	Unit	Total rate £
Q31 PLANTING – cont'd							
Market prices of trees, shrubs and plants: Notcutts Nurseries Ltd – cont'd							
Lavandula angustifolia Hidcote; 15-20 1.5 L	-	-	-	-	1.40	nr	1.40
Lavandula angustifolia Hidcote; 20-30 3 L	-	-	-	-	2.20	nr	2.20
Lavatera Barnsley; 45-60 3 L	-	-	-	-	2.45	nr	2.45
Leucothoe walteri Rainbow; 30-45 3 L	-	-	-	-	4.50	nr	4.50
Leycesteria formosa; 30-45 3 L	-	-	-	-	2.80	nr	2.80
Ligustrum ovalifolium; 45-60 2 L	-	-	-	-	1.40	nr	1.40
Ligustrum ovalifolium Aureum; 30-45 2 L	-	-	-	-	1.80	nr	1.80
Ligustrum nitida Baggesens Gold; 30-45 3 L	-	-	-	-	1.90	nr	1.90
Mahonia japonica; 45-60 3 L	-	-	-	-	2.60	nr	2.60
Olearia haastii; 30-45 3 L	-	-	-	-	2.20	nr	2.20
Osmanthus delavayi; 30-45 3 L	-	-	-	-	3.20	nr	3.20
Pachysandra terminalis; 15-20 1.5 L	-	-	-	-	1.60	nr	1.60
Pernettya mucronata Male; 20-30 3 L	-	-	-	-	2.60	nr	2.60
Philadelphus Belle Etoile; 45-60 3 L	-	-	-	-	1.80	nr	1.80
Philadelphus coronarius Aureus; 30-45 3 L	-	-	-	-	2.80	nr	2.80
Philadelphus Virginal; 45-60 3 L	-	-	-	-	2.40	nr	2.40
Phormium tenax; 30-45 3 L	-	-	-	-	3.50	nr	3.50
Photinia x fraseri Red Robin; 30-45 3 L	-	-	-	-	3.60	nr	3.60
Potentilla arbuscula; 20-30 3 L	-	-	-	-	1.60	nr	1.60
Potentilla fruticosa Pretty Polly; 20-30 3 L	-	-	-	-	2.45	nr	2.45
Prunus lusitanica; 45-60 3 L	-	-	-	-	1.80	nr	1.80
Pyracantha Orange Glow; 60-90 3 L	-	-	-	-	1.80	nr	1.80
Pyracantha Teton; 30-45 3 L	-	-	-	-	2.20	nr	2.20
Ribes sanguineum King Edward VII; 45-60 3 L	-	-	-	-	2.10	nr	2.10
Rosmarinus officinalis; 20-30 3 L	-	-	-	-	2.20	nr	2.20
Rubus Betty Ashburner; 45-60 2 L	-	-	-	-	1.30	nr	1.30
Rubus Tridel Benenden; 45-60 3 L	-	-	-	-	2.40	nr	2.40
Salix lanata; 45-60 3 L	-	-	-	-	2.60	nr	2.60
Salvia officinalis Icterina; 20-30 3 L	-	-	-	-	2.60	nr	2.60
Sambucus nigra Aurea; 45-60 3 L	-	-	-	-	1.40	nr	1.40
Sambucus racemosa Plumosa Aurea; 45-60 3 L	-	-	-	-	3.95	nr	3.95
Santolina chamaecyparissus; 20-30 3 L	-	-	-	-	1.60	nr	1.60
Sarcococca confusa; 20-30 2 L	-	-	-	-	2.80	nr	2.80
Senecio monroi; 20-25 2 L	-	-	-	-	2.60	nr	2.60
Senecio Sunshine; 30-45 3 L	-	-	-	-	1.60	nr	1.60
Skimmia japonica; 30-45 3 L	-	-	-	-	2.60	nr	2.60
Spiraea x arguta; 45-60 3 L	-	-	-	-	1.80	nr	1.80
Spiraea bumalda x Goldflame; 20-30 2 L	-	-	-	-	1.60	nr	1.60
Stephanandra incisa Crispa; 30-45 2 L	-	-	-	-	1.60	nr	1.60
Symphoricarpos albus; 45-60 3 L	-	-	-	-	1.60	nr	1.60
Symphoricarpos x chenaultii Hancock; 45-60 2 L	-	-	-	-	1.20	nr	1.20
Tamarix tetrandra; 45-60 3 L	-	-	-	-	2.40	nr	2.40
Viburnum x bodnantense Dawn; 45-60 3 L	-	-	-	-	3.40	nr	3.40
Viburnum opulus; 60-90 3 L	-	-	-	-	1.60	nr	1.60
Viburnum opulus Sterile; 45-60 3 L	-	-	-	-	2.40	nr	2.40
Viburnum plicatum Pink Beauty; 45-60 3 L	-	-	-	-	4.30	nr	4.30
Viburnum tinus; 30-45 3 L	-	-	-	-	1.95	nr	1.95
Viburnum tinus Variegatum; 20-30 3 L	-	-	-	-	2.90	nr	2.90
Container grown climbers							
Clematis Jackmanii; 60-90 2 L	-	-	-	-	3.25	nr	3.25
Clematis montana Elizabeth; 60-90 2 L	-	-	-	-	3.25	nr	3.25
Hydrangea petiolaris; 45-60 3 L	-	-	-	-	3.55	nr	3.55
Jasminum nudiflorum; 45-60 2 L	-	-	-	-	2.60	nr	2.60

Prices for Measured Works
Q PAVING/PLANTING/FENCING/SITE FURNITURE

Item Excluding site overheads and profit	PC £	Labour hours	Labour £	Plant £	Material £	Unit	Total rate £
Lonicera japonica Halliana; 60-90 3 L	-	-	-	-	2.60	nr	2.60
Parthenocissus quinquefolia; 60-90 2 L	-	-	-	-	2.40	nr	2.40
Passiflora caerulea; 60-90 2 L	-	-	-	-	3.50	nr	3.50
Vitis coignetiae; 60-90 2 L	-	-	-	-	5.50	nr	5.50
Wisteria sinensis; 60-90 3 L	-	-	-	-	6.50	nr	6.50
Container grown specimen shrubs							
Amelanchier lamarckii; 90-120 10 L	-	-	-	-	8.50	nr	8.50
Aralia elata; 120-150 10 L	-	-	-	-	20.00	nr	20.00
Aucuba japonica; 45-60 10 L	-	-	-	-	9.50	nr	9.50
Berberis darwinii; 45-60 10 L	-	-	-	-	9.50	nr	9.50
Choisya ternata; 45-60 10 L	-	-	-	-	9.50	nr	9.50
Cotoneaster cornubia; 90-120 10 L	-	-	-	-	8.50	nr	8.50
Cotoneaster Exburiensis; 90-120 10 L	-	-	-	-	9.50	nr	9.50
Cotoneaster watereri; 90-120 10 L	-	-	-	-	8.50	nr	8.50
Elaeagnus pungens Maculata; 45-60 10 L	-	-	-	-	12.50	nr	12.50
Elaeagnus ebbingei; 60-90 10 L	-	-	-	-	8.50	nr	8.50
Elaeagnus ebbingei Limelight; 60-90 10 L	-	-	-	-	12.50	nr	12.50
Escallonia rubra Crimson Spire; 60-90 10 L	-	-	-	-	8.50	nr	8.50
Euonymus fortunei Emerald N Gold; 30-45 7.3 L	-	-	-	-	8.00	nr	8.00
Fatsia japonica; 90-120 15 L	-	-	-	-	12.50	nr	12.50
Ilex aquifolium; 90-120 10 L	-	-	-	-	9.50	nr	9.50
Ilex aquifolium Golden King; 90-120 10 L	-	-	-	-	15.00	nr	15.00
Ilex aquifolium Argentea Marginata; 90-120 10 L	-	-	-	-	15.00	nr	15.00
Mahonia japonica; 60-90 10 L	-	-	-	-	12.50	nr	12.50
Phormium tenax; 60-90 10 L	-	-	-	-	14.00	nr	14.00
Phormium tenax Purpureum; 60-90 10 L	-	-	-	-	14.00	nr	14.00
Phormium Yellow Wave; 60-90 10 L	-	-	-	-	18.00	nr	18.00
Photinia x Fraseri Red Robin; 60-90 10 L	-	-	-	-	12.50	nr	12.50
Prunus x cistena Crimson Dwarf; 45-60 10 L	-	-	-	-	12.50	nr	12.50
Prunus laurocerasus; 90-120 10 L	-	-	-	-	8.50	nr	8.50
Prunus laurocerasus Otto Luyken; 45-60 10 L	-	-	-	-	8.50	nr	8.50
Prunus laurocerasus Zabeliana; 45-60 10 L	-	-	-	-	8.50	nr	8.50
Prunus lusitanica; 45-60 10 L	-	-	-	-	9.50	nr	9.50
Pyracantha Mohave; 45-60 10 L	-	-	-	-	8.50	nr	8.50
Pyracantha Orange Glow; 60-90 10 L	-	-	-	-	8.50	nr	8.50
Pyracantha Red Column; 60-90 10 L	-	-	-	-	8.50	nr	8.50
Pyracantha Soleil d'Or; 60-90 10 L	-	-	-	-	8.50	nr	8.50
Pyracantha Teton; 45-60 10 L	-	-	-	-	8.50	nr	8.50
Rhus typhina; 90-120 10 L	-	-	-	-	8.50	nr	8.50
Rosmarinus officinalis; 30-45 10 L	-	-	-	-	10.00	nr	10.00
Rubus cockburnianus Golden Vale; 90-120 10 L	-	-	-	-	10.50	nr	10.50
Skimmia Rubella; 30-45 7.5 L	-	-	-	-	9.50	nr	9.50
Spiraea japonica Little Princess; 30-45 7.5 L	-	-	-	-	8.50	nr	8.50
Spiraea Snowmound; 45-60 10 L	-	-	-	-	8.50	nr	8.50
Spiraea thunbergii; 45-60 10 L	-	-	-	-	8.50	nr	8.50
Bamboos							
Arundinaria auricoma; 40-50 7.5 L	-	-	-	-	18.50	nr	18.50
Container grown conifers							
X Cupressocyparis leylandii; 120-150 10 L	-	-	-	-	7.50	nr	7.50
X Cupressocyparis leylandii Castlewellan; 120-150 10 L	-	-	-	-	7.50	nr	7.50
Juniperus communis Repanda; 30-45 10 L	-	-	-	-	8.50	nr	8.50
Juniperus x media pfitzeriana Aurea; 45-60 10L	-	-	-	-	8.50	nr	8.50
Juniperus sabina Tamariscifolia; 45-60 10 L	-	-	-	-	8.50	nr	8.50
Pinus mugo; 45-60 10 L	-	-	-	-	12.50	nr	12.50
Pinus sylvestris; 90-120 10 L	-	-	-	-	8.50	nr	8.50
Pinus sylvestris; 150-180 50 L	-	-	-	-	60.00	nr	60.00
Pinus sylvestris; 180-210 80 L	-	-	-	-	105.00	nr	105.00

Prices for Measured Works
Q PAVING/PLANTING/FENCING/SITE FURNITURE

Item Excluding site overheads and profit	PC £	Labour hours	Labour £	Plant £	Material £	Unit	Total rate £
Q31 PLANTING – cont'd							
Market prices of trees, shrubs and plants: Notcutts Nurseries Ltd – cont'd							
Chamaecyparis lawsoniana Ellwoodii; 30-45 2 L	-	-	-	-	2.20	nr	2.20
Chamaecyparis lawsoniana Ellwoods Gold; 30-45 2 L	-	-	-	-	2.90	nr	2.90
Chamaecyparis lawsoniana Ellwoods Pillar; 30-45 2 L	-	-	-	-	2.90	nr	2.90
X Cupressocyparis leylandii; 60-90 3 L	-	-	-	-	1.80	nr	1.80
X Cupressocyparis leylandii Castlewellan; 60-90 3 L	-	-	-	-	1.80	nr	1.80
Juniperus communis Compressa; 30-45 2 L	-	-	-	-	2.60	nr	2.60
Juniperus communis Repanda; 30-45 3 L	-	-	-	-	2.60	nr	2.60
Juniperus conferta; 30-45 3 L	-	-	-	-	2.80	nr	2.80
Juniperus media Mint Julep; 30-45 3 L	-	-	-	-	2.90	nr	2.90
Juniperus media pfitzeriana; 30-45 3 L	-	-	-	-	2.70	nr	2.70
Juniperus media pfitzeriana Aurea; 30-45 3 L	-	-	-	-	2.90	nr	2.90
Juniperus sabina Tamariscifolia; 30-45 3 L	-	-	-	-	2.70	nr	2.70
Juniperus squamata Blue Star; 30-45 2 L	-	-	-	-	3.20	nr	3.20
Larix decidua; 60-90 3 L	-	-	-	-	1.60	nr	1.60
Picea albertiana Conica; 20-30 3 L	-	-	-	-	2.80	nr	2.80
Pinus mugo; 20-30 2 L	-	-	-	-	2.20	nr	2.20
Pinus nigra; 45-60 3 L	-	-	-	-	1.90	nr	1.90
Pinus nigra austriaca; 45-60 3 L	-	-	-	-	1.60	nr	1.60
Pinus nigra maritima; 45-60 3 L	-	-	-	-	1.60	nr	1.60
Pinus sylvestris; 45-60 3 L	-	-	-	-	1.60	nr	1.60
Transplants 45-60							
Acer campestre	-	-	-	-	0.19	nr	0.19
Acer platanoides	-	-	-	-	0.19	nr	0.19
Acer pseudoplatanus	-	-	-	-	0.21	nr	0.21
Alnus cordata	-	-	-	-	0.21	nr	0.21
Alnus glutinosa	-	-	-	-	0.21	nr	0.21
Alnus incana	-	-	-	-	0.21	nr	0.21
Betula pendula	-	-	-	-	0.21	nr	0.21
Carpinus betulus	-	-	-	-	0.24	nr	0.24
Cornus sanguinea	-	-	-	-	0.24	nr	0.24
Corylus avellana	-	-	-	-	0.24	nr	0.24
Crataegus monogyna	-	-	-	-	0.16	nr	0.16
Fagus sylvatica	-	-	-	-	0.26	nr	0.26
Fraxinus excelsior	-	-	-	-	0.24	nr	0.24
Hippophae rhamnoides	-	-	-	-	0.28	nr	0.28
Prunus avium	-	-	-	-	0.24	nr	0.24
Quercus robur	-	-	-	-	0.28	nr	0.28
Rosa rugosa	-	-	-	-	0.26	nr	0.26
Rosa rugosa Alba	-	-	-	-	0.26	nr	0.26
Sambucus nigra	-	-	-	-	0.24	nr	0.24
Sorbus aria	-	-	-	-	0.30	nr	0.30
Sorbus intermedia	-	-	-	-	0.32	nr	0.32
Viburnum opulus	-	-	-	-	0.32	nr	0.32
Transplants 60-90							
Acer campestre	-	-	-	-	0.21	nr	0.21
Acer platanoides	-	-	-	-	0.24	nr	0.24
Alnus cordata	-	-	-	-	0.24	nr	0.24
Amelanchier canadensis	-	-	-	-	0.28	nr	0.28
Cornus alba	-	-	-	-	0.24	nr	0.24
Crataegus monogyna	-	-	-	-	0.24	nr	0.24

Prices for Measured Works
Q PAVING/PLANTING/FENCING/SITE FURNITURE

Item Excluding site overheads and profit	PC £	Labour hours	Labour £	Plant £	Material £	Unit	Total rate £
Fagus sylvatica	-	-	-	-	0.32	nr	0.32
Fraxinus excelsior	-	-	-	-	0.26	nr	0.26
Populus alba	-	-	-	-	0.26	nr	0.26
Populus nigra italica	-	-	-	-	0.28	nr	0.28
Populus trichocarpa	-	-	-	-	0.28	nr	0.28
Prunus avium	-	-	-	-	0.28	nr	0.28
Salix alba	-	-	-	-	0.21	nr	0.21
Salix caprea	-	-	-	-	0.26	nr	0.26
Salix fragilis	-	-	-	-	0.24	nr	0.24
Salix rosmarinifolia	-	-	-	-	0.24	nr	0.24
Salix viminalis	-	-	-	-	0.24	nr	0.24
Salix vitellina	-	-	-	-	0.24	nr	0.24
Sorbus aucuparia	-	-	-	-	0.26	nr	0.26
Herbaceous							
Achillea fil. Cloth of Gold; 10-15 2 L	-	-	-	-	1.80	nr	1.80
Alchemilla mollis; 10-15 2 L	-	-	-	-	1.80	nr	1.80
Anemone hybrida Honorine Jobert; 10-15 3 L	-	-	-	-	2.20	nr	2.20
Artemisia absin. Lambrook Silver; 10-15 3 L	-	-	-	-	2.45	nr	2.45
Artemisia Powis Castle; 10-15 3 L	-	-	-	-	2.45	nr	2.45
Bergenia Bressingham Salmon; 20-30 2 L	-	-	-	-	2.25	nr	2.25
Convallaria majalis; 10-15 2 L	-	-	-	-	1.80	nr	1.80
Euphorbia amygdaloides Purpurea; 15-20 2 L	-	-	-	-	2.20	nr	2.20
Euphorbia wulfenii; 10-15 2 L	-	-	-	-	2.60	nr	2.60
Geranium oxonianum Wargrave Pink; 10-15 3 L	-	-	-	-	1.80	nr	1.80
Geranium Johnson Blue; 10-15 3 L	-	-	-	-	1.80	nr	1.80
Hosta August Moon; 10-15 2 L	-	-	-	-	2.50	nr	2.50
Hosta sieboldii; 10-15 2 L	-	-	-	-	2.50	nr	2.50
Iris foetidissima; 10-15 2 L	-	-	-	-	2.20	nr	2.20
Papaver orientale; 10-15 3 L	-	-	-	-	1.80	nr	1.80
Penstemon Garnet; 10-15 3 L	-	-	-	-	2.25	nr	2.25
Pulmonaria angustifolia; 10-15 2 L	-	-	-	-	1.80	nr	1.80
Salvia East Friesland; 10-15 3 L	-	-	-	-	1.80	nr	1.80
Sedum spectabilis Autumn Joy; 10-15 3 L	-	-	-	-	1.80	nr	1.80
Tiarella cordifolia; 10-15 2 L	-	-	-	-	1.80	nr	1.80
Waldsteinia ternata; 10-15 2 L	-	-	-	-	1.80	nr	1.80
Bare root trees 1.50 - 1.80 feathered							
Acer campestre	-	-	-	-	2.00	nr	2.00
Acer platanoides	-	-	-	-	1.50	nr	1.50
Acer pseudoplatanus	-	-	-	-	1.50	nr	1.50
Alnus cordata	-	-	-	-	1.50	nr	1.50
Alnus glutinosa	-	-	-	-	1.50	nr	1.50
Alnus incana	-	-	-	-	3.00	nr	3.00
Amelanchier canadensis	-	-	-	-	6.00	nr	6.00
Betula pendula	-	-	-	-	1.50	nr	1.50
Carpinus betulus	-	-	-	-	2.50	nr	2.50
Fagus sylvatica	-	-	-	-	3.50	nr	3.50
Fagus sylvatica purpurea	-	-	-	-	5.50	nr	5.50
Fraxinus excelsior	-	-	-	-	2.50	nr	2.50
Populus alba	-	-	-	-	1.50	nr	1.50
Populus nigra italica	-	-	-	-	1.50	nr	1.50
Populus tremula	-	-	-	-	5.50	nr	5.50
Prunus avium	-	-	-	-	2.50	nr	2.50
Prunus avium 'Plena'	-	-	-	-	6.00	nr	6.00
Quercus robur	-	-	-	-	9.50	nr	9.50
Quercus rubra	-	-	-	-	4.50	nr	4.50
Sorbus lutescens	-	-	-	-	3.00	nr	3.00
Sorbus aucuparia	-	-	-	-	2.00	nr	2.00

Prices for Measured Works
Q PAVING/PLANTING/FENCING/SITE FURNITURE

Item Excluding site overheads and profit	PC £	Labour hours	Labour £	Plant £	Material £	Unit	Total rate £
Q31 PLANTING – cont'd							
Market prices of trees, shrubs and plants: Notcutts Nurseries Ltd – cont'd							
Bare root trees 1.80 - 2.10 feathered							
Acer campestre	-	-	-	-	2.00	nr	2.00
Alnus cordata	-	-	-	-	3.00	nr	3.00
Acer platanoides	-	-	-	-	3.00	nr	3.00
Acer pseudoplatanus	-	-	-	-	3.00	nr	3.00
Alnus glutinosa	-	-	-	-	3.00	nr	3.00
Alnus incana	-	-	-	-	2.50	nr	2.50
Amelanchier canadensis	-	-	-	-	12.00	nr	12.00
Betula pendula	-	-	-	-	3.00	nr	3.00
Carpinus betulus	-	-	-	-	4.50	nr	4.50
Fagus sylvatica	-	-	-	-	5.50	nr	5.50
Fagus sylvatica purpurea	-	-	-	-	9.50	nr	9.50
Fraxinus excelsior	-	-	-	-	4.00	nr	4.00
Populus alba	-	-	-	-	3.50	nr	3.50
Populus nigra italica	-	-	-	-	3.50	nr	3.50
Populus tremula	-	-	-	-	3.50	nr	3.50
Prunus avium	-	-	-	-	4.00	nr	4.00
Prunus avium 'Plena'	-	-	-	-	8.00	nr	8.00
Quercus robur	-	-	-	-	9.50	nr	9.50
Quercus rubra	-	-	-	-	10.50	nr	10.50
Sorbus aucuparia	-	-	-	-	4.50	nr	4.50
Sorbus lutescens	-	-	-	-	9.00	nr	9.00
Bare root trees 8-10 cm							
Acer campestre	-	-	-	-	11.00	nr	11.00
Acer platanoides	-	-	-	-	9.00	nr	9.00
Acer pseudoplatanus	-	-	-	-	9.00	nr	9.00
Alnus cordata	-	-	-	-	9.00	nr	9.00
Alnus glutinosa	-	-	-	-	9.00	nr	9.00
Alnus incana	-	-	-	-	9.00	nr	9.00
Amelanchier canadensis	-	-	-	-	25.00	nr	25.00
Betula pendula	-	-	-	-	8.00	nr	8.00
Carpinus betulus	-	-	-	-	12.00	nr	12.00
Fagus sylvatica	-	-	-	-	12.00	nr	12.00
Fraxinus excelsior	-	-	-	-	13.50	nr	13.50
Fagus sylvatica Purpurea	-	-	-	-	20.00	nr	20.00
Populus tremula	-	-	-	-	11.00	nr	11.00
Populus alba	-	-	-	-	9.50	nr	9.50
Populus nigra Italica	-	-	-	-	9.50	nr	9.50
Prunus avium	-	-	-	-	10.00	nr	10.00
Prunus avium Plena	-	-	-	-	13.00	nr	13.00
Quercus robor	-	-	-	-	12.00	nr	12.00
Quercus rubra	-	-	-	-	17.00	nr	17.00
Sorbus aucuparia	-	-	-	-	11.00	nr	11.00
Bare root trees 10-12 cm sel. standard							
Acer campestre	-	-	-	-	14.00	nr	14.00
Alnus cordata	-	-	-	-	10.00	nr	10.00
Acer platanoides	-	-	-	-	10.00	nr	10.00
Acer pseudoplatanus	-	-	-	-	10.00	nr	10.00
Amelanchier canadensis	-	-	-	-	40.00	nr	40.00
Betula pendula	-	-	-	-	10.00	nr	10.00
Carpinus betulus	-	-	-	-	27.50	nr	27.50
Fagus sylvatica	-	-	-	-	18.00	nr	18.00
Fagus sylvatica purpurea	-	-	-	-	30.00	nr	30.00

Prices for Measured Works
Q PAVING/PLANTING/FENCING/SITE FURNITURE

Item Excluding site overheads and profit	PC £	Labour hours	Labour £	Plant £	Material £	Unit	Total rate £
Fraxinus excelsior	-	-	-	-	12.00	nr	12.00
Populus alba	-	-	-	-	12.50	nr	12.50
Populus nigra italica	-	-	-	-	12.50	nr	12.50
Populus tremula	-	-	-	-	16.00	nr	16.00
Prunus avium	-	-	-	-	12.00	nr	12.00
Prunus avium 'Plena'	-	-	-	-	16.00	nr	16.00
Quercus robur	-	-	-	-	21.00	nr	21.00
Quercus rubra	-	-	-	-	21.00	nr	21.00
Sorbus aucuparia	-	-	-	-	15.00	nr	15.00
Sorbus lutescens	-	-	-	-	18.00	nr	18.00
Bare root trees 12-14 cm heavy standard							
Acer campestre	-	-	-	-	24.00	nr	24.00
Alnus cordata	-	-	-	-	20.00	nr	20.00
Acer platanoides	-	-	-	-	20.00	nr	20.00
Acer pseudoplatanus	-	-	-	-	20.00	nr	20.00
Alnus glutinosa	-	-	-	-	20.00	nr	20.00
Alnus incana	-	-	-	-	20.00	nr	20.00
Amelanchier canadensis	-	-	-	-	70.00	nr	70.00
Betula pendula	-	-	-	-	20.00	nr	20.00
Carpinus betulus	-	-	-	-	28.00	nr	28.00
Fagus sylvatica	-	-	-	-	28.00	nr	28.00
Fagus sylvatica Purpurea	-	-	-	-	45.00	nr	45.00
Fraxinus excelsior	-	-	-	-	24.00	nr	24.00
Populus alba	-	-	-	-	22.00	nr	22.00
Populus nigra italica	-	-	-	-	12.50	nr	12.50
Populus tremula	-	-	-	-	16.00	nr	16.00
Prunus avium	-	-	-	-	24.00	nr	24.00
Prunus avium 'Plena'	-	-	-	-	28.00	nr	28.00
Quercus robur	-	-	-	-	32.00	nr	32.00
Quercus rubra	-	-	-	-	32.00	nr	32.00
Sorbus aucuparia	-	-	-	-	22.00	nr	22.00
Sorbus lutescens	-	-	-	-	30.00	nr	30.00
Market prices of trees; Coblands Nurseries Ltd							
Bare root trees 14-16 cm extra heavy standard							
Acer campestre	-	-	-	-	55.00	nr	55.00
Acer platanoides	-	-	-	-	40.00	nr	40.00
Acer pseudoplatanus	-	-	-	-	40.00	nr	40.00
Alnus glutinosa	-	-	-	-	40.00	nr	40.00
Alnus incana	-	-	-	-	115.00	nr	115.00
Amelanchier lamarckii	-	-	-	-	55.00	nr	55.00
Betula pendula	-	-	-	-	40.00	nr	40.00
Carpinus betulus	-	-	-	-	55.00	nr	55.00
Fagus sylvatica	-	-	-	-	55.00	nr	55.00
Fagus sylvatica Dawyck purple	-	-	-	-	115.00	nr	115.00
Fraxinus excelsior	-	-	-	-	50.00	nr	50.00
Populus tremula	-	-	-	-	50.00	nr	50.00
Prunus avium	-	-	-	-	40.00	nr	40.00
Prunus avium 'Plena'	-	-	-	-	55.00	nr	55.00
Quercus robur	-	-	-	-	50.00	nr	50.00
Quercus rubra	-	-	-	-	55.00	nr	55.00
Robinia pseudoacasia Frisia	-	-	-	-	115.00	nr	115.00
Sorbus aucuparia	-	-	-	-	40.00	nr	40.00
Sorbus aria	-	-	-	-	50.00	nr	50.00
Root ball trees; extra over for root balling and wrapping in hessian	-	-	-	-	30.00	nr	30.00

Prices for Measured Works
Q PAVING/PLANTING/FENCING/SITE FURNITURE

Item Excluding site overheads and profit	PC £	Labour hours	Labour £	Plant £	Material £	Unit	Total rate £
Q31 PLANTING – cont'd							
Market prices of aquatic plants in prepared growing medium of approved hessian bag filled with soil and fertilizer; tied, slit and weighted and placed in position in pool to receive three submerged floating leafed marginal or swamp plants; Anglo Aquarium Plant							
Water lilies							
Nymphaea in variety - white	-	-	-	-	5.80	nr	5.80
Nymphaea in variety - deep red	-	-	-	-	7.40	nr	7.40
Deep water aquatics							
Aponogeton distachyos	-	-	-	-	2.10	nr	2.10
Nymphoides peltata	-	-	-	-	1.69	nr	1.69
Marginal plants							
Acorus calamus	-	-	-	-	1.69	nr	1.69
Butomus umbellatus	-	-	-	-	1.69	nr	1.69
Caltha palustris	-	-	-	-	1.69	nr	1.69
Carex acuta	-	-	-	-	1.69	nr	1.69
Iris pseudacorus	-	-	-	-	1.69	nr	1.69
Juncus effusus	-	-	-	-	1.69	nr	1.69
Mentha aquatica	-	-	-	-	1.69	nr	1.69
Mimulus luteus	-	-	-	-	1.69	nr	1.69
Phragmites (australis)	-	-	-	-	1.69	nr	1.69
Ranunculus lingua	-	-	-	-	1.69	nr	1.69
Scirpus lacustris	-	-	-	-	1.69	nr	1.69
Sagittaria sagittifolia var. leucopetala	-	-	-	-	3.00	nr	3.00
Typha latifolia	-	-	-	-	1.69	nr	1.69
Submerged oxygenating plants							
Ceratophyllum demersum; containerised	-	-	-	-	34.00	100	34.00
Lagarosiphon major (Elodea crispa); containerised	-	-	-	-	34.00	100	34.00
Ranunculus aquatilis; containerised	-	-	-	-	34.00	100	34.00
Market Prices of Bulbs;							
Dutch Hyacinths (prepared for early flowering); 15 to 16 cm							
"Amsterdam", cerise red	-	-	-	-	161.00	100	161.00
"Anna Marie", clear pink	-	-	-	-	154.00	100	154.00
"Carnegie", white, dense spike	-	-	-	-	150.00	100	150.00
"Ostara", deep blue	-	-	-	-	157.00	100	157.00
"City of Haarlem", primrose yellow	-	-	-	-	179.00	100	179.00
Dutch Hyacinths (not prepared); 15 to 16 cm							
"Amsterdam", cerise red, Medium	-	-	-	-	129.00	100	129.00
"Anna Marie", clear pink, Early	-	-	-	-	125.00	100	125.00
"Carnegie", white, dense spike, Late	-	-	-	-	125.00	100	125.00
"Ostara", deep blue, Early	-	-	-	-	122.00	100	122.00
"City of Haarlem", primrose yellow, Late	-	-	-	-	122.00	100	122.00
Tulips, Single Early,							
"Apricot Beauty", soft salmon rose	-	-	-	-	80.00	1000	80.00
"Best Seller", bright coppery orange	-	-	-	-	91.00	1000	91.00
"Superb Mixed", blended colours	-	-	-	-	86.00	1000	86.00
Tulips, Double Early							
"Abba", glowing tomato red	-	-	-	-	70.00	1000	70.00
"Electra", deep cherry red	-	-	-	-	119.00	1000	119.00
"Murillo Sports Mixed", selected species	-	-	-	-	96.00	1000	96.00

Prices for Measured Works
Q PAVING/PLANTING/FENCING/SITE FURNITURE

Item Excluding site overheads and profit	PC £	Labour hours	Labour £	Plant £	Material £	Unit	Total rate £
Tulips, Triumph							
"Abu Hassan", deep mahogany, broad yellow edge	-	-	-	-	77.00	1000	77.00
"Arabian Mystery", deep purple violet, edged white	-	-	-	-	91.00	1000	91.00
"Coquette", creamy white	-	-	-	-	93.00	1000	93.00
"Superb Mixed", blended colours	-	-	-	-	80.00	1000	80.00
Tulips, Darwin Hybrid							
"Ad Rem", scarlet	-	-	-	-	74.00	1000	74.00
"Golden Parade", deep yellow	-	-	-	-	76.00	1000	76.00
"Superb Mixed", blended colours	-	-	-	-	77.00	1000	77.00
Tulips, Single Late or Cottage							
"Alabaster", pure white	-	-	-	-	76.00	1000	76.00
"Blushing Beauty", red, feathered rosy white	-	-	-	-	99.00	1000	99.00
"Superb Mixed", blended colours	-	-	-	-	88.00	1000	88.00
Tulips, Lily Flowered							
"Aladdin", scarlet, edged yellow	-	-	-	-	74.00	1000	74.00
"Superb Mixed", blended colours	-	-	-	-	84.00	1000	84.00
Tulips, Parrot							
"Apricot Parrot", pale apricot yellow, tinged creamy white	-	-	-	-	89.00	1000	89.00
"Rococo", carmine edged glowing red	-	-	-	-	74.00	1000	74.00
"Superb Mixed", blended colours	-	-	-	-	100.00	1000	100.00
Tulips, Double Late							
"Lilac Perfection", lilac	-	-	-	-	116.00	1000	116.00
"Upstar", pale pink and white	-	-	-	-	74.00	1000	74.00
"Superb Mixed", blended colours	-	-	-	-	93.00	1000	93.00
Tulips, Greigii							
"Ali Baba", rose	-	-	-	-	76.00	1000	76.00
"Mary Ann", carmine red, edged white	-	-	-	-	100.00	1000	100.00
"Superb Mixed", blended colours	-	-	-	-	84.00	1000	84.00
Daffodils, Yellow Trumpets; 15/17cm							
"Dutch Master", large golden yellow	-	-	-	-	55.00	1000	55.00
"King Alfred", medium golden yellow	-	-	-	-	55.00	1000	55.00
Daffodils, Bi-Coloured Trumpets; 15/17cm							
"Magnet", perianth creamy white, lemon yellow trumpet	-	-	-	-	75.00	1000	75.00
Daffodils, White Trumpets; 15/17cm							
"Mount Hood", pure white perianth, pure white frilled trumpet	-	-	-	-	90.00	1000	90.00
Narcissi, Large Cupped; 15/17cm							
"Carbineer", rich yellow perianth, bright orange cup	-	-	-	-	65.00	1000	65.00
"St. Keverne", deep golden yellow	-	-	-	-	70.00	1000	70.00
"Salome", white perianth, shaded pink trumpet	-	-	-	-	85.00	1000	85.00
Narcissi, Small Cupped; 15/17cm							
"Birma", soft yellow perianth, orange scarlet cup	-	-	-	-	70.00	1000	70.00
"Verger", white perianth, orange scarlet cup	-	-	-	-	75.00	1000	75.00
Narcissi, Double; 15/17cm							
"Flower Drift", white perianth, orange yellow cup	-	-	-	-	75.00	1000	75.00
"Texas", yellow outer petals, inter-mixed with orange scarlet	-	-	-	-	85.00	1000	85.00
Narcissi, Triandrus							
"Albus" (Angel Tears), reflexing silvery white flowers in clusters	-	-	-	-	94.00	1000	94.00
"Hawera", 2 or 3 creamy yellow flowers on a stem	-	-	-	-	40.00	1000	40.00

Prices for Measured Works
Q PAVING/PLANTING/FENCING/SITE FURNITURE

Item Excluding site overheads and profit	PC £	Labour hours	Labour £	Plant £	Material £	Unit	Total rate £
Q31 PLANTING – cont'd							
Market Prices of Bulbs; - cont'd							
Narcissi, Cyclamineus							
"Jack Snipe", long creamy white reflexing petals, short orange yellow cup	-	-	-	-	88.00	1000	88.00
Narcissi, Jonquilla							
"Jonquilla Single", buttercup yellow flowers, sweet scented	-	-	-	-	74.00	1000	74.00
"Sugar Bush", ivory white petals, orange yellow cup	-	-	-	-	98.00	1000	98.00
Narcissi, Tazetta or Poetaz; 15/17cm							
"Cheerfulness", creamy white double flowers, scented	-	-	-	-	65.00	1000	65.00
"Sir Winston Churchill", double geranium, white perianth, white trumpet	-	-	-	-	75.00	1000	75.00
Daffodils, Miniature							
"Lobularis", lemon yellow tapering perianth, darker yellow trumpet, very early	-	-	-	-	73.00	1000	73.00
"Minnow", miniature tazetta hybrid, 2-3 soft flowers on each stem	-	-	-	-	44.00	1000	44.00
"Nanus", a delightful yellow miniature	-	-	-	-	111.00	1000	111.00
Daffodils & Narcissi, Special Mixture; 15/17cm							
suitable for planting in parkland, river banks etc.	-	-	-	-	55.00	1000	55.00
Crocus, Spring Flowering Dutch; 8/9cm							
Blue	-	-	-	-	32.00	1000	32.00
Yellow	-	-	-	-	32.00	1000	32.00
Striped	-	-	-	-	40.00	1000	40.00
Mixed	-	-	-	-	32.00	1000	32.00
Crocus, Winter Flowering Species; , 5/7cm							
"Ancyrensis" (Golden Bunch), maize yellow with orange interior	-	-	-	-	27.00	1000	27.00
"Chrysanthus Advance", yellow and violet	-	-	-	-	27.00	1000	27.00
"Sieberi Violet Queen", dark blue	-	-	-	-	30.00	1000	30.00
Species Mixed, blended colours	-	-	-	-	21.00	1000	21.00
Iris, Dutch; 7/8cm							
"Golden Harvest", golden yellow	-	-	-	-	38.00	1000	38.00
"Imperator", very deep blue	-	-	-	-	34.00	1000	34.00
"Superb Mixed", blended colours	-	-	-	-	33.00	1000	33.00
Iris, Species							
"Danfordiae", sweet scented, deep lemon yellow with brown markings; dwarf variety	-	-	-	-	35.00	1000	35.00
"Reticulata", deep purple violet with yellow blotch; dwarf variety	-	-	-	-	43.00	1000	43.00
"Reticulata Harmony", royal blue with yellow rimmed white blotch	-	-	-	-	40.00	1000	40.00
"Superb Mixed", blended colours	-	-	-	-	39.00	1000	39.00
Allium							
"Giganteum", deep violet, star shaped flowers, height 80 cm	-	-	-	-	179.00	100	179.00
"Neapolitanum", pure white, height 30 cm	-	-	-	-	21.00	1000	21.00
Amaryllis Hippeastrum; 26/28cm							
"Apple Blossom", blushing pink	-	-	-	-	17.00	10	17.00
"Striped", red and white	-	-	-	-	17.00	10	17.00
Anemones; 5/6cm							
"Single De Caen", mixed shades	-	-	-	-	27.00	1000	27.00
"Blanda Pink Star", phlox purple	-	-	-	-	61.00	1000	61.00

Prices for Measured Works
Q PAVING/PLANTING/FENCING/SITE FURNITURE

Item Excluding site overheads and profit	PC £	Labour hours	Labour £	Plant £	Material £	Unit	Total rate £
Freesia							
Single Flowering Mixture	-	-	-	-	46.00	1000	46.00
Double Flowering Mixture	-	-	-	-	48.00	1000	48.00
Galanthus (Snowdrop)							
"Nivalis Double"	-	-	-	-	118.00	1000	118.00
"Nivalis Single"	-	-	-	-	53.00	1000	53.00
"Nivalis Single", supplied 'in the green'	-	-	-	-	63.00	1000	63.00
Muscari (Grape Hyacinth)							
"Armeniacum", bright blue	-	-	-	-	22.00	1000	22.00
"Botryoides Album", white	-	-	-	-	57.00	1000	57.00
Ranunculus							
Paeony Flowering Mixed	-	-	-	-	45.00	1000	45.00
Scilla							
"Campanulata Blue"	-	-	-	-	54.00	1000	54.00
Sparaxis							
"Superb Mixed"	-	-	-	-	21.00	1000	21.00
Planting - General							
Preamble: The British Standard recommendations for Nursery Stock are in BS 3936: pts 1-5. Planting of trees and shrubs as well as forestry is covered by the appropriate sections of BS 4428. Transplanting of semi-mature trees is covered by BS 4043. Prices for all planting work are deemed to include carrying out planting in accordance with BS 4428 and good horticultural practice.							
Site protection; temporary protective fencing							
Rolled chestnut pale fencing; to 100 mm diameter chestnut posts; driving into firm ground at 3 m centres; pales at 50 mm centres							
900 mm high	1.81	0.11	1.92	-	3.91	m	5.83
1200 mm high	2.43	0.11	1.92	-	4.44	m	6.36
1500 mm high	-	0.11	1.92	-	5.70	m	7.62
Chain link fencing; to BS 1722 Pt 1; including timber posts at 3m centres							
1.80 m high	6.30	0.13	2.40	-	6.30	m	8.70
2.13 m high	7.30	0.17	3.00	-	7.30	m	10.30
Extra over temporary protective fencing for removing and making good (no allowance for re-use of material)	-	0.07	1.20	0.19	-	m	1.39
Cultivation							
Treating soil with "Paraquat-Diquat" weedkiller at rate of 5 litre/ha; PC £22.20 / litre; in accordance with manufacturer's instructions; including all safety precautions							
by machine	-	-	-	0.24	1.11	100 m²	1.35
by hand	1.11	0.13	2.40	-	1.11	100 m²	3.51
Ripping up subsoil; using approved subsoiling machine; minimum depth 250 mm below topsoil; at 1.20 m centres; in							
gravel or sandy clay	-	-	-	2.31	-	100m2	2.31
soil compacted by machines	-	-	-	2.70	-	100m2	2.70
clay	-	-	-	2.89	-	100m2	2.89
chalk or other soft rock	-	-	-	5.78	-	100m2	5.78

Q PAVING/PLANTING/FENCING/SITE FURNITURE

Item Excluding site overheads and profit	PC £	Labour hours	Labour £	Plant £	Material £	Unit	Total rate £
Q31 PLANTING – cont'd							
Cultivation – cont'd							
Extra for subsoiling at 1 m centres	-	-	-	0.58	-	100m2	**0.58**
Breaking up existing ground; using pedestrian operated tine cultivator or rotavator							
100 mm deep	-	0.22	3.96	2.24	-	100m2	**6.20**
150 mm deep	-	0.28	4.95	2.80	-	100m2	**7.75**
200 mm deep	-	0.37	6.60	3.73	-	100m2	**10.33**
As above but in heavy clay or wet soils							
100 mm deep	-	0.44	7.92	4.48	-	100m2	**12.40**
150 mm deep	-	0.66	11.88	6.72	-	100m2	**18.60**
200 mm deep	-	0.83	14.85	8.40	-	100m2	**23.25**
Breaking up existing ground; using tractor drawn tine cultivator or rotavator							
100 mm deep	-	-	-	0.52	-	100 m²	**0.52**
150 mm deep	-	-	-	0.65	-	100 m²	**0.65**
200 mm deep	-	-	-	0.86	-	100 m²	**0.86**
600 mm deep	-	-	-	2.59	-	100 m²	**2.59**
Cultivating ploughed ground; using disc, drag, or chain harrow							
4 passes	-	-	-	3.10	-	100 m²	**3.10**
Rolling cultivated ground lightly; using self-propelled agricultural roller	-	0.06	1.00	0.58	-	100 m²	**1.58**
Importing only selected and approved topsoil; to BS 3882; from source not exceeding 13 km from site							
1 - 14 m³	17.32	-	-	-	51.96	m³	**51.96**
over 15 m³	17.32	-	-	-	17.32	m³	**17.32**
Spreading and lightly consolidating approved topsoil (imported or from spoil heaps); in layers not exceeding 150 mm; travel distance from spoil heaps not exceeding 100 m; by machine (imported topsoil not included)							
minimum depth 100 mm	-	1.55	27.90	35.54	-	100 m²	**63.44**
minimum depth 150 mm	-	2.33	42.00	53.49	-	100 m²	**95.49**
minimum depth 300 mm	-	4.67	84.00	106.97	-	100 m²	**190.97**
minimum depth 450 mm	-	6.99	125.82	160.29	-	100 m²	**286.11**
Spreading and lightly consolidating approved topsoil (imported or from spoil heaps); in layers not exceeding 150 mm; travel distance from spoil heaps not exceeding 100 m; by hand (imported topsoil not included)							
minimum depth 100 mm	-	20.00	360.07	-	-	100 m²	**360.07**
minimum depth 150 mm	-	30.01	540.11	-	-	100 m²	**540.11**
minimum depth 300 mm	-	60.01	1080.22	-	-	100 m²	**1080.22**
minimum depth 450 mm	-	90.02	1620.32	-	-	100 m²	**1620.32**
Extra over for spreading topsoil to slopes 15 - 30 degrees by machine or hand	-	-	-	-	-	10%	-
Extra over for spreading topsoil to slopes over 30 degrees by machine or hand	-	-	-	-	-	25%	-
Extra over spreading topsoil for travel exceeding 100 m; by machine							
100 - 150 m	-	0.02	0.36	0.14	-	m³	**0.50**
150 - 200 m	-	0.03	0.48	0.19	-	m³	**0.67**
200 - 300 m	-	0.04	0.64	0.25	-	m³	**0.89**

Prices for Measured Works
Q PAVING/PLANTING/FENCING/SITE FURNITURE

Item Excluding site overheads and profit	PC £	Labour hours	Labour £	Plant £	Material £	Unit	Total rate £
Extra over spreading topsoil for travel exceeding 100 m; by hand							
100 m	-	2.50	45.00	-	-	m³	45.00
200 m	-	3.50	63.00	-	-	m³	63.00
300 m	-	4.50	81.00	-	-	m³	81.00
Evenly grading; to general surfaces to bring to finished levels							
by machine (tractor mounted rotavator)	-	-	-	0.03	-	m²	0.03
by pedestrian operated rotavator	-	-	0.07	0.04	-	m²	0.12
by hand	-	0.01	0.18	-	-	m²	0.18
Extra over grading for slopes 15 - 30 degrees by machine or hand	-	-	-	-	-	10%	-
Extra over grading for slopes over 30 degrees by machine or hand	-	-	-	-	-	25%	-
Clearing stones; disposing off site; to distance not exceeding 13 km							
by hand; stones not exceeding 50 mm in any direction; loading to skip 5.35 m³	-	0.01	0.18	0.03	-	m²	0.21
by mechanical stone rake; stones not exceeding 50 mm in any direction; loading to 15 m³ truck by mechanical loader	-	-	0.04	0.08	-	m²	0.11
Lightly cultivating; weeding; to fallow areas; disposing debris off site; to distance not exceeding 13 km							
by hand	-	0.01	0.26	-	0.06	m²	0.32
Preparation of planting operations							
For the following topsoil improvement and planting operations add or subtract the following amounts for every £0.10 difference in the material cost price							
35 g/m²	-	-	-	-	0.35	100 m²	0.35
50 g/m²	-	-	-	-	0.50	100 m²	0.50
70 g/m²	-	-	-	-	0.70	100 m²	0.70
100 g/m²	-	-	-	-	1.00	100 m²	1.00
150 kg/ ha	-	-	-	-	15.00	ha	15.00
200 kg/ ha	-	-	-	-	20.00	ha	20.00
250 kg/ ha	-	-	-	-	25.00	ha	25.00
300 kg/ ha	-	-	-	-	30.00	ha	30.00
400 kg/ ha	-	-	-	-	40.00	ha	40.00
500 kg/ ha	-	-	-	-	50.00	ha	50.00
700 kg/ ha	-	-	-	-	70.00	ha	70.00
1000 kg/ ha	-	-	-	-	100.00	ha	100.00
1250 kg/ ha	-	-	-	-	125.00	ha	125.00
Selective herbicides; in accordance with manufacturer's instructions; PC £27.27 per litre; by machine; application rate							
30 ml/100m2	-	-	-	0.29	0.82	100 m²	1.11
35 ml/100m2	-	-	-	0.29	0.95	100 m²	1.24
40 ml/100m2	-	-	-	0.29	1.09	100 m²	1.38
50 ml/100m2	-	-	-	0.29	1.36	100 m²	1.65
3.00 l/ha	-	-	-	28.89	81.81	ha	110.70
3.50 l/ ha	-	-	-	28.89	95.44	ha	124.33
4.00 l/ha	-	-	-	28.89	109.08	ha	137.97

Prices for Measured Works
Q PAVING/PLANTING/FENCING/SITE FURNITURE

Item Excluding site overheads and profit	PC £	Labour hours	Labour £	Plant £	Material £	Unit	Total rate £
Q31 PLANTING – cont'd							
Preparation of planting operations – cont'd							
Selective herbicides; in accordance with manufacturer's instructions; PC £27.27 per litre; by hand; application rate							
30 ml/m²	-	0.17	3.00	-	0.82	100 m²	3.82
35 ml/m²	-	0.17	3.00	-	0.95	100 m²	3.96
40 ml/m²	-	0.17	3.00	-	1.09	100 m²	4.09
50 ml/m²	-	0.17	3.00	-	1.36	100 m²	4.36
3.00 l/ha	-	16.67	300.00	-	81.81	ha	381.81
3.50 l/ha	-	16.67	300.00	-	95.44	ha	395.45
4.00 l/ha	-	16.67	300.01	-	109.08	ha	409.09
General herbicides; in accordance with manufacturer's instructions; "Knapsack" spray application							
"Super Verdone" at 100 ml/100 m²	0.60	0.05	0.90	-	0.60	100 m²	1.50
"Roundup Pro" at 40 ml/100 m²	0.16	0.05	0.90	-	0.16	100 m²	1.06
"Spasor" at 50 ml/100 m²	0.70	0.05	0.90	-	0.70	100 m²	1.60
"Casoron G (residual)" at 1kg/125m2	3.13	0.05	0.90	-	3.44	100 m²	4.34
Fertilizers; in top 150 mm of topsoil; at 35g/m²							
fertilizer (18:0:0+Mg +Fe)	6.11	0.12	2.21	-	6.11	100 m²	8.32
"Enmag"	4.22	0.12	2.21	-	4.43	100 m²	6.63
fertilizer (7:7:7)	1.73	0.12	2.21	-	1.73	100 m²	3.94
"Superphosphate"	1.62	0.12	2.21	-	1.70	100 m²	3.91
fertilizer (20:10:10)	2.44	0.12	2.21	-	2.44	100 m²	4.65
"Hoof and Horn"	2.23	0.12	2.21	-	2.35	100 m²	4.55
"Bone meal"	3.10	0.12	2.21	-	3.25	100 m²	5.46
Fertilizers; in top 150 mm of topsoil at 70 g/m²							
fertilizer (18:0:0+Mg +Fe)	12.23	0.12	2.21	-	12.23	100 m²	14.43
"Enmag"	8.43	0.12	2.21	-	8.85	100 m²	11.06
fertilizer (7:7:7)	3.46	0.12	2.21	-	3.46	100 m²	5.67
"Superphosphate"	3.25	0.12	2.21	-	3.41	100 m²	5.61
fertilizer (20:10:10)	4.89	0.12	2.21	-	4.89	100 m²	7.09
"Hoof and Horn"	4.47	0.12	2.21	-	4.69	100 m²	6.90
"Bone meal"	6.20	0.12	2.21	-	6.51	100 m²	8.71
Spreading topsoil; from dump not exceeding 100 m distance; by machine (topsoil not included)							
at 1 m³ per 13 m²; 75 mm thick	-	0.56	10.00	27.06	-	100 m²	37.06
at 1 m³ per 10 m²; 100 mm thick	-	0.89	16.00	28.06	-	100 m²	44.06
at 1 m³ per 6.50 m²; 150 mm thick	-	1.11	19.98	48.91	-	100 m²	68.89
at 1 m³ per 5 m²; 200 mm thick	-	1.48	26.64	65.03	-	100 m²	91.67
Spreading topsoil; from dump not exceeding 100 m distance; by hand (topsoil not included)							
at 1 m³ per 13 m²; 75 mm thick	-	15.00	270.05	-	-	100 m²	270.05
at 1 m³ per 10 m²; 100 mm thick	-	20.00	360.07	-	-	100 m²	360.07
at 1 m³ per 6.50 m²; 150 mm thick	-	30.01	540.11	-	-	100 m²	540.11
at 1 m³ per 5 m²; 200 mm thick	-	36.67	660.13	-	-	100 m²	660.13
Imported topsoil; tipped 100 m from area of application; by machine							
at 1 m³ per 13 m²; 75 mm thick	129.90	0.56	10.00	27.06	155.88	100 m²	192.94
at 1 m³ per 10 m²; 100 mm thick	173.20	0.89	16.00	28.06	207.84	100 m²	251.90
at 1 m³ per 6.50 m²; 150 mm thick	259.80	1.11	19.98	48.91	311.76	100 m²	380.65
at 1 m³ per 5 m²; 200 mm thick	346.40	1.48	26.64	65.03	415.68	100 m²	507.35

Prices for Measured Works
Q PAVING/PLANTING/FENCING/SITE FURNITURE

Item Excluding site overheads and profit	PC £	Labour hours	Labour £	Plant £	Material £	Unit	Total rate £
Imported topsoil; tipped 100 m from area of application; by hand							
at 1 m³ per 13 m²; 75 mm thick	129.90	15.00	270.05	-	155.88	100 m²	425.93
at 1 m³ per 10 m²; 100 mm thick	173.20	20.00	360.07	-	207.84	100 m²	567.91
at 1 m³ per 6.50 m²; 150 mm thick	259.80	30.01	540.11	-	311.76	100 m²	851.87
at 1 m³ per 5 m²; 200 mm thick	346.40	36.67	660.13	-	415.68	100 m²	1075.81
Composted bark and manure soil conditioner (20 m³ loads); from not further than 25m from location; cultivating into topsoil by machine							
50 mm thick	130.00	2.86	51.43	1.87	136.50	100 m²	189.80
100 mm thick	260.00	6.05	108.88	1.87	273.00	100 m²	383.75
150 mm thick	390.00	8.90	160.29	1.87	409.50	100 m²	571.65
200 mm thick	520.00	12.90	232.29	1.87	546.00	100 m²	780.15
Composted bark soil conditioner (20 m³ loads); placing on beds by mechanical loader; spreading and rotavating into topsoil by machine							
50 mm thick	130.00	-	-	7.45	130.00	100 m²	137.45
100 mm thick	260.00	-	-	11.43	273.00	100 m²	284.43
150 mm thick	390.00	-	-	15.57	409.50	100 m²	425.07
200 mm thick	520.00	-	-	19.63	546.00	100 m²	565.63
Mushroom compost (20 m³ loads); from not further than 25m from location; cultivating into topsoil by machine							
50 mm thick	89.50	2.86	51.43	1.87	89.50	100 m²	142.80
100 mm thick	179.00	6.05	108.88	1.87	179.00	100 m²	289.75
150 mm thick	268.50	8.90	160.29	1.87	268.50	100 m²	430.65
200 mm thick	358.00	12.90	232.29	1.87	358.00	100 m²	592.15
Mushroom compost (20 m³ loads); placing on beds by mechanical loader; spreading and rotavating into topsoil by machine							
50 mm thick	89.50	-	-	7.45	89.50	100 m²	96.95
100 mm thick	179.00	-	-	11.43	179.00	100 m²	190.43
150 mm thick	268.50	-	-	15.57	268.50	100 m²	284.07
200 mm thick	358.00	-	-	19.63	358.00	100 m²	377.63
Manure (60 m³ loads); from not further than 25m from location; cultivating into topsoil by machine							
50 mm thick	110.00	2.86	51.43	1.87	110.00	100 m²	163.30
100 mm thick	220.00	6.05	108.88	1.87	231.00	100 m²	341.75
150 mm thick	330.00	8.90	160.29	1.87	346.50	100 m²	508.65
200 mm thick	440.00	12.90	232.29	1.87	462.00	100 m²	696.15
Surface applications and soil additives; pre-planting; from not further than 25 m from location; by machine							
Ground limestone soil conditioner; including turning in to cultivated ground							
0.25 kg/m² = 2.50 tonnes/ha	0.23	-	-	4.03	0.23	100 m²	4.26
0.50 kg/m² = 5.00 tonnes/ha	0.47	-	-	4.03	0.47	100 m²	4.49
0.75 kg/m² = 7.50 tonnes/ha	0.70	-	-	4.03	0.70	100 m²	4.73
1.00 kg/m² = 10.00 tonnes/ha	0.93	-	-	4.03	0.93	100 m²	4.96
Medium bark soil conditioner; A.H.S. Ltd; including turning in to cultivated ground; delivered in 15 m³ loads							
1 m³ per 40 m² = 25 mm thick	0.80	-	-	0.09	0.80	m²	0.89
1 m³ per 20 m² = 50 mm thick	1.60	-	-	0.14	1.60	m²	1.74
1 m³ per 13.33 m² = 75 mm thick	2.40	-	-	0.19	2.40	m²	2.59
1 m³ per 10 m² = 100 mm thick	3.20	-	-	0.25	3.20	m²	3.45

Prices for Measured Works
Q PAVING/PLANTING/FENCING/SITE FURNITURE

Item Excluding site overheads and profit	PC £	Labour hours	Labour £	Plant £	Material £	Unit	Total rate £
Q31 PLANTING – cont'd							
Surface applications and soil additives; pre-planting; - cont'd							
Peat soil conditioner; A.H.S. Ltd; including turning in to cultivated ground; delivered in 55 m³ loads							
1 m³ per 40 m² = 25 mm thick	0.70	-	-	0.09	0.70	m²	0.79
1 m³ per 20 m² = 50 mm thick	1.40	-	-	0.14	1.40	m²	1.54
1 m³ per 13.33 m² = 75 mm thick	2.10	-	-	0.19	2.10	m²	2.29
1 m³ per 10 m² = 100 mm thick	2.80	-	-	0.25	2.80	m²	3.05
Mushroom compost soil conditioner; A.H.S. Ltd; including turning in to cultivated ground; delivered in 20 m³ loads							
1 m³ per 40 m² = 25 mm thick	0.45	0.02	0.30	-	0.45	m²	0.75
1 m³ per 20 m² = 50 mm thick	0.90	0.03	0.55	-	0.90	m²	1.45
1 m³ per 13.33 m² = 75 mm thick	1.34	0.04	0.72	-	1.34	m²	2.06
1 m³ per 10 m² = 100 mm thick	1.79	0.05	0.90	-	1.79	m²	2.69
Mushroom compost soil conditioner; A.H.S. Ltd; including turning in to cultivated ground; delivered in 55 m³ loads							
1 m³ per 40 m² = 25 mm thick	0.10	0.02	0.30	-	0.10	m²	0.40
1 m³ per 20 m² = 50 mm thick	0.20	0.03	0.55	-	0.20	m²	0.75
1 m³ per 13.33 m² = 75 mm thick	0.30	0.04	0.72	-	0.30	m²	1.02
1 m³ per 10 m² = 100 mm thick	0.40	0.05	0.90	-	0.40	m²	1.30
Surface applications and soil additives; pre-planting; from not further than 25 m from location; by hand							
Ground limestone soil conditioner; including turning in to cultivated ground							
0.25 kg/m² = 2.50 tonnes/ha	0.23	1.20	21.60	-	0.23	100 m²	21.83
0.50 kg/m² = 5.00 tonnes/ha	0.47	1.33	24.00	-	0.47	100 m²	24.46
0.75 kg/m² = 7.50 tonnes/ha	0.70	1.50	27.00	-	0.70	100 m²	27.70
1.00 kg/m² = 10.00 tonnes/ha	0.93	1.71	30.86	-	0.93	100 m²	31.79
Medium bark soil conditioner; A.H.S. Limited; including turning in to cultivated ground; delivered in 15 m³ loads							
1 m³ per 40 m² = 25 mm thick	0.80	0.02	0.40	-	0.80	m²	1.20
1 m³ per 20 m² = 50 mm thick	1.60	0.04	0.80	-	1.60	m²	2.40
1 m³ per 13.33 m² = 75 mm thick	2.40	0.07	1.20	-	2.40	m²	3.60
1 m³ per 10 m² = 100 mm thick	3.20	0.08	1.44	-	3.20	m²	4.64
Peat soil conditioner; A.H.S. Limited; including turning in to cultivated ground; delivered in 55 m³ loads							
1 m³ per 40 m² = 25 mm thick	0.70	0.02	0.40	-	0.70	m²	1.10
1 m³ per 20 m² = 50 mm thick	1.40	0.04	0.80	-	1.40	m²	2.20
1 m³ per 13.33 m² = 75 mm thick	2.10	0.07	1.20	-	2.10	m²	3.30
1 m³ per 10 m² = 100 mm thick	2.80	0.08	1.44	-	2.80	m²	4.24
Mushroom compost soil conditioner; Melcourt. Limited; including turning in to cultivated ground; delivered in 25 m³ loads							
1 m³ per 40 m² = 25 mm thick	0.45	0.02	0.40	-	0.45	m²	0.85
1 m³ per 20 m² = 50 mm thick	0.90	0.04	0.80	-	0.90	m²	1.69
1 m³ per 13.33 m² = 75 mm thick	1.34	0.07	1.20	-	1.34	m²	2.54
1 m³ per 10 m² = 100 mm thick	1.79	0.08	1.44	-	1.79	m²	3.23

Prices for Measured Works
Q PAVING/PLANTING/FENCING/SITE FURNITURE

Item Excluding site overheads and profit	PC £	Labour hours	Labour £	Plant £	Material £	Unit	Total rate £
Mushroom compost soil conditioner; A.H.S. Limited; including turning in to cultivated ground; delivered in 55 m³ loads							
1 m³ per 40 m² = 25 mm thick	0.10	0.02	0.40	-	0.10	m²	0.50
1 m³ per 20 m² = 50 mm thick	0.20	0.04	0.80	-	0.20	m²	1.00
1 m³ per 13.33 m² = 75 mm thick	0.30	0.07	1.20	-	0.30	m²	1.50
1 m³ per 10 m² = 100 mm thick	0.40	0.08	1.44	-	0.40	m²	1.84
"Super Humus" mixed bark and manure conditioner; Melcourt Industries Ltd; delivered in 25 m³ loads							
1 m³ per 40 m² = 25 mm thick	0.65	0.02	0.40	-	0.68	m²	1.08
1 m³ per 20 m² = 50 mm thick	1.30	0.04	0.80	-	1.37	m²	2.16
1 m³ per 13.33 m² = 75 mm thick	1.95	0.07	1.20	-	2.05	m²	3.25
1 m³ per 10 m² = 100 mm thick	2.60	0.08	1.44	-	2.73	m²	4.17
Tree planting; pre-planting operations							
Excavating tree pits; depositing soil alongside pits; by machine							
600 mm x 600 mm x 600 mm deep	-	0.15	2.65	0.74	-	nr	3.39
900 mm x 900 mm x 600 mm deep	-	0.33	5.94	1.67	-	nr	7.61
1.00 m x 1.00 m x 600 mm deep	-	0.61	11.04	2.06	-	nr	13.11
1.25 m x 1.25 m x 600 mm deep	-	0.96	17.29	3.23	-	nr	20.52
1.00 m x 1.00 m x 1.00 m deep	-	1.02	18.41	3.44	-	nr	21.85
1.50 m x 1.50 m x 750 mm deep	-	1.73	31.07	5.81	-	nr	36.87
1.50 m x 1.50 m x 1.00 m deep	-	2.30	41.32	7.72	-	nr	49.04
1.75 m x 1.75 m x 1.00 mm deep	-	3.13	56.38	10.54	-	nr	66.92
2.00 m x 2.00 m x 1.00 mm deep	-	4.09	73.63	13.76	-	nr	87.39
Excavating tree pits; depositing soil alongside pits; by hand							
600 mm x 600 mm x 600 mm deep	-	0.44	7.92	-	-	nr	7.92
900 mm x 900 mm x 600 mm deep	-	1.00	18.00	-	-	nr	18.00
1.00 m x 1.00 m x 600 mm deep	-	1.13	20.25	-	-	nr	20.25
1.25 m x 1.25 m x 600 mm deep	-	1.93	34.74	-	-	nr	34.74
1.00 m x 1.00 m x 1.00 m deep	-	2.06	37.08	-	-	nr	37.08
1.50 m x 1.50 m x 750 mm deep	-	3.47	62.46	-	-	nr	62.46
1.75 m x 1.50 m x 750 mm deep	-	4.05	72.90	-	-	nr	72.90
1.50 m x 1.50 m x 1.00 m deep	-	4.63	83.34	-	-	nr	83.34
2.00 m x 2.00 m x 750 mm deep	-	6.17	111.06	-	-	nr	111.06
2.00 m x 2.00 m x 1.00 m deep	-	8.23	148.15	-	-	nr	148.15
Breaking up subsoil in tree pits; to a depth of 200 mm	-	0.03	0.60	-	-	m²	0.60
Spreading and lightly consolidating approved topsoil (imported or from spoil heaps); in layers not exceeding 150 mm; distance from spoil heaps not exceeding 100 m (imported topsoil not included); by machine							
minimum depth 100 mm	-	1.55	27.90	35.54	-	100 m²	63.44
minimum depth 150 mm	-	2.33	42.00	53.49	-	100 m²	95.49
minimum depth 300 mm	-	4.67	84.00	106.97	-	100 m²	190.97
minimum depth 450 mm	-	6.99	125.82	160.29	-	100 m²	286.11
Spreading and lightly consolidating approved topsoil (imported or from spoil heaps); in layers not exceeding 150 mm; distance from spoil heaps not exceeding 100 m (imported topsoil not included); by hand							
minimum depth 100 mm	-	20.00	360.07	-	-	100 m²	360.07
minimum depth 150 mm	-	30.01	540.11	-	-	100 m²	540.11
minimum depth 300 mm	-	60.01	1080.22	-	-	100 m²	1080.22
minimum depth 450 mm	-	90.02	1620.32	-	-	100 m²	1620.32

Prices for Measured Works
Q PAVING/PLANTING/FENCING/SITE FURNITURE

Item Excluding site overheads and profit	PC £	Labour hours	Labour £	Plant £	Material £	Unit	Total rate £
Q31 PLANTING – cont'd							
Tree planting; pre-planting operations; - cont'd							
Extra for filling tree pits with imported topsoil; PC £17.32 /m³; plus allowance for 20% settlement							
depth 100 mm	-	-	-	-	2.08	m²	2.08
depth 150 mm	-	-	-	-	3.12	m²	3.12
depth 200 mm	-	-	-	-	4.16	m²	4.16
depth 300 mm	-	-	-	-	6.24	m²	6.24
depth 400 mm	-	-	-	-	8.31	m²	8.31
depth 450 mm	-	-	-	-	9.35	m²	9.35
depth 500 mm	-	-	-	-	10.39	m²	10.39
depth 600 mm	-	-	-	-	12.47	m²	12.47
Add or deduct the following amounts for every £0.50 change in the material price of topsoil							
depth 100 mm	-	-	-	-	0.06	m²	0.06
depth 150 mm	-	-	-	-	0.09	m²	0.09
depth 200 mm	-	-	-	-	0.12	m²	0.12
depth 300 mm	-	-	-	-	0.18	m²	0.18
depth 400 mm	-	-	-	-	0.24	m²	0.24
depth 450 mm	-	-	-	-	0.27	m²	0.27
depth 500 mm	-	-	-	-	0.30	m²	0.30
depth 600 mm	-	-	-	-	0.36	m²	0.36
Structural Soils; Amsterdam Tree Sand; Metro Sand; Non compressive soil mixture for tree planting in areas to receive compressive surface treatments.							
backfilling and lightly compacting in layers; excavation, disposal, moving of material from delivery position and surface treatments not included; by machine							
individual treepits	45.37	0.50	9.00	2.52	45.37	m³	56.90
in trenches	45.37	0.42	7.50	2.10	45.37	m³	54.98
backfilling and lightly compacting in layers; excavation, disposal, moving of material from delivery position and surface treatments not included; by hand							
individual treepits	45.37	1.60	28.80	-	45.37	m³	74.18
in trenches	45.37	1.33	23.94	-	45.37	m³	69.32
Tree planting; tree staking; J Toms Ltd Extra over trees for tree stake(s); driving 500 mm into firm ground; trimming to approved height; including two tree ties to approved pattern							
one stake; 100 mm diameter x 3.00 m long	4.18	0.20	3.60	-	4.18	nr	7.78
one stake; 100 mm diameter x 2.40 m long	3.25	0.20	3.60	-	3.25	nr	6.85
two stakes; 100 mm diameter x 3.00 m long	8.36	0.30	5.40	-	8.36	nr	13.76
two stakes; 100 mm diameter x 2.40 m long	6.50	0.30	5.40	-	6.50	nr	11.90
three stakes; 100 mm diameter x 3.00 m long	12.54	0.36	6.48	-	12.54	nr	19.02
three stakes; 100 mm diameter x 2.40 m long	9.75	0.36	6.48	-	9.75	nr	16.23
Tree planting; tree anchors; Platipus Tree Anchoring Systems							
ref RF1 rootball kit; for 75 to 225 mm girth, 2 to 4.5 m high inclusive of "Plati-mat" PM1	25.31	2.50	45.00	-	25.31	nr	70.31

Prices for Measured Works
Q PAVING/PLANTING/FENCING/SITE FURNITURE

Item Excluding site overheads and profit	PC £	Labour hours	Labour £	Plant £	Material £	Unit	Total rate £
ref RF2; rootball kit; for 225 to 450 mm girth, 4.5 to 7.5 m high inclusive of "Plati-mat" PM2	41.46	2.00	36.00	-	41.46	nr	**77.46**
ref RF3; rootball kit; for 450 to 750 mm girth, 7.5 to 12 m high "Plati-mat" PM3	87.62	1.33	24.00	-	87.62	nr	**111.62**
ref CG1; guy fixing kit; 75to 225 mm girth, 2 to 4.5 m high	16.47	1.67	30.00	-	16.47	nr	**46.47**
ref CG2; guy fixing kit; 225 to 450 mm girth, 4.5 to 7.5 m high	28.99	2.00	36.00	-	28.99	nr	**64.99**
installation tools; Drive Rod for CG1 Kit	-	-	-	-	52.18	nr	**52.18**
installation tools; Drive Rod for CG2 Kit	-	-	-	-	78.63	nr	**78.63**
Extra over trees for land drain to tree pits; 100 mm diameter perforated flexible agricultural drain; including excavating drain trench; laying pipe; backfilling	0.63	1.00	18.00	-	0.69	m	**18.69**
Tree planting; tree pit additives;							
Melcourt Industries Ltd; "Topgrow"; incorporating into topsoil at 1 part "Topgrow" to 3 parts excavated topsoil; supplied in 80 l bags; pit size							
600 mm x 600 mm x 600 mm	0.37	0.02	0.30	-	0.37	nr	**0.67**
900 mm x 900 mm x 900 mm	3.22	0.05	0.90	-	3.22	nr	**4.12**
1.00 m x 1.00 m x 1.00 m	5.89	0.20	3.60	-	5.89	nr	**9.49**
1.25 m x 1.25 m x 1.25 m	11.61	0.33	6.00	-	11.61	nr	**17.61**
1.50 m x 1.50 m x 1.50 m	20.11	0.75	13.50	-	20.11	nr	**33.61**
Melcourt Industries Ltd; "Topgrow"; incorporating into topsoil at 1 part "Topgrow" to 3 parts excavated topsoil; supplied in 60 m³ loose loads; pit size							
600 mm x 600 mm x 600 mm	0.83	0.02	0.30	-	0.83	nr	**1.13**
900 mm x 900 mm x 900 mm	2.79	0.05	0.90	-	2.79	nr	**3.69**
1.00 m x 1.00 m x 1.00 m	3.83	0.20	3.60	-	3.83	nr	**7.43**
1.25 m x 1.25 m x 1.25 m	7.47	0.33	6.00	-	7.47	nr	**13.47**
1.50 m x 1.50 m x 1.50 m	12.91	0.75	13.50	-	12.91	nr	**26.41**
Alginure Products; "Alginure Root Dip"; to bare rooted plants at 1 part "Alginure Root Dip" to 3 parts water							
transplants; at 3000/15 kg bucket	1.11	0.07	1.20	-	1.11	100 nr	**2.31**
standard trees; at 112/15 kg bucket of dip	0.30	0.03	0.60	-	0.30	nr	**0.90**
medium shrubs; at 600/15 kg bucket	0.06	0.02	0.30	-	0.06	nr	**0.36**
Tree Planting; Root Barriers							
English Woodlands; "Root Director", one-piece root control planters for installation at time of planting, to divert root growth down away from pavements and out for anchorage. Excavation measured separately							
RD 1050; 1050 x 1050 mm to 1300 x 1300 mm at base	65.00	0.25	4.50	-	65.00	nr	**69.50**
RD 640; 640 x 640 mm to 870 x 870 mm at base	39.00	0.25	4.50	-	39.00	nr	**43.50**
Greenleaf; linear root deflection barriers; installed to trench measured separately							
Re-Root 2000, 1.5 mm thick	5.80	0.05	0.90	-	5.80	m	**6.70**
Re-Root 2000, 1.0 mm thick	4.88	0.05	0.90	-	4.88	m	**5.78**
Re-Root 600, 1.0 mm thick.	5.50	0.05	0.90	-	5.50	m	**6.40**

Q PAVING/PLANTING/FENCING/SITE FURNITURE

Item Excluding site overheads and profit	PC £	Labour hours	Labour £	Plant £	Material £	Unit	Total rate £
Q31 PLANTING – cont'd							
Greenleaf Tree Planting - Irrigation Systems; "Root Rain" tree pit irrigation systems							
"Metro" - "Small"; 35 mm pipe diameter 1.25 m long for specimen shrubs and standard trees							
Plastic	5.09	0.25	4.50	-	5.09	nr	9.59
Plastic with chain	6.65	0.25	4.50	-	6.65	nr	11.15
Metal with chain	8.24	0.25	4.50	-	8.24	nr	12.74
"Medium" 35 mm pipe diameter 1.75 m long for s standard and selected standard trees							
Plastic	5.40	0.29	5.14	-	5.40	nr	10.54
Plastic with chain	6.98	0.29	5.14	-	6.98	nr	12.12
Metal with chain	8.55	0.29	5.14	-	8.55	nr	13.69
"Large" 35 mm pipe diameter 2.50 m long for selected standards and extra heavy standards							
Plastic	6.24	0.33	6.00	-	6.24	nr	12.24
Plastic with chain	7.82	0.33	6.00	-	7.82	nr	13.82
Metal with chain	10.44	0.33	6.00	-	10.44	nr	16.44
"Large" 35 mm pipe diameter 2.50 m long for selected standards and extra heavy standards							
Plastic	6.24	0.33	6.00	-	6.24	nr	12.24
Plastic with chain	7.82	0.33	6.00	-	7.82	nr	13.82
Metal with chain	10.44	0.33	6.00	-	10.44	nr	16.44
"Urban" Irrigation and aeration systems; for large capacity general purpose irrigation to parkland and street verge planting							
RRUrb1; 3.0 m pipe	-	0.29	5.14	-	12.60	nr	17.74
RRUrb2; 5.0 m pipe	14.60	0.33	6.00	-	14.60	nr	20.60
RRUrb3; 8.0 m pipe	15.90	0.40	7.20	-	15.90	nr	23.10
"Precinct" Irrigation and aeration systems; heavy cast aluminium inlet for heavily trafficked locations							
"5 m pipe"	27.50	0.33	6.00	-	27.50	nr	33.50
"8 m pipe"	30.00	0.40	7.20	-	30.00	nr	37.20
Mulching of tree pits; Melcourt Industries Ltd							
Spreading mulch; to individual trees; maximum distance 25 m (mulch not included)							
50 mm thick	-	0.05	0.87	-	-	m²	0.87
75 mm thick	-	0.07	1.31	-	-	m²	1.31
100 mm thick	-	0.10	1.75	-	-	m²	1.75
Mulch; "Bark Nuggets"; to individual trees; delivered in 70 m³ loads; maximum distance 25 m							
50 mm thick	1.45	0.05	0.87	-	1.52	m²	2.40
75 mm thick	2.17	0.07	1.31	-	2.29	m²	3.60
100 mm thick	2.90	0.10	1.75	-	3.05	m²	4.79
Mulch; "Bark Nuggets"; to individual trees; delivered in 20 m³ loads; maximum distance 25 m							
50 mm thick	2.07	0.05	0.87	-	2.17	m²	3.05
75 mm thick	3.11	0.05	0.90	-	3.26	m²	4.16
100 mm thick	4.14	0.07	1.20	-	4.35	m²	5.55
Mulch; "Amenity Bark"; to individual trees; delivered in 70 m³ loads; maximum distance 25 m							
50 mm thick	0.89	0.05	0.87	-	0.93	m²	1.81
75 mm thick	1.33	0.07	1.31	-	1.40	m²	2.71
100 mm thick	1.78	0.10	1.75	-	1.86	m²	3.61

Prices for Measured Works
Q PAVING/PLANTING/FENCING/SITE FURNITURE

Item Excluding site overheads and profit	PC £	Labour hours	Labour £	Plant £	Material £	Unit	Total rate £
Mulch; "Amenity Bark"; to individual trees; delivered in 20 m³ loads; maximum distance 25 m							
50 mm thick	1.51	0.05	0.87	-	1.58	m²	2.46
75 mm thick	2.26	0.07	1.31	-	2.38	m²	3.69
100 mm thick	3.02	0.07	1.20	-	3.17	m²	4.37
Trees; planting labours only							
Bare root trees; including backfilling with previously excavated material (all other operations and materials not included)							
light standard; 6 - 8 cm girth	-	0.35	6.30	-	-	nr	6.30
standard; 8 - 10 cm girth	-	0.40	7.20	-	-	nr	7.20
selected standard; 10 - 12 cm girth	-	0.58	10.44	-	-	nr	10.44
heavy standard; 12 - 14 cm girth	-	0.83	15.00	-	-	nr	15.00
extra heavy standard; 14 - 16 cm girth	-	1.00	18.00	-	-	nr	18.00
Root balled trees; including backfilling with previously excavated material (all other operations and materials not included)							
standard; 8 - 10 cm girth	-	0.50	9.00	-	-	nr	9.00
selected standard; 10 - 12 cm girth	-	0.60	10.80	-	-	nr	10.80
heavy standard; 12 - 14 cm girth	-	0.80	14.40	-	-	nr	14.40
extra heavy standard;14 - 16 cm girth	-	1.50	27.00	-	-	nr	27.00
16 - 18 cm girth	-	1.30	23.33	20.03	-	nr	43.36
18 - 20 cm girth	-	1.60	28.80	24.73	-	nr	53.53
20 - 25 cm girth	-	4.50	81.00	69.55	-	nr	150.55
25 - 30 cm girth	-	6.00	108.00	92.73	-	nr	200.73
30 - 35 cm girth	-	11.00	198.00	170.01	-	nr	368.01
Tree planting; root balled trees; advanced nursery stock and semi-mature							
- General							
Preamble: The cost of planting semi-mature trees will depend on the size and species, and on the access to the site for tree handling machines. Prices should be obtained for individual trees and planting.							
Tree planting; bare root trees; nursery stock							
"Acer platanoides"; including backfillling with excavated material (other operations not included)							
light standard; 6 - 8 cm girth	9.25	0.35	6.30	-	9.25	nr	15.55
standard; 8 - 10 cm girth	11.25	0.40	7.20	-	11.25	nr	18.45
selected standard; 10 - 12 cm girth	16.25	0.58	10.44	-	16.25	nr	26.69
heavy standard; 12 - 14 cm girth	24.00	0.83	15.00	-	24.00	nr	39.00
extra heavy standard; 14 - 16 cm girth	40.00	1.00	18.00	-	40.00	nr	58.00
"Carpinus betulus"; including backfillling with excavated material (other operations not included)							
light standard; 6 - 8 cm girth	14.00	0.35	6.30	-	14.00	nr	20.30
standard; 8 - 10 cm girth	16.00	0.40	7.20	-	16.00	nr	23.20
selected standard; 10 - 12 cm girth	27.50	0.58	10.44	-	27.50	nr	37.94
heavy standard; 12 - 14 cm girth	40.00	0.83	15.01	-	40.00	nr	55.01
extra heavy standard; 14 - 16 cm girth	40.00	1.00	18.00	-	40.00	nr	58.00

Prices for Measured Works
Q PAVING/PLANTING/FENCING/SITE FURNITURE

Item Excluding site overheads and profit	PC £	Labour hours	Labour £	Plant £	Material £	Unit	Total rate £
Q31 PLANTING – cont'd							
Tree planting; bare root trees; nursery Stock – cont'd							
"Fraxinus excelsior"; including backfillling with excavated material (other operations not included)							
light standard; 6 - 8 cm girth	11.75	0.35	6.30	-	11.75	nr	18.05
standard; 8 - 10 cm girth	13.50	0.40	7.20	-	13.50	nr	20.70
selected standard; 10 - 12 cm girth	22.00	0.58	10.44	-	22.00	nr	32.44
heavy standard; 12 - 14 cm girth	40.00	0.83	14.94	-	40.00	nr	54.94
extra heavy standard; 14 - 16 cm girth	50.00	1.00	18.00	-	50.00	nr	68.00
"Prunus avium Plena"; including backfillling with excavated material (other operations not included)							
light standard; 6 - 8 cm girth	11.75	0.36	6.55	-	11.75	nr	18.30
standard; 8 - 10 cm girth	13.50	0.40	7.20	-	13.50	nr	20.70
selected standard; 10 - 12 cm girth	27.50	0.58	10.44	-	27.50	nr	37.94
heavy standard; 12 - 14 cm girth	45.00	0.83	15.00	-	45.00	nr	60.00
extra heavy standard; 14 - 16 cm girth	55.00	1.00	18.00	-	55.00	nr	73.00
"Quercus robur"; including backfillling with excavated material (other operations not included)							
light standard; 6 - 8 cm girth	11.75	0.35	6.30	-	11.75	nr	18.05
standard; 8 - 10 cm girth	13.50	0.40	7.20	-	13.50	nr	20.70
selected standard; 10 - 12 cm girth	22.00	0.58	10.44	-	22.00	nr	32.44
heavy standard; 12 - 14 cm girth	40.00	0.83	15.00	-	40.00	nr	55.00
extra heavy standard; 14 - 16 cm girth	50.00	1.00	18.00	-	50.00	nr	68.00
"Robinia pseudoacacia Frisia"; including backfillling with excavated material (other operations not included)							
light standard; 6 - 8 cm girth	30.00	0.35	6.30	-	30.00	nr	36.30
standard; 8 - 10 cm girth	40.00	0.40	7.20	-	40.00	nr	47.20
selected standard; 10 - 12 cm girth	55.00	0.58	10.44	-	55.00	nr	65.44
heavy standard; 12 - 14 cm girth	80.00	0.83	15.00	-	80.00	nr	95.00
extra heavy standard; 14 - 16 cm girth	115.00	1.00	18.00	-	115.00	nr	133.00
Tree planting; root balled trees; nursery stock							
"Acer platanoides"; including backfillling with excavated material (other operations not included)							
standard; 8 - 10 cm girth	-	0.48	8.64	-	21.25	nr	29.89
selected standard; 10 - 12 cm girth	-	0.56	10.09	-	36.25	nr	46.34
heavy standard; 12 - 14 cm girth	-	0.76	13.76	-	49.00	nr	62.76
extra heavy standard; 14 - 16 cm girth	-	1.20	21.60	-	70.00	nr	91.60
"Carpinus betulus"; including backfillling with excavated material (other operations not included)							
standard; 8 - 10 cm girth	-	0.48	8.64	-	26.00	nr	34.64
selected standard; 10 - 12 cm girth	-	0.56	10.09	-	47.50	nr	57.59
heavy standard; 12 - 14 cm girth	-	0.76	13.76	-	69.00	nr	82.76
extra heavy standard; 14 - 16 cm girth	-	1.20	21.60	-	85.00	nr	106.60
"Fraxinus excelsior"; including backfillling with excavated material (other operations not included)							
standard; 8 - 10 cm girth	-	0.48	8.64	-	23.50	nr	32.14

Prices for Measured Works
Q PAVING/PLANTING/FENCING/SITE FURNITURE

Item Excluding site overheads and profit	PC £	Labour hours	Labour £	Plant £	Material £	Unit	Total rate £
selected standard; 10 - 12 cm girth	-	0.56	10.09	-	42.00	nr	52.09
heavy standard; 12 - 14 cm girth	-	0.76	13.76	-	65.00	nr	78.76
extra heavy standard; 14 - 16 cm girth	-	1.20	21.60	-	80.00	nr	101.60
"Prunus avium Plena"; including backfillling with excavated material (other operations not included)							
standard; 8 - 10 cm girth	-	0.40	7.20	-	26.00	nr	33.20
selected standard; 10 - 12 cm girth	-	0.56	10.09	-	26.00	nr	36.09
heavy standard; 12 - 14 cm girth	-	0.76	13.76	-	70.00	nr	83.76
extra heavy standard; 14 - 16 cm girth	-	1.20	21.60	-	85.00	nr	106.60
"Quercus robur"; including backfillling with excavated material (other operations not included)							
standard; 8 - 10 cm girth	-	0.48	8.64	-	23.50	nr	32.14
selected standard; 10 - 12 cm girth	-	0.56	10.09	-	42.00	nr	52.09
heavy standard; 12 - 14 cm girth	-	0.76	13.76	-	65.00	nr	78.76
extra heavy standard; 14 - 16 cm girth	-	1.20	21.60	-	80.00	nr	101.60
"Robinia pseudoacacia Frisia"; including backfillling with excavated material (other operations not included)							
standard; 8 - 10 cm girth	-	0.48	8.64	-	50.00	nr	58.64
selected standard; 10 - 12 cm girth	-	0.56	10.09	-	75.00	nr	85.09
heavy standard; 12 - 14 cm girth	-	0.76	13.76	-	105.00	nr	118.76
extra heavy standard; 14 - 16 cm girth	-	1.20	21.60	-	145.00	nr	166.60
Tree planting; root balled trees; advanced nursery stock and semi-mature; Belwood Nurseries							
"Acer platanoides Emerald Queen"; including backfilling with excavated material (other operations not included)							
16 - 18 cm girth	60.00	1.30	23.33	20.03	60.00	nr	103.36
18 - 20 cm girth	70.00	1.60	28.80	24.73	70.00	nr	123.53
20 - 25 cm girth	80.00	4.50	81.00	69.55	80.00	nr	230.55
25 - 30 cm girth	140.00	6.00	108.00	92.73	140.00	nr	340.73
30 - 35 cm girth	170.00	11.00	198.00	170.01	170.00	nr	538.01
"Aesculus briotti"; including backfilling with excavated material (other operations not included)							
16 - 18 cm girth	70.00	1.30	23.33	20.03	70.00	nr	113.36
18 - 20 cm girth	78.00	1.60	28.80	24.73	78.00	nr	131.53
20 - 25 cm girth	95.00	4.50	81.00	69.55	95.00	nr	245.55
25 - 30 cm girth	160.00	6.00	108.00	92.73	160.00	nr	360.73
30 - 35 cm girth	190.00	11.00	198.00	170.01	190.00	nr	558.01
"Prunus avium Flora Plena"; including backfilling with excavated material (other operations not included)							
16 - 18 cm girth	55.00	1.30	23.33	20.03	55.00	nr	98.36
18 - 20 cm girth	65.00	1.60	28.80	24.73	65.00	nr	118.53
20 - 25 cm girth	85.00	4.50	81.00	69.55	85.00	nr	235.55
25 - 30 cm girth	125.00	6.00	108.00	92.73	125.00	nr	325.73
30 - 35 cm girth	170.00	11.00	198.00	170.01	170.00	nr	538.01

Prices for Measured Works
Q PAVING/PLANTING/FENCING/SITE FURNITURE

Item Excluding site overheads and profit	PC £	Labour hours	Labour £	Plant £	Material £	Unit	Total rate £
Q31 PLANTING – cont'd							
Tree planting; root balled trees; advanced nursery stock; - cont'd							
"Sorbus aucuparia Sheerwater Seedling"; including backfilling with excavated material (other operations not included)							
16 - 18 cm girth	59.00	1.30	23.33	20.03	59.00	nr	102.36
18 - 20 cm girth	65.00	1.60	28.80	24.73	65.00	nr	118.53
20 - 25 cm girth	80.00	4.50	81.00	69.55	80.00	nr	230.55
25 - 30 cm girth	130.00	6.00	108.00	92.73	130.00	nr	330.73
30 - 35 cm girth	170.00	11.00	198.00	170.01	170.00	nr	538.01
"Tilia cordata Green Spire"; including backfilling with excavated material (other operations not included)							
16 - 18 cm girth	55.00	1.30	23.33	20.03	55.00	nr	98.36
18 - 20 cm girth	60.00	1.60	28.80	24.73	60.00	nr	113.53
20 - 25 cm girth	65.00	4.50	81.00	69.55	65.00	nr	215.55
25 - 30 cm girth	70.00	6.00	108.00	92.73	70.00	nr	270.73
30 - 35 cm girth	110.00	11.00	198.00	170.01	110.00	nr	478.01
Tree planting; root balled trees; semi-mature and mature trees; Belwood Trees; planting and back filling; planted by tele handler or by crane; delivery included; all other operations priced separately							
Semi mature trees							
40 - 45 cm girth	260.00	8.00	144.00	50.13	260.00	nr	454.13
45 - 50 cm girth	340.00	8.00	144.00	65.92	340.00	nr	549.92
55 - 60 cm girth	460.00	10.00	180.00	80.00	460.00	nr	720.00
67 - 70 cm girth	512.00	15.00	270.00	176.73	512.00	nr	958.73
75 - 80 cm girth	590.00	18.00	324.00	174.32	590.00	nr	1088.32
over 80 cm girth	790.00	18.00	324.00	174.32	790.00	nr	1288.32
Tree planting; tree protection - General							
Preamble: Care must be taken to ensure that tree grids and guards are removed when trees grow beyond the specified diameter of guard.							
Tree planting; tree protection							
Crowders Nurseries; "Crowders Tree Tube"; olive green							
1200 mm high x 80 x 80 mm	0.71	0.07	1.20	-	0.71	nr	1.91
Crowders Nurseries; stakes; 1500 mm high for "Crowders Tree Tube"; driving into ground	0.50	0.05	0.90	-	0.50	nr	1.40
Crowders Nurseries; galvanized welded mesh tree guards; nailing to tree stakes (tree stakes not included)							
840 mm high x 200 mm diameter	10.25	0.25	4.50	-	10.25	nr	14.75
1200 mm high x 250 mm diameter	8.25	0.33	6.00	-	8.25	nr	14.25
1200 mm high x 300 mm diameter	9.00	0.33	6.00	-	9.00	nr	15.00
1800 mm high x 300 mm diameter	12.00	0.33	6.00	-	12.00	nr	18.00
Expandable plastic tree guards; including 25 mm softwood stakes							
500 mm high	0.59	0.17	3.00	-	0.59	nr	3.59
1.00 m high	0.65	0.17	3.00	-	0.65	nr	3.65

Prices for Measured Works
Q PAVING/PLANTING/FENCING/SITE FURNITURE

Item Excluding site overheads and profit	PC £	Labour hours	Labour £	Plant £	Material £	Unit	Total rate £
J. Toms; "Spiral Guard" perforated PVC guards; for trees 10 to 40 mm diameter; white, black or grey							
450 mm	0.21	0.03	0.60	-	0.21	nr	**0.81**
600 mm	0.26	0.03	0.60	-	0.26	nr	**0.86**
750 mm	0.33	0.03	0.60	-	0.33	nr	**0.93**
English Woodlands; Plastic Mesh Treeguards supplied in 50m rolls							
13 mm x 13 mm small mesh; roll width 60 cm; black	0.70	0.04	0.64	-	1.11	nr	**1.75**
13 mm x 13 mm small mesh; roll width 120 cm; black	1.32	0.06	1.06	-	1.86	nr	**2.92**
13 mm x 13 mm small mesh; roll width 60 cm; brown	0.80	0.06	1.06	-	1.21	nr	**2.27**
13 mm x 13 mm small mesh; roll width 120 cm; brown	1.41	0.06	1.06	-	1.96	nr	**3.02**
ready made tree guard in dark green fine plastic mesh, supplied flat packed	0.34	0.06	1.06	-	0.75	nr	**1.81**
Tree guards of 3 nr 3.00 m x 100 mm stakes; driving 600 mm into firm ground; bracing with timber braces at top and bottom; including 3 strands barbed wire	21.45	1.00	18.00	-	21.45	nr	**39.45**
English Woodlands; strimmer guard in heavy duty black plastic, 225 mm high	1.98	0.07	1.20	-	1.98	nr	**3.18**
Tubex Ltd; "Standard Treeshelter" inclusive of 25 mm stake; prices shown for quantities of 500 nr							
0.6 m high	0.46	0.05	0.90	-	0.64	each	**1.54**
0.75 m high	0.55	0.05	0.90	-	0.77	each	**1.67**
1.2 m high	1.84	0.05	0.90	-	2.14	each	**3.04**
1.5 m high	1.05	0.05	0.90	-	1.41	each	**2.31**
Tubex Ltd; "Shrubshelter" inclusive of 25 mm stake; prices shown for quantities of 500 nr							
"Ecostart" shelter for forestry transplants and seedlings,	0.53	0.07	1.20	-	0.53	nr	**1.73**
0.6 m high	0.80	0.07	1.20	-	0.98	nr	**2.18**
0.75 m high	1.14	0.07	1.20	-	1.36	nr	**2.56**
Tree guards; Netlon							
"Treeguard 2"; flat sheet 450 mm wide x 1.00 m high; cutting to length on site; 40 x 20 mm mesh; average 1.00 m high	0.84	0.03	0.60	-	0.84	nr	**1.44**
Extra over trees for spraying with antidessicant spray; "Wiltpruf"							
selected standards; standards; light standards	2.57	0.20	3.60	-	2.57	nr	**6.17**
standards; heavy standards	4.29	0.25	4.50	-	4.29	nr	**8.79**

Prices for Measured Works
Q PAVING/PLANTING/FENCING/SITE FURNITURE

Item Excluding site overheads and profit	PC £	Labour hours	Labour £	Plant £	Material £	Unit	Total rate £
Q31 PLANTING – cont'd							
Hedges							
Excavating trench for hedges; depositing soil alongside trench; by machine							
300 mm deep x 300 mm wide	-	0.03	0.54	0.30	-	m	0.84
300 mm deep x 450 mm wide	-	0.05	0.81	0.45	-	m	1.26
Excavating trench for hedges; depositing soil alongside trench; by hand							
300 mm deep x 300 mm wide	-	0.12	2.16	-	-	m	2.16
300 mm deep x 450 mm wide	-	0.23	4.05	-	-	m	4.05
Setting out; notching out; excavating trench; breaking up subsoil to minimum depth 300 mm							
minimum 400 mm deep	-	0.25	4.50	-	-	m	4.50
Hedge planting; including backfill with excavated topsoil; PC £0.30/nr							
single row; 200 mm centres	1.50	0.06	1.13	-	1.50	m	2.63
single row; 300 mm centres	1.00	0.06	1.00	-	1.00	m	2.00
single row; 400 mm centres	0.75	0.04	0.75	-	0.75	m	1.50
single row; 500 mm centres	0.60	0.03	0.60	-	0.60	m	1.20
double row; 200 mm centres	3.00	0.17	3.00	-	3.00	m	6.00
double row; 300 mm centres	2.00	0.13	2.40	-	2.00	m	4.40
double row; 400 mm centres	1.50	0.08	1.50	-	1.50	m	3.00
double row; 500 mm centres	1.20	0.07	1.20	-	1.20	m	2.40
Extra over hedges for incorporating manure; at 1 m³ per 30 m	0.73	0.03	0.45	-	0.73	m	1.18
Topiary; Clifton Nurseries Ltd							
Mop head trees; trimming to ball							
1.50 m high x 450 mm diameter	75.00	0.50	9.00	-	75.00	nr	84.00
Pyramid Bay trees; trimming to shape							
2.50 m high	275.00	0.50	9.00	-	275.00	nr	284.00
Shrub planting - General							
Preamble: For preparation of planting areas see "Cultivation" at the beginning of the section on planting							
Shrub planting							
Setting out; selecting planting from holding area; loading to wheel barrows; planting as plan or as directed; distance from holding area maximum 50 m; plants 2 - 3 litre containers							
plants in groups of 100 nr minimum	-	0.01	0.21	-	-	nr	0.21
plants in groups of 10 - 100 nr	-	0.02	0.30	-	-	nr	0.30
plants in groups of 3 - 5 nr	-	0.03	0.45	-	-	nr	0.45
single plants not grouped	-	0.04	0.72	-	-	nr	0.72
Forming planting holes; in cultivated ground (cultivating not included); by mechanical auger; trimming holes by hand; depositing excavated material alongside holes							
250 diameter	-	0.03	0.60	0.04	-	nr	0.64
250 x 250 mm	-	0.04	0.72	0.08	-	nr	0.80
300 x 300 mm	-	0.08	1.35	0.10	-	nr	1.45
Hand excavation; forming planting holes; in cultivated ground (cultivating not included); depositing excavated material alongside holes							
100 mm x 100 mm x 100 mm deep; with mattock or hoe	-	0.01	0.12	-	-	nr	0.12

Prices for Measured Works
Q PAVING/PLANTING/FENCING/SITE FURNITURE

Item Excluding site overheads and profit	PC £	Labour hours	Labour £	Plant £	Material £	Unit	Total rate £
250 mm x 250 mm x 300 mm deep	-	0.04	0.72	-	-	nr	0.72
300 mm x 300 mm x 300 mm deep	-	0.06	1.00	-	-	nr	1.00
400 mm x 400 mm x 400 mm deep	-	0.13	2.25	-	-	nr	2.25
500 mm x 500 mm x 500 mm deep	-	0.25	4.50	-	-	nr	4.50
600 mm x 600 mm x 600 mm deep	-	0.43	7.79	-	-	nr	7.79
900 mm x 900 mm x 600 mm deep	-	1.00	18.00	-	-	nr	18.00
1.00 m x 1.00 m x 600 mm deep	-	1.23	22.14	-	-	nr	22.14
1.25 m x 1.25 m x 600 mm deep	-	1.93	34.74	-	-	nr	34.74
Hand excavation; forming planting holes; in uncultivated ground; depositing excavated material alongside holes							
100 mm x 100 mm x 100 mm deep with mattock or hoe	-	0.03	0.45	-	-	nr	0.45
250 mm x 250 mm x 300 mm deep	-	0.06	1.00	-	-	nr	1.00
300 mm x 300 mm x 300 mm deep	-	0.06	1.13	-	-	nr	1.13
400 mm x 400 mm x 400 mm deep	-	0.25	4.50	-	-	nr	4.50
500 mm x 500 mm x 500 mm deep	-	0.33	5.86	-	-	nr	5.86
600 mm x 600 mm x 600 mm deep	-	0.55	9.90	-	-	nr	9.90
900 mm x 900 mm x 600 mm deep	-	1.25	22.50	-	-	nr	22.50
1.00 m x 1.00 m x 600 mm deep	-	1.54	27.68	-	-	nr	27.68
1.25 m x 1.25 m x 600 mm deep	-	2.41	43.42	-	-	nr	43.42
Bare root planting; to planting holes (forming holes not included); including backfilling with excavated material (bare root plants not included)							
bare root 1+1; 30 - 90 mm high	-	0.02	0.30	-	-	nr	0.30
bare root 1+2; 90 - 120 mm high	-	0.02	0.30	-	-	nr	0.30
Containerised planting; to planting holes (forming holes not inlcuded); including backfilling with excavated material (shrub or ground cover not included)							
9 cm pot	-	0.01	0.18	-	-	nr	0.18
2 litre container	-	0.02	0.36	-	-	nr	0.36
3 litre container	-	0.02	0.40	-	-	nr	0.40
5 litre container	-	0.03	0.60	-	-	nr	0.60
10 litre container	-	0.05	0.90	-	-	nr	0.90
15 litre container	-	0.07	1.20	-	-	nr	1.20
20 litre container	-	0.08	1.50	-	-	nr	1.50
Shrub planting; 2 litre containerised plants; in cultivated ground (cultivating not included); PC £1.80/nr							
average 2 plants per m²	-	0.06	1.01	-	3.60	m²	4.61
average 3 plants per m²	-	0.08	1.51	-	5.40	m²	6.91
average 4 plants per m²	-	0.11	2.02	-	7.20	m²	9.22
average 6 plants per m²	-	0.17	3.03	-	10.80	m²	13.83
Extra over shrubs for stakes	0.50	0.02	0.30	-	0.50	nr	0.80
Composted bark soil conditioners; 20 m³ loads; on beds by mechanical loader; spreading and rotavating into topsoil; by machine							
50 mm thick	130.00	-	-	7.45	130.00	100 m²	137.45
100 mm thick	260.00	-	-	11.43	273.00	100 m²	284.43
150 mm thick	390.00	-	-	15.57	409.50	100 m²	425.07
200 mm thick	520.00	-	-	19.63	546.00	100 m²	565.63

Prices for Measured Works
Q PAVING/PLANTING/FENCING/SITE FURNITURE

Item Excluding site overheads and profit	PC £	Labour hours	Labour £	Plant £	Material £	Unit	Total rate £
Q31 PLANTING – cont'd							
Shrub Planting; - cont'd							
Mushroom compost; 25 m³ loads; delivered not further than 25m from location; cultivating into topsoil by pedestrian operated machine							
50 mm thick	89.50	2.86	51.43	1.87	89.50	100 m²	142.80
100 mm thick	179.00	6.05	108.88	1.87	179.00	100 m²	289.75
150 mm thick	268.50	8.90	160.29	1.87	268.50	100 m²	430.65
200 mm thick	358.00	12.90	232.29	1.87	358.00	100 m²	592.15
Mushroom compost; 25 m³ loads; on beds by mechanical loader; spreading and rotavating into topsoil by tractor drawn rotavator							
50 mm thick	89.50	-	-	7.45	89.50	100 m²	96.95
100 mm thick	179.00	-	-	11.43	179.00	100 m²	190.43
150 mm thick	268.50	-	-	15.57	268.50	100 m²	284.07
200 mm thick	358.00	-	-	19.63	358.00	100 m²	377.63
Manure; 20 m³ loads; delivered not further than 25m from location; cultivating into topsoil by pedestrian operated machine							
50 mm thick	-	2.86	51.43	1.87	140.00	100 m²	193.30
100 mm thick	220.00	6.05	108.88	1.87	231.00	100 m²	341.75
150 mm thick	330.00	8.90	160.29	1.87	346.50	100 m²	508.65
200 mm thick	440.00	12.90	232.29	1.87	462.00	100 m²	696.15
Fertilizers (7:7:7); PC £0.49/kg; to beds; by hand							
35 g/m²	1.73	0.17	3.00	-	1.73	100 m²	4.73
50 g/m²	2.47	0.17	3.00	-	2.47	100 m²	5.47
70 g/m²	3.46	0.17	3.00	-	3.46	100 m²	6.46
Fertilizers; "Enmag" PC £1.20/kg slow release fertilizer; to beds; by hand							
35 g/m²	4.22	0.17	3.00	-	4.22	100 m²	7.22
50 g/m²	6.02	0.17	3.00	-	6.02	100 m²	9.02
70 g/m²	9.03	0.17	3.00	-	9.03	100 m²	12.03
Note: For machine incorporation of fertilizers and soil conditioners see "Cultivation".							
Herbaceous and groundcover planting							
Herbaceous plants; PC £1.00/nr; including forming planting holes in cultivated ground (cultivating not included); backfilling with excavated material; 1 litre containers							
average 4 plants per m² - 500 mm centres	-	0.09	1.68	-	4.40	m²	6.08
average 6 plants per m² - 408 mm centres	-	0.14	2.52	-	6.60	m²	9.12
average 8 plants per m² - 354 mm centres	-	0.19	3.36	-	8.80	m²	12.16
Note: For machine incorporation of fertilizers and soil conditioners see "Cultivation".							
Plant support netting; Bridport Gundry; on 50 mm diameter stakes; 750 mm long; driving into ground at 1.50 m centres							
green extruded plastic mesh; 125 mm square	0.36	0.04	0.72	-	0.36	m²	1.08

Prices for Measured Works
Q PAVING/PLANTING/FENCING/SITE FURNITURE

Item Excluding site overheads and profit	PC £	Labour hours	Labour £	Plant £	Material £	Unit	Total rate £
Bulb planting							
Bulbs; including forming planting holes in cultivated area (cultivating not included); backfilling with excavated material							
small	12.90	0.83	15.00	-	12.90	100 nr	27.90
medium	20.60	0.83	15.00	-	20.60	100 nr	35.60
large	24.50	0.91	16.36	-	24.50	100 nr	40.86
Bulbs; in grassed area; using bulb planter; including backfilling with screened topsoil or peat and cut turf plug							
small	12.90	1.67	30.00	-	12.90	100 nr	42.90
medium	20.60	1.67	30.00	-	20.60	100 nr	50.60
large	24.50	2.00	36.00	-	24.50	100 nr	60.50
Aquatic planting							
Approved hessian bags filled with soil and fertilizer; tied, slit and weighted; in pool to receive three submerged floating leafed marginal or swamp plants	-	0.10	1.80	-	0.55	nr	2.35
Aquatic plants; in prepared growing medium in pool; plant size 2 - 3 litre containerised (plants not included)	-	0.04	0.72	-	-	nr	0.72
Operations after planting							
Initial cutting back to shrubs and hedge plants; including disposal of all cuttings	-	1.00	18.00	-	-	100 m²	18.00
Mulch; Melcourt Industries Ltd; "Bark Nuggets"; to plant beds; delivered in 70 m³ loads; maximum distance 25 m							
50 mm thick	1.45	0.03	0.48	-	1.52	m²	2.01
75 mm thick	2.17	0.04	0.73	-	2.29	m²	3.01
100 mm thick	2.90	0.05	0.97	-	3.05	m²	4.02
Mulch; Melcourt Industries Ltd; "Bark Nuggets"; to plant beds; delivered in 20 m³ loads; maximum distance 25 m							
50 mm thick	2.07	0.03	0.53	-	2.17	m²	2.70
75 mm thick	3.11	0.04	0.79	-	3.26	m²	4.05
100 mm thick	4.14	0.06	1.05	-	4.35	m²	5.40
Mulch; Melcourt Industries Ltd; "Amenity Bark"; to plant beds; delivered in 70 m³ loads; maximum distance 25 m							
50 mm thick	0.89	0.03	0.53	-	0.93	m²	1.46
75 mm thick	1.33	0.04	0.79	-	1.40	m²	2.19
100 mm thick	1.78	0.04	0.66	-	1.86	m²	2.53
Mulch; Melcourt Industries Ltd; "Amenity Bark"; to plant beds; delivered in 20 m³ loads; maximum distance 25 m							
50 mm thick	1.51	0.03	0.53	-	1.58	m²	2.11
75 mm thick	2.26	0.04	0.79	-	2.38	m²	3.16
100 mm thick	3.02	0.06	1.05	-	3.17	m²	4.22
Selective herbicides; in accordance with manufacturer's instructions; PC £27.27/ litre; by machine; application rate							
30 ml/m²	-	-	-	0.29	0.82	100 m²	1.11
35 ml/m²	-	-	-	0.29	0.95	100 m²	1.24
40 ml/m²	-	-	-	0.29	1.09	100 m²	1.38
50 ml/m²	-	-	-	0.29	1.36	100 m²	1.65
3.00 l/ha	-	-	-	28.89	81.81	ha	110.70
3.50 l/ha	-	-	-	28.89	95.44	ha	124.33
4.00 l/ha	-	-	-	28.89	109.08	ha	137.97

Prices for Measured Works
Q PAVING/PLANTING/FENCING/SITE FURNITURE

Item Excluding site overheads and profit	PC £	Labour hours	Labour £	Plant £	Material £	Unit	Total rate £
Q31 PLANTING – cont'd							
Operations after planting; – cont'd							
Selective herbicides; in accordance with manufacturer's instructions; PC £27.27/ litre; by hand; application rate							
30 ml/m²	-	0.17	3.00	-	0.82	100 m²	3.82
35 ml/m²	-	0.17	3.00	-	0.95	100 m²	3.96
40 ml/m²	-	0.17	3.00	-	1.09	100 m²	4.09
50 ml/m²	-	0.17	3.00	-	1.36	100 m²	4.36
3.00 l/ha	-	16.67	300.00	-	81.81	ha	381.81
3.50 l/ha	-	16.67	300.00	-	95.44	ha	395.45
4.00 l/ha	-	16.67	300.06	-	109.08	ha	409.14
General herbicides; in accordance with manufacturer's instructions; "Knapsack" spray application							
"Casoron G (residual)"; at 1kg/125m2	3.13	0.05	0.90	-	3.44	100 m²	4.34
"Spasor"; at 50 ml/100 m²	0.70	0.05	0.90	-	0.70	100 m²	1.60
"Roundup Pro"; at 40 ml /100 m²	0.16	0.05	0.90	-	0.16	100 m²	1.06
"Super Verdone"; at 100ml/100 m²	0.60	0.05	0.90	-	0.60	100 m²	1.50
Fertilizers; in top 150 mm of topsoil at 35 g/m²							
fertilizer (18:0:0+Mg +Fe)	6.11	0.12	2.21	-	6.11	100 m²	8.32
"Enmag"	4.22	0.12	2.21	-	4.43	100 m²	6.63
fertilizer (7:7:7)	1.73	0.12	2.21	-	1.73	100 m²	3.94
fertilizer (20:10:10)	2.44	0.12	2.21	-	2.44	100 m²	4.65
"Superphosphate"	1.62	0.12	2.21	-	1.70	100 m²	3.91
"Hoof and Horn"	2.23	0.12	2.21	-	2.35	100 m²	4.55
"Bone meal"	3.10	0.12	2.21	-	3.25	100 m²	5.46
Fertilizers; in top 150 mm of topsoil at 70 g/m²							
fertilizer (18:0:0+Mg +Fe)	12.23	0.12	2.21	-	12.23	100 m²	14.43
"Enmag"	8.43	0.12	2.21	-	8.85	100 m²	11.06
fertilizer (7:7:7)	3.46	0.12	2.21	-	3.46	100 m²	5.67
fertilizer (20:10:10)	4.89	0.12	2.21	-	4.89	100 m²	7.09
"Superphosphate"	3.25	0.12	2.21	-	3.41	100 m²	5.61
"Hoof and Horn"	4.47	0.12	2.21	-	4.69	100 m²	6.90
"Bone meal"	6.20	0.12	2.21	-	6.51	100 m²	8.71
Maintenance operations (Note: the following rates apply to aftercare maintenance executed as part of a landscaping contract only)							
Weeding and hand forking planted areas; including disposing weeds and debris on site; areas maintained weekly	-	-	0.07	-	-	m²	0.07
Weeding and hand forking planted areas; including disposing weeds and debris on site; areas maintained monthly	-	0.01	0.18	-	-	m²	0.18
Extra over weeding and hand forking planted areas for disposing excavated material off site; to tip not exceeding 13 km; mechanically loaded							
slightly contaminated	-	-	-	1.55	12.50	m³	14.05
rubbish	-	-	-	1.55	12.50	m³	14.05
inert material	-	-	-	1.03	7.50	m³	8.53

Prices for Measured Works
Q PAVING/PLANTING/FENCING/SITE FURNITURE

Item Excluding site overheads and profit	PC £	Labour hours	Labour £	Plant £	Material £	Unit	Total rate £
Mulch; Melcourt Industries Ltd; "Bark Nuggets"; to plant beds; delivered in 70 m³ loads; maximum distance 25 m							
50 mm thick	1.45	0.03	0.48	-	1.52	m²	2.01
75 mm thick	2.17	0.04	0.73	-	2.29	m²	3.01
100 mm thick	2.90	0.05	0.97	-	3.05	m²	4.02
Mulch; Melcourt Industries Ltd; "Bark Nuggets"; to plant beds; delivered in 20 m³ loads; maximum distance 25 m							
50 mm thick	2.07	0.03	0.53	-	2.17	m²	2.70
75 mm thick	3.11	0.04	0.79	-	3.26	m²	4.05
100 mm thick	4.14	0.06	1.05	-	4.35	m²	5.40
Mulch; Melcourt Industries Ltd; "Amenity Bark"; to plant beds; delivered in 70 m³ loads; maximum distance 25 m							
50 mm thick	0.89	0.03	0.53	-	0.93	m²	1.46
75 mm thick	1.33	0.04	0.79	-	1.40	m²	2.19
100 mm thick	1.78	0.04	0.66	-	1.86	m²	2.53
Mulch; Melcourt Industries Ltd; "Amenity Bark"; to plant beds; delivered in 20 m³ loads; maximum distance 25 m							
50 mm thick	1.51	0.03	0.53	-	1.58	m²	2.11
75 mm thick	2.26	0.04	0.79	-	2.38	m²	3.16
100 mm thick	3.02	0.06	1.05	-	3.17	m²	4.22
Selective herbicides in accordance with manufacturer's instructions; PC £27.27 / litre; by machine; application rate							
30 ml/m²	-	-	-	0.29	0.82	100 m²	1.11
35 ml/m²	-	-	-	0.29	0.95	100 m²	1.24
40 ml/m²	-	-	-	0.29	1.09	100 m²	1.38
50 ml/m²	-	-	-	0.29	1.36	100 m²	1.65
3.00 l/ha	-	-	-	28.89	81.81	ha	110.70
3.50 l/ha	-	-	-	28.89	95.44	ha	124.33
4.00 l/ha	-	-	-	28.89	109.08	ha	137.97
Selective herbicides; in accordance with manufacturer's instructions; PC £27.27 / litre; by hand; application rate							
30 ml/m²	-	0.17	3.00	-	0.82	100 m²	3.82
35 ml/m²	-	0.17	3.00	-	0.95	100 m²	3.96
40 ml/m²	-	0.17	3.00	-	1.09	100 m²	4.09
50 ml/m²	-	0.17	3.00	-	1.36	100 m²	4.36
3.00 l/ha	-	16.67	300.00	-	81.81	ha	381.81
3.50 l/ha	-	16.67	300.00	-	95.44	ha	395.45
4.00 l/ha	-	16.67	300.01	-	109.08	ha	409.09
General herbicides; in accordance with manufacturer's instructions; "Knapsack" spray application							
"Casoron G (residual)"; at 1kg/125m2	3.13	0.05	0.90	-	3.44	100 m²	4.34
"Spasor"; at 50 ml/100 m²	0.70	0.05	0.90	-	0.70	100 m²	1.60
"Roundup Pro"; at 40 ml/100 m²	0.16	0.05	0.90	-	0.16	100 m²	1.06
"Super Verdone"; at 100 ml/100 m²	0.60	0.05	0.90	-	0.60	100 m²	1.50
Fertilizers; at 35 g/m²							
fertilizer (18:0:0+Mg +Fe)	6.11	0.12	2.21	-	6.11	100 m²	8.32
"Enmag"	4.22	0.12	2.21	-	4.43	100 m²	6.63
fertilizer (7:7:7)	1.73	0.12	2.21	-	1.73	100 m²	3.94
fertilizer (20:10:10)	2.44	0.12	2.21	-	2.44	100 m²	4.65
"Superphosphate"	1.62	0.12	2.21	-	1.70	100 m²	3.91
"Hoof and Horn"	2.23	0.12	2.21	-	2.35	100 m²	4.55
"Bone meal"	3.10	0.12	2.21	-	3.25	100 m²	5.46

Prices for Measured Works
Q PAVING/PLANTING/FENCING/SITE FURNITURE

Item Excluding site overheads and profit	PC £	Labour hours	Labour £	Plant £	Material £	Unit	Total rate £
Q31 PLANTING – cont'd							
Maintenance operations – cont'd							
Fertilizers; at 70 g/m²							
fertilizer (18:0:0+Mg +Fe)	12.23	0.12	2.21	-	12.23	100 m²	14.43
"Enmag"	8.43	0.12	2.21	-	8.85	100 m²	11.06
fertilizer (7:7:7)	3.46	0.12	2.21	-	3.46	100 m²	5.67
fertilizer (20:10:10)	4.89	0.12	2.21	-	4.89	100 m²	7.09
"Superphosphate"	3.25	0.12	2.21	-	3.41	100 m²	5.61
"Hoof and Horn"	4.47	0.12	2.21	-	4.69	100 m²	6.90
"Bone meal"	6.20	0.12	2.21	-	6.51	100 m²	8.71
Treating plants with growth retardant (maleic hydrazide); at rate of 16 l/ha; in accordance with manufacturer's instructions							
"Burtolin" spray; at 1.50 litres per 5 l water	1.06	0.03	0.60	-	1.06	nr	1.66
Watering planting; evenly; at a rate of 5 litre/m²							
using hand-held watering equipment	-	0.25	4.50	-	-	100 m²	4.50
using sprinkler equipment and with sufficient water pressure to run 1 nr 15 m radius sprinkler	-	0.14	2.50	-	-	100 m²	2.50
using movable spray lines powering 3 nr sprinkler heads with a radius of 15 m and allowing for 60% overlap (irrigation machinery costs not included)	-	0.02	0.28	-	-	100 m²	0.28
Forestry planting							
Deep ploughing rough ground to form planting ridges at							
2.00 m centres	-	0.63	11.25	7.23	-	100 m²	18.48
3.00 m centres	-	0.59	10.59	6.80	-	100 m²	17.39
4.00 m centres	-	0.40	7.20	4.62	-	100 m²	11.82
Notching plant forestry seedlings; "T" or "L" notch	26.00	0.75	13.50	-	26.00	100 nr	39.50
Turf planting forestry seedlings	26.00	2.00	36.00	-	26.00	100 nr	62.00
Selective herbicides; in accordance with manufacturer's instructions; PC £27.27 / litre; by hand; application rate							
30 ml/100m2	-	0.17	3.00	-	0.82	100 m²	3.82
35 ml/100m2	-	0.17	3.00	-	0.95	100 m²	3.96
40 ml/100m2	-	0.17	3.00	-	1.09	100 m²	4.09
50 ml/100m2	-	0.17	3.00	-	1.36	100 m²	4.36
3.00 l/ha	-	16.67	300.00	-	81.81	ha	381.81
3.50 l/ha	-	16.67	300.00	-	95.44	ha	395.45
4.00 l/ha	-	16.67	300.00	-	109.08	ha	409.08
Selective herbicides; in accordance with manufacturer's instructions; PC £27.27 / litre; by machine; application rate							
30 ml/100m2	-	-	-	0.29	0.82	100 m²	1.11
35 ml/100m2	-	-	-	0.29	0.95	100 m²	1.24
40 ml/100m2	-	-	-	0.29	1.09	100 m²	1.38
50 ml/100m2	-	-	-	0.29	1.36	100 m²	1.65
3.00 l/ha	-	-	-	28.89	81.81	ha	110.70
3.50 l/ha	-	-	-	28.89	95.44	ha	124.33
4.00 l/ha	-	-	-	28.89	109.08	ha	137.97
General herbicides; in accordance with manufacturer's instructions; "Knapsack" spray application							
"Casoron G (residual)"; at 1kg/125m2	3.13	0.05	0.90	-	3.44	100 m²	4.34
"Spasor"; at 50 ml/100 m²	0.70	0.05	0.90	-	0.70	100 m²	1.60

Prices for Measured Works
Q PAVING/PLANTING/FENCING/SITE FURNITURE

Item Excluding site overheads and profit	PC £	Labour hours	Labour £	Plant £	Material £	Unit	Total rate £
"Roundup Pro"; at 40 ml/100 m²	0.16	0.05	0.90	-	0.16	100 m²	1.06
"Super Verdone"; at 100 ml/100 m²	0.60	0.05	0.90	-	0.60	100 m²	1.50
Fertilizers; at 35 g/m²							
fertilizer (18:0:0+Mg + Fe)	6.11	0.12	2.21	-	6.11	100 m²	8.32
"Enmag"	4.22	0.12	2.21	-	4.43	100 m²	6.63
fertilizer (7:7:7)	1.73	0.12	2.21	-	1.73	100 m²	3.94
fertilizer (20:10:10)	2.44	0.12	2.21	-	2.44	100 m²	4.65
"Superphosphate"	1.62	0.12	2.21	-	1.70	100 m²	3.91
"Hoof and Horn"	2.23	0.12	2.21	-	2.35	100 m²	4.55
"Bone meal"	3.10	0.12	2.21	-	3.25	100 m²	5.46
Fertilizers; at 70 g/m²							
fertilizer (18:0:0+Mg + Fe)	12.23	0.12	2.21	-	12.23	100 m²	14.43
"Enmag"	8.43	0.12	2.21	-	8.85	100 m²	11.06
fertilizer (7:7:7)	3.46	0.12	2.21	-	3.46	100 m²	5.67
fertilizer (20:10:10)	4.89	0.12	2.21	-	4.89	100 m²	7.09
"Superphosphate"	3.25	0.12	2.21	-	3.41	100 m²	5.61
"Hoof and Horn"	4.47	0.12	2.21	-	4.69	100 m²	6.90
"Bone meal"	6.20	0.12	2.21	-	6.51	100 m²	8.71
Tree tubes; to young trees	121.00	0.30	5.40	-	121.00	100 nr	126.40
Cleaning and weeding around seedlings; once	-	0.50	9.00	-	-	100 nr	9.00
Treading in and firming ground around seedlings planted; at 2500 per ha after frost or other ground disturbance; once	-	0.33	6.00	-	-	100 nr	6.00
Beating up initial planting; once (including supply of replacement seedlings at 10% of original planting)	2.60	0.25	4.50	-	2.60	100 nr	7.10
Work to existing planting							
Control shoots lichen and suckers on street trees; Rhone Poulenc							
"Burtolin" spray at 1.50 litres per 5 l	1.06	0.03	0.60	-	1.06	nr	1.66
Cutting and trimming ornamental hedges; to specified profiles; including cleaning out hedge bottoms; hedge cut 2 occasions per annum; by hand							
up to 2.00 m high	-	0.03	0.60	-	0.19	m	0.79
2.00 - 4.00 m high	-	0.05	0.90	1.51	0.38	m	2.78
Cutting and laying field hedges; including stakes and ethering; removing or burning all debris. (Note: Rate at which work executed varies greatly with width and height of hedge. A typical hedge could be cut and laid at rate of 7 m run per man day)	-	1.00	18.00	-	1.70	m	19.70
labour charge per man day (specialist daywork)	-	1.00	18.00	-	-	m	18.00
Trimming field hedges; to specified heights and shapes							
using cutting bar	-	5.00	90.00	-	-	100 m	90.00
using flail	-	0.20	3.60	2.52	-	100 m	6.12

Prices for Measured Works
Q PAVING/PLANTING/FENCING/SITE FURNITURE

Item Excluding site overheads and profit	PC £	Labour hours	Labour £	Plant £	Material £	Unit	Total rate £
Q35 LANDSCAPE MAINTENANCE							
Preamble; Long term Landscape maintenance							
Maintenance on long term contracts differs in cost from that of maintenance as part of a landscape contract. In this section the contract period is generally 3 - 5 years. Staff are generally allocated to a single project only and therefore productivity is higher whilst overhead costs are lower. Labour costs in this section are lower than the costs used in other parts of the book. Machinery is assumed to be leased over a 5 year period and written off over the same period. The costs of maintenance and consumables for the various machinery types have been included in the information that follows. Finance costs for the machinery have not been allowed for.							
The rates shown below are for machines working in unconfined contiguous areas. Users should adjust the times and rates if working in smaller spaces or spaces with obstructions.							
Grass Cutting - tractor mounted equipment; Keith Banyard Grounds Maintenance Ltd							
Using multiple-gang mower with cylindrical cutters; contiguous areas such as playing fields and the like larger than 3000 m²							
3 gang; 2.13 m cutting width	-	0.01	0.15	0.10	-	100m2	**0.25**
5 gang; 3.40 m cutting width	-	0.01	0.13	0.13	-	100 m²	**0.26**
7 gang; 4.65 m cutting width	-	0.01	0.08	0.09	-	100 m²	**0.17**
Using multiple-gang mower with cylindrical cutters; non contiguous areas such as verges and general turf areas							
3 gang	-	0.02	0.29	0.10	-	100m2	**0.39**
5 gang	-	0.01	0.19	0.13	-	100 m²	**0.32**
Using multiple-rotary mower with vertical drive shaft and horizontally rotating bar or disc cutters; contiguous areas larger than 3000 m²							
cutting grass, overgrowth or the like using flail mower or reaper	-	0.02	0.28	0.22	-	100 m²	**0.50**
Collection of arisings							
22 cuts per year	-	0.02	0.23	0.21	-	100 m²	**0.44**
18 cuts per year	-	0.02	0.31	0.21	-	100 m²	**0.52**
12 cuts per year	-	0.03	0.47	0.21	-	100 m²	**0.68**
4 cuts per year	-	0.07	0.93	0.21	-	100 m²	**1.14**
Disposal of arisings							
22 cuts per year	-	0.01	0.12	0.04	0.40	100 m²	**0.55**
18 cuts per year	-	0.01	0.14	0.05	0.53	100 m²	**0.72**
12 cuts per year	-	0.01	0.09	0.12	0.96	100 m²	**1.18**
4 cuts per year	-	0.02	0.28	0.76	1.60	100 m²	**2.64**
Cutting grass, overgrowth or the like: using tractor-mounted side-arm flail mower: in areas inaccessible to alternative machine: on surface							
not exceeding 30 deg from horizontal	-	0.02	0.33	0.28	-	100 m²	**0.61**
30 deg to 50 deg from horizontal	-	0.05	0.65	0.28	-	100 m²	**0.93**

Prices for Measured Works
Q PAVING/PLANTING/FENCING/SITE FURNITURE

Item Excluding site overheads and profit	PC £	Labour hours	Labour £	Plant £	Material £	Unit	Total rate £
Grass Cutting - 'Ride-On' self-propelled equipment; Norris and Gardener Grounds Maintenance							
Using ride-on multiple-cylinder mower							
3 gang; 2.13 m cutting width	-	0.01	0.15	0.10	-	100m2	**0.25**
5 gang; 3.40 m cutting width	-	0.01	0.09	0.13	-	100 m²	**0.22**
Using ride-on multiple-rotary mower with horizontally rotating bar, disc or chain cutters							
cutting width 1.52 m	-	0.01	0.18	0.15	-	100 m²	**0.33**
cutting width 1.82 m	-	0.01	0.15	0.11	-	100 m²	**0.26**
cutting width 2.97 m	-	0.01	0.09	0.12	-	100 m²	**0.21**
Add for using grass box/collector for removal and depositing of arisings	-	0.04	0.53	0.05	-	100 m²	**0.58**
Grass Cutting - pedestrian operated equipment							
Using cylinder lawn mower fitted with not less than five cutting blades, front and rear rollers: on surface not exceeding 30 deg from horizontal; arisings let fly; width of cut							
51 cm	-	0.06	0.85	0.17	-	100 m	**1.02**
61 cm	-	0.05	0.71	0.15	-	100 m	**0.86**
71 cm	-	0.04	0.61	0.15	-	100 m	**0.76**
91 cm	-	0.03	0.48	0.15	-	100 m	**0.63**
Using rotary self propelled mower; width of cut							
45 cm	-	0.04	0.55	0.05	-	100 m²	**0.60**
81 cm	-	0.02	0.31	0.06	-	100 m²	**0.37**
91 cm	-	0.02	0.27	0.06	-	100 m²	**0.33**
120 cm	-	0.01	0.21	0.22	-	100 m²	**0.43**
Add for using grass box for collecting and depositing arisings							
removing and depositing arisings	-	0.05	0.70	-	-	100 m²	**0.70**
Add for 30 to 50 deg from horizontal	-	-	-	-	-	33%	-
Add for slopes exceeding 50 deg	-	-	-	-	-	100%	-
Cutting grass or light woody undergrowth: using trimmer with nylon cord or metal disc cutter: on surface							
Not exceeding 30 deg from horizontal	-	0.20	2.80	0.27	-	100 m²	**3.07**
30 -50 deg from horizontal	-	0.40	5.60	0.54	-	100 m²	**6.14**
Exceeding 50 deg from horizontal	-	0.50	7.00	0.68	-	100 m²	**7.68**
Grass Cutting - collecting arisings							
Extra over for tractor drawn and self propelled machinery using attached grass boxes; depositing arisings							
22 cuts per year	-	0.05	0.70	-	-	100 m²	**0.70**
18 cuts per year	-	0.08	1.05	-	-	100 m²	**1.05**
12 cuts per year	-	0.10	1.40	-	-	100 m²	**1.40**
4 cuts per year	-	0.25	3.50	-	-	100 m²	**3.50**
Disposing arisings							
22 cuts per year	-	0.01	0.11	0.02	0.05	100 m²	**0.18**
18 cuts per year	-	0.01	0.14	0.03	0.06	100 m²	**0.22**
12 cuts per year	-	0.01	0.20	0.04	0.09	100 m²	**0.33**
4 cuts per year	-	0.04	0.61	0.12	0.28	100 m²	**1.01**
Harrowing							
Harrowing grassed area							
drag harrow	-	0.01	0.17	0.12	-	100 m²	**0.30**
chain or light flexible spiked harrow	-	0.02	0.23	0.12	-	100 m²	**0.35**

Prices for Measured Works
Q PAVING/PLANTING/FENCING/SITE FURNITURE

Item Excluding site overheads and profit	PC £	Labour hours	Labour £	Plant £	Material £	Unit	Total rate £
Q35 LANDSCAPE MAINTENANCE – cont'd							
Scarifying							
A Plant Groundcare							
Sisis ARP4, including grass collection box, towed							
By tractor; area scarified annually	-	0.02	0.33	0.50	-	100 m²	0.83
Sisis ARP4, including grass collection box, towed							
By tractor; area scarified two years previously	-	0.03	0.39	0.61	-	100 m²	1.00
pedestrian operated self-powered equipment	-	0.07	0.98	0.68	-	100 m²	1.66
add for disposal of arisings	-	0.03	0.35	0.95	8.00	100 m²	9.30
By hand							
hand implement	-	0.50	7.00	-	-	100 m²	7.00
add for disposal of arisings	-	0.03	0.35	0.95	8.00	100 m²	9.30
Rolling							
Rolling grassed area; equipment towed by tractor; once over, using							
smooth roller	-	0.01	0.20	0.15	-	100 m²	0.35
Turf Aeration							
By machine; A Plant Groundcare							
Vertidrain turf aeration equipment towed by tractor to effect a minimum penetration of 100 to 250 mm at 100 mm centres	-	0.04	0.54	1.27	-	100 m²	1.81
Ryan GA 30; self propelled turf aerating equipment; to effect a minimum penetration of 100 mm at varying centres	-	0.04	0.54	1.47	-	100 m²	2.02
Groundsman; pedestrian operated; self powered solid or slitting tine turf aerating equipment to effect a minimum penetration of 100 mm	-	0.14	1.96	1.40	-	100 m²	3.36
Cushman core harvester; self propelled; for collection of arisings	-	0.02	0.27	0.65	-	100 m²	0.92
By hand							
Hand fork: to effect a minimum penetration of 100 mm and spaced 150 mm apart	-	1.33	18.67	-	-	100 m²	18.67
Hollow tine hand implement: to effect a minimum penetration of 100 mm and spaced 150 mm apart:	-	2.00	28.00	-	-	100 m²	28.00
Collection of arisings by hand	-	3.00	42.00	-	-	100 m²	42.00
Turf areas; Surface treatments and Top dressing; Boughton Loam Ltd							
Apply screened topdressing to grass surfaces. Spread using Tru-Lute							
Sand soil mixes 90/10 to 50/50	0.08	-	0.03	0.03	0.08	m²	0.14
Apply screened soil 3 mm, Kettering loam to goal mouths and worn areas							
20 mm thick	0.76	0.01	0.14	-	0.91	m²	1.05
10 mm thick	0.38	0.01	0.14	-	0.46	m²	0.60
Leaf Clearance; clearing grassed area of leaves and other extraneous debris							
Using equipment towed by tractor							
large grassed areas with perimeters of mature trees such as sports fields and amenity areas	-	0.01	0.17	0.29	-	100 m²	0.46
large grassed areas containing ornamental trees and shrub beds	-	0.03	0.35	0.40	-	100 m²	0.75

Prices for Measured Works
Q PAVING/PLANTING/FENCING/SITE FURNITURE

Item Excluding site overheads and profit	PC £	Labour hours	Labour £	Plant £	Material £	Unit	Total rate £
Using pedestrian operated mechanical equipment and blowers							
Grassed areas with perimeters of mature trees such as sports fields and amenity areas	-	0.04	0.56	0.10	-	100 m²	0.66
grassed areas containing ornamental trees and Shrub beds	-	0.10	1.40	0.25	-	100 m²	1.65
Verges	-	0.07	0.93	0.17	-	100 m²	1.10
By hand							
Grassed areas with perimeters of mature trees such as sports fields and amenity areas	-	0.05	0.70	0.09	-	100 m²	0.79
Grassed areas containing ornamental trees and Shrub beds	-	0.08	1.17	0.15	-	100 m²	1.32
Verges	-	1.00	14.00	1.84	-	100 m²	15.84
Removal of arisings							
Areas with perimeters of mature trees	-	0.01	0.09	0.08	0.64	100 m²	0.82
Areas containing ornamental trees and shrub beds	-	0.02	0.28	0.31	1.60	100 m²	2.19
Litter Clearance							
Collection and disposal of litter from grassed area	-	0.01	0.14	-	0.05	100 m²	0.19
Collection and disposal of litter from isolated grassed area not exceeding 1000 m²	-	0.04	0.56	-	0.05	100 m²	0.61
Edge Maintenance							
Maintain edges where lawn abuts pathway or hard surface using;							
Strimmer	-	0.01	0.07	0.01	-	m	0.08
Shears	-	0.02	0.23	-	-	m	0.23
Maintain edges where lawn abuts plant bed using							
mechanical edging tool	-	0.01	0.09	0.03	-	m	0.12
Shears	-	0.01	0.16	-	-	m	0.16
Half moon edging tool	-	0.02	0.28	-	-	m	0.28
Tree Guards, Stakes and Ties etc.							
Adjusting existing tree tie	-	0.03	0.47	-	-	nr	0.47
Taking up single or double tree stake and ties: removing and disposing	-	0.05	0.90	-	-	nr	0.90
Pruning Shrubs							
Trimming ground cover planting							
soft groundcover; vinca ivy and the like	-	1.00	14.00	-	-	100 m²	14.00
woody groundcover; cotoneaster and the like	-	1.50	21.00	-	-	100 m²	21.00
Pruning massed shrub border (measure ground area)							
Shrub beds pruned annually	-	0.01	0.14	-	-	m²	0.14
Shrub beds pruned hard every 3 years	-	0.03	0.39	-	-	m²	0.39
Cutting off dead heads							
Bush or standard rose	-	0.05	0.70	-	-	nr	0.70
Climbing rose	-	0.08	1.17	-	-	nr	1.17
Pruning Roses							
Bush or standard rose	-	0.05	0.70	-	-	nr	0.70
Climbing rose or rambling rose: tying in as required	-	0.07	0.93	-	-	nr	0.93

Prices for Measured Works
Q PAVING/PLANTING/FENCING/SITE FURNITURE

Item Excluding site overheads and profit	PC £	Labour hours	Labour £	Plant £	Material £	Unit	Total rate £
Q35 LANDSCAPE MAINTENANCE – cont'd							
Pruning Shrubs – cont'd							
Pruning Ornamental Shrub: height before pruning (increase these rates by 50% if pruning work has not been executed during the previous two years)							
Not exceeding 1m	-	0.04	0.56	-	-	nr	0.56
1 to 2 m	-	0.06	0.78	-	-	nr	0.78
exceeding 2 m	-	0.13	1.75	-	-	nr	1.75
Removing excess growth etc from face of building etc: height before pruning							
Not exceeding 2 m	-	0.03	0.40	-	-	nr	0.40
2 to 4 m	-	0.05	0.70	-	-	nr	0.70
4 to 6 m	-	0.08	1.17	-	-	nr	1.17
6 to 8 m	-	0.13	1.75	-	-	nr	1.75
8 to 10 m	-	0.14	2.00	-	-	nr	2.00
Removing epicormic growth from base of shrub or trunk and base of tree: any height: any diameter: number of growths							
Not exceeding 10	-	0.05	0.70	-	-	nr	0.70
10 to 20	-	0.07	0.93	-	-	nr	0.93
Beds Borders and Planters							
Lifting							
Bulbs	-	0.50	7.00	-	-	100 nr	7.00
Tubers or corms	-	0.40	5.60	-	-	100 nr	5.60
Established herbaceous plants; hoeing and depositing for replanting	-	2.00	28.00	-	-	100 nr	28.00
Temporary staking and tying in herbaceous plant	-	0.03	0.47	-	0.10	nr	0.57
cutting down spent growth of herbaceous plant; Clearing arisings							
Unstaked	-	0.02	0.28	-	-	nr	0.28
Staked: not exceeding 4 stakes per plant:							
Removing stakes and putting into store	-	0.03	0.35	-	-	nr	0.35
Hand weeding							
Newly planted areas	-	2.00	28.00	-	-	100 m²	28.00
Established areas	-	0.50	7.00	-	-	100 m²	7.00
removing grasses from groundcover areas	-	3.00	42.04	-	-	100 m²	42.04
Hand digging with fork: not exceeding 150 mm deep: breaking down lumps: leaving surface with a medium tilth	-	0.01	0.19	-	-	100 m²	0.19
Hand digging with fork or spade to an average depth of 230 mm: breaking down lumps: leaving surface with a medium tilth	-	0.02	0.28	-	-	100 m²	0.28
hand hoeing: not exceeding 50 mm deep: leaving surface with a medium tilth	-	0.40	5.60	-	-	100 m²	5.60
hand raking to remove stones etc: breaking down lumps: leaving surface with a fine tilth prior to planting	-	0.01	0.09	-	-	100 m²	0.09
Hand weeding; planter, window box; not exceeding 1.00 m²							
Ground level box	-	0.05	0.70	-	-	nr	0.70
Box accessed by stepladder	-	0.08	1.17	-	-	nr	1.17
Spreading compost, mulch or processed bark to a depth of 75 mm							

Prices for Measured Works
Q PAVING/PLANTING/FENCING/SITE FURNITURE

Item Excluding site overheads and profit	PC £	Labour hours	Labour £	Plant £	Material £	Unit	Total rate £
On shrub bed with existing mature planting	-	0.04	0.61	-	-	m²	0.61
recently planted areas	-	0.06	0.82	-	-	m²	0.82
Groundcover and herbaceous areas	-	0.08	1.05	-	-	m²	1.05
clearing cultivated area of leaves, litter and other extraneous debris: using hand implement							
Weekly maintenance	-	0.13	1.75	-	-	100 m²	1.75
Daily maintenance	-	0.02	0.23	-	-	100 m²	0.23
Bedding							
Lifting							
bedding plants; hoeing and depositing for disposal	-	0.33	4.67	-	-	100 m²	4.67
Hand digging with fork: not exceeding 150 mm deep: breaking down lumps: leaving surface with a medium tilth	-	0.01	0.19	-	-	100 m²	0.19
Hand weeding							
newly planted areas	-	2.00	28.00	-	-	100 m²	28.00
established areas	-	0.50	7.00	-	-	100 m²	7.00
Hand digging with fork or spade to an average depth of 230 mm: breaking down lumps: leaving surface with a medium tilth	-	0.02	0.28	-	-	100 m²	0.28
hand hoeing: not exceeding 50 mm deep: leaving surface with a medium tilth	-	0.40	5.60	-	-	100 m²	5.60
hand raking to remove stones etc: breaking down lumps: leaving surface with a fine tilth prior to planting	-	0.01	0.09	-	-	100 m²	0.09
Hand weeding; planter, window box; not exceeding 1.00 m²							
Ground level box	-	0.05	0.70	-	-	nr	0.70
Box accessed by stepladder	-	0.08	1.17	-	-	nr	1.17
Spreading only; compost, mulch or processed bark to a depth of 75 mm							
On shrub bed with existing mature planting	-	0.04	0.61	-	-	m²	0.61
recently planted areas	-	0.06	0.82	-	-	m²	0.82
Groundcover and herbaceous areas	-	0.08	1.05	-	-	m²	1.05
Collecting bedding from nursery	-	3.00	42.00	12.61	-	100 m²	54.61
Setting out							
Mass planting single variety	-	0.13	1.75	-	-	m²	1.75
Pattern	-	0.33	4.67	-	-	m²	4.67
Planting only							
massed bedding plants	-	0.20	2.80	-	-	m²	2.80
Clearing cultivated area of leaves, litter and other extraneous debris: using hand implement							
Weekly maintenance	-	0.13	1.75	-	-	100 m²	1.75
Daily maintenance	-	0.02	0.23	-	-	100 m²	0.23
Irrigation and watering							
Hand held hosepipe; flow rate 25 litres per minute; irrigation requirement							
10 litres /m²	-	0.74	10.32	-	-	100 m²	10.32
15 litres /m²	-	1.10	15.40	-	-	100 m²	15.40
20 litres /m²	-	1.46	20.48	-	-	100 m²	20.48
25 litres /m²	-	1.84	25.72	-	-	100 m²	25.72
Hand held hosepipe; flow rate 40 litres per minute; irrigation requirement							
10 litres /m²	-	0.46	6.47	-	-	100 m²	6.47
15 litres /m²	-	0.69	9.70	-	-	100 m²	9.70
20 litres /m²	-	0.91	12.78	-	-	100 m²	12.78
25 litres /m²	-	1.15	16.05	-	-	100 m²	16.05

Prices for Measured Works
Q PAVING/PLANTING/FENCING/SITE FURNITURE

Item Excluding site overheads and profit	PC £	Labour hours	Labour £	Plant £	Material £	Unit	Total rate £
Q35 LANDSCAPE MAINTENANCE – cont'd							
Hedge Cutting; field hedges cut once or twice annually							
Trimming sides and top using hand tool or hand held mechanical tools							
not exceeding 2 m high	-	0.10	1.40	0.14	-	10 m²	1.54
2 to 4 m high	-	0.33	4.67	0.45	-	10 m²	5.12
Hedge Cutting; ornamental							
Trimming sides and top using hand tool or hand held mechanical tools							
not exceeding 2 m high	-	0.13	1.75	0.17	-	10 m²	1.92
2 to 4 m high	-	0.50	7.00	0.68	-	10 m²	7.68
Hedge Cutting; reducing width; hand tool or hand held mechanical tools							
Not exceeding 2 m high							
Average depth of cut not exceeding 300 mm	-	0.10	1.40	0.14	-	10 m²	1.54
Average depth of cut 300 to 600 mm	-	0.83	11.67	1.13	-	10 m²	12.79
Average depth of cut 600 to 900 mm	-	1.25	17.50	1.69	-	10 m²	19.19
2 to 4 m high							
Average depth of cut not exceeding 300 mm	-	0.03	0.35	0.03	-	10 m²	0.38
Average depth of cut 300 to 600 mm	-	0.13	1.75	0.17	-	10 m²	1.92
Average depth of cut 600 to 900 mm	-	2.50	35.00	3.38	-	10 m²	38.38
4 to 6 m high							
Average depth of cut not exceeding 300 mm	-	0.10	1.40	0.14	-	m²	1.54
Average depth of cut 300 to 600 mm	-	0.17	2.33	0.23	-	m²	2.56
Average depth of cut 600 to 900 mm	-	0.50	7.00	0.68	-	m²	7.68
Hedge Cutting; reducing width; tractor mounted hedge cutting equipment							
Not exceeding 2 m high							
Average depth of cut not exceeding 300 mm	-	0.04	0.56	0.62	-	10 m²	1.18
Average depth of cut 300 to 600 mm	-	0.05	0.70	0.77	-	10 m²	1.47
Average depth of cut 600 to 900 mm	-	0.20	2.80	3.08	-	10 m²	5.88
2 to 4 m high							
Average depth of cut not exceeding 300 mm	-	0.01	0.17	0.19	-	10 m²	0.37
Average depth of cut 300 to 600 mm	-	0.03	0.35	0.38	-	10 m²	0.73
Average depth of cut 600 to 900 mm	-	0.02	0.28	0.31	-	10 m²	0.59
Hedge Cutting; reducing height; hand tool or hand held mechanical tools							
Not exceeding 2 m high							
Average depth of cut not exceeding 300 mm	-	0.07	0.93	0.09	-	10 m²	1.02
Average depth of cut 300 to 600 mm	-	0.13	1.87	0.18	-	10 m²	2.05
Average depth of cut 600 to 900 mm	-	0.40	5.60	0.54	-	10 m²	6.14
2 to 4 m high							
Average depth of cut not exceeding 300 mm	-	0.03	0.47	0.05	-	m²	0.51
Average depth of cut 300 to 600 mm	-	0.07	0.93	0.09	-	m²	1.02
Average depth of cut 600 to 900 mm	-	0.20	2.80	0.27	-	m²	3.07
4 to 6 m high							
Average depth of cut not exceeding 300 mm	-	0.07	0.93	0.09	-	m²	1.02
Average depth of cut 300 to 600 mm	-	0.13	1.75	0.17	-	m²	1.92
Average depth of cut 600 to 900 mm	-	0.25	3.50	0.34	-	m²	3.84

Prices for Measured Works
Q PAVING/PLANTING/FENCING/SITE FURNITURE

Item Excluding site overheads and profit	PC £	Labour hours	Labour £	Plant £	Material £	Unit	Total rate £
Hedge Cutting; removal and disposal of arisings							
Sweeping up and depositing arisings							
300 mm cut	-	0.05	0.70	-	-	10 m²	0.70
600 mm cut	-	0.20	2.80	-	-	10 m²	2.80
900 mm cut	-	0.40	5.60	-	-	10 m²	5.60
Chipping arisings							
300 mm cut	-	0.02	0.28	0.11	-	10 m²	0.39
600 mm cut	-	0.08	1.17	0.46	-	10 m²	1.62
900 mm cut	-	0.20	2.80	1.10	-	10 m²	3.90
Disposal of unchipped arisings							
300 mm cut	-	0.02	0.23	0.47	1.07	10 m²	1.77
600 mm cut	-	0.03	0.47	0.95	1.60	10 m²	3.01
900 mm cut	-	0.08	1.17	2.36	4.00	10 m²	7.53
Disposal of chipped arisings							
300 mm cut	-	-	0.05	0.16	1.60	10 m²	1.81
600 mm cut	-	0.02	0.23	0.16	3.20	10 m²	3.59
900 mm cut	-	0.03	0.47	0.31	8.00	10 m²	8.78
Q40 FENCING AND GATES							
Protective fencing							
Rolled chestnut pale fencing; fixing to 100 mm diameter chestnut posts; driving into firm ground at 3 m centres							
900 mm high	1.81	0.11	1.92	-	3.91	m	5.83
1200 mm high	2.43	0.16	2.88	-	4.56	m	7.44
1500 mm high; 3 strand	3.69	0.21	3.84	-	5.70	m	9.55
Enclosures; Earth Anchors							
"Rootfast" anchored galvanized steel enclosures post ref ADP 20-1000; 1000 mm high x 20 mm diameter with ref AA25-750 socket and padlocking ring	31.00	0.10	1.80	-	32.95	nr	34.75
steel cable, orange plastic coated	-	-	0.04	-	1.25	m	1.29
Boundary fencing; strained wire and wire mesh; H.S.Jackson & Son (Fencing)							
Strained wire fencing; concrete posts only at 2750 mm centres, 610 mm below ground; excavating holes; filling with concrete; replacing top soil; disposing surplus soil off site							
900 mm high	2.18	0.56	10.00	0.63	3.27	m	13.90
1200 mm high	2.89	0.56	10.00	0.63	3.98	m	14.61
1400 mm high	3.58	0.56	10.00	0.63	4.67	m	15.30
1800 mm high	3.95	0.78	14.00	1.57	5.04	m	20.61
2400 mm high	5.50	1.67	30.00	0.63	6.59	m	37.22
Extra over strained wire fencing for concrete straining posts with one strut; posts and struts 610 mm below ground; struts cleats, stretchers, winders, bolts, and eye bolts; excavating holes; filling to within 150 mm of ground level with concrete (1:12) - 40 mm aggregate; replacing topsoil; disposing surplus soil off site							
900 mm high	18.67	0.67	12.06	1.41	38.74	nr	52.22
1200 mm high	20.75	0.67	12.00	1.42	40.89	nr	54.31
1400 mm high	23.00	0.67	12.00	1.42	45.15	nr	58.57
1800 mm high	29.53	0.67	12.00	1.42	50.48	nr	63.90
2400 mm high	43.92	0.83	14.99	1.68	69.02	nr	85.69

Prices for Measured Works
Q PAVING/PLANTING/FENCING/SITE FURNITURE

Item Excluding site overheads and profit	PC £	Labour hours	Labour £	Plant £	Material £	Unit	Total rate £
Q40 FENCING AND GATES – cont'd							
Boundary fencing; - cont'd							
Extra over strained wire fencing for concrete straining posts with two struts; posts and struts 610 mm below ground; excavating holes; filling to within 150 mm of ground level with concrete (1:12) - 40 mm aggregate; replacing topsoil; disposing surplus soil off site							
900 mm high	26.93	0.91	16.38	1.82	56.30	nr	**74.50**
1200 mm high	30.01	0.85	15.30	1.42	60.15	nr	**76.87**
1400 mm high	33.50	0.91	16.34	1.42	69.13	nr	**86.89**
1800 mm high	42.86	0.85	15.30	1.42	76.09	nr	**92.81**
2400 mm high	43.92	0.67	12.00	1.68	81.01	nr	**94.69**
Strained wire fencing; galvanised steel angle posts only at 2750 mm centres; 610 mm below ground; driving in							
900 mm high; 40 mm x 40 mm x 5 mm	1.42	0.03	0.55	-	1.42	m	**1.96**
1200 mm high; 40 mm x 40 mm x 5 mm	1.60	0.03	0.60	-	1.60	m	**2.20**
1400 mm high; 40 mm x 40 mm x 5 mm	1.76	0.06	1.09	-	1.76	m	**2.85**
1800 mm high; 40 mm x 40 mm x 5 mm	2.32	0.07	1.31	-	2.32	m	**3.62**
2400 mm high; 45 mm x 45 mm x 5 mm	3.09	0.12	2.18	-	3.09	m	**5.27**
Galvanised steel straining posts with one strut for strained wire fencing; setting in concrete							
900 mm high; 50 mm x 50 mm x 6 mm	-	-	-	-	18.99	nr	**18.99**
1200 mm high; 50 mm x 50 mm x 6 mm	-	-	-	-	22.86	nr	**22.86**
1400 mm high; 50 mm x 50 mm x 6 mm	-	-	-	-	26.58	nr	**26.58**
1800 mm high; 50 mm x 50 mm x 6 mm	-	-	-	-	30.23	nr	**30.23**
2400 mm high; 60 mm x 60 mm x 6 mm	-	-	-	-	45.00	nr	**45.00**
Galvanised steel straining posts with two struts for strained wire fencing; setting in concrete							
900 mm high; 50 mm x 50 mm x 6 mm	-	-	-	-	27.65	nr	**27.65**
1200 mm high; 50 mm x 50 mm x 6 mm	-	-	-	-	33.67	nr	**33.67**
1400 mm high; 50 mm x 50 mm x 6 mm	-	-	-	-	40.26	nr	**40.26**
1800 mm high; 50 mm x 50 mm x 6 mm	-	-	-	-	48.44	nr	**48.44**
2400 mm high; 60 mm x 60 mm x 6 mm	-	-	-	-	65.00	nr	**65.00**
Strained wire; to posts (posts not included); 3 mm galvanized wire; fixing with galvanized stirrups							
900 mm high; 2 wire	0.14	0.03	0.60	-	0.20	m	**0.80**
1200 mm high; 3 wire	0.21	0.05	0.84	-	0.27	m	**1.11**
1400 mm high; 3 wire	0.21	0.05	0.84	-	0.27	m	**1.11**
1800 mm high; 3 wire	0.21	0.05	0.84	-	0.27	m	**1.11**
Barbed wire; to posts (posts not included); 3 mm galvanised wire; fixing with galvanised stirrups							
900 mm high; 2 wire	0.14	0.07	1.20	-	0.20	m	**1.40**
1200 mm high; 3 wire	0.21	0.09	1.68	-	0.27	m	**1.95**
1400 mm high; 3 wire	0.21	0.09	1.68	-	0.27	m	**1.95**
1800 mm high; 3 wire	0.21	0.09	1.68	-	0.27	m	**1.95**
Chain link fencing; H.S.Jackson & Son (Fencing); to strained wire and posts priced separately; 3 mm galvanised wire; 51 mm mesh; galvanized steel components; fixing to line wires threaded through posts and strained with eye-bolts; posts (not included)							
900 mm high	1.97	0.07	1.20	-	2.03	m	**3.23**
1200 mm high	2.49	0.07	1.20	-	2.55	m	**3.75**
1800 mm high	3.87	0.10	1.80	-	3.93	m	**5.73**
2400 mm high	3.87	0.10	1.80	-	3.93	m	**5.73**

Prices for Measured Works
Q PAVING/PLANTING/FENCING/SITE FURNITURE

Item Excluding site overheads and profit	PC £	Labour hours	Labour £	Plant £	Material £	Unit	Total rate £
Chain link fencing; to strained wire and posts priced separately; 3.15 mm plastic coated galvanized wire (wire only 2.50 mm); 51 mm mesh; galvanised steel components; fencing with line wires threaded through posts and strained with eye-bolts; posts (not included) (Note: plastic coated fencing can be cheaper than galvanised finish as wire of a smaller cross-sectional area can be used)							
900 mm high	1.63	0.07	1.20	-	1.75	m	2.95
1200 mm high	2.15	0.07	1.20	-	2.27	m	3.48
1400 mm high	2.47	0.07	1.20	-	2.59	m	3.79
1800 mm high	-	0.13	2.25	-	3.65	m	5.90
Extra over strained wire fencing for cranked arms and three rows of plain barbed wire							
1800 mm high	1.37	0.06	1.00	-	1.57	m	2.57
As above but galvanized barbed wire and clips in lieu of plain wire							
1 row	1.47	0.02	0.30	-	1.47	m	1.78
2 row	1.58	0.05	0.90	-	1.58	m	2.48
3 row	1.68	0.05	0.90	-	1.68	m	2.58
Field fencing; welded wire mesh; fixed to posts and straining wires measured separately							
Cattle fence; 1143 high 114 x 300 mm at bottom to 230 x 300 mm at top	0.71	0.10	1.80	-	0.91	m	2.71
Sheep fence; 910 high; 140 x 300 mm at bottom to 230 x 300 mm at top	0.56	0.10	1.80	-	0.76	m	2.56
Deer Fence; 1905 high; 89 x 150 mm at bottom to 267 x 300 mm at top	2.56	0.13	2.25	-	2.76	m	5.01
Extra for concreting in posts	-	-	-	-	1.45	nr	1.45
Extra for straining post	-	0.75	13.50	-	12.93	nr	26.43
Rabbit netting; H.S Jackson and Son; Timber stakes; peeled kiln dried pressure treated; pointed; 1.8 m posts driven 900 mm into ground at 3 m centres; (line wires and netting priced separately)							
75 -100 mm stakes	-	0.25	4.50	-	0.62	m	5.12
Corner posts or straining posts 150 mm diameter. 2.3 m high set in concrete; centres to suit local conditions or changes of direction							
1 strut	11.23	1.00	18.00	4.04	16.39	each	38.43
2 strut	13.27	1.00	18.00	4.04	18.43	each	40.47
Strained wire; to posts (posts not included); 3 mm galvanized wire; fixing with galvanized stirrups							
900 mm high; 2 wire	0.14	0.03	0.60	-	0.20	m	0.80
1200 mm high; 3 wire	0.21	0.05	0.84	-	0.27	m	1.11
Rabbit netting; 31 mm 19 gauge 1050 high netting fixed to posts line wires and straining posts or corner posts all priced separately.							
900 high turned in	0.58	0.04	0.72	-	0.58	m	1.30
900 high buried 150 mm in trench	0.58	0.08	1.50	-	0.58	m	2.08

Prices for Measured Works
Q PAVING/PLANTING/FENCING/SITE FURNITURE

Item Excluding site overheads and profit	PC £	Labour hours	Labour £	Plant £	Material £	Unit	Total rate £
Q40 FENCING AND GATES – cont'd							
Boundary fencing; strained wire and wire mesh; H.S.Jackson & Son (Fencing)							
Tubular chain link fencing; galvanized; plastic-coated; 60.3 mm diameter posts at 3.0 m centres; setting 700 mm into ground; choice of ten mesh colours; including excavating holes; backfilling and removing surplus soil; with top rail only							
900 mm high	-	-	-	-	-	m	20.00
1200 mm high	-	-	-	-	-	m	21.09
1800 mm high	-	-	-	-	-	m	23.38
2000 mm high	-	-	-	-	-	m	24.80
Tubular chain link fencing; galvanised; plastic coated; 60.3 mm diameter posts at 3.0 m centres; cranked arms and 3 lines barbed wire; setting 700 mm into ground; including excavating holes; backfilling and removing surplus soil; with top rail only							
2000 mm high	-	-	-	-	-	m	25.96
1800 mm high	!	-	-	-	-	m	25.47
Boundary fencing; Steelway Fensecure							
"Classic" 2 rail tubular fencing; top and bottom with strecher bars and straining wires in between; ; comprising 60.3 mm tubular posts at 3.00 m centres; setting in concrete; 35 mm top rail tied with aluminium and steel fittings; 50 mm x 50 mm x 355/2.5 mm PVC coated chain link; all components galvanized and coated in green nylon							
900 mm high	-	-	-	-	-	m	11.46
1200 mm high	-	-	-	-	-	m	12.63
1400 mm high	-	-	-	-	-	m	13.41
1800 mm high	-	-	-	-	-	m	14.38
2100 mm high	-	-	-	-	-	m	16.30
2400 mm high	-	-	-	-	-	m	18.02
3050 mm high	-	-	-	-	-	m	20.06
3600 mm high	-	-	-	-	-	m	22.87
End Posts; Classic range 60.3 mm diameter; setting in concrete							
900 mm high	-	-	-	-	-	nr	18.99
1200 mm high	-	-	-	-	-	nr	21.44
1400 mm high	-	-	-	-	-	nr	23.07
1800 mm high	-	-	-	-	-	nr	24.73
2100 mm high	-	-	-	-	-	nr	30.57
2400 mm high	-	-	-	-	-	nr	28.57
3050 mm high	-	-	-	-	-	nr	33.87
3600 mm high	-	-	-	-	-	nr	36.04
Corner Posts; 60.3 mm diameter; setting in concrete							
900 mm high	-	-	-	-	-	nr	11.46
1200 mm high	-	-	-	-	-	nr	12.63
1400 mm high	-	-	-	-	-	nr	13.41
1800 mm high	-	-	-	-	-	nr	14.38
2100 mm high	-	-	-	-	-	nr	16.30
2400 mm high	-	-	-	-	-	nr	18.02

Prices for Measured Works
Q PAVING/PLANTING/FENCING/SITE FURNITURE

Item Excluding site overheads and profit	PC £	Labour hours	Labour £	Plant £	Material £	Unit	Total rate £
3050 mm high	-	-	-	-	-	nr	20.06
3600 mm high	-	-	-	-	-	nr	22.87
Boundary fencing; Bekaert Fencing Products							
"Sentinel Paladin" Welded mesh colour coated green fencing; fixing to metal posts at 3 m centres with manufacturer's fixings; setting 600 mm deep in firm ground; including excavating holes; backfilling and removing surplus excavated material; includes 10 year product guarantee							
900 mm high	18.29	3.03	54.55	-	22.52	m	77.06
1200 mm high	20.95	3.03	54.55	-	25.78	m	80.32
1800 mm high	26.49	3.03	54.55	-	31.32	m	85.86
Extra over welded galvanised plastic coated mesh fencing for concreting in posts	-	-	-	-	1.45	nr	1.45
Boundary fencing; Langdons							
"Orsogril" rectangular steel bar mesh fence panels; pleione pattern; bolting to 60 mm x 80 mm uprights at 2 m centres; mesh 62 mm x 66 mm, setting in concrete							
930 mm high panels	-	-	-	-	-	m	90.79
1326 mm high panels	-	-	-	-	-	m	125.67
1722 mm high panels	-	-	-	-	-	m	151.14
Langdons; "Orsogril" rectangular steel bar mesh fence panels; pleione pattern; bolting to 80 mm x 80 mm uprights at 2 m centres; mesh 62 mm x 66 mm; setting in concrete							
930 mm high panel	-	-	-	-	-	m	93.01
1326 mm high panel	-	-	-	-	-	m	122.63
1722 mm high panel	-	-	-	-	-	m	155.85
"Orsogril" rectangular steel bar mesh fence panels; sterope pattern; bolting to 60 mm x 80 mm uprights at 2.00 m centres; mesh 62 mm x 132 mm; setting in concrete paving (paving not included)							
930 mm high panels	-	-	-	-	-	m	86.04
1326 mm high panels	-	-	-	-	-	m	112.77
1722 mm high panels	-	-	-	-	-	m	141.84
Security fencing; H.S.Jackson & Son (Fencing)							
"Barbican" galvanized steel paling fencing; on 60 mm x 60 mm posts at 3 m centres; setting in concrete							
1250 mm high	-	-	-	-	-	m	51.27
1500 mm high	-	-	-	-	-	m	54.88
2000 mm high	-	-	-	-	-	m	63.27
2500 mm high	-	-	-	-	-	m	74.34
Gates; to match "Barbican" galvanised steel paling fencing							
width 1m	-	-	-	-	-	nr	673.62
width 2 m	-	-	-	-	-	nr	681.86
width 3 m	-	-	-	-	-	nr	700.40
width 4 m	-	-	-	-	-	nr	724.09
width 8 m	-	-	-	-	-	pair	1475.99
width 9m	-	-	-	-	-	pair	1779.84
width 10 m	-	-	-	-	-	pair	2022.92

Prices for Measured Works
Q PAVING/PLANTING/FENCING/SITE FURNITURE

Item Excluding site overheads and profit	PC £	Labour hours	Labour £	Plant £	Material £	Unit	Total rate £
Q40 FENCING AND GATES – cont'd							
Security fencing; intruder guards "Viper Spike Intruder Guards"; to existing structures, including fixing bolts							
ref Viper 1; 40 mm x 5 mm x 1.2 m long with base plate	33.42	1.00	18.00	-	33.42	nr	**51.42**
ref Viper 3; 160 mm x 190 mm wide; U shape; to prevent intruders climbing pipes	33.04	0.50	9.00	-	33.04	nr	**42.04**
Security fencing; HS Jackson "Razor Barb Concertina" ; spiral wire security barriers; fixing to 600 mm steel ground stakes							
ref RBC4M/G 455 mm diameter roll - medium barb, galvanized	0.94	0.02	0.36	-	1.06	m	**1.42**
ref RBC7M/G 710 mm diameter roll - medium barb, galvanized	1.12	0.02	0.36	-	1.24	m	**1.60**
ref RBC9M/G 965 mm diameter roll - medium barb, galvanized	1.43	0.02	0.36	-	1.55	m	**1.91**
3 lines "Razor Tape" Medium Barb barbed tape; on 50 mm x 50 mm mild steel angle posts; setting in concrete							
ref RTS5M/G Razor Tape medium barb, galvanized	0.32	0.21	3.79	-	7.12	m	**10.91**
5 lines "Razor Tape" Medium Barb barbed tape; on 50 mm x 50 mm mild steel angle posts; setting in concrete							
ref RTS5M/G Razor Tape medium barb, galvanized	0.54	0.22	4.00	-	7.33	m	**11.33**
Trip rails; timber; AVS Fencing Contractors							
200 mm x 38 mm softwood rails; screwing to 100 mm x 100 mm softwood posts 700 mm long; setting in concrete; at 1.00 m centres; all treated with timber preservative	9.31	0.71	12.81	-	12.22	m	**25.03**
Extra over timber trip rails for priming in lieu of timber preservative	0.09	0.08	1.50	-	0.09	m	**1.59**
Extra over timber trip rails for bolting in lieu of screwing rails to softwood posts	1.00	0.13	2.25	-	1.00	m	**3.25**
Extra over timber trip rails for galvanised 50 mm x 50 mm mild steel posts at 1.50 m centres in lieu of 100 mm x 100 mm softwood posts	3.51	0.22	4.00	-	3.51	m	**7.51**
50 mm x 150 mm "BFC" oiled hardwood rails on 100 mm x 75 mm western red cedar posts 700 mm long; setting in concrete at 1.00 m centres	15.70	0.35	6.30	-	15.70	m	**22.00**
50 mm x 150 mm oak rails on 150 mm x 200 mm oak posts 1100 mm long; driving into firm ground at 1.00 m centres	11.32	0.35	6.30	-	11.32	m	**17.62**
Trip rails; metal; Broxap Streetscene double row 48.3 mm internal diameter mild steel tubular rails with sleeved joints; to cast iron posts 525 mm above ground mm long; setting in concrete at 1.20 m centres; all standard painted BX 243 with BX Rail C	77.31	1.00	18.00	-	81.66	m	**99.66**

Prices for Measured Works
Q PAVING/PLANTING/FENCING/SITE FURNITURE

Item Excluding site overheads and profit	PC £	Labour hours	Labour £	Plant £	Material £	Unit	Total rate £
22 mm diameter mild steel rails with ferrule joints; fixing through holes in 44 mm x 13 mm mild steel standards 700 mm long; setting in concrete at 1.20 m centres; priming	22.66	1.33	24.00	-	22.93	m	**46.93**
Double 48 mm diameter mild steel rails; to 75 mm diameter mild steel "Y" posts 700 mm long; setting in concrete at 1.20 m centres; all galvanized	45.86	1.00	18.00	-	46.13	m	**64.13**
48 mm diameter galvanized tubular rails; fixing through holes in 125 mm x 125 mm oak posts 1.20 m long; setting in firm ground at 2.00 m centres	18.25	1.18	21.18	-	18.52	m	**39.70**
48 mm diameter galvanised tubular rails; fixing through holes in 100 mm x 150 mm concrete barrel posts 1.00 m long; setting in firm ground at 2.00 m centres	12.25	1.14	20.57	-	12.57	m	**33.14**
Trip rails; Townscape Products Ltd Hollow steel section knee rails, galvanised; 500 mm high; setting in concrete							
1000 mm bays	68.53	2.00	36.00	-	74.13	m	**110.12**
1200 mm bays	17.33	2.00	36.00	-	67.25	m	**103.24**
Timber fencing; AVS Fencing Supply Palisade fencing; 22 mm x 75 mm softwood vertical palings with flat tops; nailing to 3 nr 50 mm x 125 mm horizontal softwood cant rails; housing into 100 mm x 100 mm softwood posts with weathered tops at 3.00 m centres; setting in concrete; all treated timber							
1800 mm high	15.18	0.47	8.41	-	18.49	m	**26.90**
Palisade fencing; 22 mm x 75 mm oak vertical palings with flat tops; nailing to 3 nr 50 mm x 125 mm horizontal oak cant rails; housing into 100 mm x 100 mm oak posts with weathered tops at 3.00 m centres; setting in concrete							
1800 mm high	35.43	0.47	8.40	-	38.75	m	**47.14**
Post-and-rail fencing; 3 nr 90 mm x 38 mm softwood horizontal rails; fixing with galvanized nails to 150 mm x 75 mm softwood posts; including excavating and backfilling into firm ground at 1.80 m centres; all treated timber							
1200 mm high	9.14	0.35	6.30	-	9.19	m	**15.49**
Post-and-rail fencing; 3 nr 90 mm x 38 mm oak horizontal rails; fixing with galvanised nails to 150 mm x 75 mm oak posts; including excavating and backfilling into firm ground at 1.80 m centres							
1200 mm high	15.93	0.35	6.30	-	15.98	m	**22.29**
Post-and-rail fencing; 3 nr 90 mm x 38 mm western red cedar rails; fixing with galvanised nails to 150 mm x 75 mm hardwood posts; including excavating and backfilling into firm ground at 1.80 m centres							
1200 mm high	58.56	0.35	6.30	-	58.61	m	**64.92**

Prices for Measured Works
Q PAVING/PLANTING/FENCING/SITE FURNITURE

Item Excluding site overheads and profit	PC £	Labour hours	Labour £	Plant £	Material £	Unit	Total rate £
Q40 FENCING AND GATES – cont'd							
Timber fencing; AVS Fencing Supply – cont'd							
Morticed post-and-rail fencing; 3 nr horizontal 90 mm x 38 mm softwood rails; fixing with galvanized nails; 90 mm x 38 mm softwood centre prick posts; to 150 mm x 75 mm softwood posts; including excavating and backfilling into firm ground at 2.85 m centres; all treated timber							
1200 mm high	8.07	0.35	6.30	-	8.12	m	14.42
1350 mm high five rails	8.86	0.40	7.20	-	8.91	m	16.11
Cleft rail fencing; oak or chestnut adze tapered rails 2.80 m long; morticed into joints; to 150 mm x 100 mm softwood posts 1.98 m long; including excavating and backfilling into firm ground at 2.80 m centres							
three rails	11.01	0.28	5.04	-	11.01	m	16.05
four rails	20.05	0.35	6.30	-	20.05	m	26.35
Close boarded fencing; 2 nr softwood rails; 150 x 25 mm gravel boards; 89 mm x 19 mm softwood pales lapped 13 mm; to concrete posts; including excavating and backfilling into firm ground at 3.00 m centres							
900 mm high	9.83	0.47	8.41	-	10.92	m	19.33
1050 mm high	6.86	0.47	8.41	-	12.13	m	20.54
1200 mm high	7.46	0.47	8.41	-	12.73	m	21.14
Close boarded fencing; 3 nr softwood rails; 150 x 25 mm gravel boards; 89 mm x 19 mm softwood pales lapped 13 mm; to concrete posts; including excavating and backfilling into firm ground at 3.00 m centres							
1350 mm high	13.41	0.56	10.08	-	14.50	m	24.58
1650 mm high	14.68	0.56	10.08	-	15.77	m	25.85
1800 mm high	15.16	0.56	10.08	-	16.25	m	26.33
Close boarded fencing; 3 nr softwood rails; 150 x 25 mm gravel boards; 89 mm x 19 mm softwood pales lapped 13 mm; to softwood posts 100 x 100 mm; including excavating and backfilling into firm ground at 3.00 m centres.							
1350 mm high	10.52	0.56	10.08	-	11.61	m	21.69
1650 mm high	11.60	0.56	10.08	-	12.69	m	22.77
1800 mm high	12.26	0.56	10.08	-	13.34	m	23.43
Close boarded fencing; 3 nr softwood rails; 150 x 25 mm gravel boards; 89 mm x 19 mm softwood pales lapped 13 mm; to oak posts; including excavating and backfilling into firm ground at 3.00 m centres							
1350 mm high	8.91	0.56	10.08	-	14.14	m	24.22
1650 mm high	14.69	0.56	10.08	-	15.78	m	25.86
1800 mm high	15.63	0.56	10.08	-	16.72	m	26.80
Close boarded fencing; 3 nr softwood rails; 150 x 25 mm gravel boards; 89 mm x 19 mm softwood pales lapped 13 mm; to oak posts; including excavating and backfilling into firm ground at 2.47 m centres							
1800 mm high	16.82	0.68	12.20	-	17.91	m	30.10

Prices for Measured Works
Q PAVING/PLANTING/FENCING/SITE FURNITURE

Item Excluding site overheads and profit	PC £	Labour hours	Labour £	Plant £	Material £	Unit	Total rate £
Close boarded fencing - all oak; 2 nr 76 mm x 38 mm rectangular rails; 152 mm x 25 mm gravel boards; 89 mm x 19 mm oak featheredge boards lapped 13 mm; to oak posts; including excavating and backfilling into firm ground at 2.47 m centres							
1070 mm high	7.08	0.56	10.08	-	30.46	m	40.55
1800 mm high	35.92	0.56	10.08	-	37.01	m	47.09
Extra over close boarded fencing for 65 mm x 38 mm oak cappings	4.55	0.05	0.90	-	4.55	nr	5.45
"Hit and miss" horizontal rail fencing; 87 mm x 38 mm top and bottom rails; 100 mm x 22 mm vertical boards arranged alternately on opposite side of rails; to 100 mm x 100 mm posts; including excavating and backfilling into firm ground; setting in concrete at 1.8 m centres							
treated softwood; 1800 mm high	19.08	1.33	24.00	-	21.32	m	45.32
treated softwood; 2000 mm high	19.85	1.33	24.00	-	24.67	m	48.67
primed softwood; 1800 mm high	19.08	1.67	30.00	-	21.68	m	51.68
primed softwood; 2000 mm high	19.85	1.67	30.00	-	25.03	m	55.03
Screen fencing in horizontal wavy edge overlap panels; 1.80 m wide; to 100 mm x 100 mm preservative treated softwood posts; including excavating and backfilling into firm ground							
900 mm high	6.39	0.47	8.41	-	11.14	m	19.56
1200 mm high	6.56	0.47	8.41	-	11.79	m	20.20
1500 mm high	6.71	0.56	10.11	-	13.25	m	23.36
1800 mm high	6.86	0.56	10.11	-	13.84	m	23.95
Screen fencing for oak in lieu of softwood posts							
900 mm high	4.72	-	-	-	4.72	m	4.72
1200 mm high	5.90	-	-	-	5.90	m	5.90
1500 mm high	7.08	-	-	-	7.08	m	7.08
1800 mm high	8.21	-	-	-	8.21	m	8.21
Extra over screen fencing for 300 mm high trellis tops; slats at 100 mm centres; including additional length of posts	3.78	0.10	1.80	-	3.78	m	5.58
Timber trellis; Landscapes by Design; pressure treated softwood panels Timber section 38 x 19 mm at 120 mm centres at 45, 55 or 90 deg angle; framed and weathered; panel size 1.8 x 1.8 m							
fixed to timber freestanding posts 100 x 100 mm	-	-	-	-	-	m	63.00
fixed to wall on 38 mm batten	-	-	-	-	-	m	58.80
Extra for post finials	-	-	-	-	-	nr	10.08
extra for angled top to each trellis panel at 55 deg projecting 450 mm above the top frame of each panel	-	-	-	-	-	nr	12.60
extra for scalloped radius top depression 300 mm into each panel	-	-	-	-	-	nr	15.75
extra for painting with Sadolin	-	-	-	-	-	m	12.60

Prices for Measured Works
Q PAVING/PLANTING/FENCING/SITE FURNITURE

Item Excluding site overheads and profit	PC £	Labour hours	Labour £	Plant £	Material £	Unit	Total rate £
Q40 FENCING AND GATES – cont'd							
Concrete fencing							
Panel fencing; to precast concrete posts; in 2 m bays; setting posts 600 mm into ground; sandfaced finish							
900 mm high	11.74	0.25	4.50	-	13.10	m	17.60
1200 mm high	15.65	0.25	4.50	-	17.01	m	21.51
Panel fencing; to precast concrete posts; in 2 m bays; setting posts 750 mm into ground; sandfaced finish							
1500 mm high	19.56	0.33	6.00	-	20.92	m	26.92
1800 mm high	23.47	0.36	6.54	-	24.84	m	31.38
2100 mm high	27.38	0.40	7.20	-	28.75	m	35.95
2400 mm high	32.60	0.40	7.20	-	33.96	m	41.16
Extra over concrete panel fencing for aggregate faced one side	7.44	-	-	-	7.44	m²	7.44
Windbreak fencing							
Fencing; Netlon Ltd; "Tensar" windbreak fencing; to 100 mm diameter treated softwood posts; setting 450 mm into ground; fixing with 50 mm x 25 mm treated softwood battens nailed to posts, including excavating and backfilling into firm ground; setting in concrete at 3 m centres							
1200 mm high	1.24	0.16	2.80	-	4.36	m	7.16
2000 mm high	3.78	0.16	2.80	-	6.90	m	9.70
Ball stop fencing; Steelway Fensecure							
Ball Stop Net; 30 x 30 mm netting fixed to 60.3 diameter 12 mm solid bar lattice galvanised dual posts, top, middle and bottom rails							
4.5 m high	-	-	-	-	-	m	62.58
5.0 m high	-	-	-	-	-	m	82.18
6.0 m high	-	-	-	-	-	m	90.10
7.0 m high	-	-	-	-	-	m	97.10
8.0 m high	-	-	-	-	-	m	104.45
9.0 m high	-	-	-	-	-	m	112.10
10.0 m high	-	-	-	-	-	m	120.10
Ball Stop Net; Corner posts							
4.5 m high	-	-	-	-	-	m	47.60
5.0 m high	-	-	-	-	-	m	52.10
6.0 m high	-	-	-	-	-	m	80.73
7.0 m high	-	-	-	-	-	m	72.10
8.0 m high	-	-	-	-	-	m	82.10
9.0 m high	-	-	-	-	-	m	92.10
10 m high	-	-	-	-	-	m	102.10
Ball Stop Net; End posts							
4.5 m high	-	-	-	-	-	m	39.60
5.0 m high	-	-	-	-	-	m	44.10
6.0 m high	-	-	-	-	-	m	54.10
7.0 m high	-	-	-	-	-	m	64.10
8.0 m high	-	-	-	-	-	m	74.10
9.0 m high	-	-	-	-	-	m	84.10
10 m high	-	-	-	-	-	m	95.10

Prices for Measured Works
Q PAVING/PLANTING/FENCING/SITE FURNITURE

Item Excluding site overheads and profit	PC £	Labour hours	Labour £	Plant £	Material £	Unit	Total rate £
Railings; Langdons Ltd							
Mild steel bar railings of balusters at 115 mm centres welded to flat rail top and bottom; bays 2.00 m long; bolting to 51 mm x 51 mm hollow square section posts; setting in concrete							
galvanized; 900 mm high	-	-	-	-	-	m	43.18
galvanized; 1200 mm high	-	-	-	-	-	m	52.31
galvanized; 1500 mm high	-	-	-	-	-	m	58.13
galvanized; 1800 mm high	-	-	-	-	-	m	69.75
primed; 900 mm high	-	-	-	-	-	m	40.70
primed; 1200 mm high	-	-	-	-	-	m	49.42
primed; 1500 mm high	-	-	-	-	-	m	54.64
primed; 1800 mm high	-	-	-	-	-	m	65.11
Mild steel blunt top railings of 19 mm balusters at 130 mm centres welded to bottom rail; passing through holes in top rail and welded; top and bottom rails 40 mm x 10mm; bolting to 51 mm x 51 mm hollow square section posts; setting in concrete							
galvanized; 800 mm high	-	-	-	-	-	m	46.50
galvanized; 1000 mm high	-	-	-	-	-	m	48.83
galvanized; 1300 mm high	-	-	-	-	-	m	58.13
galvanized; 1500 mm high	-	-	-	-	-	m	62.78
primed; 800 mm high	-	-	-	-	-	m	44.75
primed; 1000 mm high	-	-	-	-	-	m	46.50
primed; 1300 mm high	-	-	-	-	-	m	55.23
primed; 1500 mm high	-	-	-	-	-	m	59.30
Railings; traditional pattern; 16 mm diameter verticals at 127 mm intervals with horizontal bars near top and bottom; balusters with spiked tops; 51 x 20 mm standards; including setting 520 mm into concrete at 2.75 m centres							
primed; 1200 mm high	-	-	-	-	-	m	52.31
primed; 1500 mm high	-	-	-	-	-	m	59.30
primed; 1800 mm high	-	-	-	-	-	m	63.94
Interlaced bow-top mild steel railings; traditional park type; 16 mm diameter verticals at 80 mm intervals, welded at bottom to 50 x 10 mm flat and slotted through 38 mm x 8 mm top rail to form hooped top profile; 50 mm x 10 mm standards; setting 560 mm into concrete at 2.75 m centres							
galvanized; 900 mm high	-	-	-	-	-	m	79.06
galvanized; 1200 mm high	-	-	-	-	-	m	96.50
galvanized; 1500 mm high	-	-	-	-	-	m	115.10
galvanized; 1800 mm high	-	-	-	-	-	m	133.70
primed; 900 mm high	-	-	-	-	-	m	76.73
primed; 1200 mm high	-	-	-	-	-	m	93.01
primed; 1500 mm high	-	-	-	-	-	m	110.45
primed; 1800 mm high	-	-	-	-	-	m	127.88

Prices for Measured Works
Q PAVING/PLANTING/FENCING/SITE FURNITURE

Item Excluding site overheads and profit	PC £	Labour hours	Labour £	Plant £	Material £	Unit	Total rate £
Q40 FENCING AND GATES – cont'd							
Metal Estate Fencing; Broxap Ltd; Mild steel flat bar angular fencing; galvanized; main posts at 5.00 m centres; fixed with 1:3:6 concrete 430 deep; intermediate posts fixed with fixing claw at 1.00 m centres. "Cheshire"; curved top posts and 5 nr flat rails							
1200 high	36.71	0.28	5.00	-	38.66	m	**43.66**
"Prestbury" 5 nr plain flat or round bar							
1200 high	35.15	0.28	5.00	-	37.10	m	**42.10**
Kissing gate; Mild steel construction, flat or angular steel;							
Galvanized	298.10	2.00	36.00	-	300.05	m	**336.05**
Five bar horizontal steel fencing; 1.20 m high; 38 mm x 10 mm joiner standards at 4.50 m centres; 38 mm x 8 mm intermediate standards at 900 mm centres; including excavating and backfilling into firm ground; setting in concrete							
galvanized	-	-	-	-	-	m	**61.75**
primed	-	-	-	-	-	m	**50.75**
Extra over five bar horizontal steel fencing for 76 mm diameter end and corner posts							
galvanized	-	-	-	-	-	each	**67.50**
primed	-	-	-	-	-	each	**39.50**
Langdon Ltd; Five bar horizontal steel fencing with mild steel copings; 38 mm square hollow section mild steel standards at 1.80 m centres; including excavating and backfilling into firm ground; setting in concrete							
galvanized; 900 mm high	-	-	-	-	-	m	**80.80**
galvanized; 1200 mm high	-	-	-	-	-	m	**85.00**
galvanized; 1500 mm high	-	-	-	-	-	m	**91.00**
galvanized; 1800 mm high	-	-	-	-	-	m	**97.00**
primed; 900 mm high	-	-	-	-	-	m	**71.70**
primed; 1200 mm high	-	-	-	-	-	m	**73.60**
primed; 1500 mm high	-	-	-	-	-	m	**77.60**
primed; 1800 mm high	-	-	-	-	-	m	**83.50**
Gates - General Preamble: Gates in fences; see specification for fencing, as gates in traditional or proprietary fencing systems are usually constructed of the same materials and finished as the fencing itself.							
Gates, hardwood; Norbury (Fencing & Building Materials) Ltd "Yeoman" entrance gate, five bar diamond braced, curved hanging stile, planed keruing; fixed to 150 mm x 150 mm softwood posts; inclusive of hinges and furniture							
ref 1100 040; 0.9 m wide	107.24	5.00	90.00	-	159.97	nr	**249.97**
ref 1100 041; 1.2 m wide	116.46	5.00	90.00	-	169.19	nr	**259.19**
ref 1100 042; 1.5 m wide	125.72	5.00	90.00	-	178.45	nr	**268.45**
ref 1100 043; 1.8 m wide	181.54	5.00	90.00	-	234.27	nr	**324.27**

Prices for Measured Works
Q PAVING/PLANTING/FENCING/SITE FURNITURE

Item Excluding site overheads and profit	PC £	Labour hours	Labour £	Plant £	Material £	Unit	Total rate £
ref 1100 044; 2.1 m wide	194.41	5.00	90.00	-	247.14	nr	337.14
ref 1100 045; 2.4 m wide	201.76	5.00	90.00	-	254.49	nr	344.49
ref 1100 047; 3.0 m wide	219.82	5.00	90.00	-	272.55	nr	362.55
ref 1100 048; 3.3 m wide	227.15	5.00	90.00	-	279.88	nr	369.88
ref 1100 049; 3.6 m wide	234.46	5.00	90.00	-	287.19	nr	377.19
"Yeoman" entrance gate, five bar diamond braced, curved hanging stile, planed iroko; fixed to 150 mm x 150 mm softwood posts; inclusive of hinges and furniture							
ref 1500 040; 0.9 m wide	150.90	5.00	90.00	-	203.63	nr	293.63
ref 1500 041; 1.2 m wide	167.20	5.00	90.00	-	219.93	nr	309.93
ref 1500 042; 1.5 m wide	180.09	5.00	90.00	-	232.82	nr	322.82
ref 1500 043; 1.8 m wide	283.76	5.00	90.00	-	336.49	nr	426.49
ref 1500 044; 2.1 m wide	295.56	5.00	90.00	-	348.29	nr	438.29
ref 1500 045; 2.4 m wide	305.86	5.00	90.00	-	358.59	nr	448.59
ref 1500 046; 2.7 m wide	316.15	5.00	90.00	-	368.88	nr	458.88
ref 1500 048; 3.3 m wide	340.80	5.00	90.00	-	393.53	nr	483.53
ref 1500 049; 3.6 m wide	350.88	5.00	90.00	-	403.61	nr	493.61
Hardwood field gate, five bar diamond braced, planed keruing; fixed to 150 mm x 150 mm softwood posts; inclusive of hinges and furniture							
ref 1100 100; 0.9 m wide	67.72	4.00	72.00	-	120.45	nr	192.45
ref 1100 101; 1.2 m wide	73.45	4.00	72.00	-	126.18	nr	198.18
ref 1100 102; 1.5 m wide	86.43	4.00	72.00	-	139.16	nr	211.16
ref 1100 103; 1.8 m wide	93.99	4.00	72.00	-	146.72	nr	218.72
ref 1100 104; 2.1 m wide	103.06	4.00	72.00	-	155.79	nr	227.79
ref 1100 105; 2.4 m wide	113.48	4.00	72.00	-	166.21	nr	238.21
ref 1100 106; 2.7 m wide	122.36	4.00	72.00	-	175.09	nr	247.09
ref 1100 108; 3.3 m wide	137.29	4.00	72.00	-	190.02	nr	262.02
ref 1100 109; 3.6 m wide	143.66	4.00	72.00	-	196.39	nr	268.39
Gates, softwood; H.S.Jackson & Son (Fencing)							
Timber field gates, including wrought iron ironmongery; five bar type; diamond braced; 1.80 m high; to 200 mm x 200 mm posts; setting 750 mm into firm ground							
treated softwood; width 2400 mm	70.17	10.00	180.00	-	153.96	nr	333.96
treated softwood; width 2700 mm	76.85	10.00	180.00	-	160.64	nr	340.64
treated softwood; width 3000 mm	78.71	10.00	180.00	-	162.49	nr	342.49
treated softwood; width 3300 mm	85.75	10.00	180.00	-	169.54	nr	349.54
Featherboard garden gates, including ironmongery; to 100 mm x 120 mm posts; 1 nr diagonal brace							
treated softwood; 0.9m x 1.15m high	58.13	3.00	54.00	-	82.93	nr	136.93
treated softwood; 1.0m x 1.2m high	18.27	3.00	54.00	-	82.23	nr	136.23
treated softwood; 1.0m x 1.5m high	24.37	3.00	54.00	-	89.72	nr	143.72
treated softwood; 1.0m x 1.8m high	62.06	3.00	54.00	-	95.94	nr	149.94
Picket garden gates, including ironmongery; to match picket fence; width 1000 mm; to 100 mm x 120 mm posts; 1 nr diagonal brace							
treated softwood; 950 mm high	56.50	3.00	54.00	-	81.31	nr	135.31
treated softwood; 1200 mm high	59.09	3.00	54.00	-	83.90	nr	137.90
treated softwood; 1800 mm high	67.91	3.00	54.00	-	101.78	nr	155.78

Prices for Measured Works
Q PAVING/PLANTING/FENCING/SITE FURNITURE

Item Excluding site overheads and profit	PC £	Labour hours	Labour £	Plant £	Material £	Unit	Total rate £
Q40 FENCING AND GATES – cont'd							
Gates, tubular steel; H.S.Jackson & Son (Fencing)							
Tubular mild steel field gates, including ironmongery; diamond braced; 1.80 m high; to tubular steel posts; setting in concrete							
galvanized; width 3000 mm	71.79	5.00	90.00	-	141.90	nr	**231.89**
galvanized; width 3300 mm	75.81	5.00	90.00	-	145.92	nr	**235.91**
galvanized; width 3600 mm	79.83	5.00	90.00	-	149.94	nr	**239.93**
galvanized; width 4200 mm	88.06	5.00	90.00	-	158.17	nr	**248.16**
Gates, sliding; H.S.Jackson & Son (Fencing)							
"Sliding Gate"; including all galvanized rails and vertical rail infill panels; special guide and shutting frame posts (Note: foundations installed by suppliers)							
access width 4.00 m; 1.5m high gates	-	-	-	-	-	nr	**3229.05**
access width 4.00 m; 2.0m high gates	-	-	-	-	-	nr	**3274.37**
access width 4.00 m; 2.5m high gates	-	-	-	-	-	nr	**3318.66**
access width 6.00 m; 1.5m high gates	-	-	-	-	-	nr	**3689.46**
access width 6.00 m; 2.0m high gates	-	-	-	-	-	nr	**3744.05**
access width 6.00 m; 2.5m high gates	-	-	-	-	-	nr	**3796.58**
access width 8.00 m; 1.5m high gates	-	-	-	-	-	nr	**4204.46**
access width 8.00 m; 2.0m high gates	-	-	-	-	-	nr	**4267.29**
access width 8.00 m; 2.5m high gates	-	-	-	-	-	nr	**4327.03**
access width 10.00 m; 2.0m high gates	-	-	-	-	-	nr	**5185.02**
access width 10.00 m; 2.5m high gates	-	-	-	-	-	nr	**5262.27**
Stiles and kissing gates							
Stiles; H S Jackson & Son (Fencing); "Jacksons Nr 2007" stiles; 2 nr posts; setting into firm ground; 3 nr rails; 2 nr treads	63.92	3.00	54.00	-	73.15	nr	**127.15**
Kissing gates; Townscape Products; in mild steel bar; fixing to fencing posts (posts not included); 1.65 m x 1.30 m x 1.00 m high	138.86	5.00	90.00	-	138.86	nr	**228.86**
Pedestrian guard rails; Broxap Streetscene							
Mild steel pedestrian guard rails; to BS 3049:1976; 1.00 m high with 150 mm toe space; to posts at 2.00 m centres, galvanized finish							
vertical bar infill panel	25.10	2.00	36.00	-	30.26	m	**66.26**
vertical bar infill with 200 mm visibility gap at top	28.40	2.00	36.00	-	33.56	m	**69.56**

Prices for Measured Works
Q PAVING/PLANTING/FENCING/SITE FURNITURE

Item Excluding site overheads and profit	PC £	Labour hours	Labour £	Plant £	Material £	Unit	Total rate £
Q50 SITE/STREET FURNITURE/EQUIPMENT							
Cattle Grids - General							
Preamble: Cattle grids are not usually prefabricated, owing to their size and weight. Specification must take into account the maximum size and weight of vehicles likely to cross the grid. Drainage and regular clearance of the grid is essential. Warning signs should be erected on both approaches to the grid. For the cost of a complete grid and pit installation see "Approximate Estimates" section (book only).							
Cattle grids; H S Jackson & Son (Fencing)							
Cattle grids; supply only							
to carry 12 tonnes evenly distributed; 2.90 m x 2.36 m grid; galvanized	581.95	-	-	-	581.95	nr	**581.95**
to carry 12 tonnes evenly distributed; 3.66 m x 2.50 m grid; galvanized	1014.55	-	-	-	1014.55	nr	**1014.55**
Barriers - General							
Preamble: The provision of car and lorry control barriers may form part of the landscape contract. Barriers range from simple manual counterweighted poles to fully automated remote-control security gates, and the exact degree of control required must be specified. Complex barriers may need special maintenance and repair.							
Barriers: Autopa							
Manually operated pole barriers; counterbalance; to tubular steel supports; bolting to concrete foundation (foundation not included); aluminium boom; various finishes;							
clear opening up to 3.00 m	625.00	6.00	108.00	-	642.42	nr	**750.42**
clear opening up to 4.00 m	690.00	6.00	108.00	-	707.42	nr	**815.42**
clear opening 5.00	750.00	6.00	108.00	-	767.42	nr	**875.42**
clear opening 6.00	820.00	6.00	108.00	-	837.42	nr	**945.42**
clear opening 7.00	900.00	6.00	108.00	-	917.42	nr	**1025.42**
catch pole; arm rest for all manual barriers	105.00	-	-	-	107.18	nr	**107.18**
Electrically operated pole barriers; enclosed fan-cooled motor; double worm reduction gear; overload clutch coupling with remote controls; aluminium boom; various finishes; (exclusive of electrical connections by electrician)							
Autopa AU 3.0, 3m boom	1800.00	6.00	108.00	-	1817.42	nr	**1925.42**
Autopa AU 4.5, 4.5m boom with catchpost	1920.00	6.00	108.00	-	1937.42	nr	**2045.42**
Autopa AU 6.0, 6.0m boom with catchpost	1995.00	6.00	108.00	-	2012.42	nr	**2120.42**

Prices for Measured Works
Q PAVING/PLANTING/FENCING/SITE FURNITURE

Item Excluding site overheads and profit	PC £	Labour hours	Labour £	Plant £	Material £	Unit	Total rate £
Q50 SITE/STREET FURNITURE/EQUIPMENT - cont'd							
Vehicle crash barriers - General Preamble: See Department of Environment Technical Memorandum BE5							
Vehicle crash barriers Steel corrugated beams; untensioned Broxap Limited; effective length 3.20 m 310 mm deep x 85 mm corrugations;							
steel posts; Z-section; roadside posts	56.22	0.33	6.00	0.79	59.59	m	66.37
steel posts; Z-section; off highway	51.41	0.33	6.00	0.79	54.77	m	61.56
steel posts RSJ 760 high; for anchor fixing	63.44	0.33	6.00	-	67.29	m	73.29
steel posts RSJ 560 high; for anchor fixing to car parks	-	0.33	6.00	-	65.57	m	71.57
extra over for curved rail 6.00 m radius	18.64	-	-	-	18.64	m	18.64
Bases for street furniture Excavating; filling with no-fines cement:aggregate (1:12); bases for street furniture							
300 mm x 450 mm x 500 mm deep	-	0.75	13.50	-	5.13	nr	18.63
300 mm x 600 mm x 500 mm deep	-	1.11	19.98	-	6.84	nr	26.82
300 mm x 900 mm x 500 mm deep	-	1.67	30.06	-	10.27	nr	40.32
1750 mm x 900 mm x 300 mm deep	-	2.33	41.94	-	35.91	nr	77.85
2000 mm x 900 mm x 300 mm deep	-	2.67	48.06	-	41.04	nr	89.10
2400 mm x 900 mm x 300 mm deep	-	3.20	57.60	-	49.25	nr	106.85
2400 mm x 1000 mm x 300 mm deep	-	3.56	64.08	-	54.72	nr	118.80
Precast concrete flags; to BS 7263; to concrete bases (not included); bedding and jointing in cement:mortar (1:4)							
450 mm x 600 mm x 50 mm	7.52	1.17	21.00	-	12.98	m²	33.98
Precast concrete paving blocks; to concrete bases (not included); bedding in sharp sand; butt joints							
200 mm x 100 mm x 65 mm	7.17	0.50	9.00	-	8.88	m²	17.88
200 mm x 100 mm x 80 mm	8.00	0.50	9.00	-	9.63	m²	18.63
Engineering paving bricks; to concrete bases (not included); bedding and jointing in sulphate-resisting cement:lime:sand mortar (1:1:6)							
over 300 mm wide	10.50	0.56	10.11	-	18.47	m²	28.58
Edge restraints to pavings; haunching in concrete (1:3:6)							
200 mm x 300 mm	-	0.10	1.80	-	4.33	m	6.13
Bases; Earth Anchors Ltd							
"Rootfast" ancillary anchors; ref A1; 500 mm long 25 mm diameter and top strap ref F2; including bolting to site furniture (site furniture not included)	8.05	0.50	9.00	-	8.05	set	17.05
installation tool for above	19.50	-	-	-	19.50	nr	19.50
"Rootfast" ancillary anchors; ref A4; heavy duty 40 mm square fixed head anchors; including bolting to site furniture (site furniture not included)	-	0.33	6.00	-	-	set	6.00
"Rootfast" ancillary anchors; F4; vertical socket; including bolting to site furniture (site furniture not included)	11.15	0.05	0.90	-	11.15	set	12.05

Prices for Measured Works
Q PAVING/PLANTING/FENCING/SITE FURNITURE

Item Excluding site overheads and profit	PC £	Labour hours	Labour £	Plant £	Material £	Unit	Total rate £
"Rootfast" ancillary anchors; ref F3; horizontal socket; including bolting to site furniture (site furniture not included)	12.20	0.05	0.90	-	12.20	set	**13.10**
installation tools for the above	57.00	-	-	-	57.00	nr	**57.00**
"Rootfast" ancillary anchors; ref A3 Anchored bases; including bolting to site furniture (site furniture not included)	47.00	0.33	6.00	-	47.00	set	**53.00**
installation tools for the above	57.00	-	-	-	57.00	nr	**57.00**
Furniture/equipment - General							
Preamble: The following items include fixing to manufacturer's instructions; holding down bolts or other fittings and making good (excavating, backfilling and tarmac, concrete or paving bases not included)							
Dog waste bins; all-steel							
Earth Anchors Ltd							
HG45A, 45l, earth anchored, post mounted	152.00	0.33	6.00	-	152.00	nr	**158.00**
HG45A, 45l, as above with pedal operation	179.00	0.33	6.00	-	179.00	nr	**185.00**
Dog waste bins; cast iron							
Bins; Furnitubes International Ltd							
ref PED 701; "Pedigree"; post mounted cast iron dog waste bins; 1250 mm total height above ground; 400 mm square bin	635.00	0.75	13.50	-	640.16	nr	**653.66**
Litter bins; precast concrete in textured white or exposed aggregate finish; with wire baskets and drainage holes							
Bins; Marshalls Mono							
"Strada 710"; 710 mm diameter x 500 mm high	172.87	1.00	18.00	-	172.87	nr	**190.87**
"Newstead"; exposed aggregate finish; 760 mm high x 485 mm diameter; including wire baskets	107.83	0.67	12.00	-	107.83	nr	**119.83**
"Shirley" circular bins; 760 mm high x 1066 mm diameter; including coated wire baskets	409.12	1.00	18.00	2.40	409.12	nr	**429.52**
Bins; Neptune Outdoor Furniture							
ref SF 16 - 42l	169.00	0.50	9.00	-	169.00	nr	**178.00**
ref SF 14 - 100l	235.00	0.50	9.00	-	235.00	nr	**244.00**
Bins; Townscape Products							
"Sutton"; 750 mm high x 500 mm diameter, 70 litre capacity including GRP canopy	138.32	0.33	6.00	-	237.15	nr	**243.15**
"Braunton"; 750 mm high x 500 mm diameter, 70 litre capacity including GRP canopy	146.67	1.50	27.00	-	245.50	nr	**272.50**
Litter bins; metal; stove-enamelled perforated metal for holder and container							
Bins; Abacus Municipal Ltd							
"Model 621"; stove enamelled steel; including brackets; fixing to pole	47.00	0.25	4.50	-	47.00	nr	**51.50**
"Model 601"; stove enamelled aluminium inner containers; galvanized mild steel cradles; ragged floor mounting pieces; concreting in	85.00	0.50	9.00	-	90.16	nr	**99.16**
Bins; Townscape Products							
"Metro"; 440 x 420 x 800 high; 62 litre capacity	453.22	0.33	6.00	-	453.22	nr	**459.22**
"Voltan" Large Round; 460 diameter x 780 high; 56 litre capacity	441.98	0.33	6.00	-	441.98	nr	**447.98**
"Voltan" Small Round with pedestal; 410 diameter x 760 high; 31 litre capacity	385.50	0.33	6.00	-	385.50	nr	**391.50**

Prices for Measured Works
Q PAVING/PLANTING/FENCING/SITE FURNITURE

Item Excluding site overheads and profit	PC £	Labour hours	Labour £	Plant £	Material £	Unit	Total rate £
Q50 SITE/STREET FURNITURE/EQUIPMENT - cont'd							
Litter bins; all-steel							
Bins; Earth Anchors							
"Ranger", 100l Litter bin, 107l, pedestal mounted	376.00	0.50	9.00	-	376.00	nr	**385.00**
"Big Ben", 82l, steel frame and liner, Vyflex coated finish, earth anchored	244.00	1.00	18.00	-	244.00	nr	**262.00**
"Beau", 42l, steel frame and liner, Vyflex coated finish, earth anchored	195.00	1.00	18.00	-	195.00	nr	**213.00**
Bins; Townscape Products							
"Baltimore Major" with GRP canopy; 560 diameter x 960 high, 140 litre capacity	654.86	1.00	18.00	-	654.86	nr	**672.86**
Litter bins; cast Iron							
Bins; Furnitubes International Ltd							
"Wave Bin"; ref WVB 440; free standing 55 litre liners; 440 mm diameter x 850 mm high	643.00	0.50	9.00	-	646.78	nr	**655.78**
"Wave Bin" ref WVB 520; free standing 85 litre liners; 520 mm diameter x 850 mm high; cast iron plinth	716.00	0.50	9.00	-	719.78	nr	**728.78**
"Covent Garden"; ref COV 702; side opening, 500 mm square x 1050 mm high; 105 litre capacity	399.35	0.50	9.00	-	405.31	nr	**414.31**
"Covent Garden"; ref COV 803; side opening, 500 mm diameter x 1025 mm high; 85 litre capacity	361.63	0.50	9.00	-	367.58	nr	**376.58**
"Covent Garden"; ref COV 912; open top; 500 mm A/F octagonal x 820 mm high; 85 litre capacity	515.00	0.50	9.00	-	520.96	nr	**529.96**
"Albert"; ref ALB 800; open top; 400 mm diameter x 845 mm high; 55 litre capacity	359.65	0.50	9.00	-	365.61	nr	**374.61**
Bins; Broxap Streetscene							
"Chester"; pedestal mounted	263.17	1.00	18.00	-	269.13	nr	**287.13**
Bins; Townscape Products							
"York Major"; 650 diameter x 1060 high, 140 litre capacity	646.11	1.00	18.00	-	647.89	nr	**665.89**
Litter bins; timber faced; hardwood slatted casings with removable metal litter containers; ground or wall fixing							
Bins; SMP Playgrounds							
ref LBC22; 560 mm high	440.00	1.00	18.00	-	447.85	nr	**465.85**
Bins; Lister Lutyens							
"Monmouth"; 675 mm high x 450 mm wide; freestanding	104.00	0.33	6.00	-	104.00	nr	**110.00**
"Monmouth"; 675 mm high x 450 mm wide; bolting to ground (without legs)	96.00	1.00	18.00	-	103.85	nr	**121.85**
Bins; Woodscape Ltd							
Square, 580 mm x 580 mm x 950 mm high, with lockable lid	550.00	0.50	9.00	-	550.00	nr	**559.00**
Round, 580 mm diameter by 950 mm high, with lockable lid	545.00	0.50	9.00	-	545.00	nr	**554.00**

Prices for Measured Works
Q PAVING/PLANTING/FENCING/SITE FURNITURE

Item Excluding site overheads and profit	PC £	Labour hours	Labour £	Plant £	Material £	Unit	Total rate £
Plastic litter and grit bins; glassfibre reinforced polyester grit bins; yellow body; hinged lids							
Bins; Furnitubes International Ltd							
ref Q6; grit and salt bins; 170 litre capacity	245.46	0.25	4.50	-	245.46	nr	249.96
ref Q11; grit and salt bins; 310 litre capacity	269.82	0.25	4.50	-	269.82	nr	274.32
Bins; Wybone; Victoriana glass fibre, cast iron effect litter bins, including lockable liner							
ref LBV/2; 521 mm x 521 mm x 673 mm high; open top; square shape; 0.078 m³ capacity	185.62	1.00	18.00	-	191.58	nr	209.58
ref LVC/3; 457 mm diameter x 648 mm high; open top; drum shape; 0.084 m³ capacity; with lockable liner	347.17	1.00	18.00	-	353.13	nr	371.13
Bins; Amberol Ltd; floor standing double walled to accept stabilizing ballast							
"Enviro Bin" ; 150 litre	160.00	0.33	6.00	-	160.00	nr	166.00
"Enviro Bin" ; 90 litre	-	0.33	6.00	-	120.00	nr	126.00
"Westminster " hooded; 90 litre	105.00	0.33	6.00	-	105.00	nr	111.00
"Westminster" 90 litre	80.00	0.33	6.00	-	80.00	nr	86.00
Bins; Amberol Ltd; pole mounted							
Envirobin 50 litre top emptying with liner	70.00	0.50	9.00	-	70.00	nr	79.00
Lifebuoy stations							
Lifebuoy stations; Earth Anchors Limited							
"Rootfast" lifebuoy station complete with post ref AP44; SOLAS approved lifebouys 590 mm and lifeline	151.00	2.00	36.00	-	159.07	nr	195.07
installation tool for above	-	-	-	-	73.00	nr	73.00
Outdoor seats; concrete framed - General							
Preamble: Prices for the following concrete framed seats and benches with hardwood slats include for fixing (where necessary) by bolting into existing paving or concrete bases (bases not included) or building into walls or concrete foundations (walls and foundations not included)							
Outdoor seats; concrete framed							
Outdoor seats; Marshalls Mono							
"Kelvin" benches; free-standing; 2.00 m long	403.19	0.50	9.00	-	403.19	nr	412.19
"Kelvin" benches; below-ground fixing; 2.00 m long	387.89	2.00	36.00	-	398.21	nr	434.21
"Kelvin" seats; concrete with hardwood slats; 2.00 m long	433.61	2.00	36.00	-	443.93	nr	479.93
Outdoor seats; Townscape Products							
"Oxford" benches; 1800 mm x 430 mm x 440 mm	201.76	2.00	36.00	-	212.08	nr	248.08
"Maidstone" seats; 1800 mm x 610 mm x 785 mm	301.73	2.00	36.00	-	312.05	nr	348.05
Outdoor seats; metal framed - General							
Preamble: Metal framed seats with hardwood backs to various designs can be bolted to ground anchors.							

Prices for Measured Works
Q PAVING/PLANTING/FENCING/SITE FURNITURE

Item Excluding site overheads and profit	PC £	Labour hours	Labour £	Plant £	Material £	Unit	Total rate £
Q50 SITE/STREET FURNITURE/EQUIPMENT - cont'd							
Outdoor seats; metal framed							
Outdoor seats; Furnitubes International Ltd							
ref NS 6; "Newstead"; steel standards with iroko slats; 1.80 m long	251.50	2.00	36.00	-	269.38	nr	305.38
ref NEB 6; "New Forest Single Bench"; cast iron standards with iroko slats; 1.83 m long	370.00	2.00	36.00	-	387.88	nr	423.88
ref EA 6; "Eastgate"; cast iron standards with iroko slats; 1.86 m long	345.00	2.00	36.00	-	362.88	nr	398.88
ref NE 6; "New Forest Seat"; cast iron standards with iroko slats; 1.83 m long	495.00	2.00	36.00	-	512.88	nr	548.88
Outdoor seats; Geometric Furniture; Teak bench with back and armrests							
ref 8204; two seater; 1.38 m	286.00	2.00	36.00	-	303.88	nr	339.88
ref 8207; two seater; 1.50 m	295.10	2.00	36.00	-	312.98	nr	348.98
ref 8206; three seater; 1.85 m	534.30	2.00	36.00	-	552.18	nr	588.18
ref 8202; three seater; 1.95 m	400.40	2.00	36.00	-	418.28	nr	454.28
ref 8205; four seater; 2.40 m	642.20	2.00	36.00	-	660.08	nr	696.08
Outdoor seats; Columbia Cascade Ltd; perforated metal bench seats, range of 8 nr colours							
ref 2604-6; back-to-back metal benches; 1800 mm x 1500 mm x 860 mm high	1295.00	3.00	54.00	-	1306.92	nr	1360.92
Outdoor seats; Orchard Street Furniture; "Bramley"; broad iroko slats to steel frame							
1.20 m long	161.44	2.00	36.00	-	173.36	nr	209.36
1.80 m long	198.94	2.00	36.00	-	210.86	nr	246.86
2.40 m long	229.79	2.00	36.00	-	241.71	nr	277.71
Outdoor seats; Orchard Street Furniture; "Laxton"; narrow iroko slats to steel frame							
1.20 m long	182.45	2.00	36.00	-	194.37	nr	230.37
1.80 m long	224.20	2.00	36.00	-	236.12	nr	272.12
2.40 m long	254.79	2.00	36.00	-	266.71	nr	302.71
Outdoor seats; Orchard Street Furniture; "Lambourne"; iroko slats to cast iron frame							
1.80 m long	473.94	2.00	36.00	-	485.86	nr	521.86
2.40 m long	557.71	2.00	36.00	-	569.63	nr	605.63
Outdoor seats; SMP Playgrounds; "Datchet", steel and timber							
backless benches; ref OFDB6; 1.80 m long	301.00	2.00	36.00	-	312.92	nr	348.92
bench seats with backrests; ref OFDS6; 1.80 m long	406.00	2.00	36.00	-	417.92	nr	453.92
picnic tables with attached seats; ref OFDT6; 1.80 m long	708.00	2.00	36.00	-	719.92	nr	755.92
Outdoor seats; Broxap Streetscene; metal and timber							
"Manchester"	373.97	2.00	36.00	-	385.89	nr	421.89
"Eastgate"	352.80	2.00	36.00	-	364.72	nr	400.72
"Serpent"	421.88	2.00	36.00	-	433.80	nr	469.80
"Rotherham"	560.74	2.00	36.00	-	572.66	nr	608.66

Prices for Measured Works
Q PAVING/PLANTING/FENCING/SITE FURNITURE

Item Excluding site overheads and profit	PC £	Labour hours	Labour £	Plant £	Material £	Unit	Total rate £
Outdoor seats; Earth Anchors; "Forest-Saver", steel frame and recycled slats							
bench, 1.8m	174.00	1.00	18.00	-	174.00	nr	192.00
seat, 1.8m	270.00	1.00	18.00	-	270.00	nr	288.00
Outdoor seats; Earth Anchors; "Forest-Saver", ci frame and recycled slats							
"Evergreen" bench, 1.8m	352.00	1.00	18.00	-	352.00	nr	370.00
"Evergreen" seat, 1.8m	427.00	1.00	18.00	-	427.00	nr	445.00
Outdoor seats; all-steel							
Outdoor seats; Earth Anchors, "Ranger"							
bench, 1.8m	246.00	1.00	18.00	-	246.00	nr	264.00
seat, 1.8m	399.00	1.00	18.00	-	399.00	nr	417.00
Outdoor seats; all timber							
Outdoor seats; Lister Lutyens; "Mendip"; teak							
1.524 m long	416.00	1.00	18.00	-	437.11	nr	455.11
1.829 m long	431.00	1.00	18.00	-	452.11	nr	470.11
2.438 m long	577.00	1.00	18.00	-	598.11	nr	616.11
Outdoor seats; Lister Lutyens; "Sussex"; hardwood							
1.5 m	173.00	2.00	36.00	-	194.11	nr	230.11
Outdoor seats; Woodscape; solid hardwood							
seat type "3" with back; 2.00 m long; freestanding	610.00	2.00	36.00	-	631.11	nr	667.11
seat type "3" with back; 2.00 m long; building in	650.00	4.00	72.00	-	671.11	nr	743.11
seat type "4" with back; 2.00 m long; fixing to wall	360.00	3.00	54.00	-	371.34	nr	425.34
seat type "5" with back; 2.50 m long; freestanding	720.00	2.00	36.00	-	741.11	nr	777.11
seat type "4"; 2.00 m long; fixing to wall	285.00	2.00	36.00	-	306.11	nr	342.11
seat type "5" with back; 2.00 m long; freestanding	625.00	2.00	36.00	-	646.11	nr	682.11
seat type "4" with back; 2.50 m long; fixing to wall	450.00	3.00	54.00	-	461.34	nr	515.34
seat type "5" with back; building in	665.00	2.00	36.00	-	686.11	nr	722.11
seat type "5" with back; 2.50 m long; building in	760.00	3.00	54.00	-	781.11	nr	835.11
bench type "1"; 2.00 m long; freestanding	480.00	2.00	36.00	-	501.11	nr	537.11
bench type "1"; 2.00 m long; building in	520.00	4.00	72.00	-	541.11	nr	613.11
bench type "2"; 2.00 m long; freestanding	455.00	2.00	36.00	-	476.11	nr	512.11
bench type "2"; 2.00 m long; building in	495.00	4.00	72.00	-	516.11	nr	588.11
bench type "2"; 2.50 m long; freestanding	520.00	2.00	36.00	-	541.11	nr	577.11
bench type "2"; 2.50 m long; building in	560.00	4.00	72.00	-	581.11	nr	653.11
bench type "2"; 2.00 m long overall; curved to 5 m radius, building in	655.00	4.00	72.00	-	676.11	nr	748.11
Outdoor seats; Orchard Street Furniture; "Allington"; all iroko							
1.20 m long	254.26	2.00	36.00	-	275.37	nr	311.37
1.80 m long	292.29	2.00	36.00	-	313.40	nr	349.40
2.40 m long	355.59	2.00	36.00	-	376.70	nr	412.70
Outdoor seats; SMP Playgrounds; "Lowland range"; all timber							
bench seats with backrest; ref OFDS6; 1.50 m long	406.00	2.00	36.00	-	417.92	nr	453.92

Prices for Measured Works
Q PAVING/PLANTING/FENCING/SITE FURNITURE

Item Excluding site overheads and profit	PC £	Labour hours	Labour £	Plant £	Material £	Unit	Total rate £
Q50 SITE/STREET FURNITURE/EQUIPMENT - cont'd							
Outdoor seats; tree benches/seats Tree bench; Neptune Outdoor Furniture, "Beaufort" Hexagonal, Timber							
SF34-05A, 500 mm diameter	639.00	0.50	9.00	-	639.00	nr	648.00
SF34-15A, 1500 mm diameter	777.00	0.50	9.00	-	777.00	nr	786.00
Tree seat; Neptune Outdoor Furniture, "Beaufort" Hexagonal, Timber, with back							
SF32-10A, 720 mm diameter	932.00	0.50	9.00	-	932.00	nr	941.00
SF32-20A, 1720 mm diameter	1089.00	0.50	9.00	-	1089.00	nr	1098.00
Street furniture ranges Townscape Products; "Belgrave"; natural grey concrete							
bollards; 250 mm diameter x 500 mm high	55.00	1.00	18.00	-	57.71	nr	75.71
seats; 1800 mm x 600 mm x 736 mm high	235.00	2.00	36.00	-	235.00	nr	271.00
Abacus Municipal; "300" range of street furniture; hardwood seats without arms							
ref 304H; 1220 mm long	200.45	2.00	36.00	-	212.37	nr	248.37
ref 306H; 1830 mm long	222.30	2.00	36.00	-	234.22	nr	270.22
ref 308H; 2440 mm long	309.70	2.00	36.00	-	321.62	nr	357.62
Abacus Municipal; "300" range of street furniture; hardwood double seats without arms							
ref 304DH; 1220 mm long	328.70	2.00	36.00	-	340.62	nr	376.62
ref 306DH; 1830 mm long	397.10	2.00	36.00	-	409.02	nr	445.02
ref 308DH; 2440 mm long	553.85	2.00	36.00	-	565.77	nr	601.77
Picnic benches - General Preamble: The following items include for fixing to ground in to manufacturer's instructions or concreting in.							
Picnic tables and benches Picnic tables and benches; Abacus Municipal							
ref 306SP; softwood	397.95	2.00	36.00	-	409.87	nr	445.87
ref 306HP; hardwood	463.05	2.00	36.00	-	474.97	nr	510.97
Picnic tables and benches; SMP Playgrounds							
ref OFLTH; "Datchet" Special needs Picnic Table	939.00	2.00	36.00	-	950.92	nr	986.92
ref OFDT6; "Datchet"; 1.80 m; steel and timber; with attached seats	708.00	2.00	36.00	-	719.92	nr	755.92
Picnic tables and benches; Broxap Streetscene							
"Eastgate" picnic unit	602.80	2.00	36.00	-	614.72	nr	650.72
Picnic tables and benches; Woodscape Ltd							
Table and benches built in, 2 m long	1560.00	2.00	36.00	-	1571.92	nr	1607.92
Market Prices of Containers **Plant containers; terracotta** Capital Garden Products							
Large Pot LP63; weathered terracotta - 1170 x 1600 dia.	-	-	-	-	584.70	nr	584.70
Large Pot LP38; weathered terracotta - 610 x 970 dia.	-	-	-	-	254.85	nr	254.85
Large Pot LP23; weathered terracotta - 480 x 580 dia.	-	-	-	-	135.70	nr	135.70

Prices for Measured Works
Q PAVING/PLANTING/FENCING/SITE FURNITURE

Item Excluding site overheads and profit	PC £	Labour hours	Labour £	Plant £	Material £	Unit	Total rate £
Manhole Cover Planter 3022; 760 x 560 x 210 high	-	-	-	-	93.80	nr	93.80
Apple Basket 2215; 380 x 560 dia.	-	-	-	-	96.80	nr	96.80
Indian style Shimmer Pot 2322; 585 x 560 dia.	-	-	-	-	134.65	nr	134.65
Indian style Shimmer Pot 1717; 430 x 430 dia.	-	-	-	-	109.25	nr	109.25
Indian style Shimmer Pot 1314; 330 x 355 dia.	-	-	-	-	70.95	nr	70.95
Plant containers; faux lead							
Capital Garden Products							
Trough 2508 'Tudor Rose'; 635 x 215 x 230 high	-	-	-	-	64.90	nr	64.90
Tub 2004 'Elizabethan'; 508 mm square	-	-	-	-	89.10	nr	89.10
Tub 1513 'Elizabethan'; 380 mm square	-	-	-	-	67.50	nr	67.50
Tub 1601 'Tudor Rose'; 420 x 395 dia.	-	-	-	-	61.75	nr	61.75
Plant containers; window boxes							
Capital Garden Products							
Adam Design AD5401; 1370 x 270 x 210 h	-	-	-	-	100.45	nr	100.45
Wheatsheaf Design WH54; 1370 x 270 x 210 h	-	-	-	-	100.45	nr	100.45
Oakleaf Design OK24; 610 x 370 x 350 h	-	-	-	-	71.70	nr	71.70
Swag Design SW2402; 610 x 200 x 210 h	-	-	-	-	71.70	nr	71.70
Plant containers; timber							
Plant containers; Hardwood; Neptune Outdoor Furniture							
'Beaufort' T38-4D, 1500 x 1500 x 900 mm high	-	-	-	-	870.00	nr	870.00
'Beaufort' T38-3C, 1000 x 1500 x 700 mm high	-	-	-	-	648.00	nr	648.00
'Beaufort' T38-2A, 1000 x 500 x 500 mm high	-	-	-	-	338.00	nr	338.00
'Kara' T42-4D, 1500 x 1500 x 900 mm high	-	-	-	-	939.00	nr	939.00
'Kara' T42-3C, 1000 x 1500 x 700 mm high	-	-	-	-	699.00	nr	699.00
'Kara' T42-2A, 1000 x 500 x 500 mm high	-	-	-	-	365.00	nr	365.00
Plant containers; Hardwood; Woodscape Ltd							
Square, 900 x 900 x 420 mm high	-	-	-	-	256.80	nr	256.80
Plant containers; precast concrete							
Plant containers; Marshalls Mono							
"Strada"; 710 mm circular planter	120.85	2.00	36.00	30.91	120.85	nr	187.76
"Elba"; 970 mm diameter x 470 mm high	224.98	2.00	36.00	30.91	224.98	nr	291.89
"Shirley"; 1066 mm diameter x 760 mm high	409.12	2.00	36.00	30.91	409.12	nr	476.03
"Boulevard" bases; 2000 mm diameter x 260 mm high	575.42	2.00	36.00	15.46	575.42	nr	626.88
"Boulevard" bases; 1200 mm diameter x 260 mm high	299.90	1.00	18.00	15.46	299.90	nr	333.36
"Boulevard" rings; 1200 mm diameter x 235 mm high	201.23	1.00	18.00	15.46	201.23	nr	234.69
"Boulevard" rings; 2000 mm diameter x 235 mm high	248.90	1.50	27.00	23.18	248.90	nr	299.08
Cycle holders							
Cycle holders; George Fischer Sales; "Velop A" galvanized steel							
ref R; fixing to wall or post; making good	27.00	1.00	18.00	-	38.92	nr	56.92
ref SR(V); fixing in ground; making good	29.00	1.00	18.00	-	40.92	nr	58.92

Prices for Measured Works
Q PAVING/PLANTING/FENCING/SITE FURNITURE

Item Excluding site overheads and profit	PC £	Labour hours	Labour £	Plant £	Material £	Unit	Total rate £
Q50 SITE/STREET FURNITURE/EQUIPMENT - cont'd							
Cycle holders – cont'd							
Cycle holders; Townscape Products "Guardian" cycle holders; tubular steel frame; setting in concrete; 1250 mm x 550 mm x 775 mm high; making good	215.01	1.00	18.00	-	226.93	nr	**244.93**
"Penny" cycle stands; 600 mm diameter; exposed aggregate bollards with 8 nr cycle holders; in galvanized steel; setting in concrete; making good	450.07	1.00	18.00	-	461.99	nr	**479.99**
Cycle holders; Mawrob & Co; combined cycle racks and shelters; corrugated aluminium roof; on angle iron frame; galvanized ref MW/AV10; for 10 nr cycles; semi-vertical; 1.22 m wide x 3.05 m long x 2.15 m high	506.00	10.00	180.00	-	529.83	nr	**709.83**
ref MW/AW10; for 10 nr cycles; horizontal; 2.13 m wide x 3.05 m long x 2.15 m high	493.90	10.00	180.00	-	517.73	nr	**697.73**
Cycle holders; Broxap Mawrob cycle parking stand; BX11 6045-6, pillar mounted double sided rail of front wheel supports; holds twelve cycles	394.90	4.00	72.00	-	403.61	nr	**475.61**
Directional signage; cast aluminium Signage; Furnitubes International Ltd ref FFL 1; "Lancer"; cast aluminium finials	69.95	0.07	1.20	-	69.95	nr	**71.15**
ref FAA IS; arrow end type cast aluminium directional arms, single line; 90 mm wide	118.90	2.00	36.00	-	118.90	nr	**154.90**
ref FAA ID; arrow end type cast aluminium directional arms; double line; 145 mm wide	152.45	0.13	2.40	-	152.45	nr	**154.85**
ref FAA IS; arrow end type cast aluminium directional arms; treble line; 200 mm wide	188.50	0.20	3.60	-	188.50	nr	**192.10**
ref FCK1 211 G; "Kingston"; composite standard root columns	552.00	2.00	36.00	-	566.34	nr	**602.34**
Excavating; for bollards and barriers; by hand Holes for bollards							
400 mm x 400 mm x 400 mm; disposing off site	-	0.42	7.50	-	0.58	nr	**8.08**
600 mm x 600 mm x 600 mm; disposing off site	-	0.97	17.46	-	1.62	nr	**19.08**
Concrete bollards - General Preamble: Precast concrete bollards are available in a very wide range of shapes and sizes. The bollards listed here are the most commonly used sizes and shapes; manufacturer's catalogues should be consulted for the full range. Most manufacturers produce bollards to match their suites of street furniture, which may include planters, benches, litter bins and cycle stands. Most parallel sided bollards can be supplied in removable form, with a reduced shank, precast concrete socket and lifting hole to permit removal with a bar.							

Prices for Measured Works
Q PAVING/PLANTING/FENCING/SITE FURNITURE

Item Excluding site overheads and profit	PC £	Labour hours	Labour £	Plant £	Material £	Unit	Total rate £
Concrete bollards							
Marshalls Mono; cylinder; straight or tapered; 200 mm - 400 mm diameter; plain grey concrete; setting into firm ground (excavating and backfilling not included)							
"Richmond" 535 mm high above ground	57.93	2.00	36.00	-	66.95	nr	102.94
"Bridgford" 915 mm high above ground	52.00	2.00	36.00	-	61.02	nr	97.01
"Truro" 760 mm high above ground	132.49	2.00	36.00	-	137.34	nr	173.34
Marshalls Mono; cylinder; straight or tapered; 200 mm - 400 mm diameter; white concrete; setting into firm ground (excavating and backfilling not included)							
"Bridgford" 915 mm high above ground	86.48	2.00	36.00	-	91.33	nr	127.33
"Richmond" 535 mm high above ground	94.80	2.00	36.00	-	99.65	nr	135.65
"Truro" 760 mm high above ground	155.22	2.00	36.00	-	160.07	nr	196.07
Marshalls Mono; cylinder; straight or tapered; 200 mm - 400 mm diameter; exposed aggregate; setting into firm ground (excavating and back filling not included)							
"Worcester" 495 mm high above ground	87.86	2.00	36.00	-	92.71	nr	128.71
"Bridgford" 915 mm high above ground	65.42	2.00	36.00	-	70.27	nr	106.27
"Truro" 760 mm high above ground	149.40	2.00	36.00	-	154.25	nr	190.25
Marshalls Mono; cylinder; straight or tapered; 200 mm - 400 mm diameter; "Beadalite" reflective finish; setting into firm ground (excavating and backfilling not included)							
"Richmond" 535 mm high above ground	130.55	2.00	36.00	-	135.48	nr	171.48
"Bridgford" 915 mm high above ground	137.21	2.00	36.00	-	142.06	nr	178.06
"Wexham" 760 mm high above ground	172.96	2.00	36.00	-	181.98	nr	217.98
Marshalls Mono; "Millbank" bollards; exposed aggregate with ribbed top							
230 mm diameter x 1000 mm high	111.98	2.00	36.00	-	116.83	nr	152.83
Extra over concrete bollards for chain fixing rings	5.20	-	-	-	5.20	nr	5.20
Precast concrete verge markers; various shapes; 450 mm high							
plain grey concrete	24.91	1.00	18.00	-	28.37	nr	46.37
white concrete	30.49	1.00	18.00	-	33.95	nr	51.95
exposed aggregate	30.49	1.00	18.00	-	33.95	nr	51.95
Extra over concrete bollards for removable bollards (any size)	16.13	-	-	-	16.13	nr	16.13
Precast concrete spur stones; 450 mm diameter x 750 mm high; plain white concrete							
half-section	108.52	1.00	18.00	-	111.79	nr	129.79
three-quarter section	132.25	1.00	18.00	-	135.52	nr	153.52
Other bollards							
Hinged or removable bollards; Marshalls Mono; powder coated; steel tube with locking device; base plate; bolting to ground, or socket 267 mm deep in ground							
"Brunel" removable bollards	314.05	2.00	36.00	-	325.07	nr	361.07
Removable parking posts; Dee-Organ "Spacekeeper"; ref 3014203; folding plastic coated galvanized steel parking posts with key; reflective bands; including 300 mm x 300 mm x 300 mm concrete foundations and fixing bolts;							
850 mm high	114.22	2.00	36.00	-	125.24	nr	161.24

Prices for Measured Works
Q PAVING/PLANTING/FENCING/SITE FURNITURE

Item Excluding site overheads and profit	PC £	Labour hours	Labour £	Plant £	Material £	Unit	Total rate £
Q50 SITE/STREET FURNITURE/EQUIPMENT - cont'd							
Other bollards – cont'd							
"Spacesaver"; ref 1508101; hardwood removable bollards including key and base-plate; 150 mm x 150 mm; 600 mm high	116.86	2.00	36.00	-	120.32	nr	156.32
"Spacesaver" ref 1506101; hardwood removable bollards including key and base-plate; 150 mm x 150 mm; 800 mm high	130.04	2.00	36.00	-	133.50	nr	169.50
Cast iron bollards - General Preamble: The following bollards are particularly suitable for conservation areas. Logos for civic crests can be incorporated to order.							
Cast iron bollards Bollards; Furnitubes International Ltd (excavating and backfilling not included)							
"Doric Round"; 920 mm high x 170 mm diameter	101.00	2.00	36.00	-	105.62	nr	141.62
"Gunner round"; 750 mm high x 165 mm diameter	92.00	2.00	36.00	-	96.62	nr	132.62
"Manchester Round"; 975 mm high; 225 mm square base	112.00	2.00	36.00	-	116.62	nr	152.62
"Cannon"; 1140 mm x 210 mm diameter	146.00	2.00	36.00	-	150.62	nr	186.62
"Kenton"; heavy duty galvanized steel; 900 mm high; 350 mm diameter	330.00	2.00	36.00	-	334.62	nr	370.62
Cast iron bollards with rails - General Preamble: The following cast iron bollards are suitable for conservation areas.							
Cast iron bollards with rails Cast iron posts with steel tubular rails; Broxap Streetscene; setting into firm ground (excavating not included)							
"Sheffield"; 450 mm high; one rail, type A	53.00	1.50	27.00	-	55.18	nr	82.18
"Southbank"; 1150 mm high above ground; four rails, type C	90.10	1.00	18.00	-	92.28	nr	110.28
"Mersey"; 1085 mm high above ground; two rails, type D	97.31	1.50	27.00	-	99.49	nr	126.49
"Promenade"; 1150 mm high above ground; square; three rails, type C	105.79	1.50	27.00	-	107.97	nr	134.97
"Type A" mild steel tubular rail including connector	4.58	0.03	0.60	-	5.48	m	6.08
"Type C" mild steel tubular rail including connector	5.88	0.03	0.60	-	7.08	m	7.68
"Type D" mild steel tubular rail including connector	7.55	-	-	-	9.43	m	9.43
Timber bollards - General **Timber bollards** Woodscape Ltd; Durable Hardwood							
RP 250/1500; 250 mm diameter x 1500 mm long	181.50	1.00	18.00	-	181.70	nr	199.70
SP 250/1500; 250 mm square x 1500 mm long	181.50	1.00	18.00	-	181.70	nr	199.70
SP 150/1200; 150mm square x 1200 mm long	65.40	1.00	18.00	-	65.60	nr	83.60
SP 125/750; 125 mm square x 750 mm long	36.10	1.00	18.00	-	36.30	nr	54.30
RP 125/750; 125 mm diameter x 750 mm long	36.10	1.00	18.00	-	36.30	nr	54.30

Prices for Measured Works
Q PAVING/PLANTING/FENCING/SITE FURNITURE

Item Excluding site overheads and profit	PC £	Labour hours	Labour £	Plant £	Material £	Unit	Total rate £
Deterrent bollards							
Semi-mountable vehicle deterrent and kerb protection bollards; Furnitubes International Ltd (excavating and backfilling not included)							
"Bell decorative"	312.00	2.00	36.00	-	318.57	nr	354.57
"Half bell"	400.00	2.00	36.00	-	406.57	nr	442.57
"Full bell"	486.00	2.00	36.00	-	492.57	nr	528.57
"Three quarter bell"	496.16	2.00	36.00	-	502.73	nr	538.73
Security Bollards							
Security bollards; Furnitubes International Ltd (excavating and backfilling not included)							
"Gunner"; reinforced with steel insert and tie bars; 750 mm high above ground; 600 mm below ground	1053.00	2.00	36.00	-	1059.92	nr	1095.92
"Burr Bloc Type 6" removable steel security bollard; 750 mm high above ground; 410 mm x 285 mm	409.50	2.00	36.00	-	425.08	nr	461.08
"Burr 7 - Burr Post Type 8"; circular hollow steel post; heavy duty anti-Ram Raid; galvanised; 900 mm high above ground; 170 mm diameter	222.45	2.00	36.00	-	238.03	nr	274.03
Tree grilles; Cast Iron							
Cast iron tree grilles; Furnitubes International							
ref GS 1070 Greenwich; two part; 1000 mm square; 700 mm diameter tree hole	107.50	2.00	36.00	-	107.50	nr	143.50
ref GC 1270 Greenwich; two part, 1200 mm diameter; 700 mm diameter tree hole	155.50	2.00	36.00	-	155.50	nr	191.50
ref GCF 1026 Greenwich; steel tree grille frame for GC 1045, one part	75.40	2.00	36.00	-	75.40	nr	111.40
ref GSF 1226 Greenwich; steel tree grille frame for GC 1270 and GC 1245, one part	84.10	2.00	36.00	-	84.10	nr	120.10
Cast iron tree grilles; Marshalls Mono							
ref MSF310 grille; two part; 1200 x 1200 mm, 440 mm diameter hole; primed	333.45	3.00	54.00	-	342.96	nr	396.96
ref MSF310B grille; two part; 1200 x 1200 mm, 440 mm diameter hole ; painted black	367.66	3.00	54.00	-	377.17	nr	431.17
ref GFB310 frame; for bolt fixing; primed	147.06	1.00	18.00	-	147.06	nr	165.06
ref GFS310 frame; for spike fixing; primed	124.22	1.00	18.00	-	124.22	nr	142.22
Cast iron tree grilles; Townscape Products							
"Baltimore" 1200 mm square x 460 mm diameter tree hole	309.62	2.00	36.00	-	309.62	nr	345.62
"Baltimore" Hexagonal, maximum width 1440 mm nominal, 600 mm diameter tree hole	498.20	2.00	36.00	-	498.20	nr	534.20
Note: Care must be taken to ensure that tree grids and guards are removed when trees grow beyond the specified diameter of guard							

Prices for Measured Works
Q PAVING/PLANTING/FENCING/SITE FURNITURE

Item Excluding site overheads and profit	PC £	Labour hours	Labour £	Plant £	Material £	Unit	Total rate £
Q50 SITE/STREET FURNITURE/EQUIPMENT - cont'd							
Playground equipment - General Preamble: The range of equipment manufactured or available in the UK is so great that comprehensive coverage would be impossible, especially as designs, specifications and prices change fairly frequently. The following information should be sufficient to give guidance to anyone designing or equipping a playground. In comparing prices note that only outline specification details are given here and that other refinements which are not mentioned may be the reason for some difference in price between two apparently identical elements. The fact that a particular manufacturer does not appear under one item heading does not necessarily imply that he does not make it. Landscape designers are advised to check that equipment complies with ever more stringent safety standards before specifying.							
Playground equipment - Installation The rates below include for installation of the specified equipment by the manufacturers. Most manufactures will offer an option to install the equipment supplied by them.							
Swings - General Preamble: Prices for the following vary considerably. Those given represent the middle of the range and include multiple swings with tubular steel frames and timber or tyre seats; ground fixing and priming only.							
Swings							
Swings; Wicksteed Ltd							
traditional swings; 1850 mm high; 1 bay; 2 seat	-	-	-	-	-	nr	1505.00
traditional swings; 1850 mm high; 2 bay; 4 seat	-	-	-	-	-	nr	2411.00
traditional swings; 2450 mm high; 1 bay; 2 seat	-	-	-	-	-	nr	1450.00
traditional swings; 2450 mm high; 2 bay; 4 seat	-	-	-	-	-	nr	2268.00
traditional swings; 3050 mm high; 1 bay; 2 seat	-	-	-	-	-	nr	1536.00
traditional swings; 3050 mm high; 2 bay; 4 seat	-	-	-	-	-	nr	2378.00
single arch swings; cradle safety seats; 1850 mm high	-	-	-	-	-	nr	1182.00
single arch swings; flat rubber safety seats; 2450 mm high	-	-	-	-	-	nr	1167.00
double arch swings; flat rubber safety seats; 2450 mm high	-	-	-	-	-	nr	1427.00
Swings; Lappset UK Ltd							
ref 020414; swing frame	-	-	-	-	-	nr	919.00
Swings; Kompan Ltd							
ref M947-52; double swings	-	-	-	-	-	nr	2477.00
ref M948-52; double swings	-	-	-	-	-	nr	2098.00
ref M951; "Sunflower" swings	-	-	-	-	-	nr	1134.00

Prices for Measured Works
Q PAVING/PLANTING/FENCING/SITE FURNITURE

Item Excluding site overheads and profit	PC £	Labour hours	Labour £	Plant £	Material £	Unit	Total rate £
Slides							
Slides; Wicksteed Ltd							
"Pedestal" slides; 3.40 m	-	-	-	-	-	nr	2055.00
"Pedestal" slides; 4.40 m	-	-	-	-	-	nr	2665.00
"Pedestal" slides; 5.80 m	-	-	-	-	-	nr	3098.00
"Embankment" slides; 3.40 m	-	-	-	-	-	nr	1575.00
"Embankment" slides; 4.40 m	-	-	-	-	-	nr	2042.00
"Embankment" slides; 5.80 m	-	-	-	-	-	nr	2595.00
"Embankment" slides; 7.30 m	-	-	-	-	-	nr	3297.00
"Embankment" slides; 9.10 m	-	-	-	-	-	nr	4126.00
"Embankment" slides; 11.00 m	-	-	-	-	-	nr	4937.00
Slides; Lappset UK Ltd							
ref 142015; slide	-	-	-	-	-	nr	2129.00
ref 141115; "Jumbo" slide	-	-	-	-	-	nr	2874.00
Slides; Kompan Ltd							
ref M351; slides	-	-	-	-	-	nr	1926.00
ref M322; slide and cave	-	-	-	-	-	nr	2262.00
Moving equipment - General							
Preamble: The following standard items of playground equipment vary considerably in quality and price; the following prices are middle of the range.							
Moving equipment							
Roundabouts; Wicksteed Ltd							
"Turnstile"	-	-	-	-	-	nr	583.00
"Speedway"	-	-	-	-	-	nr	2260.00
"Spiro Whirl"	-	-	-	-	-	nr	2568.00
Seesaws							
Seesaws; Lappset UK Ltd							
ref 010300; seesaws	-	-	-	-	-	nr	639.00
ref 010236; seesaws	-	-	-	-	-	nr	1515.00
Seesaws; Wicksteed Ltd							
"Non Bump Seesaw"	-	-	-	-	-	nr	1618.00
"Jolly Gerald"	-	-	-	-	-	nr	1805.00
"Rocking Rockette"	-	-	-	-	-	nr	2415.00
"Rocking Horse"	-	-	-	-	-	nr	2746.00
Play sculptures - General							
Preamble: Many variants on the shapes of playground equipment are available, simulating spacecraft, trains, cars, houses etc., and these designs are frequently changed. The basic principles remain constant but manufacturer's catalogues should be checked for the latest styles.							
Play sculptures							
Wooden animals; SMP Playgrounds							
pig	-	-	-	-	-	nr	440.00
donkey	-	-	-	-	-	nr	460.00
rhino	-	-	-	-	-	nr	620.00
giraffe	-	-	-	-	-	nr	630.00
camel	-	-	-	-	-	nr	650.00
hippo	-	-	-	-	-	nr	950.00
elephant	-	-	-	-	-	nr	1240.00

Prices for Measured Works
Q PAVING/PLANTING/FENCING/SITE FURNITURE

Item Excluding site overheads and profit	PC £	Labour hours	Labour £	Plant £	Material £	Unit	Total rate £
Q50 SITE/STREET FURNITURE/EQUIPMENT - cont'd							
Playground equipment – cont'd							
Climbing equipment and Play structures; General							
Preamble: Climbing equipment generally consists of individually designed modules. Play structures generally consist of interlinked and modular pieces of equipment and sculptures. These may consist of climbing, play and skill based modules, nets and various other activities. Both are set into either safety surfacing or defined sand pit areas. The equipment below outlines a range from various manufacturers. Individual catalogues should be consulted in each instance. Safety areas should be allowed round all equipment.							
Climbing equipment and Play structures							
Climbing equipment and Play structures; Kompan Ltd							
ref MQ1002; "Mosaic" combination system	-	-	-	-	-	nr	4321.00
ref OK1103; "Oasis" combination system	-	-	-	-	-	nr	3912.00
Mosaiq ref MQ 1001	-	-	-	-	-	nr	3739.00
ref MQ2004; "Mosaic" combination system	-	-	-	-	-	nr	6753.00
ref OK3100; "Oasis" combination system	-	-	-	-	-	nr	15901.00
ref M480 "Castle"	-	-	-	-	-	nr	22468.00
Climbing equipment; Wicksteed Ltd							
"Junior Commando Bridge"	-	-	-	-	-	nr	880.00
"Fantasy Funrun - Under Starters Orders" set of 12 units	-	-	-	-	-	nr	12664.00
Climbing equipment; Lappset UK Ltd							
ref 120254; "Climbing Frame"	-	-	-	-	-	nr	2369.00
ref 122457; "Playhouse"	-	-	-	-	-	nr	2894.00
ref 120100; "Activity Tower"	-	-	-	-	-	nr	10434.00
ref 122124; "Tower and Climbing Frame"	-	-	-	-	-	nr	10250.00
SMP Playgrounds							
Action Pack - Cape Horn; 8 module	-	-	-	-	-	nr	7360.00
Spring equipment							
Spring based equipment for 1 - 8 year olds; Kompan Ltd							
ref M101; "Crazy Hen"	-	-	-	-	-	nr	486.00
ref M128; "Crazy Daisy"	-	-	-	-	-	nr	675.00
ref M141; "Crazy Seesaw"	-	-	-	-	-	nr	1270.00
ref M155; "Quartet Seesaw"	-	-	-	-	-	nr	960.00
ref M199; "Crazy Springboard"	-	-	-	-	-	nr	1688.00
Spring based equipment for under 12's; Lappset UK Ltd							
ref 010443; "Rocking Horse"	-	-	-	-	-	nr	749.00
Sand pits							
Market Prices							
Play Pit sand; Boughton Loam Ltd	-	-	-	-	42.57	m³	42.57
Kompan Ltd							
" Basic 500" 2900 mm x 1570 mm x 310 mm deep	-	-	-	-	-	nr	751.00

Prices for Measured Works
Q PAVING/PLANTING/FENCING/SITE FURNITURE

Item Excluding site overheads and profit	PC £	Labour hours	Labour £	Plant £	Material £	Unit	Total rate £
Flagpoles							
Flagpoles; Harrison Flagpoles; in glass fibre; smooth white finish; terylene halyards with mounting accessories; setting in concrete; to manufacturers recommendations (excavating not included)							
6.00 m high, plain	151.00	3.00	54.00	-	168.42	nr	**222.42**
6.00 m high, hinged baseplate, external halyard	192.00	4.00	72.00	-	209.42	nr	**281.42**
6.00 m high, hinged baseplate, internal halyard	289.00	4.00	72.00	-	306.42	nr	**378.42**
10.00 m high, plain	286.00	5.00	90.00	-	303.42	nr	**393.42**
10.00 m high, hinged baseplate, external halyard	699.00	5.00	90.00	-	716.42	nr	**806.42**
10.00 m high, hinged baseplate, internal halyard	407.00	5.00	90.00	-	424.42	nr	**514.42**
15.00 m high, plain	654.00	6.00	108.00	-	671.42	nr	**779.42**
15.00 m high, hinged baseplate, external halyard	699.00	6.00	108.00	-	716.42	nr	**824.42**
15.00 m high, hinged baseplate, internal halyard	880.00	6.00	108.00	-	897.42	nr	**1005.42**
Flagpoles; Harrison Flagpoles; tapered hollow steel section; base plate flange; galvanized; white painted finish; lowering gear; including bolting to concrete (excavating not included)							
7.00 m high	125.00	3.00	54.00	-	142.64	nr	**196.64**
10.00 m high	286.00	3.50	63.00	-	306.09	nr	**369.09**
13.00 m high	310.00	6.00	108.00	-	333.69	nr	**441.69**
Flagpoles; Harrison Flagpoles; in glass fibre; smooth white finish; terylene halyards for wall mounting							
3.0 m pole	171.00	2.00	36.00	-	171.00	nr	**207.00**
Flagpoles; Harrison Flagpoles; parallel galvanised steel banner flagpole - guide price							
6 m high	800.00	4.00	72.00	-	817.42	nr	**889.42**
Sports equipment; Edwards Sports Products							
Tennis courts; steel posts and fixings for hardcourt							
Round	136.50	1.00	18.00	-	136.50	set	**154.50**
Square	173.00	1.00	18.00	-	173.00	set	**191.00**
Tennis nets; not including posts or fixings							
ref 5021; "Matchplay"	59.40	3.00	54.00	-	65.21	set	**119.21**
ref 5075; "Club"	45.43	3.00	54.00	-	51.24	set	**105.24**
ref 5001; "Championship"	67.60	3.00	54.00	-	73.41	set	**127.41**
Aluminum football goal posts with sockets; senior; including nets							
ref 2026; "Club" nets	551.00	4.00	72.00	-	556.81	set	**628.81**
ref 2037; "Medium Club" nets	540.00	0.14	2.59	-	701.32	set	**703.91**
Steel football goal posts with sockets; ref 2760; including nets							
ref 2026; "Club" net	382.20	4.00	72.00	-	388.01	set	**460.01**
ref 2037; "Medium Club" net	376.19	4.00	72.00	-	382.00	set	**454.00**
Junior aluminium football goal posts with sockets; including 2 mm nets	480.00	4.00	72.00	-	485.81	set	**557.81**
Telescopic tubular steel rugby goal posts with sockets							
ref 2800; 7.30 m high	394.16	6.00	108.00	-	409.65	set	**517.65**
ref 2802; 10.70 m high	565.04	6.00	108.00	-	580.53	set	**688.53**

Prices for Measured Works
Q PAVING/PLANTING/FENCING/SITE FURNITURE

Item Excluding site overheads and profit	PC £	Labour hours	Labour £	Plant £	Material £	Unit	Total rate £
Q50 SITE/STREET FURNITURE/EQUIPMENT - cont'd							
Sports equipment; Edwards Sports Products cont'd							
Hockey goal posts with sockets; including "Club" nets							
ref 2851; steel posts	190.61	0.14	2.59	-	311.60	set	314.19
backboards 450 mm	244.61	-	-	-	244.61	set	244.61
ref 2830; wooden posts	106.87	4.00	72.00	-	112.68	set	184.68
backboards 150 mm practice	90.85	-	-	-	90.85	set	90.85
Football mini goal complete with posts, net and fittings							
12' x 6'; aluminium and steel	343.20	4.00	72.00	-	349.01	set	421.01
12' x 6'; all steel	290.68	3.00	54.00	-	296.49	set	350.49
Clothes line fittings							
Rotary outdoor clothes line; Hills Industries Ltd; mild steel tube; to concrete base (base not included)							
"Airdry"; ref 4/40	15.71	2.00	36.00	-	18.61	nr	54.61
"Airdry"; ref 3/30	12.38	2.00	36.00	-	15.28	nr	51.28
"Portadry"; ref 3/35	18.66	2.00	36.00	-	21.56	nr	57.56
"Supadry"; ref 4/50	42.50	2.00	36.00	-	45.40	nr	81.40
"Supadry"; ref 3/50	38.10	2.00	36.00	-	41.00	nr	77.00
"Builders Special"	69.26	2.00	36.00	-	72.16	nr	108.16
"Handiline"; ref 4/45	38.33	2.00	36.00	-	41.23	nr	77.23
"Handiline"; ref 3/35	24.80	2.00	36.00	-	27.70	nr	63.70
Pair of precast concrete clothes posts 2.65 m long; 125 mm x 125 mm at base tapering to 75 mm; including setting in concrete (1:2:4); base size 300 mm x 450 mm x 600 mm deep; including excavating and disposing off site	58.65	3.00	54.00	-	68.97	nr	122.97

Prices for Measured Works
R DISPOSAL SYSTEMS

Item Excluding site overheads and profit	PC £	Labour hours	Labour £	Plant £	Material £	Unit	Total rate £
R12 DRAINAGE BELOW GROUND							
Silt pits and inspection chambers							
Excavating pits; starting from ground level; by machine							
maximum depth not exceeding 1.00 m	-	0.50	9.00	15.46	-	m³	24.46
maximum depth not exceeding 2.00 m	-	0.50	9.00	23.18	-	m³	32.18
maximum depth not exceeding 4.00 m	-	0.50	9.00	46.37	-	m³	55.37
Disposal of excavated material; depositing on site in permanent spoil heaps; average 50 m	-	0.04	0.75	1.56	-	m³	2.31
Filling to excavations; obtained from on site spoil heaps; average thickness not exceeding 0.25 m	-	0.13	2.40	4.12	-	m³	6.52
Surface treatments; compacting; bottoms of excavations	-	0.05	0.90	-	-	m²	0.90
Earthwork support; distance between opposing faces not exceeding 2.00 m							
maximum depth not exceeding 1.00 m	-	0.20	3.60	-	12.39	m³	15.99
maximum depth not exceeding 2.00 m	-	0.30	5.40	-	12.39	m³	17.79
maximum depth not exceeding 4.00 m	-	0.67	12.00	-	5.60	m³	17.60
Silt pits and inspection chambers; in situ concrete							
Beds; plain in situ concrete; 11.50 N/mm2 - 40 mm aggregate							
thickness not exceeding 150 mm	-	1.00	18.00	-	80.66	m³	98.66
thickness 150 mm - 450 mm	-	0.67	12.00	-	80.66	m³	92.66
Benchings in bottoms; plain in situ concrete; 25.50 N/mm2 - 20mm aggregate							
thickness 150 mm - 450 mm	-	2.00	36.00	-	80.66	m³	116.66
Isolated cover slabs; reinforced In situ concrete; 21.00 N/mm2 - 20mm aggregate							
thickness not exceeding 150 mm	80.66	4.00	72.00	-	80.66	m³	152.66
Fabric reinforcement; BS 4483; A193 (3.02 kg/m²) in cover slabs	1.03	0.06	1.13	-	1.13	m²	2.26
Formwork to reinforced in situ concrete; isolated cover slabs							
soffits; horizontal	-	3.28	59.04	-	10.83	m²	69.87
height not exceeding 250 mm	-	0.97	17.46	-	5.81	m	23.27
Silt pits and inspection chambers; precast concrete units							
Precast concrete inspection chamber units; BS 5911; bedding, jointing and pointing in cement mortar (1:3); 600 mm x 450 mm internally							
600 mm deep	31.80	6.00	108.00	-	35.47	nr	143.47
900 mm deep	39.05	7.00	126.00	-	42.91	nr	168.91
Drainage chambers; 1200 mm x 750 mm reducing to 600 mm x 600 mm; no base unit; depth of invert							
1050 mm deep	200.60	9.00	162.00	-	210.41	nr	372.41
1650 mm deep	289.40	11.00	198.00	-	304.31	nr	502.31
2250 mm deep	378.20	12.50	225.00	-	397.25	nr	622.25
Cover slabs for chambers or shaft sections; heavy duty							
900 mm diameter internally	35.03	0.67	11.99	-	35.03	nr	47.02
1050 mm diameter internally	40.47	2.00	36.00	10.30	40.47	nr	86.77
1200 mm diameter internally	51.69	1.00	18.00	10.30	51.69	nr	79.99
1500 mm diameter internally	92.60	1.00	18.00	10.30	92.60	nr	120.90
1800 mm diameter internally	110.42	2.00	36.00	23.18	110.42	nr	169.60

Prices for Measured Works
R DISPOSAL SYSTEMS

Item Excluding site overheads and profit	PC £	Labour hours	Labour £	Plant £	Material £	Unit	Total rate £
R12 DRAINAGE BELOW GROUND - cont'd							
Brickwork							
Walls to manholes; common bricks; PC £160.00 /1000; in cement mortar (1:3)							
one brick thick	21.60	3.00	54.00	-	28.44	m²	82.44
one and a half brick thick	32.40	4.00	72.00	-	42.66	m²	114.66
two brick thick projection of footing or the like	43.20	4.80	86.40	-	56.88	m²	143.28
Walls to manholes; engineering bricks; PC £260.00 /1000; in cement mortar (1:3)							
one brick thick	31.20	3.00	54.00	-	38.52	m²	92.52
one and a half brick thick	46.80	4.00	72.00	-	57.78	m²	129.78
two brick thick projection of footing or the like	62.40	4.80	86.40	-	77.04	m²	163.44
Extra over common or engineering bricks in any mortar for fair face; flush pointing as work proceeds; English bond walls or the like	-	0.13	2.40	-	-	m²	2.40
In situ finishings; cement:sand mortar (1:3); steel trowelled; 13 mm one coat work to manhole walls; to brickwork or blockwork base; over 300 mm wide	-	0.80	14.40	-	1.92	m²	16.32
Building into brickwork; ends of pipes; making good facings or renderings							
small	-	0.20	3.60	-	-	nr	3.60
large	-	0.30	5.40	-	-	nr	5.40
extra large	-	0.40	7.20	-	-	nr	7.20
extra large; including forming ring arch cover	-	0.50	9.00	-	-	nr	9.00
Cast iron inspection chambers; to BS 437; Ashworth Ltd, drainage systems bolted flat covers; bedding in cement mortar (1:3); mechanical coupling joints							
100 mm x 100 mm							
one branch each side	175.17	2.20	39.60	-	178.02	nr	217.62
two branches each side	319.08	2.40	43.20	-	320.28	nr	363.48
100 mm x 150 mm							
one branch each side	222.25	1.80	32.40	-	222.49	nr	254.89
150 mm x 150 mm							
one branch	229.09	1.50	27.00	-	229.33	nr	256.33
one branch each side	264.14	2.75	49.50	-	264.38	nr	313.88
Step irons; to BS 1247; Ashworth Ltd. drainage systems; malleable cast iron; galvanized; building into joints							
General purpose pattern; for one brick walls	4.28	0.17	3.06	-	4.28	nr	7.34
Best quality vitrified clay half section channels; Hepworth Plc; bedding and jointing in cement:mortar (1:2)							
Channels; straight							
100 mm	3.68	0.80	14.40	-	7.08	m	21.48
150 mm	6.13	1.00	18.00	-	9.53	m	27.53
225 mm	13.77	1.35	24.30	-	17.17	m	41.47
300 mm	28.25	1.80	32.40	-	31.65	m	64.05
Bends; 15, 30, 45 or 90 degrees							
100 mm bends	3.17	0.75	13.50	-	4.87	nr	18.37
150 mm bends	5.47	0.90	16.20	-	8.03	nr	24.23
225 mm bends	21.24	1.20	21.60	-	24.64	nr	46.24
300 mm bends	64.96	1.10	19.80	-	69.22	nr	89.02

Prices for Measured Works
R DISPOSAL SYSTEMS

Item Excluding site overheads and profit	PC £	Labour hours	Labour £	Plant £	Material £	Unit	Total rate £
Best quality vitrified clay three quarter section channels; Hepworth Plc; bedding and jointing in cement:mortar (1:2)							
Branch bends; 115, 140 or 165 degrees; left or right hand							
100 mm	3.17	0.75	13.50	-	4.87	nr	18.37
150 mm	5.47	0.90	16.20	-	8.03	nr	24.23
Intercepting traps							
Vitrified clay; inspection arms; brass stoppers; iron levers; chains and staples; galvanized; staples cut and pinned to brickwork; cement:mortar (1:2) joints to vitrified clay pipes and channels; bedding and surrounding in concrete; 11.50 N/mm2 - 40 mm aggregate; cutting and fitting brickwork; making good facings							
100 mm inlet; 100 mm outlet	47.01	3.00	54.00	-	75.95	nr	129.95
150 mm inlet; 150 mm outlet	67.78	2.00	36.00	-	102.51	nr	138.51
Excavating trenches; using 3 tonne tracked excavator; to receive pipes; grading bottoms; earthwork support; filling with excavated material to within 150 mm of finished surfaces and compacting; completing fill with topsoil; disposal of surplus soil							
Services not exceeding 200 mm nominal size							
average depth of run not exceeding 0.50 m	0.78	0.12	2.16	1.14	0.94	m	4.24
average depth of run not exceeding 0.75 m	0.78	0.16	2.93	1.57	0.94	m	5.44
average depth of run not exceeding 1.00 m	0.78	0.28	5.10	2.74	0.94	m	8.78
average depth of run not exceeding 1.25 m	0.78	0.38	6.90	3.69	0.78	m	11.37
Granular beds to trenches; lay granular material, to trenches excavated separately, to receive pipes (not included)							
300 wide x 100 thick							
reject sand	-	0.05	0.90	0.25	2.09	m	3.24
reject gravel	-	0.05	0.90	0.25	2.38	m	3.53
shingle 40 mm aggregate	-	0.05	0.90	0.25	2.74	m	3.89
sharp sand	2.62	0.05	0.90	0.25	2.88	m	4.04
300 wide x 150 thick							
reject sand	-	0.08	1.35	0.38	3.13	m	4.86
reject gravel	-	0.08	1.35	0.38	3.57	m	5.30
shingle 40 mm aggregate	-	0.08	1.35	0.38	4.11	m	5.84
sharp sand	3.93	0.08	1.35	0.38	4.32	m	6.05
Excavating trenches; using 3 tonne tracked excavator; to receive pipes; grading bottoms; earthwork support; filling with imported granular material type 2 and compacting; disposal of surplus soil							
Services not exceeding 200 mm nominal size							
average depth of run not exceeding 0.50 m	4.37	0.09	1.56	0.80	5.72	m	8.09
average depth of run not exceeding 0.75 m	6.56	0.11	1.95	1.01	8.59	m	11.54
average depth of run not exceeding 1.00 m	8.75	0.14	2.51	1.31	11.45	m	15.27
average depth of run not exceeding 1.25 m	10.94	0.23	4.11	2.21	14.32	m	20.63

Prices for Measured Works
R DISPOSAL SYSTEMS

Item Excluding site overheads and profit	PC £	Labour hours	Labour £	Plant £	Material £	Unit	Total rate £
R12 DRAINAGE BELOW GROUND – cont'd							
Excavating trenches, using 3 tonne tracked excavator, to receive pipes; grading bottoms; earthwork support; filling with concrete, ready mixed ST2 ; disposal of surplus soil							
Services not exceeding 200 mm nominal size							
average depth of run not exceeding 0.50 m	9.77	0.11	1.92	0.40	11.37	m	**13.69**
average depth of run not exceeding 0.75 m	14.65	0.13	2.34	0.50	17.05	m	**19.89**
average depth of run not exceeding 1.00 m	19.54	0.17	3.00	0.67	22.73	m	**26.40**
average depth of run not exceeding 1.25 m	24.42	0.23	4.05	1.01	28.42	m	**33.47**
Earthwork Support; providing support to opposing faces of excavation; moving along as work proceeds; A Plant Acrow							
Maximum depth not exceeding 2.00 m							
distance between opposing faces not exceeding 2.00 m	-	0.80	14.40	16.11	-	m	**30.51**
Clay pipes and fittings: to BS EN295:1:1991:Hepworth Plc; Supersleeve							
100 mm clay pipes; polypropylene slip coupling; in trenches (trenches not included)							
laid straight	1.94	0.25	4.50	-	3.36	m	**7.86**
short runs under 3.00 m	1.94	0.31	5.63	-	3.36	m	**8.99**
Extra over 100 mm clay pipes for							
bends; 15-90 degree; single socket	6.17	0.25	4.50	-	6.17	nr	**10.67**
junction; 45 or 90 degree; double socket	13.48	0.25	4.50	-	13.48	nr	**17.98**
slip couplings polypropylene	2.28	0.08	1.50	-	2.28	nr	**3.78**
gully with "P" trap; 100 mm; 154 mm x 154 mm plastic grating	56.92	1.00	18.00	-	75.44	nr	**93.44**
150 mm vitrified clay pipes; polypropylene slip coupling; in trenches (trenches not included)							
laid straight	7.74	0.30	5.40	-	7.74	m	**13.14**
short runs under 3.00 m	5.96	0.33	6.00	-	5.96	m	**11.96**
Extra over 150 mm vitrified clay pipes for							
bends; 15 - 90 degree	8.62	0.28	5.04	-	12.78	nr	**17.82**
junction; 45 or 90 degree; 100 x 150	-	0.40	7.20	-	24.81	nr	**32.01**
junction; 45 or 90 degree; 150 x 150	18.10	0.40	7.20	-	26.42	nr	**33.62**
slip couplings polypropylene	4.16	0.05	0.90	-	4.16	nr	**5.06**
taper pipe 100 -150 mm	12.97	0.50	9.00	-	12.97	nr	**21.97**
taper pipe 150 -225 mm	31.86	0.50	9.00	-	31.86	nr	**40.86**
socket adaptor; connection to traditional pipes and fittings	8.37	0.33	5.94	-	12.52	nr	**18.46**
Accessories in clay							
150 mm access pipe	-	-	-	-	104.73	nr	**104.73**
150 mm rodding eye	42.38	0.50	9.00	-	45.99	nr	**54.99**
gully with "P" traps; 150 mm; 154 mm x 154 mm plastic grating	56.92	1.00	18.00	-	62.52	nr	**80.52**
PVC-u pipes and fittings; to BS EN1401; Wavin Plastics Ltd; OsmaDrain							
110 mm PVC-u pipes; in trenches (trenches not included)							
laid straight	4.97	0.08	1.44	-	4.97	m	**6.41**
short runs under 3.00 m	4.97	0.12	2.16	-	5.60	m	**7.76**

Prices for Measured Works
R DISPOSAL SYSTEMS

Item Excluding site overheads and profit	PC £	Labour hours	Labour £	Plant £	Material £	Unit	Total rate £
Extra over 110 mm PVC-u pipes for							
bends; short radius	9.70	0.25	4.50	-	9.70	nr	14.20
bends; long radius	16.81	0.25	4.50	-	16.81	nr	21.31
junctions; equal; double socket	11.57	0.25	4.50	-	11.57	nr	16.06
slip couplings	5.61	0.25	4.50	-	5.61	nr	10.11
adaptors to clay	8.38	0.50	9.00	-	8.56	nr	17.56
160 mm PVC-u pipes; in trenches (trenches not included)							
laid straight	9.68	0.08	1.44	-	9.68	m	11.12
short runs under 3.00 m	14.75	0.12	2.16	-	16.17	m	18.33
Extra over 160 mm PVC-u pipes for							
socket bend double 90 or 45 degrees	28.92	0.20	3.60	-	28.92	nr	32.52
socket bend double 15 or 30 degrees	24.73	0.20	3.60	-	24.73	nr	28.33
socket bend single 87.5 or 45 degrees	19.34	0.20	3.60	-	19.34	nr	22.94
socket bend single 15 or 30 degrees	19.12	0.20	3.60	-	19.12	nr	22.72
bends; short radius	15.20	0.25	4.50	-	24.90	nr	29.40
bends; long radius	50.22	0.25	4.50	-	50.22	nr	54.72
junctions; single	52.11	0.33	6.00	-	52.11	nr	58.10
pipe coupler	7.60	0.05	0.90	-	7.60	nr	8.50
slip couplings PVC-u	6.76	0.05	0.90	-	6.76	nr	7.66
adaptors to clay	26.18	0.50	9.00	-	26.36	nr	35.36
level invert reducer	11.18	0.50	9.00	-	11.18	nr	20.18
spiggot	18.78	0.20	3.60	-	18.78	nr	22.38
Accessories in PVC-u							
110 mm screwed access cover	-	-	-	-	7.25	nr	7.25
110 mm rodding eye	19.82	0.50	9.00	-	23.43	nr	32.43
gully with "P" traps; 110 mm; 154 mm x 154 mm grating	38.20	1.00	18.00	-	39.64	nr	57.64
inspection chambers; 450 mm diameter; 960 mm deep; heavy duty round covers and frames; double seal recessed with 6 nr 110 mm outlets/inlets	125.23	1.00	18.00	-	127.39	nr	145.39
Kerbs; to gullies; in one course Class B engineering bricks; to 4 nr sides; rendering in cement:mortar (1:3); dished to gully gratings	2.08	1.00	18.00	-	2.91	nr	20.91
Gullies Concrete; Hepworth Plc							
Concrete road gullies; to BS 5911; trapped with rodding eye and stoppers; 450 mm diameter x 1.07 m deep	31.50	6.00	108.00	-	51.50	nr	159.50
Gullies Vitrified Clay; Hepworth Plc; bedding in concrete; 11.50 N/mm2 - 40 mm aggregate							
Vitrified clay yard gullies (mud); trapped; domestic duty (up to 1 tonne)							
RGP5; 100 mm outlet; 100 mm dia; 225 mm internal width 585 mm internal depth	62.52	3.50	63.00	-	63.08	nr	126.08
RGP7; 150 mm outlet; 100 mm dia; 225 mm internal width 585 mm internal depth	62.52	3.50	63.00	-	63.08	nr	126.08
Vitrified clay yard gullies (mud); trapped; medium duty (up to 5 tonnes)							
100 mm outlet; 100 mm dia; 225 mm internal width 585 mm internal depth	88.38	3.50	63.00	-	88.94	nr	151.94
150 mm outlet; 100 mm dia; 225 mm internal width 585 mm internal depth	88.38	3.50	63.00	-	88.94	nr	151.94
Combined filter and silt bucket for Yard gullies 225 mm wide	21.98	-	-	-	21.98	nr	21.98

Prices for Measured Works
R DISPOSAL SYSTEMS

Item Excluding site overheads and profit	PC £	Labour hours	Labour £	Plant £	Material £	Unit	Total rate £
R12 DRAINAGE BELOW GROUND – cont'd							
Gullies Vitrified Clay; Hepworth Plc; - cont'd							
Vitrified clay road gullies; trapped with rodding eye;							
100 mm outlet; 300 mm internal dia, 600 mm internal depth	54.88	3.50	63.00	-	55.44	nr	**118.44**
150 mm outlet; 300 mm internal dia, 600 mm internal depth	56.19	3.50	63.00	-	56.75	nr	**119.75**
150 mm outlet; 400 mm internal dia, 750 mm internal depth	65.17	3.50	63.00	-	65.73	nr	**128.73**
150 mm outlet; 450 mm internal dia, 900 mm internal depth	-	3.50	63.00	-	88.74	nr	**151.74**
Hinged Gratings and frames for gullies; Alloy;							
135 mm for 100 mm dia gully;	-	-	-	-	8.06	nr	**8.06**
193 mm for 150 mm dia gully;	-	-	-	-	13.94	nr	**13.94**
120 mm x 120 mm	-	-	-	-	4.95	nr	**4.95**
150 mm x 150 mm	-	-	-	-	8.95	nr	**8.95**
230 mm x 230 mm	-	-	-	-	16.38	nr	**16.38**
316 mm x 316 mm	-	-	-	-	43.45	nr	**43.45**
Hinged Gratings and frames for gullies; Cast Iron;							
265 mm for 225 mm dia gully;	-	-	-	-	27.82	nr	**27.82**
150 mm x 150 mm	-	-	-	-	8.95	nr	**8.95**
230 mm x 230 mm	-	-	-	-	16.38	nr	**16.38**
316 mm x 316 mm	-	-	-	-	43.45	nr	**43.45**
Universal Gully Trap PVC-u; Wavin Plastics Ltd; OsmaDrain system; bedding in concrete; 11.50 N/mm2 - 40 mm aggregate							
Universal gully fitting; comprising gully trap only							
110 mm outlet; 110 mm dia; 205 mm internal depth	8.22	3.50	63.00	-	8.78	nr	**71.78**
Vertical inlet hopper c\w plastic grate							
272 x 183 mm	11.05	0.25	4.50	-	11.05	nr	**15.55**
Sealed access hopper							
110 x 110 mm	18.41	0.25	4.50	-	18.41	nr	**22.91**
Universal Gully PVC-u; Wavin Plastics Ltd; OsmaDrain system; Accessories to Universal gully trap							
Hoppers; backfilling with clean granular material; tamping; surrounding in lean mix concrete							
plain hopper with 110 spigot 150 mm long	9.04	0.40	7.20	-	9.34	nr	**16.54**
vertical inlet hopper with 110 spigot 150 mm long	11.05	0.40	7.20	-	11.05	nr	**18.25**
sealed access hopper with 110 spigot 150 mm long	18.41	0.40	7.20	-	18.41	nr	**25.61**
plain hopper; solvent weld to trap	6.17	0.40	7.20	-	6.17	nr	**13.37**
vertical inlet hopper; solvent wed to trap	8.13	0.40	7.20	-	8.13	nr	**15.33**
sealed access cover; PVC-U	9.38	0.10	1.80	-	9.38	nr	**11.18**

Prices for Measured Works
R DISPOSAL SYSTEMS

Item Excluding site overheads and profit	PC £	Labour hours	Labour £	Plant £	Material £	Unit	Total rate £
Gullies PVC-u; Wavin Plastics Ltd; OsmaDrain system; bedding in concrete; 11.50 N/mm2 - 40 mm aggregate							
Bottle Gully; providing access to the drainage system for cleaning							
bottle gully; 228 x 228 x 317 mm deep	18.73	0.50	9.00	-	19.16	nr	**28.16**
sealed access cover; PVC-u 217 x 217 mm	12.08	0.10	1.80	-	12.08	nr	**13.88**
grating; ductile iron 215 x 215 mm	9.32	0.10	1.80	-	9.32	nr	**11.12**
bottle gully riser; 325 mm	2.17	0.50	9.00	-	3.11	nr	**12.11**
Yard Gully; trapped 300 mm diameter 600 mm deep; including catchment bucket and ductile iron cover and frame, medium duty loading							
305 mm diameter 600 mm deep	120.94	2.50	45.00	-	123.65	nr	**168.65**
Kerbs to Gullies							
One course Class B engineering bricks to 4 nr sides; rendering in cement:mortar (1:3); dished to gully gratings							
150 mm x 150 mm	1.04	0.33	6.00	-	1.60	nr	**7.60**
Access Covers and Frames; Jones of Oswestry; bedding frame in cement mortar (1:3); cover in grease and sand; clear opening sizes							
Access Covers and frames; "Suprabloc"; to paved areas; filling with blocks cut and fitted to match surrounding paving; BS standard size manholes available							
pedestrian weight; 300 mm x 300 mm	110.43	2.50	45.00	-	116.20	nr	**161.20**
pedestrian weight; 450 mm x 450 mm	127.62	2.50	45.00	-	127.62	nr	**172.62**
pedestrian weight; 450 mm x 600 mm	137.19	2.50	45.00	-	143.20	nr	**188.20**
light vehicular weight; 300 mm x 300 mm	110.43	2.40	43.20	-	116.20	nr	**159.40**
light vehicular weight; 450 mm x 450 mm	110.43	3.00	54.00	-	117.04	nr	**171.04**
light vehicular weight; 450 mm x 600 mm	127.62	4.00	72.00	-	133.39	nr	**205.39**
heavy vehicular weight; 300 mm x 300 mm	128.08	3.00	54.00	-	134.09	nr	**188.09**
heavy vehicular weight; 450 mm x 600 mm	159.13	3.00	54.00	-	165.74	nr	**219.74**
heavy vehicular weight; 600 mm x 600 mm	178.70	4.00	72.00	-	188.31	nr	**260.31**
Extra over "Suprabloc" manhole frames and covers for							
filling recessed manhole covers with brick paviors; PC £305.00 /1000	12.20	1.00	18.00	-	13.92	m²	**31.92**
filling recessed manhole covers with vehicular paving blocks; PC £7.17 /m²	7.17	0.75	13.50	-	7.67	m²	**21.17**
filling recessed manhole covers with concrete paving flags; PC £9.36/m²	9.37	0.35	6.30	-	9.67	m²	**15.97**
Access covers and frames to BS 497; Fraser Ltd.; bedding frame in cement mortar (1:3); cover in grease and sand; light duty; clear opening sizes							
Grade A; double triangular; coated							
600 mm diameter	119.14	3.00	54.00	-	121.46	nr	**175.46**
675 mm diameter	203.83	3.00	54.00	-	206.15	nr	**260.15**
Grade A; coated; single seal solid top							
450 mm x 450 mm	65.18	1.50	27.00	-	66.98	nr	**93.98**
450 mm x 600 mm	60.18	1.80	32.40	-	62.68	nr	**95.08**
600 mm x 600 mm	127.90	2.00	36.00	-	131.10	nr	**167.10**

Prices for Measured Works
R DISPOSAL SYSTEMS

Item Excluding site overheads and profit	PC £	Labour hours	Labour £	Plant £	Material £	Unit	Total rate £
R12 DRAINAGE BELOW GROUND – cont'd							
Access covers and frames to BS 497; **Fraser Ltd.; - cont'd**							
Grade A; coated; double seal solid top							
450 mm x 450 mm	111.99	1.50	27.00	-	113.79	nr	**140.79**
600 mm x 450 mm	120.60	1.80	32.40	-	123.10	nr	**155.50**
600 mm x 600 mm	170.00	2.00	36.00	-	173.20	nr	**209.20**
Grade A; coated; single seal recessed							
450 mm x 450 mm	71.78	1.50	27.00	-	73.58	nr	**100.58**
600 mm x 600 mm	127.11	2.00	36.00	-	130.31	nr	**166.31**
Grade A; coated; double seal recessed							
450 mm x 600 mm	142.07	1.80	32.40	-	144.57	nr	**176.97**
600 mm x 600 mm	204.94	2.00	36.00	-	208.14	nr	**244.14**
Access covers and frames; to BS 497; **Fraser Ltd.; bedding frame in cement** **mortar (1:3); cover in grease and sand;** **medium duty; clear opening sizes**							
Grade B; coated							
"Standard" 600 mm diameter single seal	91.11	3.00	54.00	-	93.43	nr	**147.43**
"Standard" 600 x 450 mm single seal	96.26	3.00	54.00	-	98.58	nr	**152.58**
"Bricast" 600 x 600 mm diameter; single seal	85.15	3.00	54.00	-	87.47	nr	**141.47**
Grade B; recessed							
"Trucast" 600 x 450 mm single seal recessed	181.10	3.00	54.00	-	183.42	nr	**237.42**
R13 LAND DRAINAGE							
Ditching; clear and bottom ditch not **exceeding 1.50 m deep; trimming back** **vegetation; disposing to spoil heaps; by** **machine**							
Up to 1.50 m wide at top	-	18.00	324.00	60.55	-	100 m	**384.55**
1.50 - 2.50 m wide at top	-	26.00	468.00	90.83	-	100 m	**558.83**
2.50 - 4.00 m wide at top	-	42.00	756.00	141.29	-	100 m	**897.29**
Ditching; clear and bottom ditch not **exceeding 1.50 m deep; trimming back** **vegetation; disposing to spoil heaps; by** **hand**							
Up to 1.50 m wide at top	-	15.00	270.00	6.75	-	100 m	**276.75**
1.50 - 2.50 m wide at top	-	27.00	486.00	12.15	-	100 m	**498.15**
2.50 - 4.00 m wide at top	-	42.00	756.00	18.90	-	100 m	**774.90**
Ditching; excavating and forming ditch **and bank to given profile (normally 45** **degrees); in loam or sandy loam; by** **machine**							
Width 300 mm							
depth 600 mm	-	3.70	66.60	37.34	-	100 m	**103.94**
depth 900 mm	-	5.20	93.60	52.48	-	100 m	**146.08**
depth 1200 mm	-	7.20	129.60	72.66	-	100 m	**202.26**
depth 1500 mm	-	9.40	169.20	83.01	-	100 m	**252.21**

Prices for Measured Works
R DISPOSAL SYSTEMS

Item Excluding site overheads and profit	PC £	Labour hours	Labour £	Plant £	Material £	Unit	Total rate £
Width 600 mm							
depth 600 mm	-	-	-	154.20	-	100 m	154.20
depth 900 mm	-	-	-	228.81	-	100 m	228.81
depth 1200 mm	-	-	-	310.89	-	100 m	310.89
depth 1500 mm	-	-	-	447.68	-	100 m	447.68
Width 900 mm							
depth 600 mm	-	-	-	295.05	-	100 m	295.05
depth 900 mm	-	-	-	519.85	-	100 m	519.85
depth 1200 mm	-	-	-	698.57	-	100 m	698.57
depth 1500 mm	-	-	-	865.48	-	100 m	865.48
Width 1200 mm							
depth 600 mm	-	-	-	463.65	-	100 m	463.65
depth 900 mm	-	-	-	688.45	-	100 m	688.45
depth 1200 mm	-	-	-	911.85	-	100 m	911.85
depth 1500 mm	-	-	-	1168.40	-	100 m	1168.40
Width 1500 mm							
depth 600 mm	-	-	-	571.84	-	100 m	571.84
depth 900 mm	-	-	-	871.66	-	100 m	871.66
depth 1200 mm	-	-	-	1159.13	-	100 m	1159.13
depth 1500 mm	-	-	-	1452.77	-	100 m	1452.77
Extra for ditching in clay	-	-	-	-	-	20%	-
Ditching; excavating and forming ditch and bank to given profile (normal 45 degrees); in loam or sandy loam; by hand							
Width 300 mm							
depth 600 mm	-	36.00	648.00	-	-	100 m	648.00
depth 900 mm	-	42.00	756.00	-	-	100 m	756.00
depth 1200 mm	-	56.00	1008.00	-	-	100 m	1008.00
depth 1500 mm	-	70.00	1260.00	-	-	100 m	1260.00
Width 600 mm							
depth 600 mm	-	56.00	1008.00	-	-	100 m	1008.00
depth 900 mm	-	84.00	1512.00	-	-	100 m	1512.00
depth 1200 mm	-	112.00	2016.00	-	-	100 m	2016.00
depth 1500 mm	-	140.00	2520.00	-	-	100 m	2520.00
Width 900 mm							
depth 600 mm	-	84.00	1512.00	-	-	100 m	1512.00
depth 900 mm	-	126.00	2268.00	-	-	100 m	2268.00
depth 1200 mm	-	168.00	3024.00	-	-	100 m	3024.00
depth 1500 mm	-	210.00	3780.00	-	-	100 m	3780.00
Width 1200 mm							
depth 600 mm	-	112.00	2016.00	-	-	100 m	2016.00
depth 900 mm	-	168.00	3024.00	-	-	100 m	3024.00
depth 1200 mm	-	224.00	4032.00	-	-	100 m	4032.00
depth 1500 mm	-	280.00	5040.00	-	-	100 m	5040.00
Extra for ditching in clay							50%
Extra for ditching in compacted soil							90%

R DISPOSAL SYSTEMS

Item Excluding site overheads and profit	PC £	Labour hours	Labour £	Plant £	Material £	Unit	Total rate £
R13 LAND DRAINAGE – cont'd							
Piped ditching							
Jointed concrete pipes; to BS 5911 pt.3; including bedding, haunching and topping with 150 mm concrete; 11.50 N/mm2 - 40 mm aggregate; to existing ditch							
300 mm diameter	12.17	0.67	12.00	6.18	27.28	m	45.46
450 mm diameter	19.07	0.67	12.00	6.18	41.74	m	59.92
600 mm diameter	28.59	0.67	12.00	6.18	60.78	m	78.96
900 mm diameter	74.13	1.00	18.00	9.27	92.50	m	119.77
Jointed concrete pipes; to BS 5911 pt.1 class S; including bedding, haunching and topping with 150 mm concrete; 11.50 N/mm2 - 40 mm aggregate; to existing ditch							
1200 mm diameter	117.97	1.00	18.00	9.27	142.00	m	169.28
Extra over jointed concrete pipes for bends to 45 deg	97.36	0.67	12.00	7.73	104.69	nr	124.41
Extra over jointed concrete pipes for single junctions 300 mm dia	81.05	0.67	12.00	7.73	88.38	nr	108.10
Extra over jointed concrete pipes for single junctions 450 mm dia	90.14	0.67	12.00	7.03	97.47	nr	116.49
Extra over jointed concrete pipes for single junctions 600 mm dia	112.68	0.67	12.00	7.03	120.01	nr	139.03
Extra over jointed concrete pipes for single junctions 900 mm dia	272.40	0.67	12.00	7.03	279.73	nr	298.75
Extra over jointed concrete pipes for single junctions 1200 mm dia	403.95	0.67	12.00	7.03	411.28	nr	430.30
Concrete road gullies; to BS 5911; trapped; cement:mortar (1:2) joints to concrete pipes; bedding and surrounding in concrete; 11.50 N/mm2 - 40 mm aggregate; 450 mm diameter x 1.07 m deep; rodding eye; stoppers	31.50	6.00	108.00	-	47.94	nr	155.94
Mole drainage; White Horse Contractors Ltd							
Drain by mole plough; 50 mm diameter mole set at depth of 450 mm in parallel runs							
1.20 m centres	-	-	-	-	-	100 m²	11.19
1.20 m centres	-	-	-	-	-	ha	557.36
1.50 m centres	-	-	-	-	-	100 m²	10.28
1.50 m centres	-	-	-	-	-	ha	504.00
2.00 m centres	-	-	-	-	-	100 m²	9.35
2.00 m centres	-	-	-	-	-	ha	462.93
2.50 m centres	-	-	-	-	-	100 m²	8.31
2.50 m centres	-	-	-	-	-	ha	412.26
3.00 m centres	-	-	-	-	-	100 m²	7.27
3.00 m centres	-	-	-	-	-	ha	370.81
Drain by mole plough; 75 mm diameter mole set at depth of 450 mm in parallel runs							
1.20 m centres	-	-	-	-	-	100 m²	13.36
1.20 m centres	-	-	-	-	-	ha	667.92
1.50 m centres	-	-	-	-	-	100 m²	12.17
1.50 m centres	-	-	-	-	-	ha	601.07
2.00 m centres	-	-	-	-	-	100 m²	11.76
2.00 m centres	-	-	-	-	-	ha	569.10
2.50 m centres	-	-	-	-	-	100 m²	9.83
2.50 m centres	-	-	-	-	-	ha	492.87
3.00 m centres	-	-	-	-	-	100 m²	10.28
3.00 m centres	-	-	-	-	-	ha	428.39

R DISPOSAL SYSTEMS

Item Excluding site overheads and profit	PC £	Labour hours	Labour £	Plant £	Material £	Unit	Total rate £
Sand slitting; White Horse Contractors Ltd							
Drainage slits; at 2.00 m centres; using slitting machine; backfilling to 75 mm of surface with pea gravel; blind with sharp sand							
250 mm depth	-	-	-	-	-	100 m²	25.29
300 mm depth	-	-	-	-	-	100 m²	28.42
400 mm depth	-	-	-	-	-	100 m²	34.16
450 mm depth	-	-	-	-	-	100 m²	36.70
250 mm depth	-	-	-	-	-	ha	6500.00
300 mm depth	-	-	-	-	-	ha	6750.00
400 mm depth	-	-	-	-	-	ha	7000.00
Trenchless drainage system; White Horse Contractors Ltd; Insert perforated pipes by means of laser graded deep plough machine; backfill with gravel (not included)							
laterals 60 mm							
depth 700 mm	-	-	-	-	-	m	1.72
main 100 mm							
depth 900 mm	-	-	-	-	-	m	2.81
main 160 mm							
depth 1000 mm	-	-	-	-	-	m	3.60
Gravel backfill							
laterals 60 mm	-	-	-	-	-	m	3.82
main 110 mm	-	-	-	-	-	m	5.70
main 160 mm	-	-	-	-	-	m	7.04
Agricultural drainage; calculation table							
For calculation of drainage per hectare, the following table can be used: rates show the lengths of drains per unit and not the value.							
Lateral drains							
4.00 m centres	-	-	-	-	-	m/ha	2500.00
6.00 m centres	-	-	-	-	-	m/ha	1670.00
8.00 m centres	-	-	-	-	-	m/ha	1250.00
10.00 m centres	-	-	-	-	-	m/ha	1000.00
15.00 m centres	-	-	-	-	-	m/ha	650.00
25.00 m centres	-	-	-	-	-	m/ha	400.00
30.00 m centres	-	-	-	-	-	m/ha	330.00
Main drains (per hectare)							
1 nr (at 100 m centres)	-	-	-	-	-	m/ha	100.00
2 nr (at 50 m centres)	-	-	-	-	-	m/ha	200.00
3 nr (at 33.333 m centres)	-	-	-	-	-	m/ha	300.00
4 nr (at 25 m centres)	-	-	-	-	-	m/ha	400.00
Agricultural drainage; excavating							
Removing 150 mm depth of topsoil; 300 mm wide; depositing beside trench; by machine	-	1.50	27.00	15.14	-	100 m	42.14
Removing 150 mm depth of topsoil; 300 mm wide; depositing beside trench; by hand	-	8.00	144.00	-	-	100 m	144.00
Disposing on site; to spoil heaps; by machine							
not exceeding 100 m distance	-	0.07	1.19	2.27	-	m³	3.46
average 100 - 150 m distance	-	0.08	1.43	2.73	-	m³	4.15
average 150 - 200 m distance	-	0.09	1.67	3.19	-	m³	4.86
Removing excavated material from site to tip not exceeding 13 km; mechanically loaded							
excavated material and clean hardcore rubble	-	-	-	1.03	7.50	m³	8.53

R DISPOSAL SYSTEMS

Item Excluding site overheads and profit	PC £	Labour hours	Labour £	Plant £	Material £	Unit	Total rate £
R13 LAND DRAINAGE – cont'd							
Agricultural drainage; White Horse Contractors Ltd excavating trenches (with minimum run of 500 m) by trenching machine; including disposing subsoil to spoil heaps not exceeding 100 m							
Width 150 mm							
depth 450 mm	-	5.04	90.72	86.98	-	100 m	**177.70**
depth 600 mm	-	5.79	104.16	101.80	-	100 m	**205.96**
depth 750 mm	-	6.53	117.53	116.49	-	100 m	**234.02**
Width 225 mm							
depth 450 mm	-	4.68	84.15	81.57	-	100 m	**165.72**
depth 600 mm	-	6.60	118.80	121.41	-	100 m	**240.21**
depth 750 mm	-	7.43	133.65	139.24	-	100 m	**272.89**
Width 300 mm							
depth 600 mm	-	7.33	131.94	141.51	-	100 m	**273.45**
depth 750 mm	-	8.17	147.06	161.98	-	100 m	**309.04**
depth 900 mm	-	9.00	162.00	182.08	-	100 m	**344.08**
depth 1000 mm	-	9.72	174.96	198.00	-	100 m	**372.96**
Width 375 mm							
depth 600 mm	-	8.67	156.06	169.58	-	100 m	**325.64**
depth 750 mm	-	9.58	172.44	192.70	-	100 m	**365.14**
depth 900 mm	-	10.50	189.00	216.21	-	100 m	**405.21**
depth 1000 mm	-	11.78	212.04	242.00	-	100 m	**454.04**
Agricultural drainage; excavating for drains; by backacter excavator JCB C3X Sitemaster ; including disposing spoil to spoil heaps not exceeding 100 m							
Width 150 mm							
depth 450 mm	-	0.50	9.00	80.76	-	100 m	**89.76**
depth 600 mm	-	0.67	12.00	102.53	-	100 m	**114.53**
depth 700 mm	-	0.78	14.04	122.28	-	100 m	**136.32**
depth 900 mm	-	1.00	18.00	130.61	-	100 m	**148.61**
Width 225 mm							
depth 450 mm	-	0.75	13.50	90.23	-	100 m	**103.73**
depth 600 mm	-	1.00	18.00	115.16	-	100 m	**133.16**
depth 700 mm	-	1.17	21.06	137.05	-	100 m	**158.11**
depth 900 mm	-	1.50	27.00	165.01	-	100 m	**192.01**
depth 1000 mm	-	1.67	30.06	217.81	-	100 m	**247.87**
Width 300 mm							
depth 450 mm	-	1.00	18.00	100.63	-	100 m	**118.63**
depth 700 mm	-	1.56	28.08	156.46	-	100 m	**184.54**
depth 900 mm	-	2.00	36.00	200.95	-	100 m	**236.95**
depth 1000 mm	-	2.22	39.96	223.19	-	100 m	**263.15**
depth 1200 mm	-	2.67	48.06	268.06	-	100 m	**316.12**
depth 1500 mm	-	3.33	59.94	334.79	-	100 m	**394.73**
Width 375 mm							
depth 450 mm	-	1.25	22.50	125.55	-	100 m	**148.05**
depth 700 mm	-	1.94	34.92	195.28	-	100 m	**230.20**
depth 900 mm	-	2.50	45.00	251.11	-	100 m	**296.11**
depth 1000 mm	-	2.78	50.04	279.03	-	100 m	**329.07**
depth 1200 mm	-	3.33	59.94	334.79	-	100 m	**394.73**
depth 1500 mm	-	4.17	75.06	418.54	-	100 m	**493.60**
Width 450 mm							
depth 450 mm	-	1.50	27.00	164.08	-	100 m	**191.08**
depth 700 mm	-	2.33	41.94	255.18	-	100 m	**297.12**

R DISPOSAL SYSTEMS

Item Excluding site overheads and profit	PC £	Labour hours	Labour £	Plant £	Material £	Unit	Total rate £
depth 900 mm	-	3.00	54.00	328.16	-	100 m	382.16
depth 1000 mm	-	3.33	59.94	364.77	-	100 m	424.71
depth 1200 mm	-	4.00	72.00	437.75	-	100 m	509.75
depth 1500 mm	-	5.00	90.00	547.04	-	100 m	637.04
depth 2000 mm	-	6.67	120.06	729.61	-	100 m	849.67
depth 2500 mm	-	8.33	149.94	911.81	-	100 m	1061.75
Width 600 mm							
depth 450 mm	-	6.63	119.34	251.16	-	100 m	370.50
depth 600 mm	-	8.84	159.12	334.88	-	100 m	494.00
depth 700 mm	-	10.31	185.58	390.56	-	100 m	576.14
depth 900 mm	-	11.05	198.90	418.60	-	100 m	617.50
depth 1000 mm	-	14.73	265.14	558.00	-	100 m	823.14
depth 1200 mm	-	17.68	318.24	669.75	-	100 m	987.99
depth 1500 mm	-	22.10	397.80	837.19	-	100 m	1234.99
depth 2000 mm	-	29.46	530.28	1116.00	-	100 m	1646.28
depth 2500 mm	-	36.83	662.94	1395.19	-	100 m	2058.13
depth 3000 mm	-	36.83	662.94	1395.19	-	100 m	2058.13
Width 900 mm							
depth 450 mm	-	6.63	119.34	251.16	-	100 m	370.50
depth 600 mm	-	8.84	159.12	334.88	-	100 m	494.00
depth 700 mm	-	10.31	185.58	390.56	-	100 m	576.14
depth 900 mm	-	11.05	198.90	418.60	-	100 m	617.50
depth 1000 mm	-	14.73	265.14	558.00	-	100 m	823.14
depth 1200 mm	-	17.68	318.24	669.75	-	100 m	987.99
depth 1500 mm	-	22.10	397.80	837.19	-	100 m	1234.99
depth 2000 mm	-	29.46	530.28	1116.00	-	100 m	1646.28
depth 2500 mm	-	36.83	662.94	1395.19	-	100 m	2058.13
depth 3000 mm	-	36.83	662.94	1395.19	-	100 m	2058.13
Agricultural drainage; excavating for drains; by 7 tonne tracked excavator; including disposing spoil to spoil heaps not exceeding 100 m							
Width 600 mm							
depth 450 mm	-	4.49	80.82	142.98	-	100 m	223.80
depth 600 mm	-	5.99	107.82	190.74	-	100 m	298.56
depth 700 mm	-	6.99	125.82	222.58	-	100 m	348.40
depth 900 mm	-	8.98	161.64	285.95	-	100 m	447.59
depth 1000 mm	-	9.98	179.64	317.79	-	100 m	497.43
depth 1200 mm	-	11.98	215.64	381.48	-	100 m	597.12
depth 1500 mm	-	14.97	269.46	476.69	-	100 m	746.15
depth 2000 mm	-	19.97	359.46	635.90	-	100 m	995.36
depth 2500 mm	-	24.96	449.28	945.53	-	100 m	1394.81
depth 3000 mm	-	29.95	539.10	953.70	-	100 m	1492.80
Width 700 mm							
depth 450 mm	-	5.24	94.32	166.86	-	100 m	261.18
depth 600 mm	-	6.99	125.82	222.58	-	100 m	348.40
depth 700 mm	-	8.15	146.70	259.52	-	100 m	406.22
depth 900 mm	-	10.48	188.64	333.71	-	100 m	522.35
depth 1000 mm	-	11.65	209.70	370.97	-	100 m	580.67
depth 1200 mm	-	13.98	251.64	445.17	-	100 m	696.81
depth 1500 mm	-	17.47	314.46	556.30	-	100 m	870.76
depth 2000 mm	-	23.29	419.22	741.62	-	100 m	1160.84
depth 2500 mm	-	29.11	524.02	1103.12	-	100 m	1627.14
depth 3000 mm	-	34.94	628.92	1112.59	-	100 m	1741.51

R DISPOSAL SYSTEMS

Item Excluding site overheads and profit	PC £	Labour hours	Labour £	Plant £	Material £	Unit	Total rate £
R13 LAND DRAINAGE – cont'd							
Agricultural drainage; excavating for drains; by 7 tonne tracked excavator; - cont'd							
Width 900 mm							
depth 600 mm	-	9.89	178.02	314.93	-	100 m	**492.95**
depth 700 mm	-	11.54	207.72	367.47	-	100 m	**575.19**
depth 900 mm	-	14.84	267.12	472.80	-	100 m	**739.92**
depth 1000 mm	-	16.48	296.64	524.77	-	100 m	**821.41**
depth 1200 mm	-	19.78	356.04	629.85	-	100 m	**985.89**
depth 1500 mm	-	24.73	445.14	787.48	-	100 m	**1232.62**
depth 2000 mm	-	32.97	593.46	1049.86	-	100 m	**1643.32**
depth 2500 mm	-	41.21	741.78	1561.12	-	100 m	**2302.90**
depth 3000 mm	-	49.45	890.10	1574.64	-	100 m	**2464.74**
Width 1000 mm							
depth 600 mm	-	11.64	209.52	370.65	-	100 m	**580.17**
depth 700 mm	-	13.59	244.62	432.75	-	100 m	**677.37**
depth 900 mm	-	17.47	314.46	556.30	-	100 m	**870.76**
depth 1000 mm	-	19.41	349.38	618.07	-	100 m	**967.45**
depth 1200 mm	-	23.29	419.22	741.62	-	100 m	**1160.84**
depth 1500 mm	-	29.11	523.98	926.95	-	100 m	**1450.93**
depth 2000 mm	-	38.81	698.58	1235.83	-	100 m	**1934.41**
depth 2500 mm	-	48.52	873.36	1838.03	-	100 m	**2711.39**
depth 3000 mm	-	58.22	1047.96	1853.90	-	100 m	**2901.86**
Width 1500 mm							
depth 600 mm	-	17.47	314.46	556.30	-	100 m	**870.76**
depth 700 mm	-	20.38	366.84	648.96	-	100 m	**1015.80**
depth 900 mm	-	26.20	471.60	834.29	-	100 m	**1305.89**
depth 1000 mm	-	29.11	523.98	926.95	-	100 m	**1450.93**
depth 1200 mm	-	34.93	628.74	1112.28	-	100 m	**1741.02**
depth 1500 mm	-	43.67	786.06	1390.58	-	100 m	**2176.64**
depth 2000 mm	-	58.22	1047.96	1853.90	-	100 m	**2901.86**
depth 2500 mm	-	72.78	1310.04	2757.05	-	100 m	**4067.09**
depth 3000 mm	-	87.33	1571.94	2780.85	-	100 m	**4352.79**
Agricultural drainage; excavating for drains; by hand, including disposing spoil to spoil heaps not exceeding 100 m							
Width 150 mm							
depth 450 mm	-	22.03	396.54	-	-	100 m	**396.54**
depth 600 mm	-	29.38	528.84	-	-	100 m	**528.84**
depth 700 mm	-	34.27	616.86	-	-	100 m	**616.86**
depth 900 mm	-	44.06	793.08	-	-	100 m	**793.08**
Width 225 mm							
depth 450 mm	-	33.05	594.90	-	-	100 m	**594.90**
depth 600 mm	-	44.06	793.08	-	-	100 m	**793.08**
depth 700 mm	-	51.41	925.38	-	-	100 m	**925.38**
depth 900 mm	-	66.10	1189.80	-	-	100 m	**1189.80**
depth 1000 mm	-	73.44	1321.92	-	-	100 m	**1321.92**
Width 300 mm							
depth 450 mm	-	44.06	793.08	-	-	100 m	**793.08**
depth 600 mm	-	58.75	1057.50	-	-	100 m	**1057.50**
depth 700 mm	-	68.54	1233.72	-	-	100 m	**1233.72**
depth 900 mm	-	88.13	1586.34	-	-	100 m	**1586.34**
depth 1000 mm	-	97.92	1762.56	-	-	100 m	**1762.56**
Width 375 mm							
depth 450 mm	-	55.08	991.44	-	-	100 m	**991.44**

Prices for Measured Works
R DISPOSAL SYSTEMS

Item Excluding site overheads and profit	PC £	Labour hours	Labour £	Plant £	Material £	Unit	Total rate £
depth 600 mm	-	73.44	1321.92	-	-	100 m	1321.92
depth 700 mm	-	85.68	1542.24	-	-	100 m	1542.24
depth 900 mm	-	110.16	1982.88	-	-	100 m	1982.88
depth 1000 mm	-	122.40	2203.20	-	-	100 m	2203.20
Width 450 mm							
depth 450 mm	-	66.10	1189.80	-	-	100 m	1189.80
depth 600 mm	-	88.13	1586.34	-	-	100 m	1586.34
depth 700 mm	-	102.82	1850.76	-	-	100 m	1850.76
depth 900 mm	-	132.19	2379.42	-	-	100 m	2379.42
depth 1000 mm	-	146.88	2643.84	-	-	100 m	2643.84
Width 600 mm							
depth 450 mm	-	88.13	1586.34	-	-	100 m	1586.34
depth 600 mm	-	117.50	2115.00	-	-	100 m	2115.00
depth 700 mm	-	137.09	2467.62	-	-	100 m	2467.62
depth 900 mm	-	176.26	3172.68	-	-	100 m	3172.68
depth 1000 mm	-	195.84	3525.12	-	-	100 m	3525.12
Width 900 mm							
depth 450 mm	-	132.19	2379.42	-	-	100 m	2379.42
depth 600 mm	-	176.26	3172.68	-	-	100 m	3172.68
depth 700 mm	-	205.63	3701.34	-	-	100 m	3701.34
depth 900 mm	-	264.38	4758.84	-	-	100 m	4758.84
depth 1000 mm	-	293.76	5287.68	-	-	100 m	5287.68
Earthwork Support; moving along as work proceeds							
Maximum depth not exceeding 2.00 m							
distance between opposing faces not exceeding 2.00 m	-	0.80	14.40	16.11	-	m	30.51
Agricultural drainage; pipe laying;							
Hepworth; Agricultural clay drain pipes; to BS 1196; 300 mm length; butt joints; in straight runs							
75 mm diameter	250.42	8.00	144.00	-	256.68	100 m	400.68
100 mm diameter	431.57	9.00	162.00	-	442.36	100 m	604.36
150 mm diameter	887.11	10.00	180.00	-	909.29	100 m	1089.29
Extra over clay drain pipes for filter-wrapping pipes with "Terram" or similar filter fabric							
"Terram 700"	0.20	0.04	0.72	-	0.20	m²	0.92
"Terram 1000"	0.22	0.04	0.72	-	0.22	m²	0.94
Junctions between drains in clay pipes							
75 mm x 75 mm	9.15	0.25	4.50	-	9.57	nr	14.07
100 mm x 100 mm	11.43	0.25	4.50	-	12.11	nr	16.61
100 mm x 150 mm	14.07	0.25	4.50	-	15.09	nr	19.59
Wavin Plastics Ltd; flexible plastic perforated pipes in trenches (not included); to a minimum depth of 450 mm (couplings not included)							
"OsmaDrain"; flexible plastic perforated pipes in trenches (not included); to a minimum depth of 450 mm (couplings not included)							
60 mm diameter; available in 150 m coil	23.50	2.00	36.00	-	24.09	100 m	60.09
80 mm diameter; available in 100 m coil	35.50	2.00	36.00	-	36.39	100 m	72.39
100 mm diameter; available in 100 m coil	62.50	2.00	36.00	-	64.06	100 m	100.06
160 mm diameter; available in 35 m coil	151.00	2.00	36.00	-	154.78	100 m	190.78

Prices for Measured Works
R DISPOSAL SYSTEMS

Item Excluding site overheads and profit	PC £	Labour hours	Labour £	Plant £	Material £	Unit	Total rate £
R13 LAND DRAINAGE – cont'd							
Wavin Plastics Ltd; flexible plastic perforated pipes in trenches – cont'd							
"WavinCoil"; Plastic pipe junctions							
60 mm x 60 mm	1.36	0.05	0.90	-	1.36	nr	2.26
80 mm x 80 mm	1.48	0.05	0.90	-	1.48	nr	2.38
100 mm x 100 mm	1.66	0.05	0.90	-	1.66	nr	2.56
80 mm x 60 mm	1.43	0.05	0.90	-	1.43	nr	2.33
100 mm x 60 mm	1.52	0.05	0.90	-	1.52	nr	2.42
100 mm x 80 mm	1.58	0.05	0.90	-	1.58	nr	2.48
125 mm x 60 mm	1.68	0.05	0.90	-	1.68	nr	2.58
125 mm x 100 mm	1.74	0.05	0.90	-	1.74	nr	2.64
160 mm x 160 mm	4.28	0.05	0.90	-	4.28	nr	5.18
"WavinCoil"; couplings for flexible pipes							
60 mm diameter	0.52	0.03	0.60	-	0.52	nr	1.12
80 mm diameter	0.62	0.03	0.60	-	0.62	nr	1.21
100 mm diameter	0.68	0.03	0.60	-	0.68	nr	1.27
160 mm diameter	0.91	0.03	0.60	-	0.91	nr	1.51
Market prices of backfilling materials							
Sand	18.99	-	-	-	22.79	m³	22.79
Gravel rejects	22.68	-	-	-	24.95	m³	24.95
Topsoil; allowing for 20% settlement	11.66	-	-	-	13.99	m³	13.99
Agricultural drainage; backfilling trench after laying pipes with gravel rejects or similar; blind filling with ash or sand; topping with 150 mm topsoil from dumps not exceeding 100 m; by machine							
Width 150 mm							
depth 450 mm	-	3.30	59.40	50.15	128.46	100 m	238.02
depth 600 mm	-	4.30	77.40	65.61	182.91	100 m	325.92
depth 750 mm	-	4.96	89.28	75.81	213.51	100 m	378.60
depth 900 mm	-	6.30	113.40	96.52	281.55	100 m	491.47
Width 225 mm							
depth 450 mm	-	4.95	89.10	75.23	198.03	100 m	362.36
depth 600 mm	-	6.45	116.10	98.41	274.46	100 m	488.97
depth 750 mm	-	7.95	143.10	121.59	351.12	100 m	615.82
depth 900 mm	-	9.45	170.10	144.78	427.55	100 m	742.43
Width 375 mm							
depth 450 mm	-	8.25	148.50	125.38	329.91	100 m	603.80
depth 600 mm	-	10.75	193.50	164.02	457.37	100 m	814.89
depth 750 mm	-	13.25	238.50	202.66	585.06	100 m	1026.22
depth 900 mm	-	15.75	283.50	241.30	712.52	100 m	1237.32
Agricultural drainage; backfilling trench after laying pipes with gravel rejects or similar, blind filling with ash or sand, topping with 150 mm topsoil from dumps not exceeding 100 m; by hand							
Width 150 mm							
depth 450 mm	-	18.63	335.34	-	123.22	100 m	458.56
depth 600 mm	-	24.84	447.12	-	182.91	100 m	630.03

Prices for Measured Works
R DISPOSAL SYSTEMS

Item Excluding site overheads and profit	PC £	Labour hours	Labour £	Plant £	Material £	Unit	Total rate £
depth 750 mm	-	31.05	558.90	-	187.28	100 m	746.18
depth 900 mm	-	37.26	670.68	-	281.55	100 m	952.23
Width 225 mm							
depth 450 mm	-	27.94	502.92	-	198.03	100 m	700.95
depth 600 mm	-	37.26	670.68	-	274.46	100 m	945.14
depth 750 mm	-	46.57	838.26	-	351.12	100 m	1189.38
depth 900 mm	-	55.89	1006.02	-	427.55	100 m	1433.57
Width 375 mm							
depth 450 mm	-	46.57	838.26	-	329.91	100 m	1168.17
depth 600 mm	-	61.10	1099.80	-	457.37	100 m	1557.17
depth 750 mm	-	77.63	1397.34	-	585.06	100 m	1982.40
depth 900 mm	-	93.15	1676.70	-	712.52	100 m	2389.22
Catchwater or french drains; 100 mm diameter non-coilable perforated plastic pipes; to BS4962; including straight jointing; pipes laid with perforations uppermost; lining trench; wrapping pipes with filter fabric							
Width 300 mm							
depth 450 mm	843.62	9.10	163.80	32.46	1195.14	100 m	1391.39
depth 600 mm	847.43	9.84	177.12	43.89	1316.97	100 m	1537.99
depth 750 mm	851.42	10.60	190.80	55.64	1439.01	100 m	1685.45
depth 900 mm	855.32	11.34	204.12	67.07	1560.94	100 m	1832.14
depth 1000 mm	857.92	11.84	213.12	74.80	1642.23	100 m	1930.16
depth 1200 mm	863.12	12.84	231.12	90.26	1804.81	100 m	2126.19
Width 450 mm							
depth 450 mm	849.47	10.44	187.92	80.98	1378.04	100 m	1646.94
depth 600 mm	855.32	11.34	204.12	67.07	1560.94	100 m	1832.14
depth 750 mm	861.17	12.46	224.28	84.38	1743.84	100 m	2052.51
depth 900 mm	867.02	13.60	244.80	102.00	1926.75	100 m	2273.55
depth 1000 mm	870.92	14.34	258.12	113.44	2048.68	100 m	2420.24
depth 1200 mm	878.72	15.84	285.12	136.62	2292.55	100 m	2714.29
depth 1500 mm	890.42	18.10	325.80	171.55	2658.35	100 m	3155.70
depth 2000 mm	909.92	21.84	393.12	229.35	3268.03	100 m	3890.50
Width 600 mm							
depth 450 mm	855.32	11.24	202.32	67.07	1560.94	100 m	1830.34
depth 600 mm	863.12	12.84	231.12	90.26	1804.81	100 m	2126.19
depth 750 mm	870.92	14.34	258.12	113.44	2048.68	100 m	2420.24
depth 900 mm	878.72	15.84	285.12	136.62	2292.55	100 m	2714.29
depth 1000 mm	886.14	16.84	303.12	152.08	2446.19	100 m	2901.38
depth 1200 mm	894.32	18.84	339.12	182.99	2780.29	100 m	3302.40
depth 1500 mm	909.92	21.84	393.12	454.35	3268.03	100 m	4115.50
depth 2000 mm	935.92	19.84	357.12	414.81	4080.93	100 m	4852.86
Width 900 mm							
depth 450 mm	867.02	13.60	244.80	102.00	1926.75	100 m	2273.55
depth 600 mm	878.72	7.92	142.56	136.62	2292.55	100 m	2571.73
depth 750 mm	890.42	9.05	162.90	171.55	2658.35	100 m	2992.80
depth 900 mm	902.12	10.17	183.06	206.17	3024.16	100 m	3413.39
depth 1000 mm	909.92	10.92	196.56	229.35	3268.03	100 m	3693.94
depth 1200 mm	925.52	12.42	223.56	275.72	3755.77	100 m	4255.04
depth 1500 mm	948.92	14.67	264.06	345.26	4487.37	100 m	5096.70
depth 2000 mm	987.92	18.92	340.56	536.18	5706.72	100 m	6583.46
depth 2500 mm	1026.92	22.17	399.06	652.09	6926.07	100 m	7977.22
depth 3000 mm	1065.92	25.92	466.56	693.00	8145.41	100 m	9304.98

Prices for Measured Works
R DISPOSAL SYSTEMS

Item Excluding site overheads and profit	PC £	Labour hours	Labour £	Plant £	Material £	Unit	Total rate £
R13 LAND DRAINAGE – cont'd							
Catchwater or french drains; 160 mm diameter non-coilable perforated plastic pipes; to BS4962; including straight jointing; pipes laid with perforations uppermost; lining trench; wrapping pipes with filter fabric							
Width 600 mm							
depth 450 mm	1545.27	11.20	201.60	64.91	2248.15	100 m	**2514.66**
depth 600 mm	1553.07	12.70	228.60	88.09	2492.02	100 m	**2808.71**
depth 750 mm	1560.87	14.20	255.60	111.28	2735.89	100 m	**3102.76**
depth 900 mm	1568.67	15.70	282.60	134.46	2979.76	100 m	**3396.82**
depth 1000 mm	1573.87	16.70	300.60	149.91	3142.34	100 m	**3592.85**
depth 1200 mm	1584.27	18.70	336.60	180.91	3467.50	100 m	**3985.01**
depth 1500 mm	1599.87	21.70	390.60	227.19	3955.24	100 m	**4573.02**
depth 2000 mm	1625.87	26.70	480.60	304.46	4768.13	100 m	**5553.20**
depth 2500 mm	1651.87	31.70	570.60	381.74	5581.03	100 m	**6533.37**
depth 3000 mm	1677.87	36.70	660.60	459.01	6393.93	100 m	**7513.54**
Width 900 mm							
depth 450 mm	867.02	13.60	244.80	102.00	1926.75	100 m	**2273.55**
depth 600 mm	878.72	15.84	285.12	136.62	2292.55	100 m	**2714.29**
depth 750 mm	890.42	18.10	325.80	171.55	2658.35	100 m	**3155.70**
depth 900 mm	902.12	20.34	366.12	206.17	3024.16	100 m	**3596.45**
depth 1000 mm	909.92	21.84	393.12	229.35	3268.03	100 m	**3890.50**
depth 1200 mm	925.52	24.84	447.12	275.72	3755.77	100 m	**4478.60**
depth 1500 mm	948.92	29.34	528.12	345.26	4487.37	100 m	**5360.76**
depth 2000 mm	987.92	36.84	663.12	461.18	5706.72	100 m	**6831.02**
depth 2500 mm	1026.92	44.34	798.12	577.09	6926.07	100 m	**8301.28**
depth 3000 mm	1065.92	51.84	933.12	693.00	8145.41	100 m	**9771.54**
Catchwater or french drains; Exxon Geochemical Polymers; "Filtram" filter drain; in trenches (trenches not included); comprising filter fabric, liquid conducting core and 110 mm uPVC slitpipes; all in accordance with manufacturer's instructions; backfilling with clean broken stone 40 mm minimum; top 300 mm of fill to be gravel rejects or quarry waste							
Width 600 mm							
depth 1000 mm	883.92	16.84	303.12	152.08	2455.13	100 m	**2910.33**
depth 1200 mm	894.32	18.84	339.12	182.99	2780.29	100 m	**3302.40**
depth 1500 mm	909.92	21.84	393.12	454.35	3268.03	100 m	**4115.50**
depth 2000 mm	935.92	19.84	357.12	414.81	4080.93	100 m	**4852.86**
Width 900 mm							
depth 450 mm	867.02	13.60	244.80	102.00	1926.75	100 m	**2273.55**
depth 600 mm	878.72	15.84	285.12	136.62	2292.55	100 m	**2714.29**
depth 750 mm	890.42	18.10	325.80	171.55	2658.35	100 m	**3155.70**
depth 900 mm	902.12	20.34	366.12	206.17	3024.16	100 m	**3596.45**
depth 1000 mm	909.92	21.84	393.12	229.35	3268.03	100 m	**3890.50**
depth 1200 mm	925.52	24.84	447.12	275.72	3755.77	100 m	**4478.60**
depth 1500 mm	948.92	29.34	528.12	345.26	4487.37	100 m	**5360.76**
depth 2000 mm	987.92	37.84	681.12	536.18	5706.72	100 m	**6924.02**
depth 2500 mm	1026.92	44.34	798.12	652.09	6926.07	100 m	**8376.28**
depth 3000 mm	1065.92	51.84	933.12	768.00	8145.41	100 m	**9846.54**

Prices for Measured Works
R DISPOSAL SYSTEMS

Item Excluding site overheads and profit	PC £	Labour hours	Labour £	Plant £	Material £	Unit	Total rate £
Outfalls Reinforced concrete outfalls to water course; flank walls; for 150 mm drain outlets; overall dimensions							
900 mm x 1050 mm x 900 mm high	-	-	-	-	-	m	**424.00**
Soakaway design based on BS EN 752-4. Flat rate hourly rainfall = 50 mm /hr and assumes 100 impermeability of the run-off area. A storage capacity of the soakaway should be 1/3 of the hourly rainfall; Formulae for calculating soakaway depths are provided in the publications mentioned below and in the memoranda section of this publication. The design of soakaways is dependent on amongst other factors, soil conditions, permeability, groundwater level and runoff. The definitive documents for design of soakaways are CIRIA 156 and BRE Digest 365 dated September 1991; The suppliers of the systems below will assist through their technical divisions. Excavation earthwork support of pits and disposal not included.							
Excavating; mechanical To reduce levels							
maximum depth not exceeding 1.00 m; JCB sitemaster	-	0.05	0.90	1.55	-	m³	**2.45**
maximum depth not exceeding 1.00 m; 360 Tracked excavator	-	0.04	0.72	1.20	-	m³	**1.92**
maximum depth not exceeding 2.00 m; 360 Tracked excavator	-	0.06	1.08	1.80	-	m³	**2.88**
Disposal; mechanical Excavated material; off site; to tip not exceeding 13 km; mechanically loaded.							
inert material	-	0.03	0.60	1.03	7.50	m³	**9.13**
Insitu concrete ring beam foundations to base of soakaway; 300 mm wide x 250 deep; poured on or against earth or unblinded hardcore internal diameters of rings							
900 mm	-	4.00	72.00	-	16.83	nr	**88.83**
1200 mm	20.39	4.50	81.00	-	22.43	nr	**103.43**
1500 mm	25.49	5.00	90.00	-	28.03	nr	**118.03**
2400 mm	40.78	5.50	99.00	-	44.87	nr	**143.87**

Prices for Measured Works
R DISPOSAL SYSTEMS

Item Excluding site overheads and profit	PC £	Labour hours	Labour £	Plant £	Material £	Unit	Total rate £
R13 LAND DRAINAGE – cont'd							
Concrete soakaway rings; Milton Pipes Ltd; perforations and step irons to concrete rings at manufacturers recommended centres; placing of concrete ring soakaways to insitu concrete ring beams (1:3:6) (not included); filling and surrounding base with gravel 225 deep (not included);							
Ring diameter 900 mm							
1.00 deep; volume 636 litres	32.09	1.25	22.50	13.10	103.28	nr	138.89
1.50 deep; volume 954 litres	48.13	1.65	29.70	34.59	152.85	nr	217.14
2.00 deep; volume 1272 litres	64.18	1.65	29.70	34.59	202.41	nr	266.70
Ring diameter 1200 mm							
1.00 deep; volume 1131 litres	47.61	3.00	54.00	31.45	129.88	nr	215.33
1.50 deep; volume 1696 litres	71.42	4.50	81.00	47.17	191.14	nr	319.31
2.00 deep; volume 2261 litres	95.22	4.50	81.00	47.17	252.39	nr	380.56
2.50 deep; volume 2827 litres	119.03	6.00	108.00	62.90	286.14	nr	457.03
Ring diameter 1500 mm							
1.00 deep; volume 1767 litres	86.94	4.50	81.00	31.45	189.09	nr	301.54
1.50 deep; volume 2651 litres	130.41	6.00	108.00	44.03	277.87	nr	429.90
2.00 deep; volume 3534 litres	173.88	6.00	108.00	44.03	366.65	nr	518.68
2.50 deep; volume 4418 litres	119.03	7.50	135.00	78.62	357.11	nr	570.72
Ring diameter 2400 mm							
1.00 deep; volume 4524 litres	221.49	6.00	108.00	31.45	359.01	nr	498.46
1.50 deep; volume 6786 litres	332.24	7.50	135.00	44.03	530.79	nr	709.81
2.00 deep; volume 9048 litres	442.98	7.50	135.00	44.03	706.49	nr	885.52
2.50 deep; volume 11310 litres	553.73	9.00	162.00	78.62	878.27	nr	1118.88
Extra over for							
250 mm depth chamber ring	-	-	-	-	-	100 %	-
500 mm depth chamber ring	-	-	-	-	-	50 %	-
Cover slabs to soakaways							
heavy duty precast concrete							
900 diameter	54.85	1.00	18.00	15.72	54.85	nr	88.57
1200 diameter	70.38	1.00	18.00	15.72	70.38	nr	104.10
1500 diameter	111.78	1.00	18.00	15.72	111.78	nr	145.50
2400 diameter	396.40	1.00	18.00	15.72	396.40	nr	430.12
Step irons to concrete chamber rings.	19.88	-	-	-	19.88	m	19.88
Extra over soakaways for filter wrapping with a proprietary filter membrane							
900 mm diameter x 1.00 m deep	1.23	1.00	18.00	-	1.53	nr	19.53
900 mm diameter x 2.00 m deep	2.45	1.50	27.00	-	3.07	nr	30.07
1050 mm diameter x 1.00 m deep	1.43	1.50	27.00	-	1.79	nr	28.79
1050 mm diameter x 2.00 m deep	2.86	2.00	36.00	-	3.57	nr	39.57
1200 mm diameter x 1.00 m deep	1.63	2.00	36.00	-	2.04	nr	38.04
1200 mm diameter x 2.00 m deep	3.27	2.50	45.00	-	4.08	nr	49.08
1500 mm diameter x 1.00 m deep	2.04	2.50	45.00	-	2.55	nr	47.55
1500 mm diameter x 2.00 m deep	4.07	2.50	45.00	-	5.09	nr	50.09
1800 mm diameter x 1.00 m deep	2.45	3.00	54.00	-	3.06	nr	57.06
1800 mm diameter x 2.00 m deep	4.90	3.25	58.50	-	6.12	nr	64.62
Gravel surrounding to concrete ring soakaway							
40 mm aggregate backfilled to vertical face of soakaway wrapped with geofabric (not included)							
250 thick	26.10	0.20	3.60	6.29	26.10	m³	35.99

R DISPOSAL SYSTEMS

Item Excluding site overheads and profit	PC £	Labour hours	Labour £	Plant £	Material £	Unit	Total rate £
Backfilling to face of soakaway; **Carefully compacting as work proceeds** arising from the excavations							
average thickness exceeding 0.25 m; depositing in layers 150 mm maximum thickness	-	0.03	0.60	3.14	-	m³	3.74
"Aquacell" soakaway; Wavin Plastics Ltd; . Preformed polypropylene soakaway infiltration crate units; to trenches; surrounded by geotextile and 40 mm aggregate laid 100 thick. in trenches (Excavation, disposal and backfilling not included)							
1.00 x 500 x 400; internal volume 190 litres							
4 crates; 2.00 x 1.00 x 400; 760 litres	108.00	1.60	28.80	12.98	142.92	nr	184.70
8 crates; 2.00 x 1.00 x 800; 1520 litres	216.00	3.20	57.60	16.69	259.68	nr	333.97
12 crates; 6.00 x 500 x 800; 2280 litres	324.00	4.80	86.40	34.62	410.04	nr	531.06
16 crates; 4.00 x 1.00 x 800; 3040 litres	432.00	6.40	115.20	30.91	509.28	nr	655.39
20 crates; 5.00 x 1.00 x 800; 3800 litres	540.00	8.00	144.00	30.60	620.79	nr	795.39
30 crates; 15.00 x 1.00 x 400; 5700 litres	810.00	12.00	216.00	83.46	1015.69	nr	1315.14
60 crates; 15.00 x 1.00 x 800; 11400 litres	1620.00	20.00	360.00	109.11	1894.79	nr	2363.90
Geofabric surround to Aquacell units; **Terram Ltd** "Terram" synthetic fibre filter fabric; to face of concrete rings (not included); anchoring whilst backfilling (not included)							
"Terram 1000", 0.70 mm thick; mean water flow 50 litre/m²/s	0.43	0.05	0.90	-	0.52	m²	1.42

Prices for Measured Works
S PIPED SUPPLY SYSTEMS

Item Excluding site overheads and profit	PC £	Labour hours	Labour £	Plant £	Material £	Unit	Total rate £
S10 COLD WATER							
Blue MDPE polythene pipes; type 50; for cold water services; with compression fittings; bedding on 100 mm DOT type 1 granular fill material							
Pipes							
20 mm diameter	0.26	0.08	1.44	-	1.82	m	3.26
25 mm diameter	0.33	0.08	1.44	-	1.90	m	3.34
32 mm diameter	0.81	0.08	1.44	-	2.38	m	3.82
50 mm diameter	1.40	0.10	1.80	-	2.99	m	4.79
60 mm diameter	3.03	0.10	1.80	-	4.66	m	6.46
Hose union bib taps; to BS 5412; including fixing to wall; making good surfaces							
15 mm	10.56	0.75	13.50	-	12.34	nr	25.84
22 mm	14.90	0.75	13.50	-	17.57	nr	31.07
Stopcocks; to BS 1010; including fixing to wall; making good surfaces							
15 mm	6.26	0.75	13.50	-	8.04	nr	21.54
22 mm	9.76	0.75	13.50	-	11.54	nr	25.04
Standpipes; to existing 25mm water mains							
1.00 m high	-	-	-	-	-	nr	120.00
Hose junction bib taps; to standpipes							
19 mm	-	-	-	-	-	nr	65.00
S14 IRRIGATION							
Leaky Pipe Systems Ltd; " Leaky Pipe"; moisture leaking pipe irrigation system.							
Main supply pipe inclusive of machine excavation; exclusive of connectors							
20 mm LDPE Polytubing	0.84	0.05	0.90	0.60	0.86	m	2.36
16 mm LDPE Polytubing	0.56	0.05	0.90	0.60	0.57	m	2.07
Water filters and cartridges							
No 10; 20 mm	-	-	-	-	41.90	nr	41.90
Big Blue and RR30 cartridge; 25 mm	-	-	-	-	107.20	nr	107.20
Water filters and pressure regulator sets. Complete assemblies							
No 10; Flow rate 3.1 - 82 litres per minute	-	-	-	-	71.50	nr	71.50
Leaky pipe hose; placed 150 mm sub surface for turf irrigation							
Distance between laterals 350 mm; excavation and backfilling priced separately							
LP12L low leak	3.65	0.04	0.72	-	3.65	m²	4.37
LP12H high leak	3.16	0.04	0.72	-	3.16	m²	3.88
LP12UH ultra high leak	3.85	0.04	0.72	-	3.85	m²	4.57
Leaky pipe hose; laid to surface for landscape irrigation. Distance between laterals 600 mm							
LP12L low leak	2.12	0.03	0.45	-	2.18	m²	2.63
LP12H high leak	1.84	0.03	0.45	-	1.89	m²	2.34
LP12UH ultra high leak	2.24	0.03	0.45	-	2.30	m²	2.75
Leaky pipe hose; laid to surface for landscape irrigation. Distance between laterals 900 mm							
LP12L low leak	1.42	0.02	0.30	-	1.46	m²	1.76
LP12H high leak	1.23	0.02	0.30	-	1.26	m²	1.56
LP12UH ultra high leak	1.50	0.02	0.30	-	1.54	m²	1.84

S PIPED SUPPLY SYSTEMS

Item Excluding site overheads and profit	PC £	Labour hours	Labour £	Plant £	Material £	Unit	Total rate £
Leaky pipe hose; laid to surface for tree irrigation laid around circumference of tree pit.							
LP12L low leak	1.96	0.13	2.25	-	2.15	nr	4.40
LP12H high leak	1.70	0.13	2.25	-	1.87	nr	4.12
LP12UH ultra high leak	2.07	0.13	2.25	-	2.27	nr	4.52
Accessories							
Automatic multi-station controller stations inclusive of connections	289.00	2.00	36.00	-	289.00	nr	325.00
Solenoid valves inclusive of wiring and connections to a multi-station controller; nominal distance from controller 25 m	48.00	0.50	9.00	-	398.00	nr	407.00

S15 FOUNTAINS/WATER FEATURES

Lakes and ponds - General
Preamble: The pressure of water against a retaining wall or dam is considerable, and where water retaining structures form part of the design of water features, the landscape architect is advised to consult a civil engineer. Artificially contained areas of water in raised reservoirs over 25,000 m³ have to be registered with the local authority, and their dams will have to be covered by a civil engineer's certificate of safety.

Typical linings - General
Preamble: In addition to the traditional methods of forming the linings of lakes and ponds in puddled clay or concrete, there are a number of lining materials available. They are mainly used for reservoirs but can also help to form comparatively economic water features especially in soil which is not naturally water retentive. Information on the construction of traditional clay puddle ponds can be obtained from the British Trust for Conservation Volunteers, 36 St. Mary's Street, Wallingford, Oxfordshire. OX10 0EU. Tel: (01491) 39766. The cost of puddled clay ponds depends on the availability of suitable clay, the type of hand or machine labour that can be used, and the use to which the pond is to be put.

Lake liners; Fairwater Water Gardens Ltd; to evenly graded surface of excavations (excavating not included); all stones over 75 mm; removing debris; including all welding and jointing of liner sheets

Item	PC £	Labour hours	Labour £	Plant £	Material £	Unit	Total rate £
Geotextile underlay; inclusive of spot welding to prevent dragging							
to water features	-	-	-	-	-	m²	1.90
to lakes or large features	-	-	-	-	-	1000m2	1900.00
Butyl rubber liners; "Varnamo" inclusive of site vulcanising							
0.75 mm thick	-	-	-	-	-	m²	5.90
0.75 mm thick	-	-	-	-	-	1000m2	5900.00
1.00 mm thick	-	-	-	-	-	m²	6.55
1.00 mm thick	-	-	-	-	-	1000m2	6550.00

Prices for Measured Works
S PIPED SUPPLY SYSTEMS

Item Excluding site overheads and profit	PC £	Labour hours	Labour £	Plant £	Material £	Unit	Total rate £
S15 FOUNTAINS/WATER FEATURES – cont'd							
Lake liners; Landline Ltd; "Landflex" or "Alkorplan" geomembranes; to prepared surfaces (surfaces not included); all joints fully welded; installation by Landline employees							
"Landflex HC" polyethylene geomembranes							
0.50 mm thick	-	-	-	-	-	1000m2	2940.00
0.75 mm thick	-	-	-	-	-	1000m2	3360.00
1.00 mm thick	-	-	-	-	-	1000m2	3780.00
1.50 mm thick	-	-	-	-	-	1000m2	4300.00
2.00 mm thick	-	-	-	-	-	1000m2	4780.00
"Alkorplan PVC" geomembranes							
0.80 mm thick	-	-	-	-	-	1000 m	4720.00
1.20 mm thick	-	-	-	-	-	1000 m	7090.00
Lake liners; Rawell Marketing; to prepared graded surfaces (surfaces not included); maximum slope 2 horizontal:1 vertical; in accordance with manufacturer's instructions							
"Rawmat" bentonite sheets; 200 mm lap joints; spreading 150 mm thick screened approved topsoil; "Rawmat" to be bedded in anchor trenches round perimeter (excavation and material to trenches not included)	859.80	4.00	72.00	24.87	1031.76	100 m²	1128.63
Lake liners; Monarflex							
Polyethylene lake and reservoir lining system; welding on site by Monarflex technicians (surface preparation and backfilling not included)							
"Blackline"; 500 micron	-	-	-	-	-	100 m²	412.65
"Blackline"; 750 micron	-	-	-	-	-	100 m²	552.30
"Blackline"; 1000 micron	-	-	-	-	-	100 m²	588.00
Operations over surfaces of lake liners							
Dug ballast; evenly spread over excavation already brought to grade							
150 mm thick	421.20	2.00	36.00	49.74	463.32	100 m²	549.06
200 mm thick	561.60	3.00	54.00	74.61	617.76	100 m²	746.37
300 mm thick	842.40	3.50	63.00	87.05	926.64	100 m²	1076.69
Imported topsoil; evenly spread over excavation							
100 mm thick	173.20	1.50	27.00	37.31	207.84	100 m²	272.15
150 mm thick	259.80	2.00	36.00	49.74	311.76	100 m²	397.50
200 mm thick	346.40	3.00	54.00	74.61	415.68	100 m²	544.29
Blinding existing subsoil with 50 mm sand	131.05	1.00	18.00	49.74	144.16	100 m²	211.90
Topsoil from excavation; evenly spread over excavation							
100 mm thick	-	-	-	37.31	-	100 m²	37.31
200 mm thick	-	-	-	49.74	-	100 m²	49.74
300 mm thick	-	-	-	74.61	-	100 m²	74.61
Extra over for screening topsoil using a "Powergrid screener; removing debris	-	-	-	3.43	0.38	m³	3.81

Prices for Measured Works
S PIPED SUPPLY SYSTEMS

Item Excluding site overheads and profit	PC £	Labour hours	Labour £	Plant £	Material £	Unit	Total rate £
Waterfall construction; Fairwater Water Gardens Ltd;							
Stone placed on top of butyl liner; securing with concrete and dressing to form natural rock pools and edgings							
Portland stone m³ rate	-	-	-	-	-	m³	492.10
Portland stone tonne rate	-	-	-	-	-	tonne	273.60
Balancing tank; blockwork construction; inclusive of recirculation pump and pond level control; 110 mm balancing pipe to pond; waterproofed with butyl rubber membrane; pipework mains water top-up and overflow							
450 x 600 x 1000 mm	-	-	-	-	-	nr	876.85
extra for pump							
2000 gallons per hour; submersible	-	-	-	-	-	nr	185.25
Ornamental pools - General							
Preamble: Small pools may be lined with one of the materials mentioned under lakes and ponds, or may be in rendered brickwork, puddled clay or, for the smaller sizes, fibreglass Most of these tend to be cheaper than waterproof concrete. Basic prices for various sizes of concrete pools are given in the approximate estimates section (book only). Prices for excavation, grading, mass concrete, and precast concrete retaining walls are given in the relevant sections. The manufacturers should be consulted before specifying the type and thickness of pool liner, as this depends on the size, shape and proposed use of the pool. The manufacturer's recommendation on foundations and construction should be followed.							
Ornamental pools							
Pool liners; to 50 mm sand blinding to excavation (excavating not included); all stones over 50 mm; removing debris from surfaces of excavation; including all welding and jointing of liner sheets							
black polythene; 1000 gauge	-	-	-	-	-	m²	2.07
blue polythene; 1000 gauge	-	-	-	-	-	m²	2.19
coloured PVC; 1500 gauge	-	-	-	-	-	m²	2.74
black PVC; 1500 gauge	-	-	-	-	-	m²	2.42
black butyl; 0.75 mm thick	-	-	-	-	-	m²	10.74
black butyl; 1.00 mm thick	-	-	-	-	-	m²	14.79
black butyl; 1.50 mm thick	-	-	-	-	-	m²	23.52
Fine gravel; 100 mm; evenly spread over area of pool; by hand	2.75	0.13	2.25	-	2.75	m²	5.00
Selected topsoil from excavation; 100 mm; evenly spread over area of pool; by hand	-	0.13	2.25	-	-	m²	2.25
Extra over selected topsoil for spreading imported topsoil over area of pool; by hand	17.32	-	-	-	20.78	m³	20.78

Prices for Measured Works
S PIPED SUPPLY SYSTEMS

Item Excluding site overheads and profit	PC £	Labour hours	Labour £	Plant £	Material £	Unit	Total rate £
S15 FOUNTAINS/WATER FEATURES – cont'd							
Pool surrounds and ornament; **Haddonstone Ltd; Portland Bath or** **Terracotta cast stone** Pool surrounds; Installed to pools or water feature construction priced separately. Surrounds and copings to 112.5 mm internal brickwork							
C4HS; half small pool surround; internal diameter 1780 mm; kerb features continuous moulding enriched with ovolvo and palmette designs; inclusive of plinth and integral conch shell vases flanked by dolphins	948.95	16.00	288.00	-	960.78	nr	1248.78
C4S; small pool surround as above but with full circular construction; internal diameter 1780 mm	1729.38	48.00	864.00	-	1744.51	nr	2608.51
C4M; medium pool surround; internal diameter 2705 mm; inclusive of plinth and integral vases	2139.60	48.00	864.00	-	2171.32	nr	3035.32
C4XL; extra large pool surround; internal diameter 5450 mm	4279.20	140.00	2520.00	-	4317.02	nr	6837.02
Pool centre pieces and fountains; inclusive of plumbing and pumps							
HC350; Lotus bowl; 1830 wide with C1700 triple dolphin fountain; C180 pedestal	3398.29	8.00	144.00	-	3401.02	nr	3545.02
C251; Gothic Fountain and Gothic Upper Base A350; freestanding fountain	593.20	4.00	72.00	-	595.93	nr	667.93
HC521; Romanesque Fountain; freestanding bowl with self circulating fountain; filled with cobbles; 815 mm diameter; 348 mm high	276.60	2.00	36.00	-	335.25	nr	371.25
C300; Lion fountain 610 mm high on fountain base C305; 280 mm high	245.73	2.00	36.00	-	248.46	nr	284.46
Wall Fountains, Watertanks and Fountains Capital Garden Products							
Lion Wall Fountain F010; 965 x 940 x 520	-	-	-	-	301.05	nr	301.05
Dolphin Wall Fountain F001; 735 x 510	-	-	-	-	98.95	nr	98.95
Dutch Master Wall Fountain F012; 740 x 410	-	-	-	-	94.60	nr	94.60
Tank 2801, James II design; 725 x 725 x 760 h - 308 litres	-	-	-	-	207.45	nr	207.45
James II Tank and Fountain 4901bp; 725 x 1245 x 760 h - 557l	-	-	-	-	522.10	nr	522.10
Pool Fountain F005; Crane Catching Fish; 635 high	-	-	-	-	187.55	nr	187.55
Pool Fountain F004; Mannekin Pis; 340 high	-	-	-	-	55.70	nr	55.70
Fountain kits; typical prices of submersible units comprising fountain pumps, fountain nozzles, underwater spotlights, nozzle extension armatures, underwater terminal boxes and electrical control panels							
Single aerated white foamy water columns; ascending jet 70 mm diameter; descending water up to four times larger; jet height adjustable between 1.00 m and 1.70 m	3600.00	-	-	-	4170.00	nr	4170.00
Single aerated white foamy water columns, ascending jet 110 mm diameter; descending water up to four times larger; jet height adjustable between 1.50 m and 3.00 m	6300.00	-	-	-	7180.02	nr	7180.02

V ELEC. SUPPLY/POWER/LIGHTING SYSTEMS

Item Excluding site overheads and profit	PC £	Labour hours	Labour £	Plant £	Material £	Unit	Total rate £
V41 STREET AREA FLOODLIGHTING							
Street area floodlighting - Generally							
Preamble: There are an enormous number of luminaires available which are designed for small scale urban and garden projects. The designs are continually changing and the landscape designer is advised to consult the manufacturer's latest catalogue. Most manufacturers supply light fittings suitable for column, bracket, bulkhead, wall or soffit mounting. Highway lamps and columns for trafficked roads are not included in this section as the design of highway lighting is a very specialised subject outside the scope of most landscape contracts. The IP reference number refers to the waterproof properties of the fitting; the higher the number the more waterproof the fitting. Most items can be fitted with time clocks or PIR controls.							
Market Prices of Lamps							
Lamps							
42w TCT	-	-	-	-	16.48	nr	16.48
125w MBFU	-	-	-	-	20.11	nr	20.11
75w MBIF	-	-	-	-	57.02	nr	57.02
70w HQI-TS	-	-	-	-	57.02	nr	57.02
70w SON	-	-	-	-	54.00	nr	54.00
70w SONT	-	-	-	-	54.00	nr	54.00
100w SONT	-	-	-	-	65.79	nr	65.79
150w SONT	-	-	-	-	65.79	nr	65.79
10/13w TCD	-	-	-	-	9.83	nr	9.83
28w 2D	-	-	-	-	10.19	nr	10.19
100w GLS\E27	-	-	-	-	0.87	nr	0.87
Wall brackets and lanterns; including fixing to wall; (lamps, electric wiring, connections or related fixtures such as switch gear and time clock mechanisms not included unless otherwise indicated)							
Wall brackets and lanterns; Sugg Lighting							
762 mm "Westminster"; on large ornate bracket	628.65	0.50	9.00	-	631.32	nr	640.32
533 mm "Westminster"; on 1372 mm bow bracket	955.00	0.50	9.00	-	957.67	nr	966.67
406 mm "Classic Globe" on "Universal" plinth	384.50	0.50	9.00	-	387.17	nr	396.17
356 mm "Windsor" in "Abbey" cradle bracket	329.45	0.50	9.00	-	332.12	nr	341.12
305 mm "Windsor" in "Contract" cradle bracket	303.30	0.50	9.00	-	305.97	nr	314.97
406 mm "Windsor" on "Abbey" bracket	381.00	0.50	9.00	-	383.67	nr	392.67

Prices for Measured Works
V ELEC. SUPPLY/POWER/LIGHTING SYSTEMS

Item Excluding site overheads and profit	PC £	Labour hours	Labour £	Plant £	Material £	Unit	Total rate £
V41 STREET AREA FLOODLIGHTING – cont'd							
Bulkhead and canopy fittings; including fixing to wall and light fitting (lamp, final painting, electric wiring, connections or related fixtures such as switch gear and time clock mechanisms not included unless otherwise indicated)							
Bulkhead and canopy fittings; Louis Poulsen "Nyhavn Wall" small domed top conical shade with rings; copper wall lantern; finished untreated copper to achieve verdigris finish; also available in white aluminium; 310 mm diameter shade; with wall mounting arm; to IP 44	347.11	0.50	9.00	-	349.78	nr	358.78
Bulkhead and canopy fittings; Sugg Lighting							
"Princess" backlamp in brass	195.28	0.50	9.00	-	197.95	nr	206.95
"Victoria" backlamp in brass	237.51	0.50	9.00	-	240.18	nr	249.18
"Palace" backlamp in brass	-	0.50	9.00	-	262.67	nr	271.67
"Windsor" backlamp in brass	232.00	0.50	9.00	-	234.67	nr	243.67
Floodlighting; ground, wall or pole mounted; including fixing, (lamp, final painting, electric wiring, connections or related fixtures such as switch gear and time clock mechanisms not included unless otherwise indicated)							
Floodlighting; Outdoor Lighting (OLS)							
"Compact Flood"; ref FL-01; multi purpose ground / spike mounted wide angle floodlight; 70w HIT or SON; to IP 65	225.00	1.00	18.00	-	230.23	nr	248.23
"Morph 38"; ref SP-07; mains voltage spotlight; 60/80/120 w PAR 38; to IP 54	42.23	1.00	18.00	-	47.46	nr	65.46
Floodlight Accessories; Outdoor Lighting (OLS)							
"ES"; earth spike for Compact Flood	37.00	-	-	-	37.00	nr	37.00
"ES"; earth spike for Morph 38	5.15	-	-	-	5.15	nr	5.15
"L" Louvre for Compact Flood	44.00	-	-	-	44.00	nr	44.00
"L38"; Louvre for Morph 38	19.05	-	-	-	19.05	nr	19.05
"C "; Cowl for Compact Flood	28.00	-	-	-	28.00	nr	28.00
"BD "; Barn doors for Compact Flood	60.73	-	-	-	60.73	nr	60.73
Large area/pitch floodlighting; CU Phosco;							
ref FL444 1000w SON-T; floodlights with lamp and loose gear; narrow asymmetric beam	355.40	1.00	18.00	-	355.40	nr	373.40
ref FL444 2.0kw MBIOS; floodlight with lamp and loose gear; projector beam	600.78	1.00	18.00	-	600.78	nr	618.78
Large area floodlighting; CU Phosco;							
ref FL345/G/100S; floodlight with lamp and integral gear	240.16	1.00	18.00	-	240.16	nr	258.16
ref FL345/G/250S; floodlight with lamp and integral gear	242.47	1.00	18.00	-	242.47	nr	260.47
ref FL345/G/400 MBI; floodlight with lamp and integral gear	263.66	1.00	18.00	-	263.66	nr	281.66
Small area floodlighting; CU Phosco;							
ref FL325/G/50S; floodlight with lamp and integral gear	34.20	1.00	18.00	-	34.20	nr	52.20
ref FL325/G/70S; floodlight with lamp and integral gear	34.20	1.00	18.00	-	34.20	nr	52.20

Prices for Measured Works
V ELEC. SUPPLY/POWER/LIGHTING SYSTEMS

Item Excluding site overheads and profit	PC £	Labour hours	Labour £	Plant £	Material £	Unit	Total rate £
ref FL325/G/70 MBF; floodlight with lamp and integral gear	59.40	1.00	18.00	-	59.40	nr	77.40
Pole clamp	29.60	-	-	-	29.60	nr	29.60
ref B76; spigot cap	13.10	-	-	-	13.10	nr	13.10
"D" wall bracket	23.50	-	-	-	23.50	nr	23.50
Spotlights for uplighting and for illuminating signs and notice boards, statuary and other features); ground, wall or pole mounted; including fixing, light fitting and priming (lamp, final painting, electric wiring, connections or related fixtures such as switch gear and time clock mechanisms not included); All mains voltage (240v) unless otherwise stated							
Spotlights; Outdoor Lighting (OLS)							
Maxispotter" miniature solid copper spotlight designed to patternate and age naturally; with mounting bracket; and internal anti glare louvre; low voltage halogen reflector; 20/35/50w; to IP56;	69.78	1.00	18.00	-	75.01	nr	93.01
"WeeBee Spot"; ref SP-05 for Low voltage halogen reflector; 20/35/50w; c/w integral transformer and wall mounting box; to IP 65	78.95	1.00	18.00	-	84.18	nr	102.18
Spotlight Accessories; Outdoor Lighting (OLS)							
"ES"; earth spike for Maxispotter	6.18	-	-	-	6.18	nr	6.18
"ES"; earth spike for WeeBee Spot	14.70	-	-	-	14.70	nr	14.70
"ES"; earth spike for Morph 20	5.15	-	-	-	5.15	nr	5.15
Spotlighters, uplighters and cowl lighting; Marlin Lighting; TECHNO - SHORT ARM; head adjustable 130 deg rotation 350 deg; projection 215 mm on 210 mm base plate; 355 high; integral gear; PG16 cable gland; black							
ref SIM 3518-09 150w CDMT/HQIT	337.90	1.00	18.00	-	337.90	nr	355.90
ref SIM 3514 -09; 70w SONT	332.21	1.00	18.00	-	332.21	nr	350.21
ref SIM 3517-09; 70w CDMT/HQIT	321.42	1.00	18.00	-	321.42	nr	339.42
Recessed uplighting; including walk/drive over fully recessed uplighting; excavating, ground fixing, concreting in and making good surfaces (electric wiring, connections or related fixtures such as switch gear and time-clock mechanisms not included unless otherwise stated) (Note: Transformers will power multiple lights dependent on the distance between the light units) All mains voltage (240v) unless otherwise stated							
Recessed uplighting; Outdoor lighting (OLS); "WeeBee Up"; ultra-compact recessed halogen uplight in diecast aluminium and S/Steel top plate and toughened safety glass; 20/35w; complete with installation sleeve; low voltage requires transformer; to IP 67							
ref BU-01;	91.67	2.00	36.00	-	96.90	nr	132.90

V ELEC. SUPPLY/POWER/LIGHTING SYSTEMS

Item Excluding site overheads and profit	PC £	Labour hours	Labour £	Plant £	Material £	Unit	Total rate £
V41 STREET AREA FLOODLIGHTING – cont'd							
Recessed uplighting; Outdoor lighting (OLS); BU-Pharo; Pharo uplighters; 266 mm diameter; diecast aluminium with stainless steel top plate 10 mm toughened safety glass; 2000 kg drive over; integral control gear; to IP67							
BU-Pharo; HIT Metal Halide; white light; spot or flood or wall wash distribution	309.00	1.50	27.00	-	314.23	nr	341.23
HSE-E High pressure sodium; golden light; spot or flood or wall wash distribution	319.30	1.50	27.00	-	324.53	nr	351.53
HME-Mercury Vapour; cool white; flood distribution	278.10	1.50	27.00	-	283.33	nr	310.33
TC-T Compact fluorescent; white low power consumption; flood distribution	277.08	1.50	27.00	-	282.31	nr	309.31
Tungsten halogen 150w; flood distribution	267.80	1.50	27.00	-	273.03	nr	300.03
Accessories for Pharo uplighters; Outdoor Lighting(OLS);							
"RG" Rockguard	56.65	-	-	-	56.65	nr	56.65
"IS" Stainless steel installation sleeve	55.62	-	-	-	55.62	nr	55.62
"L" Anti glare louvre	22.66	-	-	-	22.66	nr	22.66
Recessed uplighting; Outdoor lighting (OLS); Nimbus 125-150 mm diameter uplighters; manufactured from diecast aluminium with stainless steel top plate 8 mm toughened safety glass; 2000 kg drive over; integral control gear; to IP67							
BU-02; 95 mm deep; 20/35/50w low voltage halogen; requires remote transformer	97.85	2.00	36.00	-	103.08	nr	139.08
BU-03; 172 mm deep; 20/35/50w low voltage halogen; with integral transformer	134.93	2.00	36.00	-	140.16	nr	176.16
Accessories for Nimbus uplighters; Outdoor Lighting(OLS);							
"IS" Stainless steel installation sleeve	32.45	-	-	-	32.45	nr	32.45
"BR" Brass top plate surcharge	17.51	-	-	-	17.51	nr	17.51
Wall recessed; Marlin Lighting; "EOS" range ; integral gear; asymmetric reflector; toughened reeded glass; to IP55							
ref SIM 4629-09 10w TCD "MINI EOS" 145 x 90 mm	76.83	1.00	18.00	-	76.83	nr	94.83
ref SIM 4640-09 "MEGA EOS" 80w MBFU 310 x 310 mm	355.00	1.00	18.00	-	355.00	nr	373.00
ref SIM 4639-09 "MEGA EOS" Wall recessing box	12.98	0.25	4.50	-	12.98	nr	17.48
ref SIM 4619-09 26w TCD "RECTANGULAR EOS" 270 x 145 mm	125.84	1.00	18.00	-	125.84	nr	143.84
ref SIM 4532-12 "RECTANGULAR EOS" Wall recessing box	12.98	0.25	4.50	-	12.98	nr	17.48
ref SIM 4628-09 12v 20w QT9 "MINI EOS" 145 x 90 mm	120.65	1.00	18.00	-	120.65	nr	138.65
ref SIM 4623-12 "MINI EOS" Wall recessing box	7.42	0.25	4.50	-	7.42	nr	11.92

Prices for Measured Works
V ELEC. SUPPLY/POWER/LIGHTING SYSTEMS

Item Excluding site overheads and profit	PC £	Labour hours	Labour £	Plant £	Material £	Unit	Total rate £
Underwater Lighting							
Underwater Lighting; Outdoor lighting (OLS) UW-05 "Minipower" cast bronze underwater floodlight c\w mounting bracket; 50w low voltage reflector lamp; requires remote transformer.	180.00	1.00	18.00	-	180.00	nr	198.00
Low-level lighting; positioned to provide glare free light wash to pathways steps and terraces; including forming post holes, concreting in, making good to surfaces (final painting, electric wiring, connections or related fixtures such as switch gear and time clock mechanisms not included)							
Low-level lighting; Outdoor Lighting (OLS) "Sentry"; ref AM-08; single-sided bollard type sculptural pathway light; aluminum; 670 mm high for 35w HIT; to IP 65	324.45	2.00	36.00	-	329.68	nr	365.68
"Footliter"; ref GL-05; low voltage pathlighter; 311 mm high; 6.00 m light distribution; in aluminium or solid copper; c\w earth spike; requires remote transformer; 20w halogen; to IP 44 - copper finish	69.01	2.00	36.00	-	74.24	nr	110.24
"Bricklight"; ref ST-03; for 9w PL lamp; diecast aluminium for recessing to walls; conforms to standard brick size; to IP 65	51.50	2.00	36.00	-	56.73	nr	92.73
Lighted bollards; including excavating, ground fixing, concreting in and making good surfaces (lamp, final painting, electric wiring, connections or related fixtures such as switch gear and time-clock mechanisms not included) (Note: All illuminated bollards must be earthed); heights given are from ground level to top of bollards							
Lighted bollards; Louis Poulsen "Orbiter" Vandal resistant; head of cast aluminium; domed top; anti glare rings; pole extruded aluminium; diffuser clear UV stabilised polycarbonate; powder coated; 1040 mm high; 255 mm diameter; with root or base plate; IP44	390.37	2.50	45.00	-	395.53	nr	440.53
"Waterfront" solidly proportioned; head of cast silumin; domed top; symmetrical distribution; pole extruded aluminium sandblasted or painted white; internal diffuser clear UV stabilised polycarbonate; 865 mm high; diameter 260 mm; IP55	410.00	2.50	45.00	-	415.16	nr	460.16
"Bysted" concentric louvred bollard; head cast iron; post COR-TEN steel; externally untreated to provide natural aging effect of uniform oxidised red surface finish; internal painted white; lamp diffuser rings of clear polycarbonate; 1130 mm high 280 mm diameter: to IP 44	594.89	2.50	45.00	-	600.05	nr	645.05
"Centurion" heavy duty bollard; 2 sided light distribution from stepped reflector; cast aluminium; 900 mm high; width 240 x 160 mm; to IP 44	397.58	2.50	45.00	-	402.74	nr	447.74

V ELEC. SUPPLY/POWER/LIGHTING SYSTEMS

Item Excluding site overheads and profit	PC £	Labour hours	Labour £	Plant £	Material £	Unit	Total rate £
V41 STREET AREA FLOODLIGHTING – cont'd							
Lighted bollards; - cont'd							
"Nyhavn" split level overhanging canopy type luminaire head of cast silumin; diameter 565 mm; sandblasted or powder coated; diffuser clear UV stabilized polycarbonate; post of extruded aluminium; diameter 220 mm; 1235 mm high; diameter to IP44	515.00	2.50	45.00	-	520.16	nr	565.16
Lighted bollards; Sugg Lighting							
"Cannon" bollards; flange mounted in cast aluminium with slotted apertures	335.48	2.50	45.00	-	340.64	nr	385.64
Reproduction street lanterns and columns; including excavating, concreting in, backfilling and disposing of spoil, or fixing to ground or wall, making good surfaces, light fitting and priming (final painting, electric wiring, connections or related fixtures such as switch gear and time-clock mechanisms not included) (Note: lanterns up to 14 in and columns up to 7 ft are suitable for residential lighting)							
Reproduction street lanterns and columns; Sugg Lighting; reproduction lanterns hand made to original designs; all with ES lamp holder							
762 mm "Westminster"	538.00	0.50	9.00	-	540.67	nr	549.67
533 mm "Westminster"	455.00	0.50	9.00	-	457.67	nr	466.67
432 mm "Guildhall"	422.00	0.50	9.00	-	424.67	nr	433.67
432 mm "Grosvenor"	325.00	0.50	9.00	-	327.67	nr	336.67
406 mm "Classic Globe"	310.00	0.50	9.00	-	312.67	nr	321.67
"Windsor"; large, with door, IP54	301.00	0.50	9.00	-	303.67	nr	312.67
"Windsor"; medium, with door, IP54	265.00	0.50	9.00	-	267.67	nr	276.67
"Windsor"; small, with door, IP54	224.00	0.50	9.00	-	226.67	nr	235.67
Reproduction brackets and suspensions							
Reproduction brackets and suspensions; Sugg Lighting							
"Bow" bracket	500.00	2.00	36.00	-	502.67	nr	538.67
"Large ornate"	224.65	2.00	36.00	-	227.32	nr	263.32
"Large swan neck"	194.62	2.00	36.00	-	197.29	nr	233.29
"Medium ornate"	215.59	2.00	36.00	-	218.26	nr	254.26
"Universal" bracket/pedestal	149.57	2.00	36.00	-	152.24	nr	188.24
"Medium swan neck"	155.92	2.00	36.00	-	158.59	nr	194.59
"Decorated P" bracket	90.00	2.00	36.00	-	92.67	nr	128.67
"3-arm" suspension; general	87.97	2.00	36.00	-	90.64	nr	126.64
"Plaza" bracket	73.92	2.00	36.00	-	76.59	nr	112.59
"Universal" plinth pedestal	74.50	2.00	36.00	-	77.17	nr	113.17
"Frog"; general	39.90	2.00	36.00	-	42.57	nr	78.57

V ELEC. SUPPLY/POWER/LIGHTING SYSTEMS

Item Excluding site overheads and profit	PC £	Labour hours	Labour £	Plant £	Material £	Unit	Total rate £
Reproduction lighting columns							
Reproduction lighting columns; Sugg Lighting							
2133 mm "Royal Exchange" cast aluminium columns with flanges; triple	1805.00	8.00	144.00	-	1805.00	nr	**1949.00**
3505 mm "Constitution Hill" cast aluminium columns with flanges	1701.53	8.00	144.00	-	1701.53	nr	**1845.53**
2133 mm "Royal Exchange" cast aluminium columns with flanges; single	1367.10	6.00	108.00	-	1367.10	nr	**1475.10**
3505 mm "Aylesbury" cast aluminium columns	890.93	8.00	144.00	-	890.93	nr	**1034.93**
3.00 m - 5.00 m "Harborne" combined steel/iron columns with roots	522.30	8.00	144.00	-	528.75	nr	**672.75**
2438 mm "Barley Sugar" cast aluminium columns	583.28	6.00	108.00	-	583.28	nr	**691.28**
3.00 m - 5.00 m "Cannonbury"; combined steel/iron columns with flanges	512.40	8.00	144.00	-	512.40	nr	**656.40**
3048 mm "Trafalgar" cast aluminium columns with flanges	480.38	6.00	108.00	-	480.38	nr	**588.38**
2743 mm "Napier" cast aluminium columns with flanges	421.58	6.00	108.00	-	439.00	nr	**547.00**
Precinct lighting lanterns; ground, wall or pole mounted; including fixing and light fitting (lamps, poles, brackets, final painting, electric wiring, connections or related fixtures such as switch gear and time clock mechanisms not included)							
Precinct lighting lanterns; Louis Poulsen "Nyhavn Park" side entry 90 degree curved arm mounted; steel rings over domed top; housing cast silumin sandblasted with integral gear; protected by UV stabilized clear polycarbonate diffuser; IP55	504.70	1.00	18.00	-	504.70	nr	**522.70**
"Kipp"; bottom entry pole-top; shot blasted or lacquered Hanover design award; hinged diffuser for simple maintenance; indirect lighting technique ensures low glare; IP55	272.95	1.00	18.00	-	272.95	nr	**290.95**
"Bjarne Bech"; luminaire head of galvanized sheet steel; domed head over cylindrical light shield; internal reflector surfaces painted white; IP23	279.13	1.00	18.00	-	279.13	nr	**297.13**
Precinct lighting lanterns; Marlin Lighting; "Sphereline"; clear or opal; vandal resistant; integral HPF gear; finish - black							
ref 8031 XX; 300 mm diameter opal polycarbonate globe; 10/13w TCD; to IP 44	57.63	1.00	18.00	-	57.63	nr	**75.63**
Precinct lighting lanterns; Marlin Lighting; "Sphereline"; accessories and fittings;							
ref 8030 PB26 "Boulevard" pole top 2 globe bracket	313.73	0.50	9.00	-	313.73	nr	**322.73**
Precinct lighting lanterns; Sugg Lighting							
large suspended "Superseal Rochester"; with 150W SON-T lamp	434.00	1.00	18.00	-	434.00	nr	**452.00**
medium suspended "Superseal Rochester"; with 70W SON-T lamp	351.00	1.00	18.00	-	351.00	nr	**369.00**
small suspended "Street Rochester"; with 50W SON-E lamp	289.00	1.00	18.00	-	289.00	nr	**307.00**

V ELEC. SUPPLY/POWER/LIGHTING SYSTEMS

Item Excluding site overheads and profit	PC £	Labour hours	Labour £	Plant £	Material £	Unit	Total rate £
V41 STREET AREA FLOODLIGHTING – cont'd							
Precinct lighting columns; including excavating, concreting, backfilling and disposing of spoil (final painting, electric wiring, connections or related fixtures such as switch gear and time-clock mechanisms not included)							
Precinct lighting columns; Marlin Lighting; "Skyline" tubular steel poles							
ref SPR 750 A6; waisted; 5.0 m high	179.46	6.00	108.00	-	204.55	nr	312.55
ref SPR 730 A6; waisted; 3.0 m high	131.60	6.00	108.00	-	156.69	nr	264.69
Precinct lighting columns; Sugg Lighting							
3.00 - 5.00 m steel columns with decorative embellishments	246.09	8.00	144.00	-	252.54	nr	396.54
3.00 - 5.00 m steel columns with contract embellishments	229.09	8.00	144.00	-	235.54	nr	379.54
Precinct lighting columns; CU Phosco							
"Cub" octagonal steel; 5000 mm high	105.50	8.00	144.00	-	111.95	nr	255.95
"Cub" octagonal steel; 6000 mm high	113.35	8.00	144.00	-	119.80	nr	263.80
"Cub" octagonal steel; 8000 mm high	203.20	8.00	144.00	-	209.65	nr	353.65
Transformers for low voltage lighting; Outdoor Lighting (OLS); distance from electrical supply 25 m; trenching and backfilling measured separately							
Boxed Transformer							
50 Va for maximum 50 w of lamp	48.93	2.00	36.00	-	70.04	nr	106.04
100Va for maximum 100 w of lamp	57.28	2.00	36.00	-	78.40	nr	114.40
150 Va for maximum 150 w of lamp	60.61	2.00	36.00	-	81.73	nr	117.73
200 Va for maximum 200 w of lamp	71.43	2.00	36.00	-	92.55	nr	128.55
250 Va for maximum 50 w of lamp	75.34	2.00	36.00	-	96.46	nr	132.46
Buried transformers							
50 Va for maximum 50 w of lamp	57.24	2.00	36.00	-	78.35	nr	114.35
Electrical Accessories							
Feeder pillar; ornamental bollard housing for exterior sockets and meters							
"Cannon" style; with 1 nr door	250.95	8.00	144.00	-	263.05	nr	407.05
"Fluted", with 2 nr doors	599.60	8.00	144.00	-	611.70	nr	755.70
Installation; electric cable in trench 500 mm deep (trench not included); all in accordance with IEE regulations							
Light duty 600 volt grade armoured							
3 core 2.5 mm	0.92	0.01	0.09	-	0.92	m	1.01
Twin and earth PVC cable in plastic conduit (conduit not included)							
2.50 mm2	0.14	0.01	0.18	-	0.14	m	0.32
4.00 mm2	0.43	0.01	0.18	-	0.43	m	0.61
16 mm heavy gauge high impact PVC conduit	3.08	0.03	0.60	-	3.08	m	3.68
Main switch and fuse unit; 30A	-	-	-	-	37.92	nr	37.92
Weatherproof switched socket; 13A	-	2.00	36.00	-	15.38	nr	51.38

DAVIS LANGDON & EVEREST

LONDON
Mid City Place
71 High Holborn
London WC1V 6QS
Tel: (020) 7061 7000
Fax: (020) 7061 7061
Email: neill.morrison@davislangdon-uk.com

BIRMINGHAM
75-77 Colmore Row
Birmingham
B3 2HD
Tel: (0121) 710 1100
Fax: (0121) 710 1399
Email: david.daly@davislangdon-uk.com

BRISTOL
St Lawrence House
29/31 Broad Street
Bristol
BS1 2HF
Tel: (0117) 927 7832
Fax: (0117) 925 1350
Email: alan.trolley@davislangdon-uk.com

CAMBRIDGE
36 Storey's Way
Cambridge
CB3 ODT
Tel: (01223) 351 258
Fax: (01223) 321 002
Email: stephen.bugg@davislangdon-uk.com

CARDIFF
4 Piershead Street
Capital Waterside
Cardiff CF10 4QP
Tel: (029) 20 497 497
Fax: (029) 20 497 111
Email: paul.edwards@davislangdon-uk.com

EDINBURGH
39 Melville Street
Edinburgh
EH3 7JF
Tel: (0131) 240 1350
Fax: (0131) 240 1399
Email: ian.mcandie@davislangdon-uk.com

GLASGOW
Monteith House
11 George Square
Glasgow G2 1DY
Tel: (0141) 248 0300
Fax: (0141) 248 0303
Email: hugh.fisher@davislangdon-uk.com

LEEDS
No 4 The Embankment
Victoria Wharf
Sovereign Street
Leeds LS1 4BA
Tel: (0113) 243 2481
Fax: (0113) 242 4601
Email: duncan.sissons@davislangdon-uk.com

LIVERPOOL
Cunard Building
Water Street
Liverpool L3 1JR
Tel: (0151) 236 1992
Fax: (0151) 227 5401
Email: andrew.stevenson@davislangdon-uk.com

MANCHESTER
Cloister House
Riverside
New Bailey Street
Manchester M3 5AG
Tel: (0161) 819 7600
Fax: (0161) 819 1818
Email: paul.stanion@davislangdon-uk.com

MILTON KEYNES
Everest House
Rockingham Drive
Linford Wood
Milton Keynes MK14 6LY
Tel: (01908) 304 700
Fax: (01908) 660 059
Email: kevin.sims@davislangdon-uk.com

NEWCASTLE
2nd Floor
The Old Post Office Building
St Nicholas Street
Newcastle upon Tyne NE1 1RH
Tel: (0191) 221 2012
Fax: (0191) 261 4274
Email: gary.lockey@davislangdon-uk.com

NORWICH
63 Thorpe Road
Norwich
NR1 1UD
Tel: (01603) 628 194
Fax: (01603) 615 928
Email: michael.ladbrook@davislangdon-uk.com

NOTTINGHAM
Hamilton House
Clinton Avenue
Nottingham NG5 1AW
Tel: (0115) 962 2511
Fax: (0115) 969 1079
Email: rod.exton@davislangdon-uk.com

OXFORD
Avalon House
Marcham Road
Abingdon
Oxford OX14 1TZ
Tel: (01235) 555 025
Fax: (01235) 554 909
Email: paul.coomber@davislangdon-uk.com

PETERBOROUGH
Clarence House
Minerva Business Park
Lynchwood
Peterborough PE2 6FT
Tel: (01733) 362 000
Fax: (01733) 230 875
Email: stuart.bremner@davislangdon-uk.com

PLYMOUTH
1 Ensign House
Parkway Court
Longbridge Road
Plymouth PL6 8LR
Tel: (01752) 827 444
Fax: (01752) 221 219
Email: gareth.steventon@davislangdon-uk.com

PORTSMOUTH
Unit 3, St Andrews Court
St Michaels Road
Portsmouth PO1 2PR
Tel: (023) 9281 5218
Fax: (023) 9282 7156
Email: chris.tremellen@davislangdon-uk.com

SOUTHAMPTON
Brunswick House
Brunswick Place
Southampton SO15 2AP
Tel: (023) 8033 3438
Fax: (023) 8022 6099
Email: richard.pitman@davislangdon-uk.com

DAVIS LANGDON CONSULTANCY
Mid City Place
71 High Holborn
London WC1V 6QS
Tel: (020) 7061 7007
Fax: (020) 7061 7005
Email: jim.meikle@davislangdon-uk.com

MOTT GREEN & WALL
Mid City Place
71 High Holborn
London WC1V 6QS
Tel: (020) 7061 7777
Fax: (020) 7061 7009
Email: general@mottgreenwall.co.uk

DAVIS LANGDON SCHUMANN SMITH
Southgate House
St Georges Way
Stevenage
Hertfordshire SG1 1HG
Tel: (01438) 742 642
Fax: (01438) 742 632
Email: nick.schumann@schumannsmith.com

NBW CROSHER & JAMES
LONDON
1-5 Exchange Court
Strand
London WC2R 0PQ
Tel: (020) 7845 0600
Fax: (020) 7845 0601
Email: llewellynt@nbwcrosherjames.com

BIRMINGHAM
102 New Street
Birmingham B2 4HQ
Tel: (0121) 632 3600
Fax: (0121) 632 3601
Email: Whittakerr@nbwcrosherjames.com

CARDIFF
4 Piershead Street
Capital Waterside
Cardiff CF10 4QP
Tel: (029) 20497 497
Fax: (029) 20497 111
Email: murraym@nbwcrosherjames.com

EDINBURGH
39 Melville Street
Edinburgh EH3 7JF
Tel: (0131) 220 4225
Fax: (0131) 220 4226
Email: mcfarlanei@nbwcrosherjames.com

GLASGOW
Monteith House
11 George Square
Glasgow G2 1DY
Tel: (0141) 248 0333
Fax: (0141) 248 0313
Email: fraserk@nbwcrosherjames.com

MANCHESTER
Cloister House
Riverside
New Bailey Street
Manchester M3 5AG
Tel: (0161) 819 7600
Fax: (0161) 819 1818
Email: sharmas@nbwcrosherjames.com

SOUTHAMPTON
Brunswick House
Brunswick Place
Southampton SO15 2AP
Tel: (023) 8068 2800
Fax: (023) 8033 6360
Email: reesd@nbwcrosherjames.com

with offices throughout Europe, the Middle East, Asia, Australia, Africa and the USA which together form

DAVIS LANGDON & SEAH INTERNATIONAL

DAVIS LANGDON & EVEREST
Authors of Spon's Price Books

Davis Langdon & Everest is an independent practice of Chartered Quantity Surveyors, with some 1200 staff in 23 UK offices and, through Davis Langdon & Seah International, some 2,800 staff in 95 offices worldwide.

DLE manages client requirements, controls risk, manages cost and maximises value for money, throughout the course of construction projects, always aiming to be - and to deliver - the best.

TYPICAL PROJECT STAGES, DLE INTEGRATED SERVICES AND THEIR EFFECT:

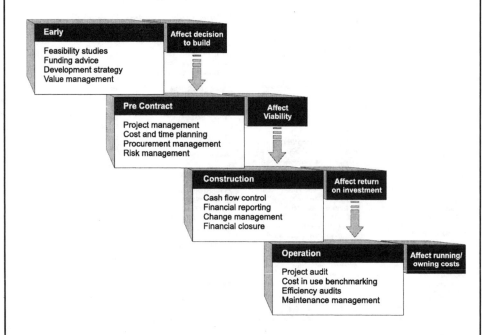

EUROPE - ASIA - AUSTRALIA - AFRICA - AMERICA
GLOBAL REACH ⇨ LOCAL DELIVERY

8
Approximate Estimates

Prices in this section are based upon the Prices for Measured Work, but allow for incidentals which would normally be measured separately in a Bill of Quantities. They do not include for Preliminaries which are priced elsewhere in this book.

Items shown as sub-contract or specialist rates would normally include the specialists overhead and profit. All other items which could fall within the scope of works of general landscape and external works contractors would not include profit

Based on current commercial rates, profits of 15% to 35% may be added to these rates to indicate the likely "with profit" values of the tasks below. The variation quoted above is dependent on the sector in which the works are taking place – domestic, public or commercial.

ESSENTIAL READING FROM SPON PRESS

Guidelines for Landscape and Visual Impact Assessment
Second Edition

The Landscape Institute and the Institute of Environmental Management and Assessment, Co-ordinated by Sue Wilson

'For any self-respecting landscape consultant, *Guidelines for Landscape and Visual Impact Assessment* is an essential reference book. The book is well-constructed, concise, compact and well-made and illustrated.'
- *The Architects' Journal* on the first edition

This new edition of these good practice guidelines provides advice on assessing the landscape and the visual impacts of development projects. Issues covered by the guidelines include: integration of landscape and visual issues into the development process; the need for a transparent approach to landscape and visual impact assessment; describing the baseline conditions; determining the magnitude and significance of impacts; reviewing the landscape and visual components of an Environmental Impact Statement.

The guidelines will enable developers and decision makers to be aware of the issues to be addressed by this type of assessment, as well as providing environmental consultants with a framework for good practice.

April 2002: 246x189 mm: 176 pp.
75 colour illustrations and 6 line drawings
HB: 0-415-23185-X: £35.00

To Order: Tel: +44 (0) 8700 768853, or +44 (0) 1264 343071 Fax: +44 (0) 1264 343005, or Post: Spon Press Customer Services, Thomson Publishing Services, Cheriton House, Andover, Hants, SP10 5BE, UK Email: book.orders@tandf.co.uk

For a complete listing of all our titles visit:
www.sponpress.com

Approximate Estimates
8.1 PRELIMINARIES

Item Excluding site overheads and profit	Unit	Total rate £
8.1 CONTRACT ADMINISTRATION AND MANAGEMENT		
Prepare and maintain Health and Safety file, prepare risk and Coshh assessments, method statements, works programmes before the start of works on site; Project value:		
£35000.00	nr	200.00
£75000.00	nr	300.00
£100000.00	nr	400.00
£200000.00 - £500000.00	nr	1000.00
SITE SETUP AND SITE ESTABLISHMENT		
Costs of initial setup of Single site cabin and 200 m² of site area for storage of vehicles plant etc. Inclusive of Site office, secure container, padlocks safety fencing electricity and water and toilet facilities.		
Site establishment	nr	498.96
Weekly hire rates on the above	nr	59.76

Approximate Estimates
8.2 DEMOLITION AND SITE CLEARANCE

Item Excluding site overheads and profit	Unit	Total rate £
Supply and erect temporary protective fencing and remove after planting		
Cleft chestnut paling 1.20 m high 75 larch posts at 3 m centres	100 m	775.00
Cleft chestnut paling 1.50 m high 75 larch posts at 3 m centres	100 m	901.00
Break up plain concrete slab and remove to licensed tip not exceeding 13 km from site:		
150 thick	m²	7.83
200 thick	m²	10.45
300 thick	m²	15.67
Clear away light fencing and gates (chain link, chestnut paling, light boarded fence or similar) and remove to licensed tip	100 m	107.34
Clear existing scrub vegetation, including shrubs and hedges and burn on site; spread ash on specified area		
by machine		
light scrub and grasses	100 m²	14.70
undergrowth brambles and heavy weed growth	100 m²	33.18
small shrubs	100 m²	33.18
As above but remove vegetation to licensed tip not exceeding 13 km from site	100 m²	35.39
by hand	100 m²	79.20
As above but remove vegetation to licensed tip not exceeding 13 km from site	100 m²	57.95
Fell trees on site, grub up roots, all by machine and remove debris to licensed tip not exceeding 13 km from site:		
trees 600 girth	each	206.12
trees 1.5 - 3.00 m girth	each	679.30
trees over 3.00 m girth	each	1505.07

Approximate Estimates
8.3 GROUNDWORK

Item Excluding site overheads and profit	Unit	Total rate £
Excavate trenches for foundations; trenches 225 mm deeper than specified foundation thickness. Pour plain concrete foundations GEN 1 10 N/mm and to thickness as described; Disposal of excavated material off site: dimensions of concrete foundations		
250 mm wide x 250 mm deep	m³	126.15
300 mm wide x 300 mm deep	m³	122.92
400 mm wide x 600 mm deep	m³	122.92
Cut and strip by machine turves 50 thick load to dumper or palette by hand, stack on site		
not exceeding 100 m travel to stack	100 m²	40.51
As above but all by hand, including barrowing	100 m²	419.99
Prices for excavating and reducing to levels are for work in light or medium soils; multiplying factors for other soils are as follows:		
clay	x 1.5	
compact Gravel	x 1.2	
soft chalk	x 2.0	
hard rock	x 3.0	
Remove topsoil average depth 300 mm, deposit in spoil heaps not exceeding 100 m travel to heap, treat once with paraquat-diquat weedkiller, turn heap once during storage, all by machine	m³	3.25
As above but all by hand	m³	30.26
Excavate to reduce levels and remove spoil to dump not exceeding 100 m travel, all by machine:		
0.25 m deep average	m²	1.71
0.30 m deep	m²	1.95
1.0 m deep using 21 tonne 360 tracked excavator	m²	5.38
1.0 m deep using JCB	m²	5.91
As above but all by hand:		
0.10 m deep average	m²	11.40
0.20 m deep average	m²	24.00
0.30 m deep average	m²	136.19
1.0 m deep average	m²	173.99
Extra for carting spoil to licensed tip offsite	m³	21.95
Spread excavated material to levels in layers not exceeding 150 mm, using scraper blade or bucket		
average thickness 100 mm	m2	0.20
As above but with imported topsoil	m2	2.41
average thickness 200 mm	m2	0.39
As above but with imported topsoil	m2	4.81
average thickness 250 mm	m2	0.49
As above but with imported topsoil	m2	6.01
Extra for work to banks exceeding 30 slope	30%	

Approximate Estimates
GROUNDWORK

Item Excluding site overheads and profit	Unit	Total rate £
Rip subsoil using approved sub-soiling machine where shown on drawings to a depth of 600 mm below topsoil at 600 mm centres, all tree roots and debris over 150 x 150 mm to be removed. Plough uncultivated ground where shown on drawings to a depths as shown		
100 mm	100 m²	3.41
200 mm	100 m²	3.54
300 mm	100 m²	3.75
400 mm	100 m²	5.47
TRENCHES		
Excavate trenches for foundations; trenches 225 mm deeper than specified foundation thickness. Pour plain concrete foundations GEN 1 10 N/mm and to thickness as described; Disposal of excavated material off site: dimensions of concrete foundations		
250 mm wide x 250 mm deep	m	9.77
300 mm wide x 300 mm deep	m	13.40
400 mm wide x 600 mm deep	m	32.51
600 mm wide x 400 mm deep	m	34.19
Excavate trench and lay foundation concrete 1:3:6 600 x 300 mm deep		
By machine	m	24.62
By hand	m	35.77

Approximate Estimates
8.4 GROUND STABILIZATION

Item Excluding site overheads and profit	Unit	Total rate £
RETAINING WALLS		
Excavate fixing trench for erosion control mat 300 x 300 mm. Place Tensar mat to trench and backfill with excavated material	m	7.96
Grade existing bank to even falls, remove all roots, stones and debris over 150 mm in any dimension. Lay 20 mm thick open texture erosion control Tensar mat with 100 mm laps, fixed with 8 x 400 mm steel pegs at 1.0 m centres. Sow with low maintenance grass at 35 g/m², spread imported topsoil to BS3882 25 mm thick incorporating medium grade sedge peat at 3 kg/m² and fertilizer at 30 g/m², water lightly.	m²	5.66
Bring bank to grade, remove all roots, stones and debris over 150 mm in any dimension. Lay Greenfix biodegradable pre-seeded erosion control mat, fixed with 6 x 300 mm steel pegs at 1.0 m centres.	100 m²	321.50
Bring surface of sub-grade to given levels and falls, remove all roots, stones and debris over 150 mm in any dimension, spread topsoil from dump on site 150 mm thick and consolidate lightly. Lay Tensar Mat erosion control mat with 100 mm laps, spread 25 mm imported topsoil with fertilizer and suitable erosion control seed.	100 m²	563.86
Bring surface of sub-grade to given levels and falls, remove all roots, stones and debris over 150 mm in any dimension, spread topsoil from dump on site and consolidate lightly. Lay Tensar Mat erosion control mat with 100 mm laps, spread 25 mm imported topsoil with fertilizer and suitable erosion control seed.	m²	5.37
Excavate trench 750 mm deep and lay concrete foundation 600 wide x 300 deep Construct Forticrete precast hollow concrete block wall with 450 mm below ground laid all in accordance with manufacturer's instructions. Fix reinforcing bar 12 mm as work proceeds Fill blocks with concrete 1:3:6 as work proceeds. 1800 high		
Type 256; 400 x 225 x 256 mm	m	165.53
Type 190; 400 x 225 x 190 mm	m	135.44
As above but 2400 high		
Type 256; 400 x 225 x 256 mm	m	229.02
Type 190; 400 x 225 x 190 mm	m	185.04

Approximate Estimates
GROUND STABILIZATION

Item Excluding site overheads and profit	Unit	Total rate £
RETAINING WALLS (cont'd)		
On foundation measured above, supply and RCC precast concrete "L" shaped units, constructed all in accordance with manufacturer's instructions. Backfill with approved excavated material compacted as the work proceeds.		
1250 mm high x 1000 mm wide	m	131.23
2400 mm high x 1000 mm wide	m	205.31
3750 mm high x 1000 mm wide	m	410.32
Excavate trench to receive foundation 300 mm deep, place plain concrete foundation 150 thick in 11.50 N/mm² concrete (sulphate-resisting cement) construct timber crib retaining wall and backfill with excavated spoil behind units.		
Keller-Comtec timber crib wall system, average 4.2 m high	m	184.42
Keller-Comtec timber crib wall system, average 1.5 m high	m	135.34
Construct bank revetment, bring bank to grade by machine, lay regulating layer of quarry scalpings, lay filter membrane, lay 100 mm drainage layer of broken stone or approved hardcore 28 - 10 mm size, blind with 25 mm sharp sand, and lay grass concrete surface (price does not include edgings or toe beams),fill with approved topsoil and fertilizer at 35 g/m², seed with dwarf rye based grass seed mix:		
Grasscrete in situ reinforced concrete surfacing GC 1, 100 thick	m²	66.57
Grasscrete in situ reinforced concrete surfacing GC 2, 150 thick	m²	71.31
Grassblock 103 open matrix blocks 406 x 406 x 103	m²	68.25
Construct revetment of Maccaferri Limited Reno mattress gabions laid on firm level ground, tightly packed with broken stone or concrete and securely wired, all in accordance with manufacturers' instructions:		
gabions 6 m x 2 m x 0.17 m, one course	m²	63.78
gabions 6 m x 2 m x 0.17 m, two courses	m²	87.73
Grade bank by mechanical scraper, lay soil-reinforcing fabric properly lapped and fixed with 200 mm steel pegs at 500 mm centres, spread approved topsoil 25 mm thick incorporating fertilizer at 35 g/m², sow with low maintenance grass at 35 g/m²:		

Approximate Estimates
GROUND STABILIZATION

Item Excluding site overheads and profit	Unit	Total rate £
Wyretex No 8 polypropylene wire reinforced fabric	m²	5.55
unseeded "Eromat Light"	m²	2.50
seeded "Covamat Standard";	m²	3.11
Netlon Erosion mat for erosion control	m²	3.87
Construct ha-ha; excavate ditch 1200 mm deep x 900 mm wide at bottom battered one side 45° slope, excavate for foundation to wall and place GEN 1 concrete foundation 150 x 500 mm, construct one and a half brick wall (Brick PC £300.00/1000) battered 10° from vertical, laid in cement:lime:sand (1:1:6) mortar in English garden wall bond, precast concrete coping weathered and throated set 150 mm above ground on high side, rake bottom and sides of ditch and seed with low maintenance grass at 35 g/m²:		
wall 600 high	m	222.54
wall 900 high	m	275.38
wall 1200 high	m	340.88
Timber Log Retaining Walls Excavate trench 300 wide to one third of the finished height of the retaining walls below. Lay 100 mm hardcore. Fix machine rounded logs set in concrete 1:3:6. Remove excavated material from site. Fix geofabric to rear of timber logs. Backfill with previously excavated material set aside in position. All works by machine		
100 mm diameter logs		
500 mm high (constructed from 1.80 m lengths)	m	70.94
1.20 mm high (constructed from 1.80 m lengths)	m	112.71
1.60 mm high (constructed from 2.40 m lengths)	m	162.94
2.00 mm high (constructed from 3.00 m lengths)	m	221.78
As above but 150 mm diameter logs		
1.20 mm high (constructed from 1.80 m lengths)	m	171.91
1.60 mm high (constructed from 2.40 m lengths)	m	239.34
2.00 mm high (constructed from 3.00 m lengths)	m	302.62
2.40 mm high (constructed from 3.60 m lengths)	m	354.82

Approximate Estimates
8.5 IN SITU CONCRETE

Item Excluding site overheads and profit	Unit	Total rate £
Mix concrete on site; aggregates delivered in 20 tonne loads; deliver mixed concrete to location by barrow distance 25 m		
1:3:6	m³	109.92
1:2:4	m³	118.46
As above but aggregates delivered in 1 tonne bags		
1:3:6	m³	137.84
1:2:4	m³	146.38
As above but ready mixed concrete		
10 N/mm²	m³	132.65
15 N/mm²	m³	137.64
Excavate foundation trench mechanically, remove spoil offsite, lay 1:3:6 site mixed concrete foundations. Distance from mixer 25 m. Depth of trench to be 225 mm deeper than foundation to allow for 3 underground brick courses priced separately. Foundation size:		
200 mm deep x 400 mm wide	m	17.07
300 mm deep x 500 mm wide	m	29.12
400 mm deep x 400 mm wide	m	29.57
400 mm deep x 600 mm wide	m	44.54
600 mm deep x 600 mm wide	m	64.13
As above but hand excavation and disposal to spoil heap 25m by barrow and off site by grab lorry		
200 mm deep x 400 mm wide	m	36.03
300 mm deep x 500 mm wide	m	63.39
400 mm deep x 400 mm wide	m	63.60
400 mm deep x 600 mm wide	m	73.18
600 mm deep x 600 mm wide	m	133.71
Excavate foundation trench mechanically, remove spoil offsite, lay ready mixed concrete GEN1 discharged directly from delivery lorry to location. Depth of trench to be 225 mm deeper than foundation to allow for 3 underground brick courses priced separately. Foundation size:		
200 mm deep x 400 mm wide	m	14.28
300 mm deep x 500 mm wide	m	23.89
400 mm deep x 400 mm wide	m	23.99
400 mm deep x 600 mm wide	m	36.17
600 mm deep x 600 mm wide	m	51.57

Approximate Estimates
8.6 BRICK/BLOCK WALLING

Item Excluding site overheads and profit	Unit	Total rate £
Excavate foundation trench 500 deep, remove spoil to dump off site, lay site mixed concrete foundations 1:3:6 350 x 150 thick, construct half brick wall with one brick piers at 2.0 m centres, laid in cement:lime:sand (1:1:6) mortar with flush joints, fair face one side, DPC two courses underground, engineering brick in cement:sand (1:3) mortar, coping of headers on end:		
Wall 900 high above DPC		
in engineering brick (class B) - £260.00/1000	m	145.99
in sandfaced facings - £300.00/1000	m	156.01
in rough stocks - £450.00/1000	m	164.47
Excavate foundation trench 400 deep, remove spoil to dump off site, lay GEN 1 concrete foundations 450 wide x 250 thick, construct one brick wall with one and a half brick piers at 3.0 m centres, all in English Garden Wall bond, laid in cement:lime:sand (1:1:6) mortar with flush joints, fair face one side, DPC two courses engineering brick in cement:sand (1:3) mortar, precast concrete coping 152 x 75 mm:		
Wall 900 high above DPC		
in engineering brick (class B) - £260.00/1000	m	245.58
in sandfaced facings - £300.00/1000	m	254.02
in rough stocks - £450.00/1000	m	270.28
Wall 1200 high above DPC		
in engineering brick (class B) - £260.00/1000	m	271.52
in sandfaced facings - £300.00/1000	m	281.52
in rough stocks - £450.00/1000	m	344.39
Wall 1800 high above DPC		
in engineering brick (class B) - £260.00/1000	m	341.48
in sandfaced facings - £300.00/1000	m	408.46
in rough stocks - £450.00/1000	m	435.73

Approximate Estimates
8.6 BRICK/BLOCK WALLING

Item Excluding site overheads and profit	Unit	Total rate £
BRICK/BLOCK WALLING (cont'd)		
Excavate foundation trench 450 deep, remove spoil to dump off site, lay GEN 1 concrete foundations 600 x 300 thick, construct one and a half brick wall with two thick brick piers at 3.0 m centres, all in English Garden Wall bond, laid in cement:lime:sand (1:1:6) mortar with flush joints, fair face one side, DPC two courses engineering brick in cement:sand (1:3) mortar, coping of headers on edge:		
Wall 900 high above DPC		
in engineering brick (class B) - £260.00/1000	m	**499.80**
in sandfaced facings - £300.00/1000	m	**254.02**
in rough stocks - £450.00/1000	m	**270.28**
Wall 1200 high above DPC		
in engineering brick (class B) - £260.00/1000	m	**271.52**
in sandfaced facings - £300.00/1000	m	**281.52**
in rough stocks - £450.00/1000	m	**344.39**
Wall 1800 high above DPC		
in engineering brick (class B) - £260.00/1000	m	**341.48**
in sandfaced facings - £300.00/1000	m	**408.46**
in rough stocks - £450.00/1000	m	**435.73**

Approximate Estimates
8.7 ROADS AND PAVINGS

Item Excluding site overheads and profit	Unit	Total rate £
KERBS AND EDGINGS		
Excavate and construct concrete foundation 300 mm wide x 150 mm deep. Lay precast concrete kerb units to BS 7263: Part 1, laid straight bedded and haunched both sides in semi-dry concrete, slump 35 mm maximum, haunching one side		
125 mm high x 125 mm thick bullnosed type BN	m	17.37
125 x 255 mm; ref HB2; SP	m	19.17
150 x 305 mm; ref HB1	m	22.98
On main paving base previously laid, supply and lay precast concrete channel units to BS 7263: Part 1, 255 mm wide x 125 mm high dished type CD, bedded in 1:3 lime:sand mortar, jointed in cement:sand mortar 1:3	m	14.45
Excavate and construct concrete foundation 600 mm wide x 200 mm deep. Lay channel of five courses of Class B engineering bricks to depths and falls bricks to be laid as headers along the channel, bedded in 1:3 lime:sand mortar bricks close jointed in cement:sand mortar 1:3	m	59.64
Excavate and lay precast concrete edging on concrete foundation and haunching one side 1:2:4 including all necessary formwork; to one side of straight path		
50 x 150	m	19.59
50 x 200	m	22.45
50 x 250	m	23.91
As above but to both sides of straight paths		
50 x 150	m	39.39
50 x 200	m	45.10
50 x 250	m	48.04
Timber edgings; Softwood		
150 x 38	m	5.06
150 x 50	m	5.32
Timber edgings hardwood		
150 x 38	m	10.06
150 x 50	m	11.92
Edge restraint to block paving; Excavate for groundbeam, lay Concrete 1:2:4 200 wide x 150mm deep. Lay butt jointed block header course on 50 mm thick sharp sand bed. Inclusive of haunching one side; butt joints		
200 x 100 x 60; laid to header course - £161.00/1000	m	12.48
200 x 100 x 60; laid to stretcher course - £161.00/1000	m	10.94
As above but clay bricks with mortar joints		
215 x 112 x 50 laid to header course - £280.00/1000	m	14.87
215 x 112 x 50 laid to stretcher course - £280.00/1000	m	14.02

Approximate Estimates
ROADS AND PAVINGS

Item Excluding site overheads and profit	Unit	Total rate £
ROADS		
Excavate weak points of excavation by hand and fill with well rammed Type 1 granular material;		
100 thick	m²	8.53
200 thick	m²	11.28
Excavate road bed and dispose excavated material off site, bring to grade and fill hollows with weak concrete, lay reinforcing mesh A142 lapped and joined, lay 150 mm thick in situ reinforced concrete roadbed 21 N/mm², with 15 impregnated fibreboard expansion joints and polysulphide based sealant at 50 m centres, 150 hardcore blinded with ash or sand, kerbs 155 x 255 to BS 7263 on both sides, including foundations haunched one side in 11.5 N/mm² concrete with all necessary formwork		
Falls, crossfalls and cambers not exceeding 15 degrees		
4.90 m wide	m	203.10
6.10 m wide	m	274.87
7.32 m wide	m	290.15
Excavate 350 mm for pathways or roadbed to receive surface 100 mm thick priced separately, bring to grade, lay 100 well-rolled hardcore; lay 150 mm Type 1 granular material; lay precast edgings kerbs BS 340; 50 x 150 on both sides, including foundations haunched one side in 11.5 N/mm² concrete with all necessary formwork and disposal off site		
Work to falls, crossfalls and cambers not exceeding 15 degrees:		
1.50 m wide	m	118.76
2.00 m wide	m	136.08
3.00 m wide	m	184.53
4.00 m wide	m	232.98
5.00 m wide	m	281.43
6.00 m wide	m	329.88
7.00 m wide	m	378.33

Approximate Estimates
ROADS AND PAVINGS

Item Excluding site overheads and profit	Unit	Total rate £
CAR PARKS		
Excavate 350 for pathways or roadbed to receive surface 100 mm thick priced separately, bring to grade, lay 100 well-rolled hardcore; lay 150 mm Type 1 granular material; lay kerbs BS 7263; 125 x 255 on both sides, including foundations haunched one side in 11.5 N/mm² concrete with all necessary formwork		
Work to falls, crossfalls and cambers not exceeding 15 degrees :		
1.50 m wide	m	116.62
2.00 m wide	m	142.48
3.00 m wide	m	190.93
4.00 m wide	m	239.38
5.00 m wide	m	287.83
6.00 m wide	m	336.28
7.00 m wide	m	384.73
As above but excavation 450 mm deep and base of Type 1 at 250 mm thick		
1.50 m wide	m	128.37
2.00 m wide	m	155.80
3.00 m wide	m	210.91
4.00 m wide	m	266.02
5.00 m wide	m	321.13
6.00 m wide	m	376.25
7.00 m wide	m	431.35
To excavated and prepared base above, lay roadbase of 40 size dense bitumen macadam to BS 4987, 70 thick, lay wearing course of 10 size dense bitumen macadam 30 thick, mark out car parking bays 5.0 m x 2.4 m with thermoplastic road paint: surfaces all mechanically laid		
per bay 5.0 m x 2.4 m	each	170.35
gangway	m2	11.64
As above but stainless steel road studs 100 x 100 to BS 873:pt.4, two per bay in lieu of thermoplastic paint:		
per bay 5.0 m x 2.4 m	each	177.28
Car park as above but with interlocking concrete blocks 200 x 100 x 80 mm; grey		
per bay 5.0 m x 2.4 m	each	207.48
gangway	m²	17.29
Car park as above but with interlocking concrete blocks 200 x 100 x 80 mm; colours		
per bay 5.0 m x 2.4 m	each	222.96
gangway	m²	18.58
Car park as above but with interlocking concrete blocks 200 x 100 x 60 mm; grey		
per bay 5.0 m x 2.4 m	each	197.04
gangway	m²	16.42
Car park as above but with interlocking concrete blocks 200 x 100 x 60 mm; colours		
per bay 5.0 m x 2.4 m	each	207.24
gangway	m²	17.27

Approximate Estimates
ROADS AND PAVINGS

Item Excluding site overheads and profit	Unit	Total rate £
CAR PARKS - cont'd		
Car park as above but with Marshalls Mono Hexpot 292 x 338 x 100 laid on 50 sand, fill with topsoil and peat (5:1) and fertilizer at 35 g/m², seed with dwarf rye grass at 35 g/m²:		
per bay 5.0 m x 2.4 m	each	406.08
gangway	m²	33.84
Car park as above but with lay Grass Concrete Grasscrete in situ continuously reinforced cellular surfacing, including expansion joints at 10 m centres, fill with topsoil and peat (5:1) and fertilizer at 35 g/m², seed with dwarf rye grass at 35 g/m²: GC2, 150 thick for HGV traffic including dust carts:		
per bay 5.0 m x 2.4 m	each	345.72
gangway	m²	28.81
GC1, 100 thick for cars and light traffic		
per bay 5.0 m x 2.4 m	each	288.84
gangway	m²	24.07
CAR PARKING FOR DISABLED PEOPLE		
To excavated and prepared base above, lay roadbase of 20 size dense bitumen macadam to BS 4987, 80 thick, lay wearing course of 10 size dense bitumen macadam 30 thick, mark out car parking bays with thermoplastic road paint:		
per bay 5.80 m x 3.25 m (ambulant)	each	219.41
per bay 6.55 m x 3.80 m (wheelchair)	each	289.72
gangway	m²	11.64
Car park as above but with interlocking concrete blocks 200 x 100 x 60 mm; grey, but mark out bays		
per bay 5.80 m x 3.25 m (ambulant)	each	315.03
per bay 6.55 m x 3.80 m (wheelchair)	each	414.91
gangway	m²	16.42
PLAY AREA FOR BALL GAMES		
Excavate for playground, bring to grade, lay 225 consolidated hardcore, blind with 100 type 1, lay macadam base of 40 size dense bitumen macadam to 75 thick, lay wearing course 30 thick		
Over 1000 m2	100 m²	3359.23
400 m2 -1000 m2	100 m²	3456.23

Approximate Estimates
ROADS AND PAVINGS

Item Excluding site overheads and profit	Unit	Total rate £
INTERLOCKING BLOCK PAVING		
Edge restraint to block paving; Excavate for groundbeam, lay Concrete 1:2:4 200 wide x 150mm deep. Lay butt jointed block header course on 50 mm thick sharp sand bed. Inclusive of haunching one side; butt joints		
200 x 100 x 60; laid to header course - £161.00/1000	m	12.48
200 x 100 x 60; laid to stretcher course - £161.00/1000	m	10.94
As above but clay bricks with mortar joints		
215 x 112 x 50 laid to header course - £280.00/1000	m	14.87
215 x 112 x 50 laid to stretcher course - £280.00/1000	m	14.02
Excavate ground and bring to levels, treat substrate with total herbicide, supply and lay granular fill Type 1 150 thick laid to falls and compacted, supply and lay brick pavers to BS 6677 in herring-bone pattern, laid on 50 compacted sharp sand, bedded in 10 loose sand, vibrated, joints filled with loose sand and vibrated, all excluding edgings or kerbs measured separately		
Blocks 200 x 100 x 60	m²	29.71
200 x 100 x 80	m²	29.86
Reduce levels, lay 150 granular material Type 1, lay 200 x 100 x 60 vehicular block paving to 90 degree herringbone pattern, on 50 compacted sand bed, vibrated, jointed in sand and vibrated, excavate and lay precast concrete edging 50 x 150 to BS 7263, on concrete foundation 1:2:4		
1.0 m wide clear width between edgings	m	64.60
1.5 m wide clear width between edgings	m	78.83
2.0 m wide clear width between edgings	m	93.09
As above but blocks laid 45 degree herringbone pattern including cutting edging blocks:		
1.0 m wide clear width between edgings	m	70.54
1.5 m wide clear width between edgings	m	84.77
2.0 m wide clear width between edgings	m	99.03
As above but all excavation by hand disposal off site by grab		
1.0 m wide clear width between edgings	m	151.15
1.5 m wide clear width between edgings	m	178.78
2.0 m wide clear width between edgings	m	206.97

Approximate Estimates
ROADS AND PAVINGS

Item Excluding site overheads and profit	Unit	Total rate £
BRICK PAVING		
Excavate ground and bring to levels, treat substrate with total herbicide, dig out weak spots and fill with compacted DoT Type 1 fill, supply and lay granular fill DoT Type 1 150 thick laid to falls and compacted, supply and lay brick pavers to BS 6677 in herring-bone pattern, laid on 50 mm compacted sharp sand, bedded in 10 mm loose sand, vibrated, joints filled with loose sand and vibrated, all excluding edgings or kerbs measured separately		
Pedestrian		
215 x 102.5 x 65 thick - £280.00/1000	m²	77.50
215 x 102.5 x 50 - £295.00/1000	m²	72.07
As above but to trafficked areas on 250mm thick Type 1 base		
215 x 102.5 x 65 thick - £280.00/1000	m²	83.35
215 x 102.5 x 50 - £295.00/1000	m²	77.92
Pedestrian areas as above but excavation by hand, barrowed to spoil heap maximum distance 25 m and removal off site by grab		
215 x 102.5 x 65 thick - £280.00/1000	m²	138.72
215 x 102.5 x 50 - £295.00/1000	m²	131.78
BASES FOR PAVING		
Excavate ground and reduce levels, to receive 38 mm thick slab and 25 mm mortar bed; dispose of excavated material off site; treat substrate with total herbicide		
Lay granular fill Type 1 150 thick laid to falls and compacted.		
All by machine	m²	10.85
All by hand except disposal by grab	m²	35.30
Lay 1:2:4 concrete base 150 thick laid to falls		
All by machine	m²	23.92
All by hand except disposal by grab	m²	54.06
As above but with concrete base 150 deep but inclusive of 150 mm compacted hardcore		
By Machine	m²	32.23
As above but inclusive of reinforcement fabric A142		
350 mm deep	m²	37.02

Approximate Estimates
ROADS AND PAVINGS

Item Excluding site overheads and profit	Unit	Total rate £
FLAG PAVING TO PEDESTRIAN AREAS		
Prices are inclusive of all mechanical excavation and disposal		
Supply and lay precast concrete flags; **Excavate ground and reduce levels,** **treat substrate with total herbicide,** **supply and lay granular fill Type 1 150** **thick laid to falls and compacted,** Standard precast concrete flags to to BS 7263 bedded and jointed in lime:sand mortar (1:3).		
450 x 450 x 70 chamfered	m²	33.75
450 x 450 x 50 chamfered	m²	32.27
600 x 300 x 50	m²	31.33
600 x 450 x 50	m²	30.41
600 x 600 x 50	m²	28.15
750 x 600 x 50	m²	27.80
900 x 600 x 50	m²	27.31
Coloured flags bedded and jointed in lime:sand mortar (1:3).		
600 x 600 x 50	m²	31.56
450 x 450 x 70 chamfered	m²	35.58
600 x 600 x 50	m²	32.05
400 x 400 x 65	m²	40.27
750 x 600 x 50	m²	31.31
900 x 600 x 50	m²	30.33
Marshalls Saxon; textured concrete flags; reconstituted Yorkstone in colours, butt jointed bedded in lime:sand mortar (1:3).		
300 x 300 x 35	m²	57.45
450 x 450 x 45	m²	49.64
600 x 300 x 35	m²	46.13
600 x 600 x 50	m²	44.15
450 x 450 x 35	m²	11.34
Tactile flags; Marshalls, Blister Tactile pavings; red or buff; for blind pedestrian guidance laid to designed pattern		
450 x 450	m²	39.62
400 x 400	m²	42.08
PEDESTRIAN DETERRENT PAVING.		
Excavate ground and bring to levels, **treat substrate with total herbicide,** **supply and lay granular fill Type 1 150** **mm thick laid to falls and compacted,** **supply and lay precast deterrent paving** **units bedded in lime:sand mortar (1:3)** **and jointed in lime:sand mortar (1:3).** Marshalls Mono		
Lambeth pyramidal paving 600 x 600 x 75	m²	53.46
Townscape		
Abbey square cobble pattern pavings; reinforced 600 x 600 x 60	m²	50.88
Geoset chamfered studs 600 x 600 x 60	m²	46.75

Approximate Estimates
ROADS AND PAVINGS

Item Excluding site overheads and profit	Unit	Total rate £
IMITATION YORKSTONE PAVINGS		
As above but imitation Yorkstone paving laid to coursed patterns		
Marshalls Heritage square or rectangular		
300 x 300 x 38	m²	57.99
600 x 300 x 30	m²	45.41
600 x 450 x 38	m²	44.76
450 x 450 x 38	m²	43.00
600 x 600 x 38	m²	39.73
As above but laid to Random rectangular patterns		
Various sizes selected from the above	m²	50.76
Imitation Yorkstone laid random rectangular as above but on concrete base 150 thick		
By machine	m²	63.83
By hand	m²	88.28
NATURAL STONE SLAB PAVING		
Excavate ground by machine and reduce levels, to receive 65 mm thick slab and 25 mm mortar bed; dispose of excavated material off site; treat substrate with total herbicide, lay granular fill Type 1 150 thick laid to falls and compacted. Lay to random rectangular pattern on 25 mm mortar bed		
new riven slabs laid random rectangular	m²	108.90
reclaimed Cathedral grade riven slabs laid random rectangular	m²	124.15
new slabs sawn 6 sides laid random rectangular	m²	95.17
new slabs, 3 sizes, sawn 6 sides laid to coursed pattern	m²	99.29
As above but excavation by hand disposal by grab		
new riven slabs laid random rectangular	m²	132.37
reclaimed Cathedral grade riven slabs laid random rectangular	m²	147.62
new slabs sawn 6 sides laid random rectangular	m²	118.64
new slabs, 3 sizes, sawn 6 sides laid to coursed pattern	m²	122.76
Excavate ground by machine and reduce levels, to receive 65 mm thick slab and 25 mm mortar bed; dispose of excavated material off site; treat substrate with total herbicide, lay granular fill Type 1 150 thick laid to falls and compacted.		
new riven slabs laid random rectangular	m²	108.90
reclaimed Cathedral grade riven slabs laid random rectangular	m²	124.15
new slabs sawn 6 sides laid random rectangular	m²	95.17
new slabs, 3 sizes, sawn 6 sides laid to coursed pattern	m²	99.29

Approximate Estimates
ROADS AND PAVINGS

Item Excluding site overheads and profit	Unit	Total rate £
As above but excavation by hand disposal by grab		
new riven slabs laid random rectangular	m²	132.37
reclaimed Cathedral grade riven slabs laid random rectangular	m²	147.62
new slabs sawn 6 sides laid random rectangular	m²	118.64
new slabs, 3 sizes, sawn 6 sides laid to coursed pattern	m²	122.76
Yorkstone paving as above but laid to 1:2:4 concrete base 150 thick, excavation and disposal by machine		
new riven slabs laid random rectangular	m²	120.99
reclaimed Cathedral grade riven slabs laid random rectangular	m²	136.24
new slabs sawn 6 sides laid random rectangular	m²	107.26
new slabs, 3 sizes, sawn 6 sides laid to coursed pattern	m²	111.38
As above but excavation by hand disposal by grab		
new riven slabs laid random rectangular	m²	151.13
reclaimed Cathedral grade riven slabs laid random rectangular	m²	166.38
new slabs sawn 6 sides laid random rectangular	m²	137.40
new slabs, 3 sizes, sawn 6 sides laid to coursed pattern	m²	141.52
GRANITE SETT PAVING - PEDESTRIAN		
Excavate ground and bring to levels, lay 100 hardcore to falls, compacted with 5 tonne roller, blind with compacted Type 1 50 mm thick, lay 100 concrete 1:2:4 , supply and lay granite setts 100 x 100 x 100 mm bedded in cement:sand mortar (1:3) 25 mm thick minimum, close butted and jointed in fine sand, all excluding edgings or kerbs measured separately		
Setts laid to bonded pattern and jointed		
new setts 100 x 100 x 100	m²	138.15
secondhand cleaned 100 x 100 x 100	m²	145.90
Setts laid in curved pattern:		
new setts 100 x 100 x 100	m²	144.09
secondhand cleaned 100 x 100 x 100	m²	151.84

Approximate Estimates
ROADS AND PAVINGS

Item Excluding site overheads and profit	Unit	Total rate £
GRANITE SETT PAVING TRAFFICKED AREAS		
Excavate ground and bring to levels, lay 100 hardcore to falls, compacted with 5 tonne roller, blind with compacted Type 1 50 mm thick, lay 150 site mixed concrete 1:2:4 reinforced with steel fabric to BS 4483 ref: A 142, supply and lay granite setts 100 x 100 x 100 mm bedded in cement:sand mortar (1:3) 25 mm thick minimum, close butted and jointed in fine sand, all excluding edgings or kerbs measured separately		
Site mixed concrete		
new setts 100 x 100 x 100	m²	144.94
secondhand cleaned 100 x 100 x 100	m²	152.69
Ready mixed concrete		
new setts 100 x 100 x 100	m²	142.66
secondhand cleaned 100 x 100 x 100	m²	150.41
CONCRETE PAVING		
Pedestrian Areas		
Excavate to reduce levels, lay 100 mm Type 1, lay 100 PAV 1 air entrained concrete, joints at max width of 6.0m cut out and sealed with sealant to BS 5212. Inclusive of all formwork and stripping	m²	65.65
Trafficked Areas		
Excavate to reduce levels, lay 150 mm Type 1 lay 150 mm PAV 2 40 N/mm² air entrained concrete, reinforced with steel mesh to BS 4483 200 x 200 square at 2.22kg/m², joints at max width of 6.0m cut out and sealed with sealant to BS 5212	m²	85.45
BEACH COBBLE PAVING		
Excavate ground and bring to levels and fill with compacted Type 1 fill 100 thick, lay GEN 1 concrete base 100 thick, supply and lay cobbles individually laid by hand bedded in cement:sand mortar (1:3) 25 thick minimum, dry grout with 1:3 cement:sand grout, brush off surplus grout and water in, sponge of cobbles as work proceeds all excluding formwork edgings or kerbs measured separately		
Scottish beach cobbles 200 - 100 mm	m²	88.78
Kidney flint cobbles 100 - 75 mm	m²	95.35
Scottish beach cobbles 75 -50 mm	m²	109.30

Approximate Estimates
ROADS AND PAVINGS

Item Excluding site overheads and profit	Unit	Total rate £
CONCRETE SETT PAVING		
Excavate ground and bring to levels, supply and lay 150 granular fill Type 1 laid to falls and compacted, supply and lay setts bedded in 50 sand and vibrated, joints filled with dry sand and vibrated, all excluding edgings or kerbs measured separately		
Marshalls Mono; Tegula precast concrete setts		
random sizes 60 thick	m²	41.77
single size 60 thick	m²	39.67
random size 80 thick	m²	44.83
single size 80 thick	m²	50.48
Concrete setts Blanc de Bierges:		
140 x 140 x 80	m²	83.27
210 x 140 x 80	m²	79.67
70 x 70 x 70	m²	88.90
GRASS CONCRETE PAVING		
Reduce levels, lay 150 mm Type 1 granular fill compacted, supply and lay precast grass concrete blocks on 20 sand and level by hand, fill blocks with 3mm sifted topsoil and pre-seeding fertilizer at 50g/m², sow with perennial ryegrass/chewings fescue see seed at 35g/m²		
Grass Concrete		
GB103 406 x 406 x 103	m²	33.77
GB83 406 x 406 x 83	m²	31.83
Extra for geotextile fabric underlayer	m²	0.73
Firepaths		
Excavate to reduce levels, lay 300 mm well rammed hardcore, blinded with 100 mm type 1, supply and lay Marshalls Mono "Grassguard 180" precast grass concrete blocks on 50 sand and level by hand, fill blocks with sifted topsoil and pre-seeding fertilizer at 50 g/m²		
firepath 3.8 m wide	m	209.72
firepath 4.4 m wide	m	242.82
firepath 5.0 m wide	m	276.14
turning areas	m²	55.23
Reduce levels, lay granular fill Type 1, 150 thick, lay precast concrete cellular blocks on 20 sharp sand, fill with topsoil and Growmore fertilizer at 50 g/m², sow with dwarf rye grass seed at 35 g/m²		
Marshalls Mono Hexpot: 360 x 415 x 100	m²	36.96
Marshalls Mono Hexpot: 292 x 338 x 100	m²	39.92
Charcon Hard Landscaping Grassgrid: 366 x 274 x 100	m²	30.83

Approximate Estimates
ROADS AND PAVINGS

Item Excluding site overheads and profit	Unit	Total rate £
SLAB/BRICK PATHS		
Stepping stone path inclusive of hand excavation and mechanical disposal, 100 mm Type 1 and sand blinding. Slabs to comply with BS 7263 laid 100 mm apart to existing turf		
600 wide; 600 x 600 x 50 slabs		
natural finish	m	39.22
coloured	m	41.06
exposed aggregate	m	48.67
900 wide; 600 x 900 x 50 slabs		
natural finish	m	43.89
coloured	m	46.66
exposed aggregate	m	58.06
Pathway inclusive of mechanical excavation and disposal, 100 mm Type 1 and sand blinding. Slabs to comply with BS 7263 close butted		
Straight butted path 900 wide; 600 x 900 x 50 slabs		
natural finish	m	17.52
coloured	m	19.36
exposed aggregate	m	26.97
Straight butted path 1200 wide; double row; 600 x 900 x 50 slabs, laid stretcher bond		
natural finish	m	30.76
coloured	m	34.85
exposed aggregate	m	51.75
Straight butted path 1200 wide, one row of 600 x 600 x 50, one row 600 x 900 x 50 slabs;		
natural finish	m	30.26
coloured	m	33.82
exposed aggregate	m	57.69
Straight butted path 1500 wide, slabs of 600 x 900 x 50, and 600 x 600 x 50 slabs, laid to bond		
natural finish	m	37.54
coloured	m	41.38
exposed aggregate	m	57.69
Straight butted path 1800 wide, two rows of 600 x 900 x 50 slabs, laid bonded		
natural finish	m	40.78
coloured	m	52.11
exposed aggregate	m	77.69
Straight butted path 1800 wide, three rows of 600 x 900 x 50 slabs, laid stretcher bond		
natural finish	m	45.28
coloured	m	56.61
exposed aggregate	m	82.19

Approximate Estimates
ROADS AND PAVINGS

Item Excluding site overheads and profit	Unit	Total rate £
Brick paved paths : Bricks 215 x 112.5 x 65 with mortar joints. Prices are inclusive of excavation, disposal off site, 100 hardcore, 100 1:2:4 concrete bed, edgings of brick 215 wide, haunched; all jointed in cement:lime:sand mortar (1:1:6).		
Path 1015 wide laid stretcher bond; edging course of headers		
rough stocks - £400.00/1000	m	117.42
engineering brick - £280.00/1000	m	112.55
Path 1015 wide laid stack bond:		
rough stocks - £400.00/1000	m	121.92
engineering brick - £280.00/1000	m	117.05
Path 1115 wide laid header bond:		
rough stocks - £400.00/1000	m	126.86
engineering brick - £280.00/1000	m	121.99
Path 1330 wide laid basketweave bond		
rough stocks - £400.00/1000	m	145.64
engineering brick - £280.00/1000	m	139.25
Path 1790 wide laid basketweave bond		
rough stocks - £400.00/1000	m	187.20
engineering brick - £280.00/1000	m	139.25
Brick paved paths; Bricks paviors 200 x 100 x 50 chamfered edge with butt joints. Prices are inclusive of excavation, 100 mm Type 1, and 50 mm sharp sand; jointing in kiln dried sand brushed in. Exclusive of edge restraints		
Path 1000 wide laid stretcher bond; edging course of headers		
rough stocks - £400.00/1000	m	112.59
engineering brick - £280.00/1000	m	112.55
Path 1330 wide laid basketweave bond		
rough stocks - £400.00/1000	m	145.64
engineering brick - £280.00/1000	m	139.25
Path 1790 wide laid basketweave bond		
rough stocks - £400.00/1000	m	187.20
engineering brick - £280.00/1000	m	139.25
GRAVEL PATHS		
Reduce levels and remove spoil to dump on site, lay 150 hardcore well rolled, lay 25 mm sand blinding and geofabric, lay 50 gravel watered and rolled, excavate and timber edge 150 x 38 to both sides of straight paths:		
1.0 m wide	m	20.10
1.5 m wide	m	27.47
2.0 m wide	m	31.26
As above but Breedon Gravel		
1.0 m wide	m	26.53
1.5 m wide	m	35.92
2.0 m wide	m	45.30

Approximate Estimates
ROADS AND PAVINGS

Item Excluding site overheads and profit	Unit	Total rate £
BARK PAVING		
Excavate to reduce levels, remove all topsoil to dump on site, treat area with herbicide. Lay 150 mm clean hardcore, blind with sand to BS 882 Grade C, lay 0.7 mm geotextile filter fabric water flow 50 l/m²/sec. Supply and fix treated softwood edging boards 50 x 150 mm to bark area on hardcore base extended 150 mm beyond the bark area, boards fixed with galvanized nails to treated softwood posts 750 x 50 x 75 mm driven into firm ground at 1.0 m centres. Edging boards to finish 25 mm above finished bark surface. Tops of posts to be flush with edging boards and once weathered. Supply and lay 100 mm Melcourt conifer walk chips 10-40 mm size.		
1.00 m wide	m	59.82
2.00 m wide	m	100.05
3.00 m wide	m	141.05
4.00 m wide	m	180.51
Extra over for wood fibre	m2	-0.05
Extra over for hardwood chips	m2	0.08
FOOTPATHS		
Excavate footpath to reduce level, remove arisings to tip on site maximum distance 25 m. Lay Type 1 granular fill 100 thick, lay base course of 28 size dense bitumen macadam 50 thick, wearing course of 10 size dense bitumen macadam 20 thick: Timber edge 150 x 38 mm.		
Areas over 1000 m2		
1.0 m wide	m	25.19
1.5 m wide	m	34.26
2.0 m wide	m	42.63
Areas 400 m2 - 1000 m2		
1.0 m wide	m	27.59
1.5 m wide	m	37.86
2.0 m wide	m	58.24

Approximate Estimates
8.8 SPECIAL SURFACES FOR SPORT/PLAYGROUNDS

Item Excluding site overheads and profit	Unit	Total rate £
BOWLING GREEN CONSTRUCTION; **Baylis Landscape Contractors**		
Bowling Green; Complete Excavate 300 mm deep and grade to level. Excavate and install 100 mm land drain to perimeter and backfill with shingle. Install 60 mm land drain to surface at 4.5 m centres backfilled with shingle. Install 50 m non perforated pipe 50 m long. Install 100 mm compacted and levelled drainage stone, blind with grit and sand. Spread 150 mm imported 70:30 sand:soil accurately compacted and levelled. Lay bowling green turf and top dress twice luted into surface. Exclusive of perimeter ditches and bowls protection		
6 rink green 38.4 x 38.4 m	each	49507.50
Install "Toro" automatic bowling green irrigation system with pump, tank, controller, electrics, pipework and 8 nr "Toro 780" sprinklers. Excluding pumphouse (optional)	each	7687.50
Supply and install to perimeter of green, "Sportsmark" preformed bowling green ditch channels		
"Ultimate Design 99" steel reinforced concrete channel, 600 mm long section	each	7295.00
"Ultimate GRC " glass reinforced concrete channel, 1.2 m long section	each	11100.00
"Ultimate Design 2001" medium density rotational moulded channel. 1 m long section with integral bowls protection	each	11100.00
Supply and fit Bowls Protection material to rear hitting face of "Ultimate" channels 1 and 2 above		
"Curl Grass" artificial grass, 0.45 m wide	each	2050.00
"Astroturf" artificial grass, 0.45 m wide	each	1860.00
50 mm rubber bowls bumper (2 rows)	each	4225.00
Bowls protection ditch liner laid loose, 300 mm wide	each	1250.00
JOGGING TRACK		
Excavate track 250 deep, treat substrate with Casoron G4, lay filter membrane, lay 100 depth gravel waste or similar, lay 100 mm compacted gravel	100 m²	2178.78
Extra for treated softwood edging 50 x 150 on 50 x 50 posts, both sides	m	7.76

Approximate Estimates
SPECIAL SURFACES FOR SPORT/PLAYGROUNDS

Item Excluding site overheads and profit	Unit	Total rate £
PLAYGROUNDS		
Excavate playground area, and dispose of arisings to tip; lay Type 1 granular fill, lay macadam surface two coat work 80 thick; base course of 28 size dense bitumen macadam 50 thick, wearing course of 10 size dense bitumen macadam 30 thick		
Areas over 1000 m2		
Excavation 180 mm base 100 mm thick	m²	20.90
Excavation 225 mm base 150 mm thick	m²	23.82
Areas 400 m2 -1000 m2		
Excavation 180 mm base 100 mm thick	m²	23.35
Excavation 225 mm base 150 mm thick	m²	26.27
Excavate playground area to given levels and falls, remove soil off site and backfill with compacted Type 1 granular fill. Lay ready mixed concrete to fall 2% in all direction		
base 150 mm thick; surface 100 mm thick	m²	22.11
base 150 mm thick; surface 150 mm thick	m²	28.07
SAFETY SURFACING; Baylis Landscape Contractors		
Excavate ground and reduce levels, to receive 38 mm thick slab and 25 mm mortar bed; dispose of excavated material off site; treat substrate with total herbicide; Lay granular fill Type 1 150 thick laid to falls and compacted. lay macadam base 40 mm thick; Supply and lay "Ruberflex" wet pour safety system to thicknesses as specified		
All by machine except macadam by hand		
Black		
15 mm thick	100 m²	4474.98
35 mm thick	100 m²	5806.98
60 mm thick	100 m²	7037.98
Coloured		
15 mm thick	100 m²	7856.98
35 mm thick	100 m²	8369.98
60 mm thick	100 m²	9189.98
All by hand except disposal by grab		
Black		
15 mm thick	100 m²	7149.37
35 mm thick	100 m²	8481.37
60 mm thick	100 m²	9712.37
Coloured		
15 mm thick	100 m²	10531.37
35 mm thick	100 m²	11044.37
60 mm thick	100 m²	11864.37

Approximate Estimates
SPECIAL SURFACES FOR SPORT/PLAYGROUNDS

Item Excluding site overheads and profit	Unit	Total rate £
Excavate playground area to 450 depth, lay 150 broken stone or clean hardcore, lay filter membrane, lay bark surface: Melcourt Industries		
Playbark 10/50 - 300 thick	100 m²	3711.55
Playbark 8/25 - 300 thick	100 m²	3506.55
SPORTSGROUND CONSTRUCTION; **Baylis Landscape Contractors** Plain sports pitches; site clearance, grading and drainage not included. **Cultivate ground and grade to levels, apply pre-seeding fertilizer at 900 kg/ha, apply pre-seeding selective weedkiller, seed in two operations with sports pitch type grass seed at 350 kg/ha, harrow and roll lightly, including initial cut. Size**		
association football, senior 114 m x 72 m	each	3256.94
association football, junior 106 m x 58 m	each	2626.56
rugby union pitch 156 m x 81 m	each	4780.34
rugby league pitch 134 m x 60 m	each	3151.88
hockey pitch 95 m x 60 m	each	2416.44
shinty pitch 186 m x 96 m	each	6724.00
men's lacrosse pitch 100 m x 55 m	each	2311.38
women's lacrosse pitch 110 m x 73 m	each	3362.00
target archery ground 150 m x 50 m	each	2731.63
cricket outfield 160 m x 142 m	each	8510.06
cycle track outfield 160 m x 80 m	each	5253.13
polo ground 330 m x 220 m	each	25109.94
Cricket square; excavate to depth of 150 mm, pass topsoil through 6 mm screen, return and mix evenly with imported marl or clay loam, bring to accurate levels, apply pre-seeding fertilizer at 50 g/m, apply selective weedkiller, seed with cricket square type g grass seed at 50 g/m², rake in and roll lightly, erect and remove temporary protective chestnut fencing, allow for initial cut and watering three times: 22.8 m x 22.8 m	each	13658.13

Approximate Estimates
8.9 PREPARATION FOR PLANTING /TURFING

Item Excluding site overheads and profit	Unit	Total rate £
CULTIVATION BY HAND		
Spread only and lightly consolidate topsoil brought from spoil heap in layers not exceeding 150, grade to specified levels, remove stones over 25mm, treat with paraquat-diquat weedkiller, all by hand		
100 mm thick	100 m²	360.07
150 mm thick	100 m²	540.11
300 mm thick	100 m²	1080.22
450 mm thick	100 m²	1620.32
As above but inclusive of loading to location by barrow maximum distance 25 m; finished topsoil depth		
100 mm thick	100 m²	792.07
150 mm thick	100 m²	1188.11
300 mm thick	100 m²	2376.22
450 mm thick	100 m²	3564.32
500 mm thick	100 m²	3958.56
600 mm thick	100 m²	4747.03
As above but loading to location by barrow maximum distance 100 m; finished topsoil depth		
100 mm thick	100 m²	810.07
150 mm thick	100 m²	1215.11
300 mm thick	100 m²	2430.22
450 mm thick	100 m²	3645.32
500 mm thick	100 m²	4048.56
600 mm thick	100 m²	4855.03
Extra to the above for incorporating mushroom compost at 50 mm/m² into the top 150 mm of topsoil; by hand	100 m²	169.00
Extra to above for imported topsoil PC 11.00 m3 allowing for 20% settlement		
100 mm thick	100 m²	139.90
150 mm thick	100 m²	209.85
300 mm thick	100 m²	419.70
450 mm thick	100 m²	629.55
500 mm thick	100 m²	699.50
600 mm thick	100 m²	839.40
750 mm thick	100 m²	1049.25
1.00 m thick	100 m²	1399.00
CULTIVATION BY MACHINE		
Treat area with systemic non selective herbicide 1 month before starting cultivation operations. Rip up subsoil using subsoiling machine to a depth of 250 below topsoil at 1.20 m centres in light to medium soils, rotavate to 200 deep in two passes, cultivate with chain harrow, roll lightly, clear stones over 50mm		
by tractor	100 m²	8.30

Approximate Estimates
PREPARATION FOR PLANTING /TURFING

Item Excluding site overheads and profit	Unit	Total rate £
As above but carrying out operations in clay or compacted gravel	100 m²	8.88
As above but ripping by tractor rotavation by pedestrian operated rotavator, clearance and raking by hand, herbicide application by knapsack sprayer	100 m²	36.95
As above but carrying out operations in clay or compacted gravel	100 m²	50.45
Spread and lightly consolidate topsoil brought from spoil heap not exceeding 100 m, in layers not exceeding 150, grade to specified levels, remove stones over 25, treat with paraquat-diquat weedkiller, all by machine		
100 mm thick	100 m²	63.44
150 mm thick	100 m²	95.49
300 mm thick	100 m²	190.97
450 mm thick	100 m²	286.11
Extra to above for imported topsoil PC £11.00 m3 allowing for 20% settlement		
100 mm thick	100 m²	139.90
150 mm thick	100 m²	209.85
300 mm thick	100 m²	419.70
450 mm thick	100 m²	629.55
500 mm thick	100 m²	699.50
600 mm thick	100 m²	839.40
750 mm thick	100 m²	1049.25
1.00 m thick	100 m²	1399.00
Extra for incorporating mushroom compost at 50 mm/m² into the top 150 mm of topsoil (compost delivered in 20 m3) loads		
manually spread, mechanically rotavated	100 m²	142.80
mechanically spread and rotavated	100 m²	96.95
Extra for incorporating manure at 50 mm/m² into the top 150 mm of topsoil loads		
manually spread, mechanically rotavated 20m3 loads	100 m²	193.30
mechanically spread and rotavated 60 m3 loads	100 m²	163.30

Approximate Estimates
8.10 SEEDING AND TURFING

Item Excluding site overheads and profit	Unit	Total rate £
Bring top 200 of topsoil to a fine tilth using tractor drawn implements, remove stones over 25 by mechanical stone rake and bring to final tilth by harrow, apply pre-seeding fertilizer at 50 g/m² and work into top 50 during final cultivation, seed with certified grass seed in two operations, roll seedbed lightly after sowing.		
general amenity grass at 35 g/m²; BSH A3	100 m²	35.10
general amenity grass at 35 g/m²; Johnsons Taskmaster	100 m²	26.73
shaded areas; BSH A6 at 50 g/m²	100 m²	42.50
Motorway and road verges; Perryfields Pro 120 25-35 g/m²	100 m²	34.42
Bring top 200 of topsoil to a fine tilth using pedestrian operated rotavator, remove stones over 25, apply pre-seeding fertilizer at 50 g/m² and work into top 50 during final hand cultivation, seed with certified grass seed in two operations, rake and roll seedbed lightly after sowing,		
general amenity grass at 35 g/m²; BSH A3	100 m²	97.62
general amenity grass at 35 g/m²; Johnsons Taskmaster	100 m²	89.25
shaded areas; BSH A6 at 50 g/m²	100 m²	105.02
Motorway and road verges; Perryfields Pro 120 25-35 g/m²	100 m²	96.94
Extra to above for using imported topsoil spread by machine		
100 minimum depth	100 m²	236.00
150 minimum depth	100 m²	354.00
Extra for slopes over 30	50 %	-
Extra to above for using imported topsoil spread by hand; maximum distance for transporting soil 100 m		
100 minimum depth	m²	568.07
150 minimum depth	m²	852.11
Extra for slopes over 30	50 %	-
Extra for mechanically screening top 25 of topsoil through 6 mm screen and spreading on seedbed, debris carted to dump on site not exceeding 100 m	m³	4.68
Bring top 200 of topsoil to a fine tilth, remove stones over 50, apply pre-emergent weedkiller in accordance with manufacturer's instructions, apply pre-seeding fertilizer at 50 g/m² and work into top 50 of topsoil during final cultivation, seed with certified grass seed in two operations, harrow and roll seedbed lightly after sowing, Apply pesticide at 100 g/m², level with mechanical lute		
outfield grass at 350 kg/ha - Perryfields Pro 40	ha	8551.00

Approximate Estimates
SEEDING AND TURFING

Item Excluding site overheads and profit	Unit	Total rate £
outfield grass at 350 kg/ha - Perryfields Pro 70	ha	8523.00
sportsfield grass at 300 kg/ha - Johnsons Sportsmaster	ha	8217.00
Areas not requiring pesticide or mechanical operations		
low maintenance grass at 350 kg/ha	ha	4535.00
verge mixture grass at 150 kg/ha	ha	3683.00
Extra for wild flora mixture at 30 kg/ha		
BSH WSF 75kg /ha	ha	3225.00
Extra for slopes over 30	50%	-
Cut existing turf to 1.0 x 1.0 x 0.5 m turves, roll up and move to stack not exceeding 100 m		
by pedestrian operated machine, roll up and stack by hand	100 m²	68.21
all works by hand	100 m²	195.00
Extra for boxing and cutting turves	100 m²	3.49
Bring top 200 of topsoil to a fine tilth in 2 passes, remove stones over 25 by and bring to final tilth, apply pre-seeding fertilizer at 50 g/m² and work into top 50 during final cultivation, roll turf bed lightly		
using tractor drawn implements and mechanical stone rake	m2	0.25
cultivation by pedestrian rotavator, all other operations by hand	m2	0.60
As above but bring turf from stack not exceeding 100 m, lay turves to stretcher bond using plank barrow runs, firm turves using wooden turf beater.		
using tractor drawn implements and mechanical stone rake	m2	1.78
cultivation by pedestrian rotavator, all other operations by hand	m2	1.99
As above but including imported turf; Rowlawn Medallion		
using tractor drawn implements and mechanical stone rake	m2	3.79
cultivation by pedestrian rotavator, all other operations by hand	m2	4.00
Extra over to all of the above for watering on two occasions and carrying out initial cut		
by ride on triple mower	100 m2	1.09
by pedestrian mower	100 m2	4.55
by pedestrian mower; box cutting	100 m2	5.09
Extra for using imported topsoil spread and graded by machine		
25 minimum depth	m2	0.66
75 minimum depth	m²	1.77
100 minimum depth	m²	2.36
150 minimum depth	m²	3.54
Extra for using imported topsoil spread and graded by hand; distance of barrow run 25 m		
25 minimum depth	m²	1.51
75 minimum depth	m²	3.77
100 minimum depth	m²	4.78
150 minimum depth	m²	7.17
Extra for work on slopes over 30 including pegging with 200 galvanized wire pins	m²	0.43

Approximate Estimates
SEEDING AND TURFING

Item Excluding site overheads and profit	Unit	Total rate £
Inturf Big Roll **Supply, deliver in one consignment, fully prepare the area and install in Big Roll format Inturf 553; a turfgrass comprising dwarf perennial ryegrass, smooth stalked rneadowgrass and fescues; installation by tracked machine**		
Preparation by machine	m²	2.99
cultivation by pedestrian rotavator, all other operations by hand	m²	3.38
Erosion control Cultivation and preparation operations by pedestrian rotavator as above by pedestrian rotavator, all other operations by hand; Laying only to prepared ground; Greenkeeper reinforced turf on banks netting including pegging with 200 pegs at 500 centres	m²	8.35
On ground previously cultivated, bring area to level, treat with herbicide. Lay 20 mm thick open texture erosion control mat with 100 mm laps, fixed with 8 x 400 mm steel pegs at 1.0 m centres. Sow with low maintenance grass suitable for erosion control on slopes at 35 g/m². Spread imported topsoil 25 mm thick and fertilizer at 35 g/m², water lightly using sprinklers on two occasions.	m²	5.55
As above but hand-watering by hose pipe maximum distance from mains supply 50 m	m²	5.63

Approximate Estimates
8.11 AFTERCARE OF TURF AREAS

Item Excluding site overheads and profit	Unit	Total rate £
Maintenance executed as part of a landscape construction contract.		
Grass cutting; fine turf, using pedestrian guided machinery; arisings boxed and disposed of off site		
per occasion	100 m²	4.92
per annum 26 cuts	m²	1.28
per annum 18 cuts	m²	0.92
Grass cutting; standard turf, using self propelled 3 gang machinery		
per occasion	100 m²	0.37
per annum 26 cuts	m²	0.10
per annum 18 cuts	m²	0.07
Maintenance for one year; recreation areas, parks, amenity grass areas, using tractor drawn machinery:		
per occasion	ha	26.47
per annum 26 cuts	ha	688.22
per annum 18 cuts	ha	476.46
Aerate ground with spiked aerator, apply weedkiller once, apply spring/summer fertilizer once, apply autumn/winter fertilizer once, cut grass, 16 cuts, sweep up leaves twice.		
as part of a landscape contract, defects liability	ha	2002.52
as part of a long term maintenance contract	ha	1553.00

Approximate Estimates
8.12 PLANTING

Item Excluding site overheads and profit	Unit	Total rate £
HEDGE PLANTING		
Excavate trench for hedge 300 wide x 450 deep by machine, deposit spoil alongside, and plant hedging plants in single row at 200 centres, and backfill with excavated material incorporating organic manure at 1 m³ per 5 m³, and carry out initial cut, including delivery of plants from nursery:		
bare root hedging plants		
PC - £0.30	100 m	548.94
PC - £0.60	100 m	698.94
PC - £1.00	100 m	898.94
PC - £1.20	100 m	998.94
PC - £1.50	100 m	1148.94
As above but two rows of hedging plants at 300 centres staggered rows:		
PC - £0.30	100 m	725.94
PC - £0.60	100 m	925.94
PC - £1.00	100 m	1192.60
PC - £1.50	100 m	1525.94
As above but all by hand two rows of hedging plants at 300 centres staggered rows		
PC - £0.30	100 m	963.00
PC - £0.60	100 m	1163.00
PC - £1.00	100 m	1429.66
PC - £1.50	100 m	1763.00
TREE PLANTING		
Excavate tree pit by hand, fork over bottom of pit, plant tree with roots well spread out, backfill with excavated material, incorporating organic manure at 1 m³ per 3 m³ of soil, one tree stake and two ties; tree pits square in sizes shown		
Light standard bare root tree in pit; PC £10.00		
600 x 600 deep	each	40.40
900 x 900 deep	each	53.93
Standard tree bare root tree in pit; PC £13.95		
600 x 600 deep	each	46.50
900 x 600 deep	each	55.22
900 x 900 deep	each	65.58
Standard root balled tree in pit; PC £16.00		
600 x 600 deep	each	57.94
900 x 600 deep	each	66.66
900 x 900 deep	each	77.02
Selected standard bare root tree, in pit; PC £18.80		
900 x 900 deep	each	77.32
1.00 m x 1.00 m deep	each	87.40
Selected standard root ball tree in pit; PC £30.00		
900 x 900 deep	each	96.97
1.00 m x 1.00 m deep	each	139.09

Approximate Estimates
PLANTING

Item Excluding site overheads and profit	Unit	Total rate £
Heavy standard bare root tree, in pit; PC £39.50		
900 x 900 deep	each	99.82
1.00 m x 1.00 m deep	each	115.27
1.20 m x 1.00 m deep	each	135.77
Heavy standard root ball tree, in pit; PC £50.00		
900 x 900 deep	each	123.64
1.00 m x 1.00 m deep	each	139.09
1.20 m x 1.00 m deep	each	159.59
Extra heavy standard bare root tree in pit; PC £49.00		
1.00 m x 1.00 deep	each	128.33
1.20 m x 1.00 m deep	each	179.21
1.50 m x 1.00 m deep	each	187.35
Extra heavy standard root ball tree in pit; PC £67.00		
1.00 m x 1.00 deep	each	161.93
1.20 m x 1.00 m deep	each	212.81
1.50 m x 1.00 m deep	each	220.95
Excavate tree pit by machine, fork over bottom of pit, plant tree with roots well spread out, backfill with excavated material, incorporating organic manure at 1 m³ per 3 m³ of soil, one tree stake and two ties; tree pits square in sizes shown		
Light standard bare root tree in pit; PC £10.00		
600 x 600 deep	each	35.87
900 x 900 deep	each	47.35
Standard tree bare root tree in pit; PC £13.95		
600 x 600 deep	each	41.97
900 x 600 deep	each	44.83
900 x 900 deep	each	50.00
Standard root balled tree in pit; PC £16.00		
600 x 600 deep	each	53.41
900 x 600 deep	each	56.27
900 x 900 deep	each	61.44
Selected standard bare root tree, in pit; PC £18.80		
900 x 900 deep	each	61.74
1.00 m x 1.00 m deep	each	72.17
Selected standard root ball tree in pit; PC £30.00		
900 x 900 deep	each	96.97
1.00 m x 1.00 m deep	each	123.86
Heavy standard bare root tree, in pit; PC £39.50		
900 x 900 deep	each	99.82
1.00 m x 1.00 m deep	each	100.04
1.20 m x 1.00 m deep	each	113.83
Heavy standard root ball tree, in pit; PC £50.00		
900 x 900 deep	each	123.64
1.00 m x 1.00 m deep	each	123.86
1.20 m x 1.00 m deep	each	137.65
Extra heavy standard bare root tree in pit; PC £49.00		
1.00 m x 1.00 deep	each	113.10
1.20 m x 1.00 m deep	each	126.89
1.50 m x 1.00 m deep	each	153.32
Extra heavy standard root ball tree in pit; PC £67.00		
1.00 m x 1.00 deep	each	146.70
1.20 m x 1.00 m deep	each	160.49
1.50 m x 1.00 m deep	each	186.92

Approximate Estimates
PLANTING

Item Excluding site overheads and profit	Unit	Total rate £
SEMI MATURE TREE PLANTING		
Excavate tree pit deep by machine, fork over bottom of pit, plant rootballed tree Acer platanoides "Emerald Queen" using telehandler where necessary, backfill with excavated material, incorporating Melcourt Topgrow bark/manure mixture at 1 m³ per 3 m³ of soil, Platypus underground guying system; tree pits 1500 x 1500 x 1500 mm deep inclusive of Platimats; excavated material not backfilled to treepit spread to surrounding area		
16 - 18 cm girth - £69.00	each	256.32
18 - 20 cm girth - £77.00	each	276.49
20 - 25 cm girth - £130.00	each	383.51
25 - 30 cm girth - £179.00	each	500.84
30 - 35 cm girth - £220.00	each	698.11
As above but treepits 2.00 x 2.00 x 1.5 m deep		
40 - 45 cm girth - £715.00	each	743.85
45 - 50 cm girth - £805.00	each	839.64
55 - 60 cm girth - £1265.00	each	1009.72
67 - 70 cm girth - £1800.00	each	1248.45
75 - 80 cm girth - £2400.00	each	1378.04
Extra to the above for imported topsoil moved 25 m from tipping area and disposal off site of excavated material		
Tree pits 1500 x 1500 x1500 m deep		
16 - 18 cm girth	each	152.76
18 - 20 cm girth	each	146.32
20 - 25 cm girth	each	140.44
25 - 30 cm girth	each	135.32
30 - 35 cm girth	each	130.78
SHRUBS GROUND COVERS AND BULBS		
Excavate planting holes on 250 x 250 mm x 300 deep to area previously ripped and rotavated; excavated material left alongside planting hole		
by mechanical auger		
250 mm centres (16 plants per m2)	m2	10.24
300 mm centres (11.11 plants per m2)	m2	7.11
400 mm centres (6.26 plants per m2)	m2	4.00
450 mm centres (4.93 plants per m2)	m2	3.16
500 mm centres (4 plants per m2)	m2	2.56
600 mm centres (2.77 plants per m2)	m2	1.77
750 mm centres (1.77 plants per m2)	m2	1.13
900 mm centres (1.23 plants per m2)	m2	0.79
1.00 m centres (1 plants per m2)	m2	0.64
1.50 m centres (0.44 plants per m2)	m2	0.28
As above but excavation by hand		
250 mm centres (16 plants per m2)	m2	10.24
300 mm centres (11.11 plants per m2)	m2	7.11
400 mm centres (6.26 plants per m2)	m2	4.00

Approximate Estimates
PLANTING

Item Excluding site overheads and profit	Unit	Total rate £
450 mm centres (4.93 plants per m2)	m2	3.16
500 mm centres (4 plants per m2)	m2	2.56
600 mm centres (2.77 plants per m2)	m2	1.77
750 mm centres (1.77 plants per m2)	m2	1.13
900 mm centres (1.23 plants per m2)	m2	0.79
1.00 m centres (1 plants per m2)	m2	0.64
1.50 m centres (0.44 plants per m2)	m2	0.28
Clear light vegetation from planting area and remove to dump on site, dig planting holes, plant whips with roots well spread out, backfill with excavated topsoil, including one 38 x 38 treated softwood stake, two tree ties, and mesh guard 750 high. Planting matrix 1.5 m x 1.5 m. Allow for beating up once at 10% of original planting, cleaning and weeding round whips once, applying weedkiller once at 35 gm/m², applying fertilizer once at 35 gm/m², using the following mix of whips, bare rooted:		
plant bare root plants average price 0.28 p each to a required matrix		
plant mix as above	100 m²	148.42
plant bare root plants average price 0.75 p each to a required matrix		
plant mix as above	100 m²	158.44
Cultivate and grade shrub bed, bring top 300 mm of topsoil to a fine tilth, incorporating Mushroom compost at 50 mm and Enmag slow release fertilizer; rake and bring to given levels, remove all stones and debris over 50 mm, dig planting holes average 300 x 300x 300 mm deep. Supply and plant specified shrubs in quantities as shown below, backfill with excavated material as above. Water to field capacity and mulch 50 mm bark chips 20-40 mm size. Water and weed regularly for 12 months and replace failed plants. Shrubs -3L PC £2.20; Ground covers - 9 cm PC £1.20		
100 % shrub area 300 centres		
300 mm centres	100 m2	4947.86
400 mm centres	100 m2	2957.27
500 mm centres	100 m2	2035.89
600 mm centres	100 m2	1535.36
100 % groundcovers		
200 mm centres	100 m2	6303.99
300 mm centres	100 m2	3021.49
400 mm centres	100 m2	1874.46
500 mm centres	100 m2	1342.89

Approximate Estimates
PLANTING

Item Excluding site overheads and profit	Unit	Total rate £
SHRUBS GROUND COVERS AND BULBS – cont'd		
Cultivate and grade shrub bed – cont'd		
Groundcover 30% / Shrubs 70% at the distances shown below		
200mm / 300 mm	100 m2	5291.61
300 mm /400 mm	100 m2	2946.99
300 mm /500 mm	100 m2	2319.99
400 mm/ 500 mm	100 m2	1967.79
Groundcover 50% / Shrubs 50% at the distances shown below		
200mm / 300 mm	100 m2	5630.79
300 mm /400 mm	100 m2	2985.09
300 mm /500 mm	100 m2	2536.59
400 mm/ 500 mm	100 m2	1950.09
Cultivate ground by machine and rake to level; plant bulbs as shown; bulbs PC 24.50 /100		
15 bulbs per m²	100 m²	641.23
25 bulbs per m²	100 m²	1049.83
50 bulbs per m²	100 m²	2071.33
Cultivate ground by machine and rake to level; plant bulbs as shown; bulbs PC 12.90 /100		
15 bulbs per m²	100 m²	446.83
25 bulbs per m²	100 m²	725.83
50 bulbs per m²	100 m²	1423.33
Form holes in grass areas and plant bulbs using bulb planter, backfill with organic manure and turf plug: Bulbs PC £13.00/100		
15 bulbs per m²	100 m²	643.50
25 bulbs per m²	100 m²	1072.50
50 bulbs per m²	100 m²	2145.00
PLANTING PLANTERS		
To brick planter, coat insides with 2 coats RIW liquid asphaltic composition; fill with 50 mm shingle and cover with geofabric; fill with screened topsoil incorporating 25% Topgrow compost and Enmag		
Planters 1.00 m deep		
1.00 x 1.00	each	105.78
1.00 x 2.00	each	179.22
1.00 x 3.00	each	253.14
Planters 1.50 m deep		
1.00 x 1.00	each	158.39
1.00 x 2.00	each	230.28
1.00 x 3.00	each	379.11

Approximate Estimates
PLANTING

Item Excluding site overheads and profit	Unit	Total rate £
TREE GRILLES Supply and fix cast iron tree grille 1200 x 1200 with 440 mm diameter tree hole complete with irrigation inlet and locking points, on matching steel angle paving support frame with flanges laid on paving base. PC £309.84	each	396.96

8.13 FENCING

Item Excluding site overheads and profit	Unit	Total rate £
BALLSTOP FENCING		
Supply and erect plastic coated 30 x 30 mm netting fixed to 60.3 diameter 12 mm solid bar lattice galvanized dual posts, top, middle and bottom rails with 3 horizontal rails on 60.3 mm dia. nylon coated tubular steel posts at 3.0 m centres and 60.3 mm dia. straining posts with struts at 50 m centres set 750 mm into FND2 concrete footings 300 x 300 x 600 mm deep. Include framed chain link gate 900 x 1800 mm high to match complete with hinges and locking latch.		
4500 high	100 m	6258.00
5000 high	100 m	8218.00
6000 high	100 m	9010.00
CATTLE GRID		
Excavate pit 4.0 m x 3.0 m x 500, dispose of spoil on site, lay concrete base (C7P) 100 thick on 150 hardcore, excavate trench and lay 100 dia. clay agricultural drain outlet, form concrete sides to pit 150 thick (C15P) including all formwork, construct supporting walls one brick thick in engineering brick laid in cement mortar (1:3) at 400 mm centres, install cattle grid by H S Jackson & Son (Fencing), erect side panels on steel posts set in concrete (1:3:6), erect 2 nr warning signs standard DoT pattern	each	2649.18
Erect stock gate, ms tubular field gate, diamond braced 1.8 m high hung on tubular steel posts set in concrete (C7P), complete with ironmongery, all galvanized		
width 3.00 m	each	231.89
width 4.20 m	each	248.16
CONCRETE FENCES		
Supply and erect precast concrete post and panel fence in 2 m bays, panels to be shiplap profile, aggregate faced one side, posts set 600 mm in ground in concrete.		
1500 mm high	m	26.92
1800 mm high	m	31.38
2100 mm high	m	35.95
Extra over for exposed aggregate faced panels	m²	7.44

Approximate Estimates
FENCING

Item Excluding site overheads and profit	Unit	Total rate £
DEER STOCK RABBIT FENCING		
Construct rabbit-stop fencing, erect galvanized wire netting, mesh 31, 900 above ground, 150 below ground turned out and buried, on 75 dia. treated timber posts 1.8 m long driven 700 into firm ground at 4.0 m centres, netting clipped to top and bottom straining wires 2.63 mm diameter, straining post 150 mm diameter, x 2.3 m long driven into firm ground at 50 m intervals		
turned in 150 mm	100 m	**723.62**
buried 150 mm	100 m	**801.62**
Construct deer-stop fencing, erect 5 no. 4 dia. plain galvanized wires and 5 no. 2 ply galvanized barbed wires at 150 spacing, on 45 x 45 x 5mm angle iron posts 2.4 m long driven into firm ground at 3.0 m centres, driven into firm ground at 3.0 m centres, with timber droppers 25 x 38 x 1.0 m long at 1.5 m centres	100 m	**1222.60**
Supply and erect forestry fencing of three lines of 3 mm plain galvanized wire tied to 1700 x 65 mm dia angle iron posts at 2750 m centres with 1850 x 100 mm dia straining posts and 1600 x 80 mm dia struts at 50.0 m centres driven into firm ground	100 m	**820.88**
CHESTNUT PALE FENCING		
Erect chestnut pale fencing, cleft chestnut pales, two lines galvanized wire, galvanized tying wire, treated softwood posts at 3.0 m centres and straining posts and struts at 50 m centres driven into firm ground:		
900 high, posts 75 dia. x 1200 long	m	**5.83**
1200 high, posts 75 dia. x 1500 long	m	**7.44**
CHAIN LINK FENCING		
Supply and erect chain link fencing. Form post holes and erect concrete posts and straining posts with struts at 50 m centres all set in 1:3:6 concrete. Fix line wires and fix 3 mm galvanized wire 50 mm chainlink fencing.		
900 mm high	m	**18.97**
1200 mm high	m	**20.56**
1800 mm high	m	**23.42**
2400 mm high	m	**45.77**

Approximate Estimates
FENCING

Item Excluding site overheads and profit	Unit	Total rate £
CHAIN LINK FENCING – cont'd		
As above but with plastic coated 3.15 gauge galvanized wire mesh		
900 mm high	m	18.69
1200 mm high	m	20.29
1800 mm high	m	23.59
Extra for additional concrete straining posts with 1 strut set in concrete		
900 high	each	52.22
1200 high	each	54.31
1400 high	each	58.57
1800 high	each	63.90
2400 high	each	85.69
Extra for additional concrete straining posts with 2 struts set in concrete		
900 high	each	74.50
1200 high	each	76.87
1400 high	each	86.89
1800 high	each	92.81
2400 high	each	94.69
Chain link fencing 3 mm galvanized as above but with mild steel angle posts and straining posts instead of concrete posts		
900 mm high	m	6.37
1200 mm high	m	7.59
1800 mm high	m	11.06
2400 mm high	m	13.01
Extra for additional angle iron straining posts with 1 strut set in concrete		
900 high	each	18.99
1200 high	each	22.86
1400 high	each	26.58
1800 high	each	30.23
2400 high	each	45.00
Extra for additional angle iron straining posts with 2 struts set in concrete		
900 high	each	27.65
1200 high	each	33.67
1400 high	each	40.26
1800 high	each	48.44
2400 high	each	65.00
TIMBER FENCING		
Construct timber rail, horizontal hit and miss type, rails 150 x 25, posts 100 x 100 at 1.8 m centres, twice stained with coloured wood preservative, including excavation for posts and concreting into ground (C7P):		
In treated softwood		
1800 mm high	m	45.32
In primed softwood		
1800 mm high	m	51.68

Approximate Estimates
FENCING

Item Excluding site overheads and profit	Unit	Total rate £
Erect close boarded timber fence in treated softwood, pales 89 x 19 lapped, 152 x 25 gravel boards, two 76 x 38 rectangular rails:		
Concrete posts 100 x 100 at 3.0 m centres set into ground in 1:3:6 concrete		
900 mm high	m	19.33
1350 mm high	m	24.58
1800 mm high	m	26.33
As above but with oak posts		
1350 mm high	m	24.22
1800 mm high	m	26.80
As above but with softwood posts		
1350 mm high	m	21.69
1650 mm high	m	22.77
1800 mm high	m	23.43
Erect post and rail fence, three horizontal rails 100 x 38, fixed with galvanized nails to 100 x 100 posts driven into firm ground:		
In treated softwood		
1200 mm high	m	15.49
In oak or chestnut		
1200 mm high	m	22.29
Construct post and rail fence to BS 1722: Part 7 Type MPR 13/4 1300 mm high with five rails 87 x 38 mm. Rails morticed into treated softwood posts 2100 75 x 150 mm at 3.0 m centres set 700 mm into firm ground with intermediate prick posts 1800 x 87 x 38		
1300 high	m	16.11
Construct cleft oak rail fence, with rails 300 mm minimum girth tennoned both ends. 125 x 100 mm treated softwood posts double mortised for rails, corner posts 125 x 125 mm, driven into firm ground at 2.5 m centres.		
3 rails	m	16.05
4 rails	m	26.35

Approximate Estimates
FENCING

Item Excluding site overheads and profit	Unit	Total rate £
RAILINGS		
Erect ms bar railing of 19 mm balusters at 115 mm centres welded to ms top and bottom rails 40 x 10, bays 2.0 m long, bolted to 51 x 51 ms hollow section posts set in C15P concrete, all metal work galvanized after fabrication:		
900 mm high	m	43.18
1200 mm high	m	59.75
1500 mm high	m	65.57
Supply and erect mild steel pedestrian guard rail Class A to BS 3049, panels 1000 mm high x 2000 mm wide with 150 mm toe space and 200 mm visibility gap at top, rails to be rectangular hollow sealed section 50 x 30 x 2.5 mm, vertical support 25 x 19 mm central between intermediate and top rail. Posts to be set 300 mm into paving base. All components factory welded and factory primed for painting on site	m	69.56
SECURITY FENCING		
Supply and erect chainlink fence, 51 mm x 3 mm mesh, with line wires and stretcher bars bolted to concrete posts at 3.0 m centres and straining posts at 10 m centres. Posts set in concrete 450 x 450 x 33% of height of post deep. Fit straight extension arms of 45 x 45 x 5 mm steel angle with three lines of barbed wire and droppers. All metalwork to be factory hot- dip galvanized for painting on site.		
900 high	m	22.35
1200 high	m	24.90
1800 high	m	33.84
As above but with PVC coated 3.15 mm mesh (diameter of wire 2.5 mm)		
900 high	m	22.07
1200 high	m	24.63
1800 high	m	34.01
Add to fences above for base of fence to be fixed with hairpin staples cast into concrete ground beam 1:3:6 site mixed concrete; mechanical excavation disposal to on site spoil heaps		
125 x 225 mm deep.	m	4.48
Add to fences above for straight extension arms of 45 x 45 x 5 mm steel angle with three lines of barbed wire and droppers.	m	2.57

Approximate Estimates
FENCING

Item Excluding site overheads and profit	Unit	Total rate £
Supply and erect palisade security fence Jacksons "Barbican" 2500 mm high with rectangular hollow section steel pales at 150 mm centres on three 50 x 50 x 6 mm rails. Rails bolted to 80 x 60 mm posts set in concrete 450 x 450 x 750 mm deep at 2750 mm centres, tops of pales to pointes and set at 45 degree angle. All metalwork to be hot-dip factory galvanized for painting on site.	m	74.34
Supply and erect single gate to match above complete with welded hinges and lock.		
1000 mm wide	each	673.62
4.0 mm wide	each	724.09
8.00 mm wide	pair	1475.99
Supply and erect Orsogril proprietary welded steel mesh panel fencing on steel posts set 750 mm deep in concrete foundations 600 x 600 mm. Supply and erect proprietary single gate 2.0 m wide to match fencing		
930 high	100 m	9079.00
1326 high	100 m	12567.00
1722 high	100 m	15114.00
TRIP RAILS		
Erect trip-rail, treated softwood rail 200 x 38, screwed to 100 x 100 oak posts with non-corroding screws, including excavating and setting posts in concrete (C7P) with central drainage hole, posts at 1.5 m centres, top of rail 450 above ground	m	25.03
Erect trip rail of galvanized steel tube 38 internal dia. with sleeved joint fixed to 38 dia. steel posts as above 700 long, set in C7P concrete at 1.20 m centres, metalwork primed and painted two coats metal preservative paint	m	103.24

Approximate Estimates
8.14 STREET FURNITURE

Item Excluding site overheads and profit	Unit	Total rate £
BENCHES SEATS BOLLARDS		
Supply and install proprietary powder coated steel parking posts and bases to specified colour as shown on drawing, fold down top locking type, complete with keys and instructions. Posts and bases set in concrete footing 200 x 200 x 300 mm deep.	each	169.32
Supply and install 10 no. proprietary precast concrete "Millbank" bollards 355 mm high above ground x 405 mm deep, 500 mm dia. where shown on drawing, white spar exposed aggregate finish, bedded in concrete base 750 dia, x 300 mm deep.	10 nr	1609.10
In grassed area excavate for base 2500 x 1575 mm and lay 100 mm hardcore, 100 mm concrete, brick pavers in stack bond bedded in 25 mm cement:lime:sand mortar. Supply and fix where shown on drawing proprietary seat, hardwood slats on black powder coated steel frame, bolted down with 4 no. 24 x 90 mm recessed hex-head stainless steel anchor bolts set into concrete.	set	1283.44
Supply and fix to wall where shown on drawing wall-mounted proprietary litter bins powder coated timber slatted, open top with metal liner, fixed with 2 no. stainless steel recessed hex-head anchor bolts 16 x 60 mm.	each	465.85
Supply and fix cycle stand 1250 m long x 550 mm long of 60.3 mm black powder coated hollow steel sections to BS 4948:Part 2, one-piece with rounded top corners, set 250 mm into paving.	each	244.93
Supply and locate in position proprietary precast concrete planters, 970 mm diameter planter 470 high with white spar exposed aggregate finish. Drainage holes to be positioned away from footway in direction of paving fall with white spar exposed aggregate finish. Fill with topsoil placed over 50 mm shingle and terram; plant with assorted 5 litre and 3 litre shrubs at to provide instant effect.	each	382.57

Approximate Estimates
8.15 DRAINAGE

Item Excluding site overheads and profit	Unit	Total rate £
AGRICULTURAL DRAINAGE		
Excavate and form ditch and bank with 45 degree sides in light to medium soils; all widths taken at bottom of ditch:		
300 wide x 600 deep	100 m	103.94
600 wide x 900 deep	100 m	228.81
1.20 m wide x 900 deep	100 m	688.45
1.50 m wide x 1.20 m deep	100 m	1159.13
Clear and bottom existing ditch average 1.50 m deep, trim back vegetation and remove debris to licensed tip not exceeding 13 km, lay jointed concrete pipes; to BS 5911 pt.1 class S; including bedding, haunching and topping with 150 mm concrete; 11.50 N/mm2 - 40 mm aggregate; back fill with approved spoil from site.		
pipes 300 dia.	100 m	5356.83
pipes 450 dia.	100 m	6785.53
pipes 600 dia.	100 m	8664.34
SUB SOIL DRAINAGE. BY MACHINE		
Main drain; Remove 150 topsoil and deposit alongside trench, excavate drain trench by machine and lay flexible perforated drain, lay bed of gravel rejects 100 mm. backfill with gravel rejects or similar to within 150 of finished ground level, complete fill with topsoil, remove surplus spoil to approved dump on site:		
Main drain 160 mm in supplied in 35 m lengths		
450 mm deep	100 m	574.67
600 mm deep	100 m	692.00
900 mm deep	100 m	1167.36
Extra for couplings	each	1.51
As above but with 100 mm main drain supplied in 100 m lengths		
450 mm deep	100 m	469.98
600 mm deep	100 m	582.65
900 mm deep	100 m	782.28
Extra for couplings	each	1.27
Laterals to mains above; herringbone pattern; excavation and back filling as above; inclusive of connecting lateral to main drain.		
160 mm pipe to 450 mm deep trench		
laterals at 1.0 m centres	100 m²	574.67
laterals at 2.0 m centres	100 m²	287.34
laterals at 3.0 m centres	100 m²	189.65
laterals at 5.0 m centres	100 m²	114.94
laterals at 10.0 m centres	100 m²	57.46

Approximate Estimates
DRAINAGE

Item Excluding site overheads and profit	Unit	Total rate £
SUB SOIL DRAINAGE. BY MACHINE – cont'd		
160 mm pipe to 600 mm deep trench		
laterals at 1.0 m centres	100 m²	692.00
laterals at 2.0 m centres	100 m²	346.00
laterals at 3.0 m centres	100 m²	228.36
laterals at 5.0 m centres	100 m²	138.40
laterals at 10.0 m centres	100 m²	69.20
160 mm pipe to 900 mm deep trench		
laterals at 1.0 m centres	100 m²	1167.36
laterals at 2.0 m centres	100 m²	583.67
laterals at 3.0 m centres	100 m²	385.23
laterals at 5.0 m centres	100 m²	233.48
laterals at 10.0 m centres	100 m²	116.73
Extra for 160/160 mm couplings connecting laterals to main drain		
laterals at 1.0 m centres	10 m	51.80
laterals at 2.0 m centres	10 m	25.90
laterals at 3.0 m centres	10 m	17.25
laterals at 5.0 m centres	10 m	10.36
laterals at 10.0 m centres	10 m	5.18
100 mm pipe to 450 mm deep trench		
laterals at 1.0 m centres	100 m²	469.98
laterals at 2.0 m centres	100 m²	234.99
laterals at 3.0 m centres	100 m²	155.10
laterals at 5.0 m centres	100 m²	93.99
laterals at 10.0 m centres	100 m²	47.00
100 mm pipe to 600 mm deep trench		
laterals at 1.0 m centres	100 m²	582.65
laterals at 2.0 m centres	100 m²	291.33
laterals at 3.0 m centres	100 m²	192.27
laterals at 5.0 m centres	100 m²	116.53
laterals at 10.0 m centres	100 m²	58.26
100 mm pipe to 900 mm deep trench		
laterals at 1.0 m centres	100 m²	782.28
laterals at 2.0 m centres	100 m²	391.15
laterals at 3.0 m centres	100 m²	258.16
laterals at 5.0 m centres	100 m²	156.45
laterals at 10.0 m centres	100 m²	78.23
80 mm pipe to 450 mm deep trench		
laterals at 1.0 m centres	100 m²	442.31
laterals at 2.0 m centres	100 m²	221.16
laterals at 3.0 m centres	100 m²	145.97
laterals at 5.0 m centres	100 m²	88.46
laterals at 10.0 m centres	100 m²	44.23
80 mm pipe to 600 mm deep trench		
laterals at 1.0 m centres	100 m²	554.98
laterals at 2.0 m centres	100 m²	277.50
laterals at 3.0 m centres	100 m²	183.14
laterals at 5.0 m centres	100 m²	111.00
laterals at 10.0 m centres	100 m²	55.49
80 mm pipe to 900 mm deep trench		
laterals at 1.0 m centres	100 m²	754.61
laterals at 2.0 m centres	100 m²	377.32
laterals at 3.0 m centres	100 m²	249.03
laterals at 5.0 m centres	100 m²	150.92
laterals at 10.0 m centres	100 m²	75.46

Approximate Estimates
DRAINAGE

Item Excluding site overheads and profit	Unit	Total rate £
Extra for 100/80 mm junctions connecting laterals to main drain		
laterals at 1.0 m centres	10 m	24.80
laterals at 2.0 m centres	10 m	12.40
laterals at 3.0 m centres	10 m	8.26
laterals at 5.0 m centres	10m	4.96
laterals at 10.0 m centres	10m	2.48
SUB SOIL DRAINAGE; BY HAND		
Main drain; Remove 150 topsoil and deposit alongside trench, excavate drain trench by machine and lay flexible perforated drain, lay bed of gravel rejects 100 mm. backfill with gravel rejects or similar to within 150 of finished ground level, complete fill with topsoil, remove surplus spoil to approved dump on site:		
Main drain 160 mm in supplied in 35 m lengths		
450 mm deep	100 m	1630.63
600 mm deep	100 m	2073.00
900 mm deep	100 m	2867.43
Extra for couplings	each	1.51
As above but with 100 mm main drain supplied in 100 m lengths		
450 mm deep	100 m	1099.16
600 mm deep	100 m	1402.93
900 mm deep	100 m	1989.37
Extra for couplings	each	1.27
Laterals to mains above; herringbone pattern; excavation and back filling as above; inclusive of connecting lateral to main drain.		
160 mm pipe to 450 mm deep trench		
laterals at 1.0 m centres	100 m²	1630.63
laterals at 2.0 m centres	100 m²	815.32
laterals at 3.0 m centres	100 m²	538.11
laterals at 5.0 m centres	100 m²	326.13
laterals at 10.0 m centres	100 m²	163.07
160 mm pipe to 600 mm deep trench		
laterals at 1.0 m centres	100 m²	2073.00
laterals at 2.0 m centres	100 m²	1036.50
laterals at 3.0 m centres	100 m²	684.10
laterals at 5.0 m centres	100 m²	414.61
laterals at 10.0 m centres	100 m²	207.30
160 mm pipe to 900 mm deep trench		
laterals at 1.0 m centres	100 m²	2867.43
laterals at 2.0 m centres	100 m²	1433.71
laterals at 3.0 m centres	100 m²	946.25
laterals at 5.0 m centres	100 m²	573.48
laterals at 10.0 m centres	100 m²	286.75

Approximate Estimates
DRAINAGE

Item Excluding site overheads and profit	Unit	Total rate £
SUB SOIL DRAINAGE; BY HAND – cont'd		
Laterals to mains – cont'd		
Extra for 160/160 mm couplings connecting laterals to main drain		
laterals at 1.0 m centres	10 m	51.80
laterals at 2.0 m centres	10 m	25.90
laterals at 3.0 m centres	10 m	17.25
laterals at 5.0 m centres	10 m	10.36
laterals at 10.0 m centres	10 m	5.18
100 mm pipe to 450 mm deep trench		
laterals at 1.0 m centres	100 m²	1099.16
laterals at 2.0 m centres	100 m²	549.58
laterals at 3.0 m centres	100 m²	362.72
laterals at 5.0 m centres	100 m²	219.83
laterals at 10 m centres	100 m²	109.92
100 mm pipe to 600 mm deep trench		
laterals at 1.0 m centres	100 m²	1402.93
laterals at 2.0 m centres	100 m²	701.46
laterals at 3.0 m centres	100 m²	462.97
laterals at 5.0 m centres	100 m²	280.59
laterals at 10.0 m centres	100 m²	140.29
100 mm pipe to 900 mm deep trench		
laterals at 1.0 m centres	100 m²	1989.37
laterals at 2.0 m centres	100 m²	994.69
laterals at 3.0 m centres	100 m²	656.50
laterals at 5.0 m centres	100 m²	397.88
laterals at 10.0 m centres	100 m²	198.94
80 mm pipe to 450 mm deep trench		
laterals at 1.0 m centres	100 m²	1071.49
laterals at 2.0 m centres	100 m²	535.75
laterals at 3.0 m centres	100 m²	353.59
laterals at 5.0 m centres	100 m²	214.30
laterals at 10.0 m centres	100 m²	107.15
80 mm pipe to 600 mm deep trench		
laterals at 1.0 m centres	100 m²	1375.26
laterals at 2.0 m centres	100 m²	687.63
laterals at 3.0 m centres	100 m²	453.84
laterals at 5.0 m centres	100 m²	275.06
laterals at 10.0 m centres	100 m²	137.52
80 mm pipe to 900 mm deep trench		
laterals at 1.0 m centres	100 m²	1961.70
laterals at 2.0 m centres	100 m²	980.86
laterals at 3.0 m centres	100 m²	647.37
laterals at 5.0 m centres	100 m²	392.35
laterals at 10.0 m centres	100 m²	196.17
Extra for 100/80 mm couplings connecting laterals to main drain		
laterals at 1.0 m centres	10 m	24.80
laterals at 2.0 m centres	10 m	12.40
laterals at 3.0 m centres	10 m	8.26
laterals at 5.0 m centres	10 m	4.96
laterals at 10.0 m centres	10 m	2.48

Approximate Estimates
DRAINAGE

Item Excluding site overheads and profit	Unit	Total rate £
DRAIN BY MOLE PLOUGH		
50 dia. mole at 450 deep		
1.20 m centres	100 m²	**11.19**
1.50 m centres	100 m²	**10.28**
2.00 m centres	100 m²	**9.35**
2.50 m centres	100 m²	**8.31**
3.00 m centres	100 m²	**7.27**
75 dia. mole at 600 deep		
1.20 m centres	100 m²	**13.36**
1.50 m centres	100 m²	**12.17**
2.00 m centres	100 m²	**11.76**
2.50 m centres	100 m²	**9.83**
3.00 m centres	100 m²	**10.28**
SURFACE WATER DRAINAGE		
Excavate trench by excavator to 450 deep, lay 100 vitrified clay drain with butt joints, bedding Class B, backfill with excavated material screened to remove stones over 40, backfill to be laid in layers not exceeding 150, top with 150 mm topsoil remove surplus material to approved dump on site not exceeding 100m, final level of fill to allow for settlement.	100 m	**1413.36**
Excavate hole, supply and set in concrete (C10P) vitrified clay trapped mud (dirt) gully with rodding eye to BS 65, complete with galvanized bucket and cast iron hinged locking grate and frame, flexible joint to pipe. Connect to drainage system with flexible joints	each	**151.94**
Excavate pit for inspection chamber 1500 x 1500 x 1500 deep, including earthwork support and disposal of spoil to dump on site not exceeding 100m, lay concrete (1:2:4) base 1500 dia. x 200 thick, 110 vitrified clay channels, benching in concrete (1:3:6) allowing one outlet and two inlets for 110 dia pipe, construct inspection chamber with 1200 x 1200 x 1200 mm 1 brick thick walls of engineering brick Class B, backfill with excavated material, complete with 2 no. cast iron step irons, supply and fix light duty precast concrete cover slab.	each	**838.48**
Supply and fix vitrified clay interceptor trap 100 mm inlet; 100 mm outlet; to manhole bedded in 10 aggregate concrete (C10P), complete with brass stopper and chain.	each	**129.95**
Connect to drainage system with flexible joints.	each	**20.22**
Excavate hole and lay 100 concrete base (1:3:6) 150 x150 to suit given invert level of drain, supply and connect trapped precast concrete road gully 450 dia x 1.07 m deep with 160 outlet to BS 5911, set in concrete surround, connect to vitrified clay drainage system with flexible joints, supply and fix straight bar dished top cast iron grating and frame, bedded in cement:sand mortar (1:3)	each	**275.18**

Approximate Estimates
DRAINAGE

Item Excluding site overheads and profit	Unit	Total rate £
SURFACE WATER DRAINAGE – cont'd		
Excavate trench by excavator 600 deep, lay 110 PVC-U drainpipe to BS 4660, Type 2 bedding, backfill to 150 mm above pipe with gravel rejects, lay non woven geofabric and fill with topsoil to ground level.	100 m	1129.97
Excavate pit for inspection chamber, supply and install cast iron inspection chamber unit 650 deep 100 x 150 mm one brach each side, bedded in Type 1 granular material.	each	279.34
Excavate hole and lay 100 concrete (C20P) base 150 x 150 to suit given invert level of drain, supply and connect trapped PVC-U gully complete with cast iron grate and frame, connect to drainage system, backfill with DoT Type 1 granular fill.	each	69.86
SOAKAWAYS		
Construct soakaway from perforated concrete rings; Excavation, casting insitu concrete ring beam base; and filling with gravel 250mm deep; placing perforated concrete rings; surrounding with geofabric and bacfilling with 250 mm granular surround and excavated material; step irons and cover slab; inclusive of all earthwork retention and disposal offsite of surplus material		
900 mm diameter		
1.00 m deep	nr	476.90
2.00 m deep	nr	657.07
1200 mm diameter		
1.00 m deep	nr	607.43
2.00 m deep	nr	920.20

Approximate Estimates
8.16 IRRIGATION

Item Excluding site overheads and profit	Unit	Total rate £
LEAKY PIPE IRRIGATION		
Works by machine; Main supply and connection to laterals; Excavate trench for main or ring main 450 mm deep; Supply and lay pipe; backfill and lightly compact trench;		
20 mm LDPE	100 m	180.06
16 mm LDPE	100 m	179.77
Works by hand; Main supply and connection to laterals; Excavate trench for main or ring main 450 mm deep; Supply and lay pipe; backfill and lightly compact trench;		
20 mm LDPE	100 m	1271.03
16 mm LDPE	100 m	1257.03
Turf area irrigation; laterals to mains; to cultivated soil; excavate trench 150 deep using hoe or mattock; lay moisture leaking pipe laid 150 mm below ground at centres of 350 mm		
low leak	100 m²	509.00
high leak	100 m²	460.00
Landscape area irrigation; laterals to mains; moisture leaking pipe laid to the surface of irrigated areas at 600 mm centres		
low leak	100 m²	263.00
high leak	100 m²	234.00
Landscape area irrigation; laterals to mains; moisture leaking pipe laid to the surface of irrigated areas at 900 mm centres		
low leak	100 m²	176.00
high leak	100 m²	156.00
Multistation controller	each	325.00
Solenoid valves connected to automatic controller.	each	407.00

Approximate Estimates
8.17 WATER FEATURES

Item Excluding site overheads and profit	Unit	Total rate £
LAKES AND PONDS		
Excavate for lake average depth 1.0 m, allow for bringing to specified levels, reserve topsoil, remove spoil to approved dump on site, remove all stones and debris over 75, lay 500 micron sheet including welding all joints and seams by specialist. Screen and replace topsoil 200 mm thick.		
Prices are for lakes of regular shape:		
under 1 ha	ha	100489.00
1-5 ha	ha	97799.00
As above but with 1000 micron		
under 1 ha	ha	118024.00
1-5 ha	ha	156599.00
As above but with butyl rubber sheet. Prices are for lakes of regular shape:		
0.75 mm thick	ha	118224.00
1.00 mm thick	ha	124724.00
Extra for removing spoil to tip not exceeding 13 km	m³	21.95
Extra for 25 sand blinding to lake bed	100 m²	105.95
Extra for screening topsoil	m²	0.76
Extra for spreading imported topsoil	m²	544.29
Plant aquatic plants in lake topsoil:		
Aponogeton distachyum - £210.00/100	100	282.00
Acorus calamus - £169.00/100	100	282.00
Butomus umbellatus - £169.00/100	100	282.00
Typha latifolia - £169.00/100	100	282.00
Nymphaea - £425.00/100	100	282.00

Approximate Estimates
8.18 TIMBER DECKING

Item Excluding site overheads and profit	Unit	Total rate £
WYCKHAM-BLACKWELL LTD		
Timber decking; support structure of timber joists for decking laid on blinded base (measured separately)		
joists 50 x 150 mm	10 m2	264.70
joists 50 x 200 mm	10 m2	291.70
joists 50 x 250 mm	10 m2	365.90
As above but inclusive of timber decking boards in yellow cedar 142 mm wide x 42 mm thick		
joists 50 x 150 mm	10 m2	671.60
joists 50 x 200 mm	10 m2	698.60
joists 50 x 250 mm	10 m2	772.80
As above but boards 141 x 26 mm thick		
joists 50 x 150 mm	10 m2	496.60
joists 50 x 200 mm	10 m2	291.70
joists 50 x 250 mm	10 m2	365.90
As above but decking boards in red cedar 131 mm wide x 42 mm thick		
joists 50 x 150 mm	10 m2	481.10
joists 50 x 200 mm	10 m2	508.10
joists 50 x 250 mm	10 m2	582.30
Add to all of the above for handrails fixed to posts 100 x 100 x 1370 high		
square balusters at 100 centres	m	112.77
square balusters at 300 centres	m	87.22
turned balusters at 100 centres	m	131.97
turned balusters at 300 centres	m	93.56

Approximate Estimates
8.19 LIGHTING

Item Excluding site overheads and profit	Unit	Total rate £
Excavate trench 600 mm deep; supply and install 3 core 2.5 mm armoured cable and backfill with excavated material		
by machine	m	5.49
by hand	m	28.37
Supply and install Marlin Sphereline lighting units at locations shown on drawings: 3.0 m tapered aluminium columns to BS 5649, set 300 mm into paving, 500 mm dia. clear acrylic globe luminaires to IP54 complete with control gear and electrical supply, installed all in accordance with IEE regulations.	each	691.42
Supply and install 10 no. floodlighting units at locations 20 m apart; 5.0 m galvanized painted tubular steel columns to BS 5649, set 450 mm into paving, low light pollution aluminium luminaires to BS 4533:102.5 to IP65 complete with control gear and electrical supply, installed all in accordance with IEE regulations inclusive of electrical connections	each	515.38
Supply and install 30 no. illuminated bollards at locations shown on drawings: 900 mm high blue powder coated steel bollard to BS 873: Part 3 set 450 mm into ground, louvered polycarbonate luminaire to IP54, complete with control gear and electrical supply, installed all in accordance with IEE regulations.		
PC of lighting unit £395.00	each	473.02
PC of lighting unit £610.00	each	677.54
Supply and install 4 no. 12 volt ground mounted recessed brick lights at locations 5 metres apart; black powder coated diecast aluminium casing 220 mm dia. to IP55 recessed to paving, polycarbonate diffuser, complete with control gear and electrical supply, installed all in accordance with IEE regulations.	set	2169.39

9
Tables and Memoranda

QUICK REFERENCE CONVERSION TABLES

	Imperial		Metric	
1. LINEAR				
0.039	in	1	mm	25.4
3.281	ft	1	metre	0.305
1.094	yd	1	metre	0.914
2. WEIGHT				
0.020	cwt	1	kg	50.802
0.984	ton	1	tonne	1.016
2.205	lb	1	kg	0.454
3. CAPACITY				
1.760	pint	1	litre	0.568
0.220	gal	1	litre	4.546
4. AREA				
0.002	in^2	1	mm^2	645.16
10.764	ft^2	1	m^2	0.093
1.196	yd^2	1	m^2	0.836
2.471	acre	1	ha	0.405
0.386	mile^2	1	km^2	2.59
5. VOLUME				
0.061	in^3	1	cm^3	16.387
35.315	ft^3	1	m^3	0.028
1.308	yd^3	1	m^3	0.765
6. POWER				
1.310	HP	1	kW	0.746

CONVERSION FACTORS - METRIC TO IMPERIAL

Multiply Metric	Unit	By	To obtain Imperial	Unit
Length				
kilometre	km	0.6214	statute mile	ml
metre	m	1.0936	yard	yd
centimetre	cm	0.0328	foot	ft
millimetre	mm	0.0394	inch	in
Area				
hectare	ha	2.471	acre	
square kilometre	km²	0.3861	square mile	sq ml
square metre	m²	10.764	square foot	sq ft
square metre	m²	1550	square inch	sq in
square centimetre	cm²	0.155	square inch	sq in
Volume				
cubic metre	m³	35.336	cubic foot	cu ft
cubic metre	m³	1.308	cubic yard	cu yd
cubic centimetre	cm³	0.061	cubic inch	cu in
cubic centimetre	cm³	0.0338	fluid ounce	fl oz
Liquid volume				
litre	l	0.0013	cubic yard	cu yd
litre	l	61.02	cubic inch	cu in
litre	l	0.22	Imperial gallon	gal
litre	l	0.2642	US gallon	US gal
litre	l	1.7596	pint	pt
Mass				
metric tonne	t	0.984	long ton	lg ton
metric tonne	t	1.102	short ton	sh ton
kilogram	kg	2.205	pound, avoirdupois	lb
gram	g or gr	0.0353	ounce, avoirdupois	oz

Unit mass

kilograms/cubic metre	kg/m³	0.062	pounds/cubic foot	lbs/cu ft
kilograms/cubic metre	kg/m³	1.686	pounds/cubic yard	lbs/cu yd
tonnes/cubic metre	t/m³	1692	pound/cubic yard	lbs/cu yd
kilograms/sq centimetre	kg/cm²	14.225	pounds/square inch	lbs/sq in
kilogram-metre	kg.m	7.233	foot-pound	ft-lb

Force

meganewton	MN	9.3197	tons force	tonf
kilonewton	kN	225	pounds force	lbf
newton	N	0.225	pounds force	lbf

Pressure and stress

meganewton per square metre	MN/m²	9.3197	tons force/square foot	tonf/ft²
kilopascal	kPa	0.145	pounds/square inch	psi
bar		14.5	pounds/square inch	psi
kilograme metre	kgm	7.2307	foot-pound	ft-lb

Energy

kilocalorie	kcal	3.968	British thermal unit	Btu
metric horsepower	CV	0.9863	horse power	hp
kilowatt	kW	1.341	horse power	hp

Speed

kilometres/hour	km/h	0.621	miles/hour	mph

CONVERSION FACTORS - IMPERIAL TO METRIC

Multiply Imperial	Unit	By	To obtain metric	Unit
Length				
statute mile	ml	1.609	kilometre	km
yard	yd	0.9144	metre	m
foot	ft	0.3048	centimetre	cm
inch	in	25.4	millimetre	mm
Area				
acre	acre	0.4047	hectare	ha
square mile	sq ml	2.59	square kilometre	km²
square foot	sq ft	0.0929	square metre	m²
square inch	sq in	0.0006	square metre	m²
square inch	sq in	6.4516	square centimetre	cm²
Volume				
cubic foot	cu ft	0.0283	cubic metre	m³
cubic yard	cu yd	0.7645	cubic metre	m³
cubic inch	cu in	16.387	cubic centimetre	cm³
fluid ounce	fl oz	29.57	cubic centimetre	cm³
Liquid volume				
cubic yard	cu yd	764.55	litre	l
cubic inch	cu in	0.0164	litre	l
Imperial gallon	gal	4.5464	litre	l
US gallon	US gal	3.785	litre	l
US gallon	US gal	0.833	Imperial gallon	gal
pint	pt	0.5683	litre	l
Mass				
long ton	lg ton	1.016	metric tonne	tonne
short ton	sh ton	0.907	metric tonne	tonne
pound	lb	0.4536	kilogram	kg
ounce	oz	28.35	gram	g

Unit mass

pounds/cubic foot	lb/ cu ft	16.018	kilogram's/cubic metre	kg/m³
pounds/cubic yard	lb/cu yd	0.5933	cubic/cubic metre	kg/m³
pounds/cubic yard	lb/cu yd	0.0006	tonnes/cubic metre	t/m³
foot-pound	ft-lb	0.1383	kilogram-metre	kg.m

Force

tons force	tonf	0.1073	meganewton	MN
pounds force	lbf	0.0045	kilonewton	kN
pounds force	lbf	4.45	newton	N

Pressure and stress

pounds/square inch	psi	0.1073	kilogram/sq. centimetre	kg/cm²
pounds/square inch	psi	6.89	kilopascal	kPa
pounds/square inch	psi	0.0689	bar	
foot-pound	ft-lb	0.1383	kilogram metre	kgm

Energy

British Thermal Unit	Btu	0.252	kilocalorie	kcal
horsepower (hp)	hp	1.014	metric horsepower	CV
horsepower (hp)	hp	0.7457	kilowatt	kW

Speed

miles/hour	mph	1.61	kilometres/hour	km/h

CONVERSION TABLES – cont'd

Length

Millimetre	mm	1 in	=	25.4 cm	1 mm	=	0.0394 in
Centimetre	cm	1 in	=	2.54 cm	1 cm	=	0.3937 in
Metre	m	1 ft	=	0.3048 m	1 m	=	3.2808 ft
		1 yd	=	0.9144 m	1 m	=	1.0936 yd
Kilometre	km	1 mile	=	1.6093 km	1 km	=	0.6214 mile

Note:
1 cm =	10 mm	1 ft	=	12 in
1 m =	100 cm	1 yd	=	3 ft
1 km =	1,000 m	1 mile	=	1,760 yd

Area

Square millimetre	mm²	1 in²	=	645.2 mm²	1 mm²	=	0.0016 in²
Square centimetre	cm²	1 in²	=	6.4516 cm²	1 cm²	=	1.1550 in²
Square metre	m²	1 ft²	=	0.0929 m²	1 m²	=	10.764 ft²
		1 yd²	=	0.8361 m²	1 m²	=	1.1960 yd²
Square Kilometre	km²	1 mile2	=	2.590 km²	1 km²	=	0.3861 mile2

Note:
1 cm²	=	100 m²	1 ft²	=	144 in²
1 m²	=	10,000 cm²	1 yd²	=	9 ft²
1 km²	=	100 hectares	1 mile²	=	640 acres
			1 acre	=	4,840 yd²

Volume

Cubic Centimetre		cm³	1 cm³ =	0.0610 in³	1 in³	=	16.387 cm³
Cubic Decimetre		dm³	1 dm³ =	0.0353 ft³	1 ft³	=	28.329 dm³
Cubic metre		m³	1 m³ =	35.3147 ft³	1 ft³	=	0.0283 m³
			1 m³ =	1.3080 yd³	1 yd³	=	0.7646 m³
Litre	L		1 L =	1.76 pint	1 pint	=	0.5683 L
			=	2.113 US pt			0.4733 US L

Note:
1 dm³	=	1,000 cm³	1 ft³	=	1.728 in³
1 m³	=	1,000 dm³	1 yd³	=	27 ft³
1 L	=	1 dm³	1 pint	=	20 fl oz
1 HL	=	100 L	1 gal	=	8 pints

Mass

Milligram	mg	1 mg	=	0.0154 grain	1 grain	=	64.935 mg
Gram	g	1 g	=	0.0353 oz	1 oz	=	28.35 g
Kilogram	kg	1 kg	=	2.2046 lb	1 lb	=	0.4536 kg
Tonne	t	1 t	=	0.9842 ton	1 ton	=	1.016 t

Note:
- 1 g = 1,000 mg
- 1 kg = 1000 g
- 1 t = 1,000 kg

- 1 oz = 437.5 grains
- 1 lb = 16 oz
- 1 stone = 14 lb
- 1 cwt = 112 lb
- 1 ton = 20 cwt

Force

Newton	N	1 lbf	=	4.448 N	1 kgf	=	9.807 N
Kilonewton	kN	1 lbf	=	0.00448 kN	1 ton f	=	9.964 kN
Meganewton	MN	100 tonf	=	0.9964 MN			

Pressure and stress

Kilonewton per square metre	kN/m²	1 lbf/in² =	6.895 kN/m²
		1 bar =	100 kN/m²
Meganewton per square metre	MN/m²	1 tonf/ft² =	107.3 kN/m² = 0.1073 MN/m²
		1 kgf/cm² =	98.07 kN/m²
		1 lbf/ft² =	0.04788 kN/m²

Temperature

Degree Celcius °C

$$°C = \frac{5 \times (°F - 32)}{9}$$

$$°F = \frac{(9 \times °C)}{5} + 32$$

CONVERSION TABLES – cont'd

Metric Equivalents

1 km	=	1000 m
1 m	=	100 cm
1 cm	=	10 mm
1 km²	=	100 ha
1 ha	=	10,000 m²
1 m²	=	10,000 cm²
1 cm²	=	100 mm²
1 m³	=	1,000 litres
1 litre	=	1,000 cm³
1 metric tonne	=	1,000 kg
1 quintal	=	100 kg
1 N	=	0.10197 kg
1 kg	=	1000 g
1 g	=	1000 mg
1 bar	=	14.504 psi
1 cal	=	427 kg.m
1 cal	=	0.0016 cv.h
torque unit	=	0.00116 kw.h
1 CV	=	75 kg.m/s
1 kg/cm²	=	0.97 atmosph

English Unit Equivalents

1 mile	=	1760 yd
1 yd	=	3 ft
1 ft	=	12 in
1 sq mile	=	640 acres
1 acre	=	43,560 sq ft
1 sq ft	=	144 sq in
1 cu ft	=	7.48 gal liq
1 gal	=	231 cu in
	=	4 quarts liq
1 quart	=	32 fl oz
1 fl oz	=	1.80 cu in
	=	437.5 grains
1 stone	=	14 lb
1 cwt	=	112 lb
1 sh ton	=	2000 lb
1 lg ton	=	2240 lb
	=	20 cwt
1 lb	=	16 oz, avdp
1 Btu	=	778 ft lb
	=	0.000393 hph
	=	0.000293 kwh
1 hp	=	550 ft-lb/sec
1 atmosph	=	14.7 lb/in^2

CONVERSION TABLES - cont'd

Power Units

kW	=	Kilowatt
HP	=	Horsepower
CV	=	Cheval Vapeur (Steam Horsepower)
	=	French designation for Metric Horsepower
PS	=	Pferderstarke (Horsepower)
	=	German designation of Metric Horsepower
1 HP	=	1.014 CV = 1.014 PS
	=	0.7457 kW
1 PS	=	1 CV = 0.9863 HP
	=	0.7355 kW
1 kW	=	1.341 HP
	=	1.359 CV
	=	1.359 PS

SPEED CONVERSION

km/h	m/min	mph	fpm
1	16.7	0.6	54.7
2	33.3	1.2	109.4
3	50.0	1.9	164.0
4	66.7	2.5	218.7
5	83.3	3.1	273.4
6	100.0	3.7	328.1
7	116.7	4.3	382.8
8	133.3	5.0	437.4
9	150.0	5.6	492.1
10	166.7	6.2	546.8
11	183.3	6.8	601.5
12	200.0	7.5	656.2
13	216.7	8.1	710.8
14	233.3	8.7	765.5
15	250.0	9.3	820.2
16	266.7	9.9	874.9
17	283.3	10.6	929.6
18	300.0	11.2	984.3
19	316.7	11.8	1038.9
20	333.3	12.4	1093.6
21	350.0	13.0	1148.3
22	366.7	13.7	1203.0
23	383.3	14.3	1257.7
24	400.0	14.9	1312.3
25	416.7	15.5	1367.0
26	433.3	16.2	1421.7
27	450.0	16.8	1476.4
28	466.7	17.4	1531.1
29	483.3	18.0	1585.7
30	500.0	18.6	1640.4
31	516.7	19.3	1695.1
32	533.3	19.9	1749.8
33	550.0	20.5	1804.5
34	566.7	21.1	1859.1
35	583.3	21.7	1913.8

SPEED CONVERSION – cont'd

km/h	m/min	mph	fpm
36	600.0	22.4	1968.5
37	616.7	23.0	2023.2
38	633.3	23.6	2077.9
39	650.0	24.2	2132.5
40	666.7	24.9	2187.2
41	683.3	25.5	2241.9
42	700.0	26.1	2296.6
43	716.7	26.7	2351.3
44	733.3	27.3	2405.9
45	750.0	28.0	2460.6
46	766.7	28.6	2515.3
47	783.3	29.2	2570.0
48	800.0	29.8	2624.7
49	816.7	30.4	2679.4
50	833.3	31.1	2734.0

FORMULAE

Two dimensional figures

Figure	Diagram of figure	Surface area	Perimeter
Square		a^2	$4a$
Rectangle		ab	$2(a+b)$
Triangle		½ch ½$ab \sin C$ $\sqrt{s(s-a)(s-b)(s-c)}$ where $s = ½(a+b+c)$	$a+b+c = 2s$
Circle		πr^2 ¼πd^2 where $2r = d$	$2\pi r$ πd
Parallelogram		ah	$2(a+b)$
Trapezium		½$h(a+b)$	$a+b+c+d$
Ellipse		Approximately πab	$\pi(a+b)$
Hexagon		$2.6 \times a^2$	
Octagon		$4.83 \times a^2$	
Sector of circle		½rb or $\dfrac{q}{360}\pi r^2$ note: $b = $ angle $\dfrac{q}{360} \times \pi 2r$	
Segment of a circle		$S - T$ where $S = $ area of sector $T = $ area of triangle	
Bellmouth		$\dfrac{3}{14} \times r^2$	

FORMULAE - continued

Three dimensional figures

Figure	Diagram of figure	Surface area	Volume
Cube		$6a^2$	a^3
Cuboid/ rectangular block		$2(ab + ac + bc)$	abc
Prism/ triangular block		$bd + hc + dc + ad$	$½hcd$ $½ab \sin C \, d$ $d\sqrt{s(s-a)(s-b)(s-c)}$ where $s = ½(a + b + c)$
Cylinder		$2\pi rh + 2\pi r^2$ $\pi dh + ½\pi d^2$	$\pi r^2 h$ $¼\pi d^2 h$
Sphere		$4\pi r^2$	$^4/_3 \pi r^3$
Segment of sphere		$2\pi Rh$	$^1/_6 \pi h (3r^2 + h^2)$ $^1/_3 \pi h^2 (3R - h)$
Pyramid		$(a + b)l + ab$	$^1/_3 abh$
Frustrum of a pyramid		$l(a+b+c+d) + \sqrt{(ab+cd)}$ [regular figure only]	$^h/_3 (ab + cd + \sqrt{abcd})$
Cone		$\pi rl + \pi r^2$ $½\pi dh + ¼\pi d^2$	$^1/_3 \pi r^2 h$ $^1/_{12} \pi d^2 h$
Frustrum of a cone		$\pi r^2 + \pi R^2 + \pi h (R + r)$	$^1/_3 \pi h (R^2 + Rr + r^2)$

FORMULAE

Formula	Description
Pythagoras theorem	$A^2 = B^2 + C^2$ where A is the hypotenuse of a right-angled triangle and B and C are the two adjacent sides
Simpsons Rule	The Area is divided into an even number of strips of equal width, and therefore has an odd number of ordinates at the division points area = $\dfrac{S(A + 2B + 4C)}{3}$ where S = common interval (strip width) A = sum of first and last ordinates B = sum of remaining odd ordinates C = sum of the even ordinates The Volume can be calculated by the same formula, but by substituting the area of each co-ordinate rather than its length.
Trapezoidal Rule	A given trench is divided into two equal sections, giving three ordinates, the first, the middle and the last. volume = $\dfrac{S \times (A + B + 2C)}{2}$ where S = width of the strips A = area of the first section B = area of the last section C = area of the rest of the sections
Prismoidal Rule	A given trench is divided into two equal sections, giving three ordinates, the first, the middle and the last. volume = $\dfrac{L \times (A + 4B + C)}{6}$ where L = total length of trench A = area of the first section B = area of the middle section C = area of the last section

EARTHWORK

Weights of typical materials handled by excavators.
The weight of the material is that of the state in its natural bed and includes moisture.
Adjustments should be made to allow for loose or compacted states.

Material	kg/m³	lb/cu yd
Adobe	1914	3230
Ashes	610	1030
Asphalt, rock	2400	4050
Basalt	2933	4950
Bauxite: alum ore	2619	4420
Borax	1730	2920
Caliche	1440	2430
Carnotite	2459	4150
Cement	1600	2700
Chalk (hard)	2406	4060
Cinders	759	1280
Clay: dry	1908	3220
Clay: wet	1985	3350
Coal: bituminouos	1351	2280
Coke	510	860
Conglomerate	2204	3720
Dolomite	2886	4870
Earth: dry	1796	3030
Earth: moist	1997	3370
Earth: wet	1742	2940
Feldspar	2613	4410
Felsite	2495	4210
Fluorite	3093	5220
Gabbro	3093	5220
Gneiss	2696	4550
Granite	2690	4540
Gravel, dry	1790	3020
Gypsum	2418	4080

Material	kg/m³	lb/cu yd
Lignite broken	1244	2100
Limestone	2596	4380
Magnesite, magnesium ore	2993	5050
Marble	2679	4520
Marl	2216	3740
Peat	700	1180
Potash	2193	3700
Pumice	640	1080
Quartz	2584	4360
Rhyolite	2400	4050
Sand: dry	1707	2880
Sand: wet	1831	3090
Sand and gravel - dry	1790	3020
- wet	2092	3530
Sandstone	2412	4070
Schist	2684	4530
Shale	2637	4450
Slag (blast)	2868	4840
Slate	2667	4500
Taconite	3182	5370
Topsoil	1440	2430
Trachyte	2400	4050
Traprock	2791	4710

Material	kg/m³	lb/cu yd
Snow - dry	130	220
- wet	510	860
Water	1000	62
Quarry waste	1438	90
Hardcore (consolidated)	1928	120

EARTHWORK – cont'd

Transport Capacities

Type of vehicle	Capacity of vehicle	
	Payload	Heaped capacity
Wheelbarrow	150	0.10
1 tonne dumper	1250	1.00
2.5 tonne dumper	4000	2.50
Articulated dump truck (Volvo A20 6x4)	18500	11.00
Articulated dump truck (Volvo A35 6x6)	32000	19.00
Large capacity rear dumper (Euclid R35)	35000	22.00
Large capacity rear dumper (Euclid R85)	85000	50.00

Bulkage of soils (After excavation)

Type of soil	Approximate bulking of 1 m³ after excavation
Vegetable soil and loam	25 - 30%
Soft clay	30 - 40%
Stiff clay	10 - 20%
Gravel	20 - 25%
Sand	40 - 50%
Chalk	40 - 50%
Rock, weathered	30 - 40%
Rock, unweathered	50 - 60%

Shrinkage of materials (On being deposited)

Type of soil	Approximate bulking of 1 m³ after excavation
Clay	10%
Gravel	8%
Gravel and sand	9%
Loam and light sandy soils	12%
Loose vegetable soils	15%

Voids in material used as sub bases or beddings

Material	m³ of voids/m³
Alluvium	0.37
River grit	0.29
Quarry sand	0.24
Shingle	0.37
Gravel	0.39
Broken stone	0.45
Broken bricks	0.42

EARTHWORK – cont'd

Angles of repose

Type of soil		degrees
Clay	- dry	30
	- damp, well drained	45
	- wet	15 - 20
Earth	- dry	30
	- damp	45
Gravel	- moist	48
Sand	- dry or moist	35
	- wet	25
Loam		40

Slopes and angles

Ratio of base to height	Angle in degrees
5 : 1	11
4 : 1	14
3 : 1	18
2 : 1	27
1½ : 1	34
1 : 1	45
1 : 1½	56
1 : 2	63
1 : 3	72
1 : 4	76
1 : 5	79

Grades (In Degrees and Percents)

Degrees	Percent	Degrees	Percent
1	1.8	24	44.5
2	3.5	25	46.6
3	5.2	26	48.8
4	7.0	27	51.0
5	8.8	28	53.2
6	10.5	29	55.4
7	12.3	30	57.7
8	14.0	31	60.0
9	15.8	32	62.5
10	17.6	33	64.9
11	19.4	34	67.4
12	21.3	35	70.0
13	23.1	36	72.7
14	24.9	37	75.4
15	26.8	38	78.1
16	28.7	39	81.0
17	30.6	40	83.9
18	32.5	41	86.9
19	34.4	42	90.0
20	36.4	43	93.3
21	38.4	44	96.6
22	40.4	45	100.0

EARTHWORK – cont'd

Bearing powers

Ground conditions		Bearing Power		
		km/mý	lb/iný	metric t/m²
Rock (broken)		483	70	50
Rock (solid)		2,415	350	240
Clay,	dry or hard	380	55	40
	medium dry	190	27	20
	soft or wet	100	14	10
Gravel,	cemented	760	110	80
Sand,	compacted	380	55	40
	clean dry	190	27	20
Swamp and alluvial soils		48	7	5

Earthwork support

Maximum depth of excavation in various soils without the use of earthwork support:

Ground conditions	Feet (ft)	Metres (m)
Compact soil	12	3.66
Drained loam	6	1.83
Dry sand	1	0.3
Gravelly earth	2	0.61
Ordinary earth	3	0.91
Stiff clay	10	3.05

It is important to note that the above table should only be used as a guide. Each case must be taken on its merits and, as the limited distances given above are approached, careful watch must be kept for the slightest signs of caving in.

CONCRETE WORK

Weights of concrete and concrete elements

Type of Material	kg/m³	lb/cu ft
Ordinary concrete (dense aggregates)		
Non-reinforced plain or mass concrete		
Nominal weight	2305	144
Aggregate - limestone	2162 to 2407	135 to 150
- gravel	2244 to 2407	140 to 150
- broken brick	2000 (av)	125 (av)
- other crushed stone	2326 to 2489	145 to 155
Reinforced concrete		
Nominal weight	2407	150
Reinforcement - 1%	2305 to 2468	144 to 154
- 2%	2356 to 2519	147 to 157
- 4%	2448 to 2703	153 to 163
Special concretes		
Heavy concrete		
Aggregates - barytes, magnetite	3210 (min)	200 (min)
steel shot, punchings	5280	330
Lean mixes		
Dry-lean (gravel aggregate)	2244	140
Soil-cement (normal mix)	1601	100

CONCRETE WORK – cont'd

Type of material		kg/m² per mm thick	lb/sq ft per inch thick
Ordinary concrete (dense aggregates)			
Solid slabs (floors, walls etc.)			
Thickness:	75 mm or 3 in	184	37.5
	100 mm or 4 in	245	50
	150 mm or 6 in	378	75
	250 mm or 10 in	612	125
	300 mm or 12 in	734	150
Ribbed slabs			
Thickness:	125 mm or 5 in	204	42
	150 mm or 6 in	219	45
	225 mm or 9 in	281	57
	300 mm or 12 in	342	70
Special concretes			
Finishes etc			
Rendering, screed etc			
Granolithic, terrazzo		1928 to 2401	10 to 12.5
Glass-block (hollow) concrete		1734 (approx)	9 (approx)
Prestressed concrete		Weights as for reinforced concrete (upper limits)	
Air-entrained concrete		Weights as for plain or reinforced concrete	

Average weight of aggregates

Materials	Voids %	Weight kg/m³
Sand	39	1660
Gravel 10 - 20 mm	45	1440
Gravel 35 - 75 mm	42	1555
Crushed stone	50	1330
Crushed granite (over 15 mm)	50	1345
(n.e. 15 mm)	47	1440
'All-in' ballast	32	1800 - 2000

Material	kg/m³	lb/cu yd
Vermiculite (aggregate)	64-80	108-135

Material	kg/m³	lb/cu ft
All-in aggregate	1999	125

CONCRETE WORK – cont'd

Common mixes (per m³)

Recom-mended mix	Class of work suitable for: -	Cement (kg)	Sand (kg)	Coarse aggregate (kg)	No 50kg bags cement per m³ of combined aggregate
1:3:6	Roughest type of mass concrete such as footings, road haunching over 300mm thick	208	905	1509	4.00
1:2.5:5	Mass concrete of better class than 1:3:6 such as bases for machinery, walls below ground etc.	249	881	1474	5.00
1:2:4	Most ordinary uses of concrete, such as mass walls above ground, road slabs etc. and general reinforced concrete work	304	889	1431	6.00
1:1.5:3	Watertight floors, pavements and walls, tanks, pits, steps, paths, surface of 2 course roads, reinforced concrete where where extra strength is required	371	801	1336	7.50
1:1:2	Work of thin section such as fence posts and small precast work	511	720	1206	10.50

Prescribed mixes for ordinary structural concrete

Weights of cement and total dry aggregates in kg to produce approximately one cubic metre of fully compacted concrete together with the percentages by weight of fine aggregate in total dry aggregates.

Conc. grade	Nominal max. size of aggregate (mm) Workability	40		20		14		10	
		Med.	High	Med.	High	Med.	High	Med.	High
	Limits to slump that may be expected (mm)	50-100	100-150	25-75	75-125	10-50	50-100	10-25	25-50
7	Cement (kg)	180	200	210	230	-	-	-	-
	Total aggregate (kg)	1950	1850	1900	1800	-	-	-	-
	Fine aggregate (%)	30-45	30-45	35-50	35-50	-	-	-	-
10	Cement (kg)	210	230	240	260	-	-	-	-
	Total aggregate (kg)	1900	1850	1850	1800	-	-	-	-
	Fine aggregate (%)	30-45	30-45	35-50	35-50	-	-	-	-
15	Cement (kg)	250	270	280	310	-	-	-	-
	Total aggregate (kg)	1850	1800	1800	1750	-	-	-	-
	Fine aggregate (%)	30-45	30-45	35-50	35-50	-	-	-	-
20	Cement (kg)	300	320	320	350	340	380	360	410
	Total aggregate (kg)	1850	1750	1800	1750	1750	1700	1750	1650
	Sand Zone 1 (%)	35	40	40	45	45	50	50	55
	Zone 2 (%)	30	35	35	40	40	45	45	50
	Zone 3 (%)	30	30	30	35	35	40	40	45
25	Cement (kg)	340	360	360	390	380	420	400	450
	Total aggregate (kg)	1800	1750	1750	1700	1700	1650	1700	1600
	Sand Zone 1 (%)	35	40	40	45	45	50	50	55
	Zone 2 (%)	30	35	35	40	40	45	45	50
	Zone 3 (%)	30	30	30	35	35	40	40	45
30	Cement (kg)	370	390	400	430	430	470	460	510
	Total aggregate (kg)	1750	1700	1700	1650	1700	1600	1650	1550
	Sand Zone 1 (%)	35	40	40	45	45	50	50	55
	Zone 2 (%)	30	35	35	40	40	45	45	50
	Zone 3 (%)	30	30	30	35	35	40	40	45

CONCRETE WORK – cont'd

Weights of bar reinforcement

Nominal sizes (mm)	Cross-sectional area (mm^2)	Mass kg/m	Length of bar m/tonne
6	28.27	0.222	4505
8	50.27	0.395	2534
10	78.54	0.617	1622
12	113.10	0.888	1126
16	201.06	1.578	634
20	314.16	2.466	405
25	490.87	3.853	260
32	804.25	6.313	158
40	1265.64	9.865	101
50	1963.50	15.413	65

Weights of bars (at specific spacings)

Weights of metric bars in kilogrammes per square metre

Size (mm)	Spacing of bars in millimetres									
	75	100	125	150	175	200	225	250	275	300
6	2.96	2.220	1.776	1.480	1.27	1.110	0.99	0.89	0.81	0.74
8	5.26	3.95	3.16	2.63	2.26	1.97	1.75	1.58	1.44	1.32
10	8.22	6.17	4.93	4.11	3.52	3.08	2.74	2.47	2.24	2.06
12	11.84	8.88	7.10	5.92	5.07	4.44	3.95	3.55	3.23	2.96
16	21.04	15.78	12.63	10.52	9.02	7.89	7.02	6.31	5.74	5.26
20	32.88	24.66	19.73	16.44	14.09	12.33	10.96	9.87	8.97	8.22
25	51.38	38.53	30.83	25.69	22.02	19.27	17.13	15.41	14.01	12.84
32	84.18	63.13	50.51	42.09	36.08	31.57	28.06	25.25	22.96	21.04
40	131.53	98.65	78.92	65.76	56.37	49.32	43.84	39.46	35.87	32.88
50	205.51	154.13	123.31	102.76	88.08	77.07	68.50	61.65	56.05	51.38

Basic weight of steelwork taken as 7850 kg/m³

Basic weight of bar reinforcement per metre run = 0.00785 kg/mm²

The value of PI has been taken as 3.141592654

CONCRETE WORK – cont'd
Fabric reinforcement

Preferred range of designated fabric types and stock sheet sizes

Fabric reference	Longitudinal wires			Cross wires			Mass
	Nominal wire size (mm)	Pitch (mm)	Area (mm/m²)	Nominal wire size (mm)	Pitch (mm)	Area (mm/m²)	(kg/m²)
Square mesh							
A393	10	200	393	10	200	393	6.16
A252	8	200	252	8	200	252	3.95
A193	7	200	193	7	200	193	3.02
A142	6	200	142	6	200	142	2.22
A98	5	200	98	5	200	98	1.54
Structural mesh							
B1131	12	100	1131	8	200	252	10.90
B785	10	100	785	8	200	252	8.14
B503	8	100	503	8	200	252	5.93
B385	7	100	385	7	200	193	4.53
B283	6	100	283	7	200	193	3.73
B196	5	100	196	7	200	193	3.05
Long mesh							
C785	10	100	785	6	400	70.8	6.72
C636	9	100	636	6	400	70.8	5.55
C503	8	100	503	5	400	49.00	4.34
C385	7	100	385	5	400	49.00	3.41
C283	6	100	283	5	400	49.00	2.61
Wrapping mesh							
D98	5	200	98	5	200	98	1.54
D49	2.5	100	49	2.5	100	49	0.77
Stock sheet size	Length 4.8 m			Width 2.4 m		Sheet area 11.52m²	

Wire

SWG	6g	5g	4g	3g	2g	1g	1/0g	2/0g	3/0g	4/0g	5/0g
diameter											
in	0.192	0.212	0.232	0.252	0.276	0.300	0.324	0.348	0.372	0.400	0.432
mm	4.9	5.4	5.9	6.4	7.0	7.6	8.2	8.8	9.5	0.2	1.0
area											
in^2	0.029	0.035	0.042	0.050	0.060	0.071	0.082	0.095	0.109	0.126	0.146
mm^2	19	23	27	32	39	46	53	61	70	81	95

Average weight kg/m³ of steelwork reinforcement in concrete for various building elements

	kg/m³ concrete
Substructure	
Pile caps	110 - 150
Tie beams	130 - 170
Ground beams	230 - 330
Bases	90 - 130
Footings	70 - 110
Retaining walls	110 - 150
Superstructure	
Slabs - one way	75 - 125
Slabs - two way	65 - 135
Plate slab	95 - 135
Cantilevered slab	90 - 130
Ribbed floors	80 - 120
Columns	200 - 300
Beams	250 - 350
Stairs	130 - 170
Walls - normal	30 - 70
Walls - wind	50 - 90

Note: For exposed elements add the following % :

Walls 50%, Beams 100%, Columns 15%

CONCRETE WORK – cont'd

Formwork stripping times

Normal Curing Periods

Conditions under which concrete is maturing	Minimum periods of protection for different types of cement					
	Number of days (where the average surface temperature of the concrete exceeds 10°C during the whole period)			Equivalent maturity (degree hours) calculated as the age of the concrete in hours multiplied by the number of degrees Celsius by which the average surface temperature of the concrete exceeds -10°C		
	Other	SRPC	OPC or RHPC	Other	SRPC	OPC or RHPC
1. Hot weather or drying winds	7	4	3	3500	2000	1500
2. Conditions not covered by 1	4	3	2	2000	1500	1000

KEY

OPC - Ordinary Portland Cement

RHPC - Rapid-hardening Portland cement

SRPC - Sulphate-resisting Portland cement

Minimum period before striking formwork

	Minimum period before striking		
	Surface temperature of concrete		
	16 °C	17 °C	t °C(0-25)
Vertical formwork to columns, walls and large beams	12 hours	18 hours	$\frac{300}{t+10}$ hours
Soffit formwork to slabs	4 days	6 days	$\frac{100}{t+10}$ days
Props to slabs	10 days	15 days	$\frac{250}{t+10}$ days
Soffit formwork to beams	9 days	14 days	$\frac{230}{t+10}$ days
Props to beams	14 days	21 days	$\frac{360}{t+10}$ days

MASONRY

Weights of bricks and blocks

Walls and components of walls	kg/m² per mm thick	lb/sq ft per inch thick
Blockwork		
Hollow clay blocks; average)	1.15	6
Common clay blocks	1.90	10
Brickwork		
Engineering clay bricks	2.30	12
Refactory bricks	1.15	6
Sand-lime (and similar) bricks	2.02	10.5

Weights of stones

Type of stone	kg/m3	lb/cu ft
Natural stone (solid)		
Granite	2560 to 2927	160 to 183
Limestone - Bath stone	2081	130
- Marble	2723	170
- Portland stone	2244	140
Sandstone	2244 to 2407	140 to 150
Slate	2880	180
Stone rubble (packed)	2244	140

Quantities of bricks and mortar

Materials per m² of wall:		
Thickness	**No. of Bricks**	**Mortar m³**
Half brick (112.5 mm)	58	0.022
One brick (225 mm)	116	0.055
Cavity, both skins (275 mm)	116	0.045
1.5 brick (337 mm)	174	0.074
Mass brickwork per m³	464	0.36

Mortar mixes: Quantities of Dry Materials

	Imperial cu yd			Metric m³		
Mix	Cement cwts	Lime cwts	Sand cu yds	Cement tonnes	Lime tonnes	Sand cu m
1:3	7.0	-	1.04	0.54	-	1.10
1:4	6.3	-	1.10	0.40	-	1.20
1:1:6	3.9	1.6	1.10	0.27	0.13	1.10
1:2:9	2.6	2.1	1.10	0.20	0.15	1.20
0:1:3	-	3.3	1.10	-	0.27	1.00

MASONRY – cont'd

Mortar mixes for various uses

Mix	Use
1:3	Construction designed to withstand heavy loads in all seasons
1:1:6	Normal construction not designed for heavy loads. Sheltered and moderate conditions in spring and summer. Work above d:p:c - sand, lime bricks, clay blocks etc.
1:2:9	Internal partitions with blocks which have high drying shrinkage, pumicblocks, etc. any periods
0:1:3	Hydraulic lime only should be used in this mix and may be used for construction not designed for heavy loads and above d:p:c spring and summer.

Quantities of bricks and mortar required per m² of walling

Description	Unit	Nr of bricks required	Mortar required (m³)		
			No frogs	Single frogs	Double frogs
Standard bricks **Brick size** **215 x 102.5 x 50 mm**					
half brick wall (103 mm) (103 mm)	m²	72	0.022	0.027	0.032
2 x half brick cavity wall (270 mm)	m²	144	0.044	0.054	0.064
one brick wall (215 mm)	m²	144	0.052	0.064	0.076
one and a half brick wall (328 mm)	m²	216	0.073	0.091	0.108
mass brickwork	m³	576	0.347	0.413	0.480
Brick size **215 x 102.5 x 65 mm**					
half brick wall (103 mm)	m²	58	0.019	0.022	0.026
2 x half brick cavity wall (270 mm)	m²	116	0.038	0.045	0.055
one brick wall (215 mm)	m²	116	0.046	0.055	0.064
one and a half brick wall (328 mm)	m²	174	0.063	0.074	0.088
mass brickwork	m³	464	0.307	0.360	0.413
Metric modular bricks		**Perforated**			
Brick co-ordinating size **200 x 100 x 75 mm**					
90 mm thick	m²	67	0.016	0.019	
190 mm thick	m²	133	0.042	0.048	
290 mm thick	m²	200	0.068	0.078	
Brick co-ordinating size **200 x 100 x 100 mm**					
90 mm thick	m²	50	0.013	0.016	
190 mm thick	m²	100	0.036	0.041	
290 mm thick	m²	150	0.059	0.067	
Brick co-ordinating size **300 x 100 x 75 mm**					
90 mm thick	m²	33	-	0.015	
Brick co-ordinating size **00 x 100 x 100 mm**					
90 mm thick	m²	44	0.015	0.018	
Note: Assuming 10 mm deep joints.					

MASONRY – cont'd

Mortar required per m² blockwork (9.88 blocks/m²)							
Wall thickness	75	90	100	125	140	190	215
Mortar m³/m²	0.005	0.006	0.007	0.008	0.009	0.013	0.014

Mortar Mixes

Mortar Group	Cement: lime: sand	Masonry cement: sand	Cement: sand with plasticiser
1	1 : 0-0.25:3		
2	1 : 0.5 :4-4.5	1 : 2.5-3.5	1 : 3-4
3	1 : 1:5-6	1 : 4-5	1 : 5-6
4	1 : 2:8-9	1 : 5.5-6.5	1 : 7-8
5	1 : 3:10-12	1 : 6.5-7	1 : 8

Group 1: strong inflexible mortar
Group 5: weak but flexible.

All mixes within a group are of approximately similar strength.
Frost resistance increases with the use of plasticisers.
Cement:lime:sand mixes give the strongest bond and greatest resistance to rain penetration.
Masonry cement equals ordinary Portland cement plus a fine neutral mineral filler and an air entraining agent.

Calcium Silicate Bricks

Type	Strength	Location
Class 2 crushing strength	14.0N/mm2	not suitable for walls
Class 3	20.5N/mm2	walls above dpc
Class 4	27.5N/mm2	cappings and copings
Class 5	34.5N/mm2	retaining walls
Class 6	41.5N/mm2	walls below ground
Class 7	48.5N/mm2	walls below ground

The Class 7 calcium silicate bricks are therefore equal in strength to Class B bricks.
Calcium silicate bricks are not suitable for DPCs

Durability of Bricks

FL	Frost resistant with low salt content
FN	Frost resistant with normal salt content
ML	Moderately frost resistant with low salt content
MN	Moderately frost resistant with normal salt content

Brickwork Dimensions

No. of Horizontal Bricks	Dimensions mm	No. of Vertical courses	No. of Vertical courses
1/2	112.5	1	75
1	225.0	2	150
1 1/2	337.5	3	225
2	450.0	4	300
2 1/2	562.5	5	375
3	675.0	6	450
3 1/2	787.5	7	525
4	900.0	8	600
4 1/2	1012.5	9	675
5	1125.0	10	750
5 1/2	1237.5	11	825
6	1350.0	12	900
6 1/2	1462.5	13	975
7	1575.0	14	1050
7 1/2	1687.5	15	1125
8	1800.0	16	1200
8 1/2	1912.5	17	1275
9	2025.0	18	1350
9 1/2	2137.5	19	1425
10	2250.0	20	1500
20	4500.0	24	1575
40	9000.0	28	2100
50	11250.0	32	2400
60	13500.0	36	2700
75	16875.0	40	3000

MASONRY – cont'd

Standard available block sizes			
Block	**Length x height**		
	Co-ordinating size	Work size	Thicknesses (work size)
A	400 x 100	390 x 90	(75, 90, 100, 140 &
	400 x 200	440 x 190	(190 mm
	450 x 225	440 x 215	(75, 90, 100, 140,
			(190 & 215 mm
B	400 x 100	390 x 90	(75, 90, 100, 140 &
	400 x 200	390 x 190	(190 mm
	450 x 200	440 x 190	(
	450 x 225	440 x 215	(75, 90, 100, 140,
	450 x 300	440 x 290	(190 & 215 mm
	600 x 200	590 x 190	(
	600 x 225	590 x 215	(
C	400 x 200	390 x 190	(
	450 x 200	440 x 190	(
	450 x 225	440 x 215	(60 & 75 mm
	450 x 300	440 x 290	(
	600 x 200	590 x 190	(
	600 x 225	590 x 215	(

TIMBER

Weights of timber

Material	kg/m³	lb/cu ft
General	806 (avg)	50 (avg)
Douglas fir	479	30
Yellow pine, spruce	479	30
Pitch pine	673	42
Larch, elm	561	35
Oak (English)	724 to 959	45 to 60
Teak	643 to 877	40 to 55
Jarrah	959	60
Greenheart	1040 to 1204	65 to 75
Quebracho	1285	80

Material	kg/m² per mm thickness	lb/sq ft per inch thickness
Wooden boarding and blocks		
Softwood	0.48	2.5
Hardwood	0.76	4
Hardboard	1.06	5.5
Chipboard	0.76	4
Plywood	0.62	3.25
Blockboard	0.48	2.5
Fibreboard	0.29	1.5
Wood-wool	0.58	3
Plasterboard	0.96	5
Weather boarding	0.35	1.8

TIMBER - cont'd

Conversion tables (for sawn timber only)

Inches >	Millimetres	Feet >	Metres
1	25	1	0.300
2	50	2	0.600
3	75	3	0.900
4	100	4	1.200
5	125	5	1.500
6	150	6	1.800
7	175	7	2.100
8	200	8	2.400
9	225	9	2.700
10	250	10	3.000
11	275	11	3.300
12	300	12	3.600
13	325	13	3.900
14	350	14	4.200
15	375	15	4.500
16	400	16	4.800
17	425	17	5.100
18	450	18	5.400
19	475	19	5.700
20	500	20	6.000
21	525	21	6.300
22	550	22	6.600
23	575	23	6.900
24	600	24	7.200

Planed softwood

The finished end section size of planed timber is usually 3/16" less than the original size from which it is produced. This however varies slightly dependant upon availability of material and origin of species used.

Standard (timber) to cubic metres and cubic metres to standards (timber)

m³	m³/Standards	Standard
4.672	1	0.214
9.344	2	0.428
14.017	3	0.642
18.689	4	0.856
23.361	5	1.070
28.033	6	1.284
32.706	7	1.498
37.378	8	1.712
42.05	9	1.926
46.722	10	2.140
93.445	20	4.281
140.167	30	6.421
186.890	40	8.561
233.612	50	10.702
280.335	60	12.842
327.057	70	14.982
373.779	80	17.122
420.502	90	19.263
467.224	100	21.403

1 cu metre = 35.3148
1 cu ft = 0.028317 cu metres
1 std = 4.67227 cu metres

TIMBER – cont'd

Standards (timber) to cubic metres and cubic metres to standards (timber)

1 cu metre	=	35.3148 cu ft =	0.21403 std
1 cu ft	=	0.028317 cu metres	
1 std	=	4.67227 cu metres	

Basic sizes of sawn softwood available (cross sectional areas)

Thickness (mm)	Width (mm)								
	75	100	125	150	175	200	225	250	300
16	x	x	x	x					
19	x	x	x	x					
22	x	x	x	x					
25	x	x	x	x	x	x	x	x	x
32	x	x	x	x	x	x	x	x	x
36	x	x	x	x					
38	x	x	x	x	x	x	x		
44	x	x	x	x	x	x	x	x	x
47*	x	x	x	x	x	x	x	x	x
50	x	x	x	x	x	x	x	x	x
63	x	x	x	x	x	x	x		
75		x	x	x	x	x	x	x	x
100		x		x		x		x	x
150				x		x			x
200						x			
250								x	
300									x

* This range of widths for 47 mm thickness will usually be found to be available in construction quality only.

Note: The smaller sizes below 100 mm thick and 250 mm width are normally but not exclusively of European origin. Sizes beyond this are usually of North and South American origin.

Basic lengths of sawn softwood available (metres)

1.80	2.10	3.00	4.20	5.10	6.00	7.20
2.40	3.30	4.50	5.40	6.30		
2.70	3.60	4.80	5.70	6.60		
	3.90			6.90		

Note: Lengths of 6.00 m and over will generally only be available from North American species and may have to be recut from larger sizes.

Reductions from basic size to finished size of timber by planing of two opposed faces

Purpose	15 - 35 mm	36 - 100 mm	101 - 150 mm	over 150 mm
a) constructional timber	3 mm	3 mm	5 mm	6 mm
b) matching interlocking boards	4 mm	4 mm	6 mm	6 mm
c) wood trim not specified in BS 584	5 mm	7 mm	7 mm	9 mm
d) joinery and cabinet work	7 mm	9 mm	11 mm	13 mm

Note: The reduction of width or depth is overall the extreme size and is exclusive of any reduction of the face by the machining of a tongue or lap joints.

Weights of metalwork

Material	kg/m³	lb/cu ft
Metals, steel construction, etc		
Iron		
- cast	7207	450
- wrought	7687	480
- ore - general	2407	150
- (crushed) Swedish	3682	230
Steel	7854	490
Copper		
- cast	8731	545
- wrought	8945	558
Brass	8497	530
Bronze	8945	558
Aluminium	2774	173
Lead	11322	707
Zinc (rolled)	7140	446
	g/mm² per metre	lb/sq ft per foot
Steel bars	7.85	3.4

Structural steelwork	Net weight of member @ 7854 kg/m³
rivetted	+ 10% for cleats, rivets, bolts, etc
welded	+ 1.25% to 2.5% for welds, etc
Rolled sections	
beams	+ 2.5%
stanchions	+ 5% (extra for caps and bases)
Plate	
web girders	+ 10% for rivets or welds, stiffeners, etc

	kg/m	lb/ft
Steel stairs : industrial type		
1 m or 3ft wide	84	56
Steel tubes		
50 mm or 2 in bore	5 to 6	3 to 4
Gas piping		
20 mm or 3/4 in	2	1¼

KERBS/EDGINGS/CHANNELS

Precast Concrete Kerbs to BS 7263
Straight kerb units: length from 450 to 915 mm

150mm high x 125mm thick		
bullnosed	type BN	
half battered	type HB3	
255mm high x 125mm thick		
45 degree splayed	type SP	
half battered	type HB2	
305mm high x 150mm thick		
half battered	type HB1	
Quadrant kerb units		
150 mm high x 305 and 455 mm radius to match	type BN	type QBN
150 mm high x 305 and 455 mm radius to match	type HB2, HB3	type QHB
150 mm high x 305 and 455 mm radius to match	type SP	type QSP
255 mm high x 305 and 455 mm radius to match	type BN	type QBN
255 mm high x 305 and 455 mm radius to match	type HB2, HB3	type QHB
225 mm high x 305 and 455 mm radius to match	type SP	type QSP
Angle kerb units		
305 x 305 x 225 mm high x 125 mm thick		
bullnosed external angle	type XA	
splayed external angle to match type SP	type XA	
bullnosed internal angle	type IA	
splayed internal angle to match type SP	type IA	

KERBS/EDGINGS/CHANNELS – cont'd

Channels
    255 mm wide x 125 mm high flat   type CS1
    150 mm wide x 125 mm high flat type   CS2
    255 mm wide x 125 mm high dished   type CD

Transition kerb units

from kerb type SP to HB	left handed	type TL
	right handed	type TR
from kerb type BN to HB	left handed	type DL1
	right handed	type DR1
from kerb type BN to SP	left handed	type DL2
	right handed	type DR2

Radial kerbs and channels

All profiles of kerbs and channels

External radius	Internal radius
1000 mm	3000
2000	4500
3000	6000
4500	7500
6000	9000
7500	1050
9000	1200
1050	
1200	

Precast Concrete Edgings to BS 7263

Round top type ER	Flat top type EF	Bullnosed top type EBN
150 x 50 mm	150 x 50 mm	150 x 50 mm
200 x 50	200 x 50	200 x 50
250 x 50	250 x 50	250 x 50

BASES

Cement Bound Material for Bases and Sub-bases

CBM1: very carefully graded aggregate from 37.5 - 75ym, with a 7-day strength of 4.5N/mm2

CBM2: same range of aggregate as CBM1 but with more tolerance in each size of ggregate with a 7-day strength of 7.0N/mm2

CBM3: crushed natural aggregate or blastfurnace slag, graded from 37.5mm - 150ym for 40mm aggregate, and from 20 - 75ym for 20mm aggregate, with a 7-day strength of 10N/mm2

CBM4: crushed natural aggregate or blastfurnace slag, graded from 37.5mm - 150ym for 40mm aggregate, and from 20 - 75ym for 20mm aggregate, with a 7-day strength of 15N/mm2

INTERLOCKING BRICK/BLOCK ROADS/PAVINGS

Sizes of Precast Concrete Paving Blocks to BS 6717: Part 1

Type R blocks
200 x 100 x 60 mm
200 x 100 x 65
200 x 100 x 80
200 x 100 x 100

Type S
Any shape within a 295 mm space

Sizes of Clay Brick Pavers to BS 6677: Part 1
200 x 100 x 50 mm thick
200 x 100 x 65
210 x 105 x 50
210 x 105 x 65
215 x 102.5 x 50
215 x 102.5 x 65

Type PA: 3 kN
Footpaths and pedestrian areas, private driveways, car parks, light vehicle traffic and over-run.

Type PB: 7 kN
Residential roads, lorry parks, factory yards, docks, petrol station forecourts, hardstandings, bus stations.

PAVING AND SURFACING

Weights and sizes of paving and surfacing

Description of Item		Quantity per tonne
Paving 50 mm thick	900 x 600 mm	15
Paving 50 mm thick	750 x 600 mm	18
Paving 50 mm thick	600 x 600 mm	23
Paving 50 mm thick	450 x 600 mm	30
Paving 38 mm thick	600 x 600 mm	30
Path edging	914 x 50 x 150 mm	60
Kerb (including radius and tapers)	125 x 254 x 914 mm	15
Kerb (including radius and tapers)	125 x 150 x 914 mm	25
Square channel	125 x 254 x 914 mm	15
Dished channel	125 x 254 x 914 mm	15
Quadrants	300 x 300 x 254 mm	19
Quadrants	450 x 450 x 254 mm	12
Quadrants	300 x 300 x 150 mm	30
Internal angles	300 x 300 x 254 mm	30
Fluted pavement channel	255 x 75 x 914 mm	25
Corner stones	300 x 300 mm	80
Corner stones	360 x 360 mm	60
Cable covers	914 x 175 mm	55
Gulley kerbs	220 x 220 x 150 mm	60
Gulley kerbs	220 x 200 x 75 mm	120

Material	kg/m³	lb/cu yd
Tarmacadam	2306	3891
Macadam (waterbound)	2563	4325
Vermiculite (aggregate)	64-80	108-135
Terracotta	2114	3568
Cork - compressed	388	24
	kg/m²	lb/sq ft
Clay floor tiles, 12.7mm	27.3	5.6
Pavement lights	122	25
Damp proof course	5	1
	kg/m² per mm thickness	lb/sq ft per inch thickness
Paving Slabs (stone)	2.3	12
Granite setts	2.88	15
Asphalt	2.30	12
Rubber flooring	1.68	9
Poly-vinylchloride	1.94 (avg)	10 (avg)

Coverage (m²) per cubic metre of materials used as sub bases or capping layers

Consolidated thickness laid in (mm)	Square metre coverage		
	Gravel	Sand	Hardcore
50	15.80	16.50	-
75	10.50	11.00	-
100	7.92	8.20	7.42
125	6.34	6.60	5.90
150	5.28	5.50	4.95
175	-	-	4.23
200	-	-	3.71
225	-	-	3.30
300	-	-	2.47

Approximate rate of spreads

Average thickness of course mm	Description	Approximate rate of spread			
		Open Textured		Dense, Medium & Fine Textured	
		kg/m²	m²/t	kg/m²	m²/t
35	14 mm open textured or dense wearing course	60-75	13-17	70-85	12-14
40	20 mm open textured or dense base course	70-85	12-14	80-100	10-12
45	20 mm open textured or dense base course	80-100	10-12	95-100	9-10
50	20 mm open textured or dense, or 28 mm dense base course	85-110	9-12	110-120	8-9
60	28 mm dense base course, 40 mm open textured of dense base course or 40 mm single course as base course		8-10	130-150	7-8
65	28 mm dense base course, 40 mm open textured or dense base course or 40 mm single course	100-135	7-10	140-160	6-7
75	40 mm single course, 40 mm open textured or dense base course, 40 mm dense roadbase	120-150	7-8	165-185	5-6
100	40 mm dense base course or roadbase	-	-	220-240	4-4.5

Surface Dressing Roads: Coverage (m²) per tonne of Material

Size in mm	Sand	Granite chips	Gravel	Limestone Chips
Sand	168	-	-	-
3	-	148	152	165
6	-	130	133	144
9	-	111	114	123
13	-	85	87	95
19	-	68	71	78

Sizes of Flags to BS 7263

Reference	Nominal Size	Thickness
A	600 x 450 mm	50 and 63 mm
B	600 x 600	50 and 63
C	600 x 750	50 and 63
D	600 x 900	50 and 63
E	450 x 450	50 and 70 chamfered top surface
F	400 x 400	50 and 65 chamfered top surface
G	300 x 300	50 and 60 chamfered top surface

Sizes of Natural Stone Setts to BS 435

Width		Length		Depth
100 mm	x	100 mm	x	100 mm
75	x	150 to 250	x	125
75	x	150 to 250	x	150
100	x	150 to 250	x	100
100	x	150 to 250	x	150

SPORTS
Sizes of Sports Areas

Sizes in <u>metres</u> given include clearances:

Association football	Senior		114	x	72
	Junior		108	x	58
	International		100 - 110	x	64 - 75
Football	American	Pitch	109.80	x	48.80
		Overall	118.94	x	57.94
	Australian	Overall	135 - 185	x	110 - 155
	Canadian	Overall	145.74	x	59.47
	Gaelic		128 - 146.4	x	76.8 - 91.50
Handball			91 - 110	x	55 - 65
Hurling			137	x	82
Rugby	Union pitch		56	x	81
	League		134	x	80
Hockey pitch			100.5	x	61
Men's lacrosse pitch			106	x	61
Women's lacrosse pitch			110	x	60
Target archery ground			150	x	50
Archery (Clout)			7.3m	firing area	
			Range 109.728 (Women), 146.304 (Men).		
			182.88 (Normal range)		
400m Running Track			115.61	bend length x 2	
6 lanes			84.39	straight length x 2	
		Overall	176.91 long	x	92.52 wide
Baseball		Overall	60m	x	70
Basketball			14.0	x	26.0
Camogie			91 - 110	x	54 - 68

Discus and Hammer	Safety cage 2.74m square			
	Landing area 45 arc (65° safety) 70 m radius			
Javelin	Runway	36.5 x	4.27	
	Landing area	80 - 95	x	48
Jump High	Running area	38.8 x	19	
	Landing area	5	x	4
Long	Runway	45	x	1.22
	Landing area	9	x	2.750
Triple	Runway	45	x	1.22
	Landing area	7.3	x	2.75
Korfball		90	x	40
Netball		15.25	x	30.48
Pole Vault	Runway	45	x	1.22
	Landing area	5	x	5
Polo		275	x	183
Rounders	Overall	19	x	17
Shot Putt	Base	2.135	dia	
	Landing area	65° arc,	25m radius from base	
Shinty		128 -183	x	64 - 91.5
Tennis	Court	23.77	x	10.97
	Overall minimum	36.27	x	18.29
Tug-of-war		46	x	5

SEEDING/TURFING AND PLANTING

BS 3882: 1994 Topsoil Quality

Topsoil grade	Properties
Premium	natural topsoil, high fertility, loamy texture, good soil structure, suitable for intensive cultivation
General Purpose	natural or manufactured topsoil of lesser quality than Premium, suitable for agriculture or amenity landscape, may need fertilizer or soil structure improvement.
Economy	selected subsoil, natural mineral deposit such as river silt or greensand. The grade comprises two subgrades; "Low clay" and "High clay" which is more liable to compaction in handling. This grade is suitable for low production agricultural land and amenity woodland or conservation planting areas.

Forms of Trees to BS 3936: 1992

Standards: shall be clear with substantially straight stems. Grafted and budded trees shall have no more than a slight bend at the union. Standards shall be designated as Half, Extra light, Light, Standard, Selected standard, Heavy, and Extra heavy.

Sizes of Standards
- Heavy standard — 12-14 cm girth x 3.50 to 5.00 m high
- Extra Heavy standard — 14-16 cm girth x 4.25 to 5.00 m high
- Extra Heavy standard — 16-18 cm girth x 4.25 to 6.00 m high
- Extra Heavy standard — 18-20 cm girth x 5.00 to 6.00 m high

Semi-mature trees: between 6.0m and 12.0 m tall with a girth of 20 to 75cm at 1.0 m above ground.

Feathered trees: shall have a defined upright central leader, with stem furnished with evenly spread and balanced lateral shoots down to or near the ground.

Whips: shall be without significant feather growth as determined by visual inspection.

Multi-stemmed trees: shall have two or more main stems at, near, above or below ground.

Seedlings grown from seed and not transplanted shall be specified when ordered for sale as:

1+0 one year old seedling		
2+0 two year old seedling		
1+1 one year seed bed,	one year transplanted	= two year old seedling
1+2 one year seed bed,	two years transplanted	= three year old seedling
2+1 two year seed bed,	one year transplanted	= three year old seedling
1u1 two years seed bed,	undercut after 1 year	= two year old seedling
2u2 four years seed bed,	undercut after 2 years	= four year old seedling

Cuttings.	The age of cuttings (plants grown from shoots, stems, or roots of the mother plant) shall be specified when ordered for sale. The height of transplants and undercut seedlings/cuttings (which have been transplanted or undercut at least once) shall be stated in centimetres. The number of growing seasons before and after transplanting or undercutting shall be stated.
0+1	one year cutting
0+2	two year cutting
0+1+1	one year cutting bed, one year transplanted = two year old seedling
0+1+2	one year cutting bed, two years transplanted = three year old seedling

SEEDING/TURFING AND PLANTING

Grass Cutting Capacities in m2 per Hour

Speed mph	Width Of Cut in metres												
	0.5	0.7	1.0	1.2	1.5	1.7	2.0	2.0	2.1	2.5	2.8	3.0	3.4
1.0	724	1127	1529	1931	2334	2736	3138	3219	3380	4023	4506	4828	5472
1.5	1086	1690	2293	2897	3500	4104	4707	4828	5069	6035	6759	7242	8208
2.0	1448	2253	3058	3862	4667	5472	6276	6437	6759	8047	9012	9656	10944
2.5	1811	2816	3822	4828	5834	6840	7846	8047	8449	10058	11265	12070	13679
3.0	2173	3380	4587	5794	7001	8208	9415	9656	10139	12070	13518	14484	16415
3.5	2535	3943	5351	6759	8167	9576	10984	11265	11829	14082	15772	16898	19151
4.0	2897	4506	6115	7725	9334	10944	12553	12875	13518	16093	18025	19312	21887
4.5	3259	5069	6880	8690	10501	12311	14122	14484	15208	18105	20278	21726	24623
5.0	3621	5633	7644	9656	11668	13679	15691	16093	16898	20117	22531	24140	27359
5.5	3983	6196	8409	10622	12834	15047	17260	17703	18588	22128	24784	26554	30095
6.0	4345	6759	9173	11587	14001	16415	18829	19312	20278	24140	27037	28968	32831
6.5	4707	7322	9938	12553	15168	17783	20398	20921	21967	26152	29290	31382	35566
7.0	5069	7886	10702	13518	16335	19151	21967	22531	23657	28163	31543	33796	38302

Number of plants per m2 : For plants planted on an evenly spaced grid
Planting distances

mm	0.10	0.15	0.20	0.25	0.35	0.40	0.45	0.50	0.60	0.75	0.90	1.00	1.20	1.50
0.10	100.00	66.67	50.00	40.00	28.57	25.00	22.22	20.00	16.67	13.33	11.11	10.00	8.33	6.67
0.15	66.67	44.44	33.33	26.67	19.05	16.67	14.81	13.33	11.11	8.89	7.41	6.67	5.56	4.44
0.20	50.00	33.33	25.00	20.00	14.29	12.50	11.11	10.00	8.33	6.67	5.56	5.00	4.17	3.33
0.25	40.00	26.67	20.00	16.00	11.43	10.00	8.89	8.00	6.67	5.33	4.44	4.00	3.33	2.67
0.35	28.57	19.05	14.29	11.43	8.16	7.14	6.35	5.71	4.76	3.81	3.17	2.86	2.38	1.90
0.40	25.00	16.67	12.50	10.00	7.14	6.25	5.56	5.00	4.17	3.33	2.78	2.50	2.08	1.67
0.45	22.22	14.81	11.11	8.89	6.35	5.56	4.94	4.44	3.70	2.96	2.47	2.22	1.85	1.48
0.50	20.00	13.33	10.00	8.00	5.71	5.00	4.44	4.00	3.33	2.67	2.22	2.00	1.67	1.33
0.60	16.67	11.11	8.33	6.67	4.76	4.17	3.70	3.33	2.78	2.22	1.85	1.67	1.39	1.11
0.75	13.33	8.89	6.67	5.33	3.81	3.33	2.96	2.67	2.22	1.78	1.48	1.33	1.11	0.89
0.90	11.11	7.41	5.56	4.44	3.17	2.78	2.47	2.22	1.85	1.48	1.23	1.11	0.93	0.74
1.00	10.00	6.67	5.00	4.00	2.86	2.50	2.22	2.00	1.67	1.33	1.11	1.00	0.83	0.67
1.20	8.33	5.56	4.17	3.33	2.38	2.08	1.85	1.67	1.39	1.11	0.93	0.83	0.69	0.56
1.50	6.67	4.44	3.33	2.67	1.90	1.67	1.48	1.33	1.11	0.89	0.74	0.67	0.56	0.44

Grass Clippings Wet: Based on 3.5 m3 /ton

Annual Kg/100 m2	Average 20 cuts Kg/100m2	m2 /tonne	m2 /m3
32.0	1.6	61162.1	214067.3

Nr of Cuts		22	20	18	16	12	4
Kg/cut		1.45	1.60	1.78	2.00	2.67	8.00
Area capacity of 3 tonne vehicle per load m2		206250	187500	168750	150000	112500	37500
Load m3		100 m2 units / m3 of vehicle space					
	1	196.4	178.6	160.7	142.9	107.1	35.7
	2	392.9	357.1	321.4	285.7	214.3	71.4
	3	589.3	535.7	482.1	428.6	321.4	107.1
	4	785.7	714.3	642.9	571.4	428.6	142.9
	5	982.1	892.9	803.6	714.3	535.7	178.6

FENCING AND GATES

Types of Preservative to BS 5589:1989

Creosote (tar oil) can be "factory" applied

	by pressure to BS 144: pts 1&2
	by immersion to BS 144: pt 1
	by hot and cold open tank to BS 144: pts 1&2
Copper/chromium/arsenic (CCA)	by full cell process to BS 4072 pts 1&2
Organic solvent (OS)	by double vacuum (vacvac) to BS 5707 pts 1&3
	by immersion to BS 5057 pts 1&3
Pentachlorophenol (PCP)	by heavy oil double vacuum to BS 5705 pts 2&3

Boron diffusion process (treated with disodium octaborate to BWPA Manual 1986.
Note: Boron is used on green timber at source and the timber is suppled dry.

Cleft Chestnut Pale Fences to BS 1722:Part 4:1986

Pales	Pale spacing	Wire lines	
900 mm long	75 mm	2	temporary protection
1050	75 or 100	2	light protective fences
1200	75	3	perimeter fences
1350	75	3	perimeter fences
1500	50	3	narrow perimeter fences
1800	50	3	light security fences

FENCING & GATES - cont'd

Close-boarded Fences to BS 1722 :Pt 5: 1986

Close-boarded fences 1.05 to 1.8m high
Type BCR (recessed) or BCM (morticed) with concrete posts 140 x 115 mm tapered and Type BW
with timber posts:

Palisade Fences to BS 1722:pt 6:1986.

Wooden palisade fences
Type WPC with concrete posts 140 x 115 mm tapered and Type WPW with timber posts.

For both types of fence:

Height of fence 1050 mm:	two rails
Height of fence 1200 mm:	two rails
Height of fence 1500 mm:	three rails
Height of fence 1650 mm:	three rails
Height of fence 1800 mm:	three rails
Height of fence 1800 mm:	three rails

Post and Rail Fences to BS 1722: part 7

Wooden post and rail fences
Type MPR 11/3 morticed rails and Type SPR 11/3 nailed rails
Height to top of rail 1100 mm
Rails: three rails 87 mm 38 mm

Type MPR 11/4 morticed rails and Type SPR 11/4 nailed rails
Height to top of rail 1100 mm
Rails: four rails 87 mm 38 mm.

Type MPR 13/4 morticed rails and Type SPR 13/4 nailed rails
Height to top of rail 1300 mm
Rail spacing 250 mm, 250 mm, and 225 mm from top
Rails: four rails 87 mm 38 mm.

Steel Posts to BS 1722: Part 1.

Rolled steel angle iron posts for chain link fencing:

Posts	Fence height	Strut	Straining post
1500 x 40 x 40 x 5 mm	900 mm	1500 x 40 x 40 x 5 mm	1500 x 50 x 50 x 6 mm
1800 x 40 x 40 x 5 mm	1200 mm	1800 x 40 x 40 x 5 mm	1800 x 50 x 50 x 6 mm
2000 x 45 x 45 x 5 mm	1400 mm	2000 x 45 x 45 x 5 mm	2000 x 60 x 60 x 6 mm
2600 x 45 x 45 x 5 mm	1800 mm	2600 x 45 x 45 x 5 mm	2600 x 60 x 60 x 6 mm
3000 x 50 x 50 x 6 mm with arms	1800 mm	2600 x 45 x 45 x 5 mm	3000 x 60 x 60 x 6 mm

Concrete Posts to BS 1722: Part 1.

Concrete posts for chain link fencing:

Posts and straining posts	Fence height	Strut
1570 mm 100 x 100 mm	900 mm	1500 mm x 75 x 75 mm
1870 mm 125 x 125 mm	1200 mm	1830 mm x 100 x 75 mm
2070 mm 125 x 125 mm	1400 mm	1980 mm x 100 x 75 mm
2620 mm 125 x 125 mm	1800 mm	2590 mm x 100 x 85 mm
3040 mm 125 x 125 mm	1800 mm	2590 mm x 100 x 85 mm (with arms)

FENCING AND GATES – cont'd

Rolled Steel Angle Posts to BS 1722: Part 2.

Rolled steel angle posts for rectangular wire mesh (field) fencing

Posts	Fence height	Strut	Straining post
1200 x 40 x 40 x 5 mm	600 mm	1200 x 75 x 75 mm	1350 x 100 x 100 mm
1400 x 40 x 40 x 5 mm	800 mm	1400 x 75 x 75 mm	1550 x 100 x 100 mm
1500 x 40 x 40 x 5 mm	900 mm	1500 x 75 x 75 mm	1650 x 100 x 100 mm
1600 x 40 x 40 x 5 mm	1000 mm	1600 x 75 x 75 mm	1750 x 100 x 100 mm
1750 x 40 x 40 x 5 mm	1150 mm	1750 x 75 x 100 mm	1900 x 125 x 125 mm

Concrete Posts to BS 1722: Part 2.

Concrete posts for rectangular wire mesh (field) fencing

Posts	Fence height	Strut	Straining post
1270 x 100 x 100 mm	600 mm	1200 x 75 x 75 mm	1420 x 100 x 100 mm
1470 x 100 x 100 mm	800 mm	1350 x 75 x 75 mm	1620 x 100 x 100 mm
1570 x 100 x 100 mm	900 mm	1500 x 75 x 75 mm	1720 x 100 x 100 mm
1670 x 100 x 100 mm	600 mm	1650 x 75 x 75 mm	1820 x 100 x 100 mm
1820 x 125 x 125 mm	1150 mm	1830 x 75 x 100 mm	1970 x 125 x 125 mm

Cleft Chestnut Pale Fences to BS 1722: part 4: 1986

Timber Posts to BS 1722: Part 2.

Timber posts for wire mesh and hexagonal wire netting fences.
Round timber for general fences

Posts	Fence height	Strut	Straining post
1300 x 65 mm dia.	600 mm	1200 x 80 mm dia	1450 x 100 mm dia
1500 x 65 mm dia.	800 mm	1400 x 80 mm dia	1650 x 100 mm dia
1600 x 65 mm dia.	900 mm	1500 x 80 mm dia	1750 x 100 mm dia
1700 x 65 mm dia.	1050 mm	1600 x 80 mm dia	1850 x 100 mm dia
1800 x 65 mm dia.	1150 mm	1750 x 80 mm dia	2000 x 120 mm dia

Squared timber for general fences

Posts	Fence height	Strut	Straining post
1300 x 75 x 75 mm	600 mm	1200 x 75 x 75 mm	1450 x 100 x 100 mm
1500 x 75 x 75 mm	800 mm	1400 x 75 x 75 mm	1650 x 100 x 100 mm
1600 x 75 x 75 mm	900 mm	1500 x 75 x 75 mm	1750 x 100 x 100 mm
1700 x 75 x 75 mm	1050 mm	1600 x 75 x 75 mm	1850 x 100 x 100 mm
1800 x 75 x 75 mm	1150 mm	1750 x 75 x 75 mm	2000 x 125 x 100 mm

Steel Fences to BS 1722: Pt 9: 1992

Mild steel fences: round or square verticals; flat standards and horizontals.
Tops of vertical bars may be bow-top, blunt, or pointed

Round or square bar railings.

	Fence height	Top/bottom rails and flat posts	Vertical bars
Light	1000 mm	40 x 10 mm 450 mm in ground	12 mm dia at 115 mm cs
	1200 mm	40 x 10 mm 550 mm in ground	12 mm dia at 115 mm cs
	1400 mm	40 x 10 mm 550 mm in ground	12 mm dia at 115 mm cs
Light	1000 mm	40 x 10 mm 450 mm in ground	16 mm dia at 120 mm cs
	1200 mm	40 x 10 mm 550 mm in ground	16 mm dia at 120 mm cs
	1400 mm	40 x 10 mm 550 mm in ground	16 mm dia at 120 mm cs
Medium	1200 mm	50 x 10 mm 550 mm in ground	20 mm dia at 125 mm cs
	1400 mm	50 x 10 mm 550 mm in ground	20 mm dia at 125 mm cs
	1600 mm	50 x 10 mm 600 mm in ground	22 mm dia at 145 mm cs
	1800 mm	50 x 10 mm 600 mm in ground	22 mm dia at 145 mm cs
Heavy	1600 mm	50 x 10 mm 600 mm in ground	22 mm dia at 145 mm cs
	1800 mm	50 x 10 mm 600 mm in ground	22 mm dia at 145 mm cs
	2000 mm	50 x 10 mm 600 mm in ground	22 mm dia at 145 mm cs
	2200 mm	50 x 10 mm 600 mm in ground	22 mm dia at 145 mm cs

Timber Field Gates to BS 3470: 1975

Gates made to this standard are designed to open one way only.
All timber gates are 1100 mm high.
Width over stiles 2400, 2700, 3000, 3300, 3600, and 4200 mm.
Gates over 4200 mm should be made in two leaves.

Steel Field Gates to BS 3470: 1975

Heavy duty: width over stiles 2400, 3000, 3600 and 4500 mm
Light duty: width over stiles 2400, 3000, and 3600 mm
All steel gates are 1100 mm high.

FENCING AND GATES – cont'd

Domestic Front Entrance Gates to BS 4092:part 1: 1966

Metal gates: Single gates are 900 mm high minimum, 900 mm, 1000 mm and 1100 mm wide

Domestic Front Entrance Gates to BS 4092:part 2: 1966.

Wooden gates: All rails shall be tenoned into the stiles
Single gates are 840 mm high minimum, 801 mm and 1020 mm wide
Double gates are 840 mm high minimum, 2130, 2340 and 2640 mm wide

Timber Bridle Gates to BS 5709:1979 (Horse Or Hunting Gates)

Gates open one way only
Minimum width between posts   1525 mm
Minimum height   1100 mm

Timber Kissing Gates to BS 5709:1979

Minimum width	700 mm
Minimum height	1000 mm
Minimum distance between shutting posts	600 mm
Minimum clearance at mid-point	600 mm

Metal Kissing Gates to BS 5709:1979

Sizes are the same as those for timber kissing gates.
Maximum gaps between rails 120 mm.

Categories of Pedestrian Guard Rail to BS 3049:1976

Class A for normal use
Class B where vandalism is expected
Class C where crowd pressure is likely

DRAINAGE

Weights and dimensions - vitrified clay pipes

Product	Nominal diameter (mm)	Effective length (mm)	BS 65 limits of tolerance		Crushing Strength (kN/m)	Weight	
			min (mm)	max (mm)		kg/pipe	kg/m
Supersleve	100	1600	96	105	35.00	14.71	9.19
	150	1750	146	158	35.00	29.24	16.71
Hepsleve	225	1850	221	236	28.00	84.03	45.42
	300	2500	295	313	34.00	193.05	77.22
	150	1500	146	158	22.00	37.04	24.69
Hepseal	225	1750	221	236	28.00	85.47	48.84
	300	2500	295	313	34.00	204.08	81.63
	400	2500	394	414	44.00	357.14	142.86
	450	2500	444	464	44.00	454.55	181.63
	500	2500	494	514	48.00	555.56	222.22
	600	2500	591	615	57.00	796.23	307.69
	700	3000	689	719	67.00	1111.11	370.45
	800	3000	788	822	72.00	1351.35	450.45
Hepline	100	1600	95	107	22.00	14.71	9.19
	150	1750	145	160	22.00	29.24	16.71
	225	1850	219	239	28.00	84.03	45.42
	300	1850	292	317	34.00	142.86	77.22
Hepduct (Conduit)	90	1500	-	-	28.00	12.05	8.03
	100	1600	-	-	28.00	14.71	9.19
	125	1750	-	-	28.00	20.73	11.84
	150	1750	-	-	28.00	29.24	16.71
	225	1850	-	-	28.00	84.03	45.42
	300	1850	-	-	34.00	142.86	77.22

DRAINAGE - cont'd

Weights and dimensions - vitrified clay pipes

Nominal internal diameter (mm)	Nominal wall thickness (mm)	Approximate weight kg/m
150	25	45
225	29	71
300	32	122
375	35	162
450	38	191
600	48	317
750	54	454
900	60	616
1200	76	912
1500	89	1458
1800	102	1884
2100	127	2619

Wall thickness, weights and pipe lengths vary, depending on type of pipe required.

The particulars shown above represent a selection of available diameters and are applicable to strength class 1 pipes with flexible rubber ring joints.

Tubes with Ogee joints are also available.

Weights and dimensions - PVC-U pipes

	Nominal size	Mean outside diameter (mm)		Wall thickness (mm)	Weight kg/m
		min	max		
Standard pipes	82.4	82.4	82.7	3.2	1.2
	110.0	110.0	110.4	3.2	1.6
	160.0	160.0	160.6	4.1	3.0
	200.0	200.0	200.6	4.9	4.6
	250.0	250.0	250.7	6.1	7.2
Perforated pipes					
- heavy grade	As above	As above	As above	As above	As above
- thin wall	82.4	82.4	82.7	1.7	-
	110.0	110.0	110.4	2.2	-
	160.0	160.0	160.6	3.2	-

Width of trenches required for various diameters of pipes

Pipe diameter (mm)	Trench n.e. 1.5 m deep (mm)	Trench over 1.5 m deep (mm)
n.e.100	450	600
100-150	500	650
150-225	600	750
225-300	650	800
300-400	750	900
400-450	900	1050
450-600	1100	1300

DRAINAGE BELOW GROUND AND LAND DRAINAGE

Flow of Water Which Can Be Carried by Various Sizes of Pipe

Clay or concrete pipes

Gradient of pipeline

Pipe size		1:10	1:20	1:30	1:40	1:50	1:60	1:80	1:100
		Flow in litres per second							
DN 100	15.0	8.5	6.8	5.8	5.2	4.7	4.0	3.5	
DN 150	28.0	19.0	16.0	14.0	12.0	11.0	9.1	8.0	
DN 225	140.0	95.0	76.0	66.0	58.0	53.0	46.0	40.0	

Plastic pipes

Gradient of pipeline

Pipe size	1:10	1:20	1:30	1:40	1:50	1:60	1:80	1:100
	Flow in litres per second							
82.4mm i/dia	12.0	8.5	6.8	5.8	5.2	4.7	4.0	3.5
110mm i/dia	28.0	19.0	16.0	14.0	12.0	11.0	9.1	8.0
160mm i/dia	76.0	53.0	43.0	37.0	33.0	29.0	25.0	22.0
200mm i/dia	140.0	95.0	76.0	66.0	58.0	53.0	46.0	40.0

Vitrified (Perforated) Clay Pipes and Fittings to BS En 295-5 1994

Length not specified

75 mm bore	250 mm bore	600 mm bore
100	300	700
125	350	800
150	400	1000
200	450	1200
225	500	

Pre-cast Concrete Pipes: Pre-stressed Non-pressure Pipes and Fittings: Flexible Joints to BS 5911 :Pt.103: 1994

Rationalized metric nominal sizes: 450, 500

Length:	500	-	1000 by 100 increments
	1000	-	2200 by 200 increments
	2200	-	2800 by 300 increments

Angles: length: 450 - 600 angles 45, 22.5, 11.25 °
600 or more angles 22.5, 11.25 °

Pre-cast Concrete Pipes: Un-reinforced and Circular Manholes and Soakaways to BS 5911 :Pt.200: 1994

Nominal Sizes:

Shafts: 675, 900 mm

Chambers: 900, 1050, 1200, 1350, 1500, 1800, 2100, 2400, 2700, 3000 mm.

Large chambers: To have either tapered reducing rings or a flat reducing slab in order to accept the standard cover.

Ring depths:
1. 300 - 1200 mm by 300 mm increments except for bottom slab and rings below cover slab, these are by 150 mm increments.
2. 250 - 1000 mm by 250 mm increments except for bottom slab and rings below cover slab, these are by 125 mm increments.

Access hole: 750 x 750 mm for DN 1050 chamber
1200 x 675 mm for DN 1350 chamber

Calculation of Soakaway Depth

The following formula determines the depth of concrete ring soakaway that would be required for draining given amounts of water.

$$h = \frac{4ar}{3\pi D^2}$$

h = depth of the chamber below the invert pipe
A = The are to be drained
r = The hourly rate of rainfall (50 mm per hour)
π = pi
D = internal diameter of the soakaway

This table shows the depth of chambers in each ring size which would be required to contain the volume of water specified. These allow a recommended storage capacity of 1/3 (one third of the hourly rain fall figure).

DRAINAGE BELOW GROUND AND LAND DRAINAGE – cont'd

Table showing required depth of concrete ring chambers in metres

AREA m2 Ring Size	50	100	150	200	300	400	500
0.9	1.31	2.62	3.93	5.24	7.86	10.48	13.10
1.1	0.96	1.92	2.89	3.85	5.77	7.70	9.62
1.2	0.74	1.47	2.21	2.95	4.42	5.89	7.37
1.4	0.58	1.16	1.75	2.33	3.49	4.66	5.82
1.5	0.47	0.94	1.41	1.89	2.83	3.77	4.72
1.8	0.33	0.65	0.98	1.31	1.96	2.62	3.27
2.1	0.24	0.48	0.72	0.96	1.44	1.92	2.41
2.4	0.18	0.37	0.55	0.74	1.11	1.47	1.84
2.7	0.15	0.29	0.44	0.58	0.87	1.16	1.46
3.0	0.12	0.24	0.35	0.47	0.71	0.94	1.18

Pre-cast Concrete Inspection Chambers and Gullies to BS 5911 :Pt.230: 1994

Nominal sizes: 375 diameter, 750, 900 mm deep
  450 diameter, 750, 900, 1050, 1200 mm deep

Depths: from the top for trapped or un-trapped units:
  centre of outlet 300 mm
  invert (bottom) of the outlet pipe 400 mm

Depth of water seal for trapped gullies:
  85 mm, rodding eye int. diam. 100 mm

Cover slab: 65 mm min.

Ductile Iron Pipes to BS En 598 : 1995

Type K9 with flexible joints should be used for surface water drainage.
5500 mm or 8000 mm long

80 mm bore	400 mm bore	1000 mm bore
100	450	1100
150	500	1200
200	600	1400
250	700	1600
300	800	
350	900	

Bedding Flexible Pipes: Pvc-u Or Ductile Iron

Type 1	=	100mm fill below pipe, 300mm above pipe: single size material
Type 2	=	100mm fill below pipe, 300mm above pipe: single size or graded material
Type 3	=	100mm fill below pipe, 75mm above pipe with concrete protective slab over
Type 4	=	100mm fill below pipe, fill laid level with top of pipe
Type 5	=	200mm fill below pipe, fill laid level with top of pipe
Concrete	=	25mm sand blinding to bottom of trench, pipe supported on chocks, 100mm concrete under the pipe, 150mm concrete over the pipe.

Bedding Rigid Pipes: Clay Or Concrete

(for vitrified clay pipes the manufacturer should be consulted)

Class D:	Pipe laid on natural ground with cut-outs for joints, soil screened to remove stones over 40mm and returned over pipe to 150mm min depth. Suitable for firm ground with trenches trimmed by hand.
Class N:	Pipe laid on 50mm granular material of graded aggregate to Table 4 of BS 882, or 10mm aggregate to Table 6 of BS 882, or as dug light soil (not clay) screened to remove stones over 10mm.
Class B:	As Class N, but with granular bedding extending half way up the pipe diameter.
Class F:	Pipe laid on 100mm granular fill to BS 882 below pipe, minimum 150mm granular fill above pipe: single size material. Suitable for machine dug trenches.
Class A:	Concrete 100mm thick under the pipe extending half way up the pipe, backfilled with the appropriate class of fill. Used where there is only a very shallow fall to the drain. Class A bedding allows the pipes to be laid to an exact gradient.
Concrete surround:	25mm sand blinding to bottom of trench, pipe supported on chocks, 100mm concrete under the pipe, 150mm concrete over the pipe. It is preferable to bed pipes under slabs or wall in granular material.

PIPED SUPPLY SYSTEMS

Identification of Service Tubes From Utility to Dwellings

Utility	Colour	Size	Depth
British Telecom	grey	54 mm od	450 mm
Electricity	black	38 mm od	450 mm
Gas	yellow	42 mm od rigid	450 mm
		60 mm od convoluted	
Water	may be blue	(normally untubed)	750 mm

ELECTRICAL SUPPLY/POWER/LIGHTING SYSTEMS

Electrical Insulation Class En 60.598 BS 4533

Class 1:   luminaires comply with class 1 (I) earthed electrical requirements
Class 2:   luminaires comply with class 2 (II) double insulated electrical requirements
Class 3:   luminaires comply with class 3 (III) electrical requirements

Protection to Light Fittings

BS EN 60529:1992 Classification for degrees of protection provided by enclosures.
(IP Code - International or ingress Protection)

1st characteristic:		against ingress of solid foreign objects.
The figure	2	indicates that fingers cannot enter
	3	that a 2.5 mm diameter probe cannot enter
	4	that a 1.0 mm diameter probe cannot enter
	5	the fitting is dust proof (no dust around live parts)
	6	the fitting is dust tight (no dust entry)
2nd characteristic:		ingress of water with harmful effects:
The figure	0	indicates unprotected
	1	vertically dripping water cannot enter
	2	water dripping 15° (tilt) cannot enter
	3	spraying water cannot enter
	4	splashing water cannot enter
	5	jetting water cannot enter
	6	powerful jetting water cannot enter
	7	proof against temporary immersion
	8	proof against continuous immersion
Optional additional codes:		A-D protects against access to hazardous parts;
	H	High voltage apparatus
	M	fitting was in motion during water test
	S	fitting was static during water test
	W	protects against weather
Marking code arrangement:		(example) IPX5S = IP (International or Ingress Protection);
		X (denotes omission of firstcharacteristic);
		5 = jetting;
		S = static during water test.

LANDFILL TAX

Waste liable at the lower rate

Group	Description of material	Conditions	
1	Rocks and soils	Naturally occurring	includes clay, sand, gravel, sandstone, limestone, crushed stone, china clay, construction stone, stone from the demolition of buildings or structures, slate, topsoil, peat, silt and dredgings glass includes fritted enamel, but excludes glass fibre and glass reinforced plastics.
2	Ceramic or concrete materials		ceramics includes bricks, bricks and mortar, tiles, clay ware, pottery, china and refractories concrete includes reinforced concrete, concrete blocks, breeze blocks and aircrete blocks, but excludes concrete plant washings
3	Minerals	Processed or prepared, not used	moulding sands excludes sands containing organic binders clays includes moulding clays and clay absorbents, including Fuller's earth and bentonite man-made mineral fibres includes glass fibres, but excludes glass-reinforced plastic and asbestos silica, mica and mineral abrasives
4	Furnace slags		vitrified wastes and residues from thermal processing of minerals where, in either case, the residue is both fused and insoluble slag from waste incineration
5	Ash		comprises only bottom ash and fly ash from wood, coal or waste combustion excludes fly ash from municipal, clinical, and hazardous waste incinerators and sewage sludge incinerators

LANDFILL TAX – cont'd

6	Low activity inorganic compound		comprises only titanium dioxide, calcium carbonate, magnesium carbonate, magnesium oxide, magnesium hydroxide, iron oxide, ferric hydroxide, aluminium oxide, aluminium hydroxide & zirconium dioxide
7	Calcium sulphate	Disposed of either at a site not licensed to take putrescible waste or in a containment cell which takes only calcium sulphate	includes gypsum and calcium sulphate based plasters, but excludes plasterboard
8	Calcium hydroxide and brine	Deposited in brine cavity	
9	Water	Containing other qualifying material in suspension	

Volume to weight conversion factors

Waste category	Typical waste types	Cubic metres to tonne - multiply by:	Cubic yards to tonne - multiply by:
Inactive or inert waste	Largely water insoluble and non or very slowly biodegradable: e.g. sand, subsoil, concrete, bricks, mineral fibres, fibreglass etc.	1.5	1.15

ESSENTIAL READING FROM SPON PRESS

External Works, Roads and Drainage
A Practical Guide
Phil Pitman

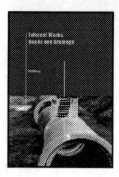

`The text provides practical input which can be of benefit to both student and professional alike.' - Building Engineer

External Works Roads and Drainage: A Practical Guide bridges the gap between theory and practice in building, construction and civil engineering, providing practical guidance, and the knowledge required 'on the job'.

This comprehensive book includes sections on legislation, environmental issues, surface water, highway and foul drainage design, road and pavement design, and external works.

The book is well illustrated, with flow charts to help guide you through the design process, bullet points and tables. Written in an accessible style, this guide is an essential reference for engineers, architects, designers, technicians and students looking to increase their design and project planning knowledge.

May 2001: 234x156 mm: 328 pp.
PB: 0-419-25760-8: £29.99

To Order: Tel: +44 (0) 8700 768853, or +44 (0) 1264 343071 Fax: +44 (0) 1264 343005, or Post: Spon Press Customer Services, Thomson Publishing Services, Cheriton House, Andover, Hants, SP10 5BE, UK Email: book.orders@tandf.co.uk

For a complete listing of all our titles visit:
www.sponpress.com

Index

Access Covers and Frames, 249, 250
Agricultural Drainage, 253-258, 327
Amsterdam Tree Sand, 188
Anchors, 85,188, 211, 226-231
Aquatic Plants, 178, 199, 334
Artificial Grass, 146
Asphalt, 110, 111, 128

Backfilling Materials, 258
Ball Stop Fencing, 220
Balusters, 108, 221, 335
Bar Railings, 221
Bark
    mulch, 169, 190, 191, 199, 201, 209, 316, 317
    surfaces, 132, 145, 147, 152, 304, 307
Barriers, 216, 225, 226
Bases, 92, 106, 125, 226, 296, 326
Bedding, 209
Beds, 208
Blocks, 83, 84,104, 123, 133-134, 139, 285, 291-295
Bollards, 232, 234-237, 273, 326, 336
Boundary Fencing, 211-215
Brick,
    copings, 104
    pavings, 123, 134-135, 326
    market prices, 100
Brick/Block Walling, 101-103
Brickwork, 90, 110, 111, 116, 118, 224, 287-295, 302-303, 305, 320
Bulbs, 316
    lifting, 208
    market prices, 178
    planting, 199

Cast Stone,
    cladding/features, 109
    ornamental garden buildings, 81
    pool surrounds, 268
    walling, 105
Cattle Grids, 225
Cement, 96, 100-103, 116
Ceramic Tiles, 116
Chambers, 115, 243, 244, 247
Channels, 120-125, 244, 245, 331
Chestnut Pale Fencing, 181, 211, 321
Clay,
    brick pavings, 135, 291, 295
    channels, 244, 248, 257
    drainage, 320, 331
    lake liners, 265, 267
    pipes, 245, 246, 257
    soil, 92, 93, 150, 151, 160-163, 181, 182, 283, 307
    tennis court surface, 146
    tiles, 116
Clearing Site Vegetation, 89
Cleft Rail Fencing, 218
Climbers, 172
Compacting, 83, 94, 131, 243
Concrete
    blocks, 83, 84, 104, 123, 133, 139, 133, 144, 226, 285, 293, 295
    bollards, 232, 234-235
    fencing, 220
    grass concrete, 83, 86, 143, 244,286, 301
    in situ, 86, 95-99, 120-124, 126, 127, 143, 243, 286, 294
    mixes, 95-96
    outfalls, 261
    paviors, 139
    pipes, 252
    reinforced, 86, 90, 92, 95, 96, 286, 292, 308
    revetments, 85, 88
    screeds, 111,116
    setts, 122, 140, 301
Conveyor, 73, 79
Copings, 104, 107, 222, 268
Crane, 74, 80, 194
Crash Barriers, 226
Crib Walls, 82
Cricket Squares, 157, 163
Cutting,
    curved cutting, 135
    grass cutting, 163-167, 204-205, 311, 313
    hedge cutting, 203, 210
    raking cutting, 135
    shrub cutting, 199, 207-208
Cycle holders, 233, 234

Damp Proof Courses, 107
Damp Proofing, 110, 111
Decking, 108, 335
Disposal, 92-93, 119, 243
    of arisings, 204, 206-207, 211
    systems, 211, 243, 296, 298

Disposal Systems, 243
Ditching, 250, 251
Drainage,
    channels, 120-125, 244, 245,
        291, 305, 331
    systems, 114, 124, 243, 327
    land drainage, 250
    mole drainage, 252
    slits, 253
    agricultural drainage calculation table, 253
Dry Stone Walling, 104
Dwarf Walls, 105

Earthwork Support, 119, 243-246, 257, 261, 331
Edge
    Brick on, 104, 123
    maintenance, 207
    restraints, 134, 139, 226
Edgings, 120 - 124, 131, 133, 135, 141, 142, 147, 291-296, 303
Edging Boards, 131, 147, 304
Electrical Accessories, 276
Embankments, 84, 164
Erosion, 82, 85, 87, 162, , 312
Erosion Control Mats, 84, 87, 88, 285, 287, 312
Excavation and Path Preparation, 130
Expansion Joints, 127, 143, 292, 294

Fencing, 78, 181, 211-225, 282, 307, 320-325
Fertilizers, 148, 155, 162, 168 - 170, 184, 198, 200 - 203
Filling, 82-84, 87, 89-94,
    104, 125, 131, 188, 243
Filling To Make Up Levels, 90, 94, 125, 131
Finishes, 116, 145
Flagpoles, 241
Flexible Sheet Materials, 87 - 88
Flexible Sheet Tanking, 111
Folly, 81
Forestry Planting, 202
Formwork, 98, 120, 127, 243, 291 - 293, 300, 320
Fountains, 265 - 268
Fungicide, 148

Gabions, 82 - 85, 286
Gates, 76, 211 - 225, 282
Glacial Boulders, 105

Grading, 83, 94, 1149, 127, 151 - 152, 183, 245 - 246, 267, 307
Granite
    aggregate, 128 - 130
    kerbs, 121 - 122
    paving, 74 - 75, 138, 140
    setts, 122 - 123, 141, 299 - 300
    walls, 104
Granolithic Paving, 116
Grass
    cutting, 204 - 205, 313
    fertilizers, 148
    maintenance, 148
    reinforcement, 87
    seed, 87 - 88, 143, 156, 160 - 161, 301, 307
Grass Concrete, 83, 86, 143, 286, 301
Grassblocks, 134
Grasscrete, 86, 143, 286, 294
Grassroad, 134
Green Roof Systems, 111
Grit Bins, 229
Groundcover Planting, 198
Growth Retardant, 202
Gullies, 124, 247 - 248, 252

Ha-ha walls, 287
Handrails, 108, 335
Hardcore, 91, 96-98, 125, 131
Harrowing, 154, 161, 163, 168, 205
Health and Safety, 77, 281
Hedges, 196, 282
    hedge cutting, 203, 210, 211
Herbaceous and Groundcover Planting, 198
Herbicides, 125, 131, 170, 171, 183, 184, 199 - 202
Hose, 162, 209, 264, 265, 312
Hydroplanting, 113

In Situ concrete, see Concrete
Intercepting Traps, 245
Intruder Guards, 216
Irrigation, 115, 146, 168, 202, 209, 319, 333
Irrigation Systems, 75, 190, 264, 305

Kerb Drainage Systems, 124
Kerbs, 120 - 124, 249, 291 - 301
Kidney Flint Cobbles, 143, 300
Kiln Dried Sand Joints, 135, 139

Index

Lakes, 265-267, 334
Lamps, 269, 275
Land Drainage, 250-262
Landscape Maintenance, 204-210
Leaf Clearance, 166, 206
Leaky pipe, 264, 265, 333
Liability, 111, 313
Lifebuoy Stations, 229
Lighting Columns, 275, 276
Liquid Applied Tanking, 110
Liquid Sod, 162
Litter and Grit Bins, 229
Litter Clearance, 207
Log Retaining Walls, 86, 287

Macadam, 90, 128-130, 146, 293-294, 304, 306
   resin bound, 132
Maintenance Operations, 166, 200, 204, 313
Management, 74, 77, 78, 281
Manhole Covers, 124, 249
Market Prices of,
   aquatic plants, 178
   backfilling materials, 258
   bulbs, 178
   containers, 232
   grass maintenance materials, 148
   grass seed, 156
   lamps, 269
   masonry, 100
   mulching materials, 169
   planting materials, 169
   pre-seeding materials, 148
   sand pits, 240
   surfacing materials, 145
   topsoil, 89
   trees, shrubs & plants, 171
Marking,
   car parks, 130
   pitches, 167
Masonry, 100
Mastic Asphalt Tanking, 110
Method Statements, 77
Mole Drainage, 252
Movement of materials, 73, 74, 76, 134
Moving Equipment, 239
Mulch, 169, 199, 201, 208, 209, 317
Mulching Materials, 169
Mulching Of Tree Pits, 190

Mushroom Compost, 169, 185-187, 198, 309, 317

Natural Stone
   cladding, 109
   kerbs, 121
   paving, 90, 140, 141, 298
   rubble, 104

Office, temporary, 78, 281
Operations after Planting, 199, 200
Ornamental
   garden buildings, 81
   hedges, 203, 210
   landscaping, 164
   masonry, 106
   pools, 267
   railings, 117
   trees, 166, 206, 207
   turf, 157

Outfalls, 261

Painting/Clear Finishing, 117, 118
Parking Posts, 235, 326
Paths, 96, 291, 302
   firepaths, 86, 134, 301
   footpaths, 127, 128, 304
   gravel paths, 303
   preparation, 130
Pavilion, 81
Paving, 74, 106, 123, 130-144, 291-297
   radial, 137, 141
Paving Blocks, 123, 133, 144, 226
Paviors, 90, 134, 135, 139, 249, 303
Pedestrian Guard Rails, 224
Pesticides, 155
Picnic Benches, 232
Piers, 102, 103, 106, 289
Pier Caps, 106, 107
Pipes,
   clay, 245
   concrete, 252
   leaky, 264
   pipe laying, 257, 259, 327
   piped ditching, 252
   perforated, 257-259, 327
   polypropylene, 246
   polythene, 264
   pvc-U, 246-247
   trenches for, 119, 245, 327-331

Pits, 90, 91, 96, 119, 170, 187-190, 240, 243, 261, 314-316
Planters, 111, 189, 208, 234, 318, 326
Plant Containers, 232, 233
Planting
　bulbs, 199
　hedges, 196
　herbaceous, 198
　materials, 169
　operations, 183
　operations after, 199
　shrubs, 196
　trees, 75, 187-195
Plants, 171
Plastered/rendered/roughcast, 116
Play
　climbing equipment, 240
　playgrounds, 146, 147, 238, 294, 305-307
　playsand, 240
　roundabouts, 239
　sculptures, 239
　seesaws, 239
　slides, 239
　spring equipment, 240
　structures, 240
　swings, 238
Ponds, 111, 265, 267, 334
Pools, 267, 268
　pool surrounds, 268
Precast Concrete
　blocks, 133
　bollards, 234
　channels, 123, 291
　copings, 107, 287
　drainage, 243, 267, 331
　edgings, 120, 291, 295
　fencing, 220
　grass blocks, 86, 144, 301
　kerbs, 120, 291
　litter bins, 227
　pavings, 136-138, 226, 297, 301
　plant containers, 233
　retaining walls, 83, 286
Prefabricated Buildings, 78, 81
Programmes, 77, 281
Protective Fencing, 181, 211
Pruning Shrubs, 207, 208

Quarry Tiles, 116

Rabbit Fencing, 321
Rabbit Netting, 213
Railings,
　mild steel railings, 221, 324
　pedestrian guard rails, 224, 324
　trip rails, 216, 325
Reinforced Concrete, 90, 95-99, 126-127, 296
Reinforcement, 98-99
Rendering, 110
Retaining Walls, 82-86, 111, 267, 287
Revetments, 85, 88
Risk Assessments, 77
Road Repairs, 130
Rockery Stone, 105
Rolling, 94, 128, 131, 132, 151, 161-167, 182, 206
Root Barriers, 114, 191
Roundabouts, 239
Rubble Walling, 104

Sand,
　Amsterdam tree sand, 190
　blinding, 86, 97, 110, 123, 131, 266, 267, 302, 303, 334
　market prices, 100, 145, 240, 258
　playsand, 145
　reject sand, 245
　sand pits, 240
　slitting, 253
Scarifying, 167, 168, 206
Screeds, 113, 116
Screen Fencing, 219
Seats & Benches, 229- 234, 326
Secure Storage, 78
Security, 78
　bollards, 237
　fencing, 215, 216, 225, 324, 325
Seesaws, 239
Setting Out, 78, 83, 167, 196, 209
Setts, 122, 123, 140, 141, 299-301
Shrubs,
　market prices, 171-177
　planting, 181, 189, 190, 195, 197, 316-318
　pruning, 207
Signage, 234
Signs, 225, 271, 320
Sika, 110
Sills, 107

Silt
  box, 124
  bucket, 247
  pits, 243
Silts, 162
Site
  accommodation, 78
  preparation, 89
  protection, 181
Slate, 107, 141
Slides, 239
Slip Layers, 112
Slip Resistant, 116, 139
Slotted Grating, 124
Soakaway(s), 261-263, 332
Soil,
  additives, 153, 185
  conditioners, 152
  filling, 94
  preparation, 153, 185
  stabilisation, 82-88
Sports
  areas, 145-146
  equipment, 241, 242
  pitches, 157, 158, 164, 166, 206, 207, 305, 307, 311
Stakes, 188, 194, 195, 197, 198, 203, 207, 208, 213, 216
Standpipes, 264
Step Irons, 244, 262, 331, 332
Stepping Stones, 137, 302
Steps, 76, 96, 273
Stiles and Kissing Gates, 224
Stopcocks, 264
Street Area Floodlighting, 269
Street Furniture, 225, 326
Stump Grinding, 89
Subsoiling, 93, 150, 181, 182, 308
Surface,
  applications, 152, 153, 185, 186
  erosion, 85-88
  finishes, 116
  treatments, 94, 125, 131, 139, 188, 206
Surfacing Materials, 145
Swings, 238
Synthaprufe, 110

Tanking, 110, 111
Temporary,
  office, 78
  protective fencing, 181, 282, 307
  roads, 80
  toilets, 79

Tender and Contract Documents, 146, 167, 241
Tennis Courts, 146, 167, 241
Tensar, 83, 84, 87, 164, 220, 285
Terram, 87, 131, 257, 263, 326
Ties
  tree, 188, 207, 314, 315
  wall, 102
Tiles, 116, 117, 141, 147, 164
Timber
  bollards, 236
  decking, 108, 335
  edgings, 147, 291
  fencing, 216-218, 322
  gates, 223
  litter bins, 228
  log retaining walls, 86, 287
  outdoor seats, 230-232
  plant containers, 233
  retaining walls, 82, 85, 286
  trellis, 219
Timbercrib, 82
Toilet facilities, 79
Top Dressing, 148, 168
Topiary, 196
Topsoil, 89, 148, 169
Transformers, 271, 276
Transplants, 170, 174, 189, 198
Traps, 245-247
Tree(s),
  anchors, 188
  market prices, 171-177
  planting, 75, 181, 187-194, 314
  removing, 89, 282, 284
  tree guards, 194, 195, 207
  tree grilles, 194, 237, 319
  tree tubes, 203
Trellis, 219
Trenches, 90-91, 119-120, 245-247, 257, 260, 263, 283
Trenchless Drainage System, 253
Trip Rails, 216, 217, 325
Turf,
  aeration, 167, 206
  areas, 206, 333
  industrially grown, 164
  irrigation, 264
  lifting, 89, 163
  maintenance materials, 148-150
  preparation of turf beds, 163
  protection, 80
  reinforcement, 165
Turfing, 148, 164, 308-313

Unclimbable fencing, 78
Underwater Lighting, 268, 273
Unreinforced Concrete, 126

Vehicle Crash Barriers, 226
Vehicular Paving, 133, 295

Walkways, 80
Walls
   block, 104
   brick, 102
   brick/block walling, 101, 289
   cast stone, 105
   crib, 82, 286
   dry stone, 104
   dwarf, 105
   granite, 104
   ha ha, 287
   natural stone/rubble, 104
   retaining, 82-86, 285-287
Water Features, 265, 334
Waterfall Construction, 267
Waterproof Rendering, 110
Watertanks, 268
Watertight Floors, 96
Weedkiller(s), 150, 155, 156, 168, 181, 283, 307-317
Wild Flora Seed Mixtures, 159, 160, 311
Willow Walling, 87
Windbreak Fencing, 220
Window Box(es), 208, 209, 233
Work To Existing Planting, 203

York Stone
   cladding, 109
   edging, 141
   kerbs, 122
   paving, 137, 140-142, 297-299

CD-Rom Single-User Licence Agreement

We welcome you as a user of this Spon Press CD-ROM and hope that you find it a useful and valuable tool. Please read this document carefully. **This is a legal agreement** between you (hereinafter referred to as the "Licensee") and Taylor and Francis Books Ltd., under the imprint of Spon Press (the "Publisher"), which defines the terms under which you may use the Product. **By breaking the seal and opening the package containing the CD-ROM you agree to these terms and conditions outlined herein. If you do not agree to these terms you must return the Product to your supplier intact, with the seal on the CD case unbroken.**

1. Definition of the Product
The product which is the subject of this Agreement, *Spon's Landscape and External Works Price Book 2004 on CD-ROM* (the "Product") consists of:

1.1 Underlying data comprised in the product (the "Data")
1.2 A compilation of the Data (the "Database")
1.3 Software (the "Software") for accessing and using the Database
1.4 A CD-ROM disk (the "CD-ROM")

2. Commencement and licence
2.1 This Agreement commences upon the breaking open of the package containing the CD-ROM by the Licensee (the "Commencement Date").
2.2 This is a licence agreement (the "Agreement") for the use of the Product by the Licensee, and not an agreement for sale.
2.3 The Publisher licenses the Licensee on a non-exclusive and non-transferable basis to use the Product on condition that the Licensee complies with this Agreement. The Licensee acknowledges that it is only permitted to use the Product in accordance with this Agreement.

3. Installation and Use
3.1 The Licensee may provide access to the Product for individual study in the following manner: The Licensee may install the Product on a secure local area network on a single site for use by one user. For more than one user or for a wide area network or consortium, use is only permissible with the express permission of the Publisher in writing and requires payment of the appropriate fee as specified by the Publisher, and signature by the Licensee of a separate multi-user licence agreement.
3.2 The Licensee shall be responsible for installing the Product and for the effectiveness of such installation.
3.3 Text from the Product may be incorporated in a coursepack. Such use is only permissible with the express permission of the Publisher in writing and requires the payment of the appropriate fee as specified by the Publisher and signature of a separate licence agreement.
3.4 The CD-ROM is a free addition to the book and no technical support will be provided.

4. Permitted Activities
4.1 The Licensee shall be entitled:
4.1.1 to use the Product for its own internal purposes;
4.1.2 to download onto electronic, magnetic, optical or similar storage medium reasonable portions of the Database provided that the purpose of the Licensee is to undertake internal research or study and provided that such storage is temporary;
4.1.3 to make a copy of the Database and/or the Software for back-up/archival/disaster recovery purposes.
4.2 The Licensee acknowledges that its rights to use the Product are strictly set out in this Agreement, and all other uses (whether expressly mentioned in Clause 5 below or not) are prohibited.

5. Prohibited Activities
The following are prohibited without the express permission of the Publisher:

5.1 The commercial exploitation of any part of the Product.
5.2 The rental, loan, (free or for money or money's worth) or hire purchase of this product, save with the express consent of the Publisher.
5.3 Any activity which raises the reasonable prospect of impeding the Publisher's ability or opportunities to market the Product.
5.4 Any networking, physical or electronic distribution or dissemination of the product save as expressly permitted by this Agreement.
5.5 Any reverse engineering, decompilation, disassembly or other alteration of the Product save in accordance with applicable national laws.
5.6 The right to create any derivative product or service from the Product save as expressly provided for in this Agreement.
5.7 Any alteration, amendment, modification or deletion from the Product, whether for the purposes of error correction or otherwise.

6. General Responsibilities of the License

6.1 The Licensee will take all reasonable steps to ensure that the Product is used in accordance with the terms and conditions of this Agreement.

6.2 The Licensee acknowledges that damages may not be a sufficient remedy for the Publisher in the event of breach of this Agreement by the Licensee, and that an injunction may be appropriate.

6.3 The Licensee undertakes to keep the Product safe and to use its best endeavours to ensure that the product does not fall into the hands of third parties, whether as a result of theft or otherwise.

6.4 Where information of a confidential nature relating to the product of the business affairs of the Publisher comes into the possession of the Licensee pursuant to this Agreement (or otherwise), the Licensee agrees to use such information solely for the purposes of this Agreement, and under no circumstances to disclose any element of the information to any third party save strictly as permitted under this Agreement. For the avoidance of doubt, the Licensee's obligations under this sub-clause 6.4 shall survive the termination of this Agreement.

7. Warrant and Liability

7.1 The Publisher warrants that it has the authority to enter into this agreement and that it has secured all rights and permissions necessary to enable the Licensee to use the Product in accordance with this Agreement.

7.2 The Publisher warrants that the CD-ROM as supplied on the Commencement Date shall be free of defects in materials and workmanship, and undertakes to replace any defective CD-ROM within 28 days of notice of such defect being received provided such notice is received within 30 days of such supply. As an alternative to replacement, the Publisher agrees fully to refund the Licensee in such circumstances, if the Licensee so requests, provided that the Licensee returns the Product to the Publisher. The provisions of this sub-clause 7.2 do not apply where the defect results from an accident or from misuse of the product by the Licensee.

7.3 Sub-clause 7.2 sets out the sole and exclusive remedy of the Licensee in relation to defects in the CD-ROM.

7.4 The Publisher and the Licensee acknowledge that the Publisher supplies the Product on an "as is" basis. The Publisher gives no warranties:

7.4.1 that the Product satisfies the individual requirements of the Licensee; or

7.4.2 that the Product is otherwise fit for the Licensee's purpose; or

7.4.3 that the Data are accurate or complete of free of errors or omissions; or

7.4.4 that the Product is compatible with the Licensee's hardware equipment and software operating environment.

7.5 The Publisher hereby disclaims all warranties and conditions, express or implied, which are not stated above.

7.6 Nothing in this Clause 7 limits the Publisher's liability to the Licensee in the event of death or personal injury resulting from the Publisher's negligence.

7.7 The Publisher hereby excludes liability for loss of revenue, reputation, business, profits, or for indirect or consequential losses, irrespective of whether the Publisher was advised by the Licensee of the potential of such losses.

7.8 The Licensee acknowledges the merit of independently verifying Data prior to taking any decisions of material significance (commercial or otherwise) based on such data. It is agreed that the Publisher shall not be liable for any losses which result from the Licensee placing reliance on the Data or on the Database, under any circumstances.

7.9 Subject to sub-clause 7.6 above, the Publisher's liability under this Agreement shall be limited to the purchase price.

8. Intellectual Property Rights

8.1 Nothing in this Agreement affects the ownership of copyright or other intellectual property rights in the Data, the Database of the Software.

8.2 The Licensee agrees to display the Publishers' copyright notice in the manner described in the Product.

8.3 The Licensee hereby agrees to abide by copyright and similar notice requirements required by the Publisher, details of which are as follows:

"© 2004 Spon Press. All rights reserved. All materials in *Spon's Architects' and Builders' Price Book 2004* are copyright protected. © 2000 Adobe Systems Incorporated. All rights reserved. No such materials may be used, displayed, modified, adapted, distributed, transmitted, transferred, published or otherwise reproduced in any form or by any means now or hereafter developed other than strictly in accordance with the terms of the licence agreement enclosed with the CD-ROM. However, text and images may be printed and copied for research and private study within the preset program limitations. Please note the copyright notice above, and that any text or images printed or copied must credit the source."

8.4 This Product contains material proprietary to and copyedited by the Publisher and others. Except for the licence granted herein, all rights, title and interest in the Product, in all languages, formats and media throughout the world, including copyrights therein, are and remain the property of the Publisher or other

9. Non-assignment
This Agreement and the licence contained within it may not be assigned to any other person or entity without the written consent of the Publisher.

10. Termination and Consequences of Termination.
10.1　The Publisher shall have the right to terminate this Agreement if:
10.1.1　the Licensee is in material breach of this Agreement and fails to remedy such breach (where capable of remedy) within 14 days of a written notice from the Publisher requiring it to do so; or
10.1.2　the Licensee becomes insolvent, becomes subject to receivership, liquidation or similar external administration; or
10.1.3　the Licensee ceases to operate in business.
10.2　The Licensee shall have the right to terminate this Agreement for any reason upon two month's written notice. The Licensee shall not be entitled to any refund for payments made under this Agreement prior to termination under this sub-clause 10.2.
10.3　Termination by either of the parties is without prejudice to any other rights or remedies under the general law to which they may be entitled, or which survive such termination (including rights of the Publisher under sub-clause 6.4 above).
10.4　Upon termination of this Agreement, or expiry of its terms, the Licensee must:
10.4.1　destroy all back up copies of the product; and
10.4.2　return the Product to the Publisher.

11. General
11.1　Compliance with export provisions
The Publisher hereby agrees to comply fully with all relevant export laws and regulations of the United Kingdom to ensure that the Product is not exported, directly or indirectly, in violation of English law.
11.2　Force majeure
The parties accept no responsibility for breaches of this Agreement occurring as a result of circumstances beyond their control.
11.3　No waiver
Any failure or delay by either party to exercise or enforce any right conferred by this Agreement shall not be deemed to be a waiver of such right.
11.4　Entire agreement
This Agreement represents the entire agreement between the Publisher and the Licensee concerning the Product. The terms of this Agreement supersede all prior purchase orders, written terms and conditions, written or verbal representations, advertising or statements relating in any way to the Product.
11.5　Severability
If any provision of this Agreement is found to be invalid or unenforceable by a court of law of competent jurisdiction, such a finding shall not affect the other provisions of this Agreement and all provisions of this Agreement unaffected by such a finding shall remain in full force and effect.
11.6　Variations
This agreement may only be varied in writing by means of variation signed in writing by both parties.
11.7　Notices
All notices to be delivered to: Spon Press, an imprint of Taylor & Francis Books Ltd., 11 New Fetter Lane, London, EC4P 4EE, UK.
11.8　Governing law
This Agreement is governed by English law and the parties hereby agree that any dispute arising under this Agreement shall be subject to the jurisdiction of the English courts.

If you have any queries about the terms of this licence, please contact:
Spon's Price Books
Spon Press
an Imprint of Taylor & Francis Books Ltd.
11 New Fetter Lane
London EC4P 4EE
Tel: +44 (0) 20 7842 2428
Fax: +44 (0) 20 7842 2300
www.sponpress.com

CD-ROM Installation Instructions

System requirements

Minimum

- 66 MHz processor
- 12MB of RAM
- 10MB available hard disk space
- Quad speed CD-ROM drive
- Microsoft Windows 95/98/2000/NT/ME/XP
- VGA or SVGA (256 colours)
- Mouse

Recommended

- 133 MHz (or better) processor
- 16 MB of RAM
- 10 MB available hard disk space or more
- 12x speed CD-ROM drive
- Microsoft Windows 95/98/2000/NT
- SVGA monitor (256 colours) or better
- Mouse

Microsoft ® is a registered trademark and Windows™ is a trademark of the Microsoft Corporation.

Installation

How to install *Spon's Landscape and External Works Price Book 2004 CD-ROM*

Windows 95/98/2000/NT

Spon's Landscape and External Works Price Book 2004 CD-ROM should run automatically when inserted into the CD-ROM drive. If it fails to run, follow the instructions below.

- Click the **Start** button and choose **Run**.
- Click the **Browse** button.
- Select your CD-ROM drive.
- Select the Setup file (setup.exe) then click **Open**.
- Click the OK button.
- Follow the instructions on screen.
- The installation process will create a folder containing an icon for *Spon's Landscape and External Works Price Book 2004 CD-ROM* and also an icon on your desktop.

How to run the *Spon's Landscape and External Works Price Book 2004 CD-ROM*

- Double click the icon (from the folder or desktop) installed by the Setup program.
- Follow the instructions on screen.

The CD-ROM is a free addition to the book and no technical support will be provided. For help with the use of the CD-ROM please visit www.pricebooks.co.uk

© **COPYRIGHT ALL RIGHTS RESERVED**

All materials in *Spon's Landscape and External Works Price Book 2004 CD-ROM* are copyright protected. No such materials may be used, displayed, modified, adapted, distributed, transmitted, transferred, published or otherwise reproduced in any form or by any means now or hereafter developed other than strictly in accordance with the terms of the above licence agreement.

The software used in *Spon's Landscape and External Works Price Book 2004 CD-ROM* is furnished under a licence agreement. The software may be used only in accordance with the terms of the licence agreement.